Studies in Computational Intelligence

Volume 625

Series editor

Janusz Kacprzyk, Polish Academy of Sciences, Warsaw, Poland
e-mail: kacprzyk@ibspan.waw.pl

About this Series

The series "Studies in Computational Intelligence" (SCI) publishes new developments and advances in the various areas of computational intelligence—quickly and with a high quality. The intent is to cover the theory, applications, and design methods of computational intelligence, as embedded in the fields of engineering, computer science, physics and life sciences, as well as the methodologies behind them. The series contains monographs, lecture notes and edited volumes in computational intelligence spanning the areas of neural networks, connectionist systems, genetic algorithms, evolutionary computation, artificial intelligence, cellular automata, self-organizing systems, soft computing, fuzzy systems, and hybrid intelligent systems. Of particular value to both the contributors and the readership are the short publication timeframe and the worldwide distribution, which enable both wide and rapid dissemination of research output.

More information about this series at http://www.springer.com/series/7092

Anis Koubaa
Editor

Robot Operating System (ROS)

The Complete Reference (Volume 1)

 Springer

Editor
Anis Koubaa
Prince Sultan University
Riyadh
Saudi Arabia

and

CISTER/INESC-TEC, ISEP
Polytechnic Institute of Porto
Porto
Portugal

Studies in Computational Intelligence
ISBN 978-3-319-79884-4   ISBN 978-3-319-26054-9   (eBook)
DOI 10.1007/978-3-319-26054-9

Springer Cham Heidelberg New York Dordrecht London
© Springer International Publishing Switzerland 2016
Softcover reprint of the hardcover 1st edition 2016

Printed on acid-free paper

Springer International Publishing AG Switzerland is part of Springer Science+Business Media
(www.springer.com)

Preface

ROS is an open-source robotic middleware for the large-scale development of complex robotic systems. Although the research community is quite active in developing applications with ROS and extending its features, the amount of references does not translate the huge amount of work being done.

The objective of the book is to provide the reader with a comprehensive coverage of the Robot Operating Systems (ROS) and the latest related systems, which is currently considered as the main development framework for robotics applications.

There are 27 chapters organized into eight parts. Part I presents the basics and foundations of ROS. In Part II, four chapters deal with navigation, motion and planning. Part III provides four examples of service and experimental robots. Part IV deals with real-world deployment of applications. Part V presents signal-processing tools for perception and sensing. Part VI provides software engineering methodologies to design complex software with ROS. Simulations frameworks are presented in Part VII. Finally, Part VIII presents advanced tools and frameworks for ROS including multi-master extension, network introspection, controllers and cognitive systems.

I believe that this book will be a valuable companion for ROS users and developers to learn more about ROS capabilities and features.

January 2016   Anis Koubaa

Acknowledgements

The editor would like to acknowledge the support of King Abdulaziz City for Science and Technology (KACST) through the funded research project entitled "MyBot: A Personal Assistant Robot Case Study for Elderly People Care" under the grant number 34-75, and also the support of Prince Sultan University.

Acknowledgements to Reviewers

The Editor would like to thank the following reviewers for their great contributions in the review process of the book by providing a quality feedback to authors.

Yasir	Javed	Prince Sultan University
André S. De	Oliveira	Universidade Tecnológica Federal do Paraná
Bence	Magyar	PAL Robotics
Alfredo	Soto	Freescale Semiconductors
Huimin	Lu	National University of Defense Technology
Dirk	Thomas	Open Source Robotics Foundation
Walter	Fetter Lages	Universidade Federal do Rio Grande do Sul
Roberto	Guzman	Robotnik
Joao	Fabro	Universidade Tecnológica Federal do Paraná
Ugo	Cupcic	SHADOW ROBOT COMPANY LTD.
Rafael	Bekrvens	University of Antwerp
Andreas	Bihlmaier	Karlsruhe Institute of Technologie (KIT)
Timo	Röhling	Fraunhofer FKIE
Maram	Alajlan	Al-Imam Mohamed bin Saud University
Timm	Linder	Social Robotics Lab, University of Freiburg
Fadri	Furrer	Autonomous Systems Lab, ETH Zurich
Péter	Fankhauser	ETH Zurich
William	Woodall	Open Source Robotics Foundation
Jenssen	Chang	Gaitech International Ltd.
William	Morris	Undefined
Markus	Achtelik	Autonomous Systems Lab, ETH Zurich
Chienliang	Fok	University of Texas at Austin
Mohamedfoued	Sriti	Al-Imam Muhammad Ibn Saud Islamic University
Steven	Peters	Open Source Robotics Foundation

Contents

Part I ROS Basics and Foundations

MoveIt!: An Introduction 3
Sachin Chitta

Hands-on Learning of ROS Using Common Hardware 29
Andreas Bihlmaier and Heinz Wörn

Threaded Applications with the roscpp API 51
Hunter L. Allen

Part II Navigation, Motion and Planning

Writing Global Path Planners Plugins in ROS: A Tutorial 73
Maram Alajlan and Anis Koubâa

A Universal Grid Map Library: Implementation and Use Case
for Rough Terrain Navigation 99
Péter Fankhauser and Marco Hutter

ROS Navigation: Concepts and Tutorial 121
Rodrigo Longhi Guimarães, André Schneider de Oliveira,
João Alberto Fabro, Thiago Becker and Vinícius Amilgar Brenner

Localization and Navigation of a Climbing Robot Inside a LPG
Spherical Tank Based on Dual-LIDAR Scanning of Weld Beads 161
Ricardo S. da Veiga, Andre Schneider de Oliveira, Lucia Valeria Ramos de
Arruda and Flavio Neves Junior

Part III Service and Experimental Robots

People Detection, Tracking and Visualization Using ROS
on a Mobile Service Robot 187
Timm Linder and Kai O. Arras

A ROS-Based System for an Autonomous Service Robot 215
Viktor Seib, Raphael Memmesheimer and Dietrich Paulus

Robotnik—Professional Service Robotics Applications with ROS 253
Roberto Guzman, Roman Navarro, Marc Beneto and Daniel Carbonell

Standardization of a Heterogeneous Robots Society Based on ROS 289
Igor Rodriguez, Ekaitz Jauregi, Aitzol Astigarraga, Txelo Ruiz
and Elena Lazkano

Part IV   Real-World Applications Deployment

ROS-Based Cognitive Surgical Robotics 317
Andreas Bihlmaier, Tim Beyl, Philip Nicolai, Mirko Kunze,
Julien Mintenbeck, Luzie Schreiter, Thorsten Brennecke,
Jessica Hutzl, Jörg Raczkowsky and Heinz Wörn

ROS in Space: A Case Study on Robonaut 2 343
Julia Badger, Dustin Gooding, Kody Ensley, Kimberly Hambuchen
and Allison Thackston

**ROS in the MOnarCH Project: A Case Study in Networked Robot
Systems** 375
João Messias, Rodrigo Ventura, Pedro Lima and João Sequeira

Case Study: Hyper-Spectral Mapping and Thermal Analysis 397
William Morris

Part V   Perception and Sensing

**A Distributed Calibration Algorithm for Color and Range Camera
Networks** 413
Filippo Basso, Riccardo Levorato, Matteo Munaro
and Emanuele Menegatti

Acoustic Source Localization for Robotics Networks 437
Riccardo Levorato and Enrico Pagello

Part VI   Software Engineering with ROS

ROS Web Services: A Tutorial 463
Fatma Ellouze, Anis Koubâa and Habib Youssef

rapros: **A ROS Package for Rapid Prototyping** 491
Luca Cavanini, Gionata Cimini, Alessandro Freddi, Gianluca Ippoliti
and Andrea Monteriù

**HyperFlex: A Model Driven Toolchain for Designing and
Configuring Software Control Systems for Autonomous Robots** 509
Davide Brugali and Luca Gherardi

Integration and Usage of a ROS-Based Whole Body Control Software Framework. 535
Chien-Liang Fok and Luis Sentis

Part VII ROS Simulation Frameworks

Simulation of Closed Kinematic Chains in Realistic Environments Using *Gazebo* 567
Michael Bailey, Krystian Gebis and Miloš Žefran

RotorS—A Modular Gazebo MAV Simulator Framework 595
Fadri Furrer, Michael Burri, Markus Achtelik and Roland Siegwart

Part VIII Advanced Tools for ROS

The ROS Multimaster Extension for Simplified Deployment of Multi-Robot Systems. 629
Alexander Tiderko, Frank Hoeller and Timo Röhling

Advanced ROS Network Introspection (ARNI). 651
Andreas Bihlmaier, Matthias Hadlich and Heinz Wörn

Implementation of Real-Time Joint Controllers 671
Walter Fetter Lages

LIDA Bridge—A ROS Interface to the LIDA (Learning Intelligent Distribution Agent) Framework 703
Thiago Becker, André Schneider de Oliveira, João Alberto Fabro
and Rodrigo Longhi Guimarães

Part I
ROS Basics and Foundations

MoveIt!: An Introduction

Sachin Chitta

Abstract MoveIt! is state of the art software for mobile manipulation, incorporating the latest advances in motion planning, manipulation, 3D perception, kinematics, control and navigation. It provides an easy-to-use platform for developing advanced robotics applications, evaluating new robot designs and building integrated robotics products for industrial, commercial, R&D and other domains. MoveIt! is the most widely used open-source software for manipulation and has been used on over 65 different robots. This tutorial is intended for both new and advanced users: it will teach new users how to integrate MoveIt! with their robots while advanced users will also be able to get information on features that they may not be familiar with.

1 Introduction

Robotics has undergone a transformational change over the last decade. The advent of new open-source frameworks like ROS and MoveIt! has made robotics more accessible to new users, both in research and consumer applications. In particular, ROS has revolutionized the *developers* community, providing it with a set of tools, infrastructure and best practices to build new applications and robots (like the Baxter research robot). A key pillar of the ROS effort is the notion of *not re-inventing the wheel* by providing easy to use libraries for different capabilities like navigation, manipulation, control (and more).

MoveIt! provides the core functionality for manipulation in ROS. MoveIt! builds on multiple pillars:

- **A library of capabilities**: MoveIt! provides a library of robotic capabilities for manipulation, motion planning, control and mobile manipulation.
- **A strong community**: A strong community of users and developers that help in maintaining and extending MoveIt! to new applications.

S. Chitta (✉)
Kinema Systems Inc., Menlo Park, CA 94025, USA
e-mail: robot.moveit@gmail.com
URL: http://moveit.ros.org

© Springer International Publishing Switzerland 2016
A. Koubaa (ed.), *Robot Operating System (ROS)*, Studies in Computational
Intelligence 625, DOI 10.1007/978-3-319-26054-9_1

3

Fig. 1 Robots using MoveIt!

- **Tools**: A set of tools that allow new users to integrate MoveIt! with their robots and advanced users to deploy new applications.

Figure 1 shows a list of robots that MoveIt! has been used with. The robots range from industrial robots from all the leading vendors and research robots from all over the world. The robots include single arm, dual-armed robots, mobile manipulation systems, and humanoid robots. MoveIt! has been used in applications ranging from search and rescue (the DARPA robotics challenge), unstructured autonomous pick and place (with industrial robots like the UR5), mobile manipulation (with the PR2 and other robots), process tasks like painting and welding, with the (simulated) Robonaut robot for target applications in the space station. MoveIt! has been used or will be used by teams in the DARPA Robotics Challenge, the ROS-Industrial Consortium, the upcoming Amazon Picking Challenge, the NASA sample retrieval challenge.

2 A Brief History

MoveIt! evolved from the Arm Navigation framework [1, 2] in ROS. The Arm Navigation framework was developed after the development of the base navigation stack in ROS to provide the same functionality that was now available for base navigation in ROS. It combined kinematics, motion planning, 3D perception and an interface to control to provide the base functionality of *moving an arm* in unstructured environments. The central node in the Arm Navigation framework, called *move_arm*, was designed to be robot agnostic, i.e. usable with any robot. It connected to several other nodes, for kinematics, motion planning, 3D perception and other capabilities, to generate collision-free trajectories for robot arms. The Arm Navigation framework was further combined with the ROS grasping pipeline to create, for the first time, a general grasping and manipulation framework that could (and was) ported onto several different robots with different kinematics and grippers.

The Arm Navigation framework had several shortcomings. Each capability in the framework was designed as a separate ROS node. This required sharing data (particularly environment data) across several processes. The need for synchronization between the nodes led to several issues: (a) a mis-match of state between separate nodes often resulted in motion plans that were invalid, (b) communication bottlenecks because of the need to send expensive 3D data to several different nodes and (c) difficulty in extending the types of services offered by *move_arm* since it required changing the structure of the node itself. MoveIt! was designed to address all of these issues.

3 MoveIt! Architecture

The architecture of MoveIt! is shown in Fig. 2. The central node in MoveIt! is called *move_group*. It is intended to be light-weight, managing different capabilities and integrating kinematics, motion planning and perception. It uses a plugin-based architecture (adopted from ROS)—dramatically improving MoveIt!'s extensibility when compared to Arm Navigation. The plugin architecture allows users to add and share *capabilities* easily, e.g. a new implementation of pick and place or motion planning. The use of plugins is a central feature of MoveIt! and differentiates it from Arm Navigation.

Users can access the actions and services provided by *move_group* in one of three ways:

- **In C++**: using the *move_group_interface* package that provides an easy to setup interface to *move_group* using a C++ API. This API is primarily meant for advanced users and is useful when creating higher-level capabilities.

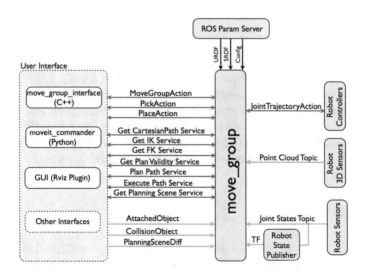

Fig. 2 MoveIt! high-level architecture

- **In Python**: using the *moveit_commander* package. This API is recommended for scripting demos and for building applications.
- **Through a GUI**: using the Motion Planning plugin to Rviz (the ROS visualizer). This API is recommended for visualization, initial interaction with robots through MoveIt! and for quick demonstrations.

One of the primary design principles behind MoveIt! is to expose an easy to use API for beginners to use while retaining access to the entire underlying API for more advanced users. MoveIt! users can access any part of the functionality directly if desired, allowing custom users to modify and architect their own applications. MoveIt! builds on several component technologies, each of which we will describe in brief detail.

3.1 Collision Checking

MoveIt! relies on the FCL [3] package for native collision checking. The collision checking capabilities are implemented using a plugin architecture, allowing any collision checker to be integrated with MoveIt!. FCL provides a state of the art implementation of collision checking, including the ability to do continuous collision checking. Collision checking is often the most expensive part of motion planning, consuming almost 80–90 % of the time for generating a motion plan. The use of an ALLOWED COLLISION MATRIX allows a user to specify which pairs of bodies do not need to be checked against each other, saving significant time. The ALLOWED COLLISION MATRIX is automatically configured by the MoveIt! Setup Assistant but can also be modified online by the user.

3.2 Kinematics

MoveIt! utilizes a plugin-based architecture for solving inverse kinematics while providing a native implementation of forward kinematics. Natively, MoveIt! uses a numerical solver for inverse kinematics for any robot. Users are free to add their own custom solvers, in particular analytic solvers are much faster than the native solver. Examples of analytic solvers that are integrated with MoveIt! include the solver for the PR2 robot. A popular plugin-based solver for MoveIt! is based on IKFast [4] and offers analytic solvers for industrial arms that are generated (in code).

3.3 Motion Planning

MoveIt! works with motion planners through a plugin interface. This allows MoveIt! to communicate with and use different motion planners from multiple libraries, making MoveIt! easily extensible. The interface to the motion planners is through a ROS Action or service (offered by the move_group node). The default motion planners for move_group are configured using the MoveIt! Setup Assistant. OMPL (Open Motion Planning Library) is an open-source motion planning library that primarily implements randomized motion planners. MoveIt! integrates directly with OMPL and uses the motion planners from that library as its primary/default set of planners. The planners in OMPL are abstract; i.e. OMPL has no concept of a robot. Instead, MoveIt! configures OMPL and provides the back-end for OMPL to work with problems in Robotics.

3.4   Planning Scene

The planning scene is used to represent the world around the robot and also stores the state of the robot itself. It is maintained by the planning scene monitor inside the move group node. The planning scene monitor listens to:

- **Robot State Information**: on the *joint_states* topic and using transform information from the ROS TF transform tree.
- **Sensor Information**: using a world geometry monitor that integrates 3D occupancy information and other object information.
- **World Geometry Information**: from user input or other sources, e.g. from an object recognition service.

The planning scene interface provides the primary interface for users to modify the state of the world that the robot operates in.

3.5   3D Perception

3D perception in MoveIt! is handled by the occupancy map monitor. The Occupancy map monitor uses an Octomap to maintain the occupancy map of the environment. The Octomap can actually encode probabilistic information about individual cells although this information is not currently used in MoveIt!. The Octomap can directly be passed into FCL, the collision checking library that MoveIt! uses. Input to the occupancy map monitor is from depth images, e.g. from an ASUS Xtion Pro Sensor or the Kinect 2 sensor. The depth image occupancy map updater includes its own self-filter, i.e. it will remove visible parts of the robot from the depth map. It uses current information about the robot (the robot state) to carry out this operation. Figure 3 shows the architecture corresponding to the 3D perception components in MoveIt!.

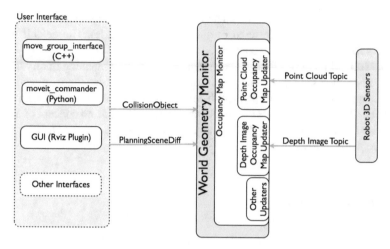

Fig. 3 The 3D perception pipeline in MoveIt!: architecture

3.6 Trajectory Processing

MoveIt! includes a trajectory processing component. Motion planners will typically only generate paths, i.e. there is no timing information associated with the paths. MoveIt! includes trajectory processing routines that can work on these paths and generate trajectories that are properly time-parameterized accounting for the maximum velocity and acceleration limits imposed on individual joints. These limits are read from a seperate file specified for each robot.

3.7 Using This Tutorial

MoveIt! is a large package and it is impossible to cover it in its entirety in a book chapter. This document serves as a reference for the tutorial that users can use but must be used in conjunction with the online documentation on the MoveIt! website. The online resource will remain the most up to date source of information on MoveIt!. This paper will introduce the most important concepts in MoveIt! and also provide helpful hints for new users. We assume that the user is already familiar with ROS. Readers should go through the ROS Tutorials—in particular, they should learn about ROS topics, services, using the ROS parameter server, ROS actions, the ROS build system and the ROS transform infrastructure (TF).

The example URDFs and MoveIt! config packages used in this tutorial for the Fanuc M10ia robot can be found in the examples repository.

3.8 Installing MoveIt!

MoveIt! can easily be installed on a Ubuntu 14.04 distribution using ROS Indigo. The most updated instructions for installing MoveIt! can be found on the MoveIt! installation page. It is recommended that most users follow the instructions for installing from binaries. There are three steps to installing MoveIt!:

1. Install ROS—follow the latest instructions on the ROS installation page.
2. Install MoveIt!:

```
sudo apt-get install ros-indigo-moveit-full
```

3. Setup your environment:

```
source /opt/ros/indigo/setup.bash
```

4 Starting with MoveIt!: The Setup Assistant

The first step in working with MoveIt! is to use the MoveIt! Setup Assistant.[1] The setup assistant is designed to allow users to import new robots and create a MoveIt! package for interacting, visualizing and simulating their robot (and associated workcell). The primary function of the setup assistant is to generate a Semantic Robot Description Format (SRDF) file for the robot. It also generates a set of files that allow the user to start a visualized demonstration of the robot instantly. We will not describe the Setup Assistant in detail (the latest instructions can always be found on the MoveIt! website [5]). We will instead focus on the parts of the process that creates the most confusion for new users.

4.1 Start

To start the setup assistant:

```
rosrun moveit_setup_assistant moveit_setup_assistant
```

This will bring up a startup screen with two choices: Create New MoveIt! Configuration Package or Edit Existing MoveIt! Configuration Package. Users should select Create New MoveIt! Configuration Package for any new robot or workcell (even if the robots in the workcells already have their own configuration package). Figure 4 illustrates this for

---

[1]This tutorial assumes that the user is using ROS Indigo on a Ubuntu 14.04 distribution.

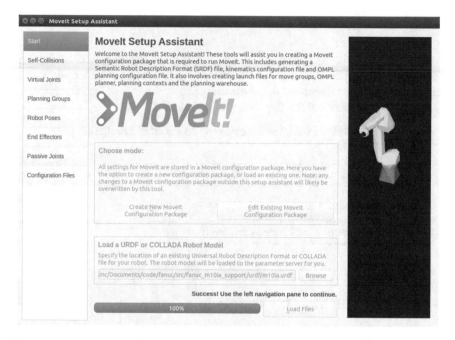

Fig. 4 Loading a Robot into the Setup Assistant

a Fanuc M10ia robot. Note that users can select either a URDF file or a *xacro* file (often used to put together multiple robots).

The Setup Assistant is also capable of editing an existing configuration. The primary reason to edit an existing configuration is to regenerate the Allowed Collision Matrix (ACM). This matrix needs to be re-generated when any of the following happens:

- The geometric description of your robot (URDF) has changed—i.e., the mesh representation being used for the robot has changed. Note here that the collision mesh representation is the key component of the URDF that MoveIt! uses. Changing the visual description of the robot while keeping the collision representation unchanged will not require the MoveIt! Setup Assistant to be run again.
- The joint limits specified for the robot have changed—this changes the limits that the Setup Assistant uses in sampling states for the Allowed Collision Matrix (ACM). Failing to run the Setup Assistant again may result in a state where the robot is allowed to move into configurations where it could be in collision with itself or with other parts of the environment.

4.2 Generating the Self-Collision Matrix

The key choice in generating the self-collision matrix is the number of random samples to be generated. Using a higher number results in more samples being generated but also slows down the process of generating the MoveIt! config package. Selecting a lower number implies that fewer samples are generated and there is a possibility that some collision checks may be wrongly disabled. We have found in practice, that generating at least 10,000 samples (the default value) is a good practice. Figure 5 shows what you should expect to see at the end of this step (Remember to press the SAVE button!).

4.3 Add Virtual Joints

Virtual joints are sometimes required to specify where the robot is in the *world*. A virtual joint could be related to the motion of a mobile base or it could be fixed, e.g. for an industrial robot bolted to the ground. Virtual joints are not always required—you can work with the default URDF model of the robot for most robots. If you do add a virtual joint, remember that there has to be a source of transform information for it (e.g. a localization module for a mobile base or a TF static transform publisher

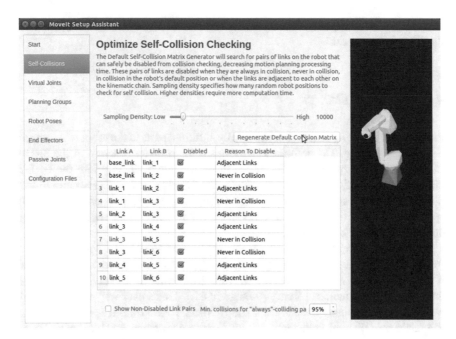

Fig. 5 Generating the self-collision matrix

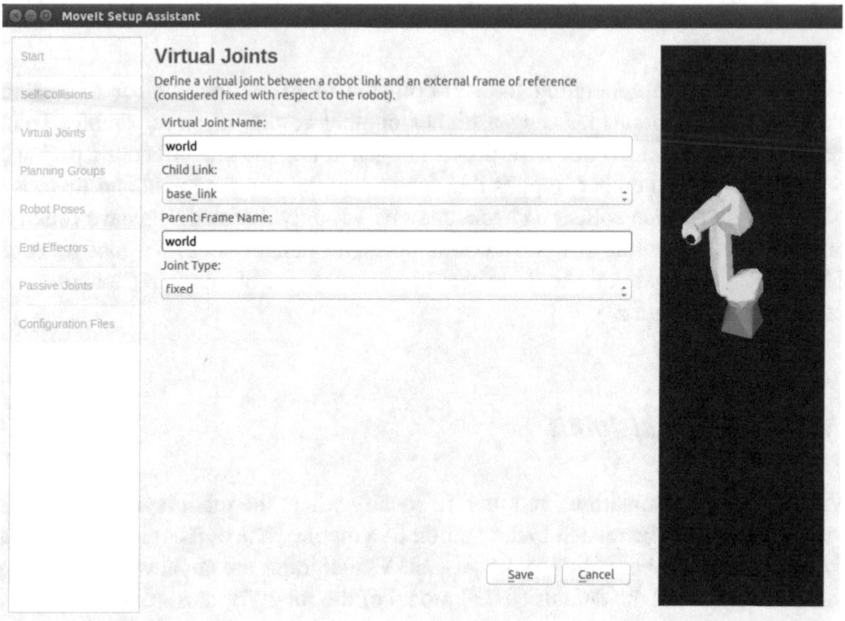

Fig. 6 Adding virtual joints

for a fixed robot). Figure 6 illustrates the process of adding a *fixed* joint that attaches the robot to the world.

4.4 Planning Groups

Planning groups bring together, semantically, different parts of the robot into a group, e.g. an arm or a leg. The definition of groups is the primary function of the SRDF. In the future, it is hoped that this information will move directly into the URDF. Groups are typically defined by grouping a set of joints together. Every child link of the joints is now a member of the group. Groups can also be defined as a chain by specifying the first link and the last link in the chain—this is more convenient when defining an arm or a leg.

In defining a group, you also have the opportunity to define a *kinematic solver* for the group (note that this choice is optional). The default kinematic solver that is always available for a group is the MoveIt! KDL Kinematics solver built around the Kinematics Dynamics Library package (KDL). This solver will only work with *chains*. It automatically checks (at startup) whether the group it is configured for is a chain or a disjoint collection of joints. Custom kinematics solvers can also be integrated into MoveIt! using a plugin architecture and will show up in the list of

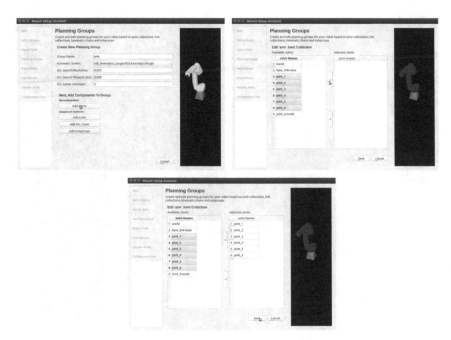

Fig. 7 Adding planning groups

choices for choosing a kinematics solver. Note that you may (and should) elect not to initialize a kinematics solver for certain groups (e.g. a parallel jaw gripper).

Figure 7 show an example where a Fanuc robot arm is configured to have a group that represents its six joints. The joints are added to the group using the "Add Joints" button (which is the recommended button). You can also define a group using just a link, e.g. to define an end-effector for the Fanuc M10ia robot, you would use the *tool0* link to define an end-effector group.

4.5 Robot Poses

The user may also add fixed poses of the robot into the SRDF. These poses are often used to describe configurations that are useful in different situations, e.g. a home position. These poses are then easily accessible using the internal C++ API of MoveIt!. Figure 8 shows a pose defined for the Fanuc M10ia. Note that these poses are user-defined and do not correspond to a native zero or home pose for the robot.

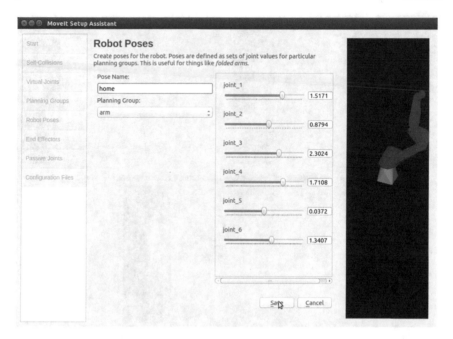

Fig. 8 Adding robot poses

4.6 Passive Joints

Certain joints in the robot can be designated as *passive joints*. This allows the various components of MoveIt! to know that such joints cannot be used for planning or control.

4.7 Adding End-Effectors (Optional)

Certain groups in the robot can be designated as *end-effectors*. This allows users to interact through these groups using the Rviz interface. Figure 9 shows the *gripper* group being designated as an end-effector.

4.8 Configuration Files

The last step in the MoveIt! Setup Assistant is to generate the configuration files that MoveIt! will use (Fig. 10). Note that it is convention to name the generated MoveIt! config package as robot_name_moveit_config. E.g. for the Fanuc robot used in our

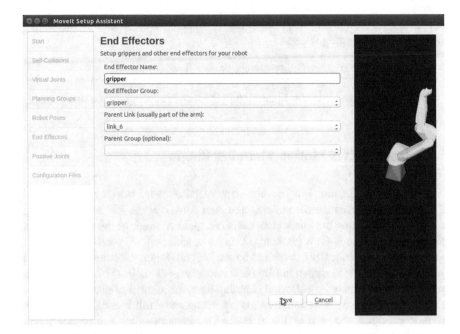

Fig. 9 Adding end-effectors

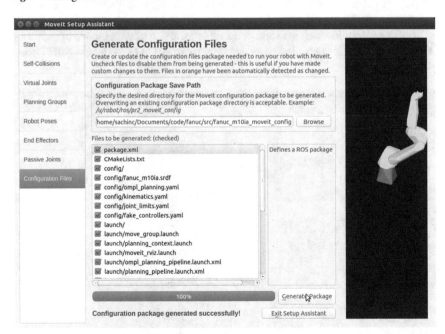

Fig. 10 Generating configuration files

example, we would name the package fanuc_m10ia_moveit_config. Now, running the initial demonstration is easy:

```
roslaunch <moveit_config_package_name> demo.launch
```

This entire process is better illustrated in the movie accompanying this paper.

5   Using the Rviz Motion Planning Plugin

The Rviz motion planning plugin is the primary interface for working with MoveIt!. It allows users to create environments and plan collision-free motions. It is shown in Fig. 11. Here, using the Context tab, users can make a choice of the type of motion planner to use. The default planning library available with MoveIt! is the OMPL library—it is automatically configured by the MoveIt! Setup Assistant. The next tab (labeled PLANNING) is shown in Fig. 12. It allows users to adjust parameters for the planning process, including allowed planning time, the number of planning attempts and the speed of the desired motion (as a percentage of full speed). It also allows users to configure the start and goal states for planning—they can be configured randomly, to match the current state of a simulated (or real) robot or to a *named* state setup in the MoveIt! Setup Assistant. The PLANNING tab also allows the user to plan and execute motions for the robot.

The Scene tab allows users to add or delete objects into the workspace of the robot. Objects or workcells are typically loaded directly from CAD models, e.g. STL files.

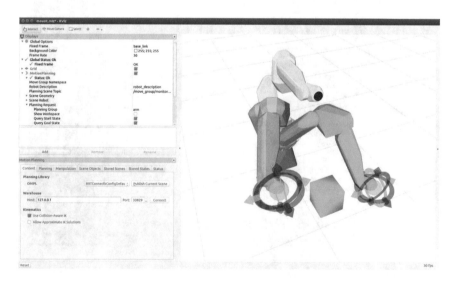

Fig. 11 MoveIt! Rviz Plugin

Fig. 12 MoveIt! Rviz Plugin: the planning interface (*left*) and the scene tab (*right*)

Fig. 13 MoveIt! Rviz
Plugin: an imported model

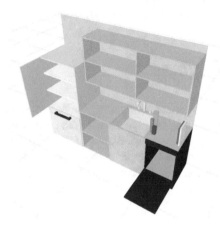

To explore this functionality, first download two files: a simple STL file representing a box and a representation of a complete *scene* for the robot to work in (in the *.scene* format). These files can be imported into the Planning Scene using the import buttons in the SCENE tab. To import STL files, use the IMPORT FILE button. To import Scene files, use the Import From Text button. Figure 13 shows a countertop model imported into MoveIt!. Once you import a model, make sure to return to the CONTEXT tab and press the Publish Current Scene button to indicate to MoveIt! that it should be working with an updated model of the environment.

5.1 Visualization and Interaction

The Rviz motion planning plugin is also the primary tool for visualization of MoveIt! plans and for interaction with MoveIt! robot models. Figure 14 shows the robot interaction models that the user can interact with. A blue ball attached to the end-effector allows the user to easily drag the end-effector around in any environment while ROS interactive markers are used for finer position and orientation control of

Fig. 14 MoveIt! Rviz
Plugin: interaction and
visualization. The *blue ball*
allows direct position
interaction with the
end-effector of a robot. Path
traces can be visualized in
MoveIt!

the end-effector. Planned paths are visualized using the plugin and can be cycled
continuously for better introspection.

5.2   Useful Hints

It is important, while using the MoveIt! Rviz Plugin, to make sure that the start state
for planning always matches the current state of the robot. It is also important to set
the speed to a conservative limit initially, especially with industrial robots, to make
sure that the robot is moving safely in its environment. User should also check the
collision model of their workcell and ensure that it is an accurate representation of
their workcell.

6   The move_group_interface

The primary recommended code API to MoveIt! is through the *move_group_interface*.
It provides both a C++ and Python API to the *move_group* node. The interface
abstracts the ROS API to MoveIt! and makes it easier to use. The ROS API is configured primarily using *constraints*, e.g. a position constraint for the end-effector
or joint constraints for the entire robot. A position constraint is typically specified
using a box volume in space for the end-effector. The *move_group_interface* allows
users to specify these constraints directly as a desired position, orientation or joint
configuration for the robot.

6.1   Planning to a Pose Goal

The code below shows how to plan to move the robot to a pose goal using the
C++ API.

```
moveit::planning_interface::MoveGroup group(``right_arm'');

geometry_msgs::Pose target_pose1;
target_pose1.orientation.w = 1.0;
target_pose1.position.x = 0.28;
target_pose1.position.y = -0.7;
target_pose1.position.z = 1.0;
group.setPoseTarget(target_pose1);

moveit::planning_interface::MoveGroup::Plan my_plan;
bool success = group.plan(my_plan);
```

The first line sets the group that the interface is working with (using the group name specified in the SRDF). Note the use of a ROS message to specify the pose target.

6.2 Planning to a Joint Goal

The code below shows how to plan to move the robot to a joint goal using the C++ API.

```
std::vector<double> group_variable_values;
group.getCurrentState()->copyJointGroupPositions(group.getCurrentState()
->getRobotModel()->getJointModelGroup(group.getName()),
    group_variable_values);

group_variable_values[0] = -1.0;
group.setJointValueTarget(group_variable_values);

moveit::planning_interface::MoveGroup::Plan my_plan;
success = group.plan(my_plan);
```

First, we get the current joint values for the group. Then, we modify one of the joint values to specify a new target. If the target is outside joint limits, no plan will be generated.

6.3 Move to Joint or Pose Goals

The process for moving to a joint or pose goal is the same as for planning to these goals. We will only use a different function call: move().

```
success = group.move(my_plan);
```

6.4 Adding Objects into the Environment

Objects can be easily added using both the C++ and Python API and through the
Rviz motion planning plugin. The motion planning plugin allows users to directly
import .STL files (e.g. a representation of the workcell). The main parameters that
can be adjusted are the collision model for the added object and the location of the
object. MoveIt! allows different types of collision models including primitives (box,
cylinder, sphere, cone) and mesh models. Mesh models should be simplified as far
as possible to minimize the number of traingles in them.

Workcell components can also be added directly using the PlanningSceneInterface
class.

```
moveit::planning_interface::PlanningSceneInterface
    planning_scene_interface;

moveit_msgs::CollisionObject collision_object;
collision_object.header.frame_id = group.getPlanningFrame();
collision_object.id = ''box1'';

/* Define a box to add to the world. */
shape_msgs::SolidPrimitive primitive;
primitive.type = primitive.BOX;
primitive.dimensions.resize(3);
primitive.dimensions[0] = 0.4;
primitive.dimensions[1] = 0.1;
primitive.dimensions[2] = 0.4;

geometry_msgs::Pose box_pose;
box_pose.orientation.w = 1.0;
box_pose.position.x = 0.6;
box_pose.position.y = -0.4;
box_pose.position.z = 1.2;

collision_object.primitives.push_back(primitive);
collision_object.primitive_poses.push_back(box_pose);
collision_object.operation = collision_object.ADD;

std::vector<moveit_msgs::CollisionObject> collision_objects;
collision_objects.push_back(collision_object);

planning_scene_interface.addCollisionObjects(collision_objects);
```

Attaching and detaching the collision object from the environment is also simple.
It is only important to make sure that the object has already been added to the
environment before attaching an object.

```
/* Attach the object */
group.attachObject(collision_object.id);

/* Detach the object */
group.detachObject(collision_object.id);
```

6.5   Helpful Hints

There are several debugging steps that a user can follow in case things don't go as planned. Here are some helpful hints for what can go wrong and how to fix it.

- **Robot won't move**: If the joint limits for the robot are not set properly, the robot may not be able to move. Check the URDF of the robot and make sure that each joint has a range of joint values to move through. Check to make sure that the maximum joint value is greater than the minimum joint value.
- **Robot won't move when I define soft limits**: If soft limits are defined for the robot in the URDF, they will be used by MoveIt!. Make sure that they are valid.
- **Motion plans are not being generated successfully**: Check that no two parts of the robot are in self-collision at all joint configurations. This can especially happen if you add new parts to the robot in the URDF but have not run the robot again through the MoveIt! Setup Assistant. If any robot parts appear red, they are in collision—run the MoveIt! Setup Assistant with the complete robot model to make sure that all collision checks that need to be disabled are labeled correctly in the SRDF.
- **GUI-based Interaction is not working properly**: Robots with 6 or more degrees of freedom do well with Rviz interfaction. Robots with less than 6 DOFs are harder to interact with through the plugin.
- **The motion plans are moving into collision**: MoveIt! checks each motion plan segment for collisions at a certain discretization. If this discretization value is too large, motion segments will not be checked at a fine discretization and the resulting motions may actually pass through parts of the environment. The discretization value can be adjusted using the *longest_valid_segment* parameter in the *ompl_planning.yaml* file.
- **The motion plans are moving into collision**: If using Rviz, make sure that you have pressed the Publish Planning Scene button before planning.

6.6   Additional Resources

Additional tutorials are available on the MoveIt! website from the tutorials page.

- Tutorial for the Move Group Interface (C++, Python)—this tutorial described the main user interface to MoveIt! for users who would like to interact through a ROS interface. It is the recommended interface for all beginners.
- Tutorial for using the kinematics API (C++)—this tutorial delves deeper into the use of kinematics (forward and inverse kinematics) using the programmatic C++ API. This tutorial is only recommended for more advanced users of MoveIt!.
- Tutorial for the Planning Scene API (C++, ROS)—this tutorial explains the structure and interface to the planning scene API.

- Loading motion planners from plugins, using kinematic constraints: (C++)—this tutorial explains how to load and use motion planners directly from the C++ interface and also how to specify and use some types of kinematic constraints. It is intended for advanced users.
- Using the motion planning pipeline and planning request adapters: (C++)—this tutorial explains how to use the planning request adapters as part of a motion planning pipeline to change the input to and output from the motion planners. It is intended for advanced users.

7 Connecting to a Robot

MoveIt! can connect directly to a robot through a ROS interface. The requirements on a ROS interface include a source of joint information, transform information and an interface to a trajectory controller:

- **Joint States**: A source of joint state information is needed. This source must publish the state information (at least the position of each joint) at a reasonable rate on the *joint_states* topic. A typical rate is 100 Hz. Different components can publish on the same topic and all the information will be combined internally by MoveIt! to maintain the right state. Note that MoveIt! can account for *mimic* joints, i.e. coupled joints where a single actuator or motor controls the motion of two joints.
- **Transform Information**: MoveIt! uses the joint state information to maintain its own transform tree internally. However, joint state information does not contain information about external virtual joints, e.g. the position of the robot in an external map. This transform information must exist on the ROS transform server using the TF package in ROS.
- **Trajectory Action Controller**: MoveIt! also requires the existence of a trajectory action controller that supports a ROS *action* interface using the FOLLOWJOINT-TRAJECTORY action in the *control_msgs* package.
- **Gripper Command Action Interface–OPTIONAL**: The gripper command action interface, allows for easy control of a gripper. It differs from the Trajectory Action interface in allowing the user to set a maximum force that can be applied by the gripper.

7.1 Configuring the Controller Interface

Configuring the controller interface requires generating the right controller configuration YAML file. An example of a controller configuration for two arms of a robot and a gripper is given below:

```
controller_list:
 - name: r_arm_controller
   action_ns: follow_joint_trajectory
   type: FollowJointTrajectory
   default: true
   joints:
     - r_shoulder_pan_joint
     - r_shoulder_lift_joint
     - r_upper_arm_roll_joint
     - r_elbow_flex_joint
     - r_forearm_roll_joint
     - r_wrist_flex_joint
     - r_wrist_roll_joint
 - name: l_arm_controller
   action_ns: follow_joint_trajectory
   type: FollowJointTrajectory
   default: true
   joints:
     - l_shoulder_pan_joint
     - l_shoulder_lift_joint
     - l_upper_arm_roll_joint
     - l_elbow_flex_joint
     - l_forearm_roll_joint
     - l_wrist_flex_joint
     - l_wrist_roll_joint
 - name: gripper_controller
   action_ns: gripper_action
   type: GripperCommand
   default: true
   joints:
     - l_gripper_joint
     - r_gripper_joint
```

This configuration should go into a YAML file (called controllers.yaml) inside the MoveIt! config package. Now, create the controller launch file (call it robot_moveit_controller_manager.launch where *robot* is the name of your robot—the robot name needs to match the name specified when you created your MoveIt! config directory).

Add the following lines to this file:

```
<launch>
 <!-- Set the param that trajectory_execution_manager needs to find the
      controller plugin -->
 <arg name=''moveit_controller_manager''
      default=''moveit_simple_controller_manager/MoveItSimpleControllerManager''/>
 <param name=''moveit_controller_manager'' value=''$(arg
      moveit_controller_manager)''/>
 <!-- load controller_list -->
 <rosparam file=''$(find my_robot_name_moveit_config)/config/controllers.yaml''/>
</launch>
```

MAKE SURE to replace *my_robot_name_moveit_config* with the correct path for your MoveIt! config directory.

Now, you should be ready to have MoveIt! talk to your robot.

7.2 Debugging Hints

The FollowJointTrajectory or GripperCommand interfaces on your robot must be communicating in the correct namespace. In the above example, you should be able to see the following topics (using rostopic list) on your robot:

```
/r_arm_controller/follow_joint_trajectory/goal
/r_arm_controller/follow_joint_trajectory/feedback
/r_arm_controller/follow_joint_trajectory/result
/l_arm_controller/follow_joint_trajectory/goal
/l_arm_controller/follow_joint_trajectory/feedback
/l_arm_controller/follow_joint_trajectory/result
/gripper_controller/gripper_action/goal
/gripper_controller/gripper_action/feedback
/gripper_controller/gripper_action/result
```

You should also be able to see (using *rostopic info topic_name*) that the topics are published/subscribed to by the controllers on your robot and also by the *move_group* node.

7.3 Integrating 3D Perception

MoveIt! also integrated 3D perception for building an updated representation of the environment. This representation is based on the OCTOMAP package. The 3D perception elements include components that perform *self-filtering*, i.e. filter out parts of the robot from sensor streams. Input to the Octomap is from any source of 3D information in the form of a depth map. The resolution of the OCTOMAP is adjustable. Using a coarser resolution ensures that the 3D perception pipeline can run at a reasonable rate. A 10 cm resolution should result in an update rate of close to 10 Hz. The 3D perception can be integrated using a YAML configuratoin file.

```
sensors:
  - sensor_plugin: occupancy_map_monitor/DepthImageOctomapUpdater
    image_topic: /head_mount_kinect/depth_registered/image_raw
    queue_size: 5
    near_clipping_plane_distance: 0.3
    far_clipping_plane_distance: 5.0
    shadow_threshold: 0.2
    padding_scale: 4.0
    padding_offset: 0.03
    filtered_cloud_topic: filtered_cloud
```

You will now need to update the *moveit_sensor_manager.launch* file in the launch directory of your MoveIt! configuration directory with this sensor information (this file is auto-generated by the Setup Assistant but is empty). You will need to add the following line into that file to configure the set of sensor sources for MoveIt! to use:

```
\begin{verbatim}
<rosparam command=''load'' file=''$(find
    my_moveit_config)/config/sensors_kinect.yaml'' />
\end{verbatim}
```

Note that you will need to input the path to the right file you have created above. You will also need to configure the Octomap by adding the following lines into the moveit_sensor_manager.launch:

```
<param name=''octomap_frame'' type=''string''
    value=''odom_combined'' />
<param name=''octomap_resolution'' type=''double''
    value=''0.05'' />
<param name=''max_range'' type=''double'' value=''5.0'' />
```

7.4 Helpful Hints

- **Controller Configuration**: MoveIt! does not implement any controllers on its own. Users will have to implement their own controller or, preferably, use ROS-Control [6] as a controller framework. The ROS-Control package includes trajectory interpolation and smoothing that is essential for smooth operation on industrial and other robots.
- **Sensor Configuration**: The self-filtering routines function best if the URDF is an accurate representation of the robot. Padding parameters can also be used to adjust the size of the meshes used for self-filtering. It is always best to add a little padding to the meshes since uncertainty in the motion of the robot can cause the *self-filtering* to fail.

8 Building Applications with MoveIt!

MoveIt! is a platform for robotic manipulation. It forms an ideal base to build large-scale applications. Examples of such applications include:

- **Pick and Place**: MoveIt! includes a pick and place pipeline. The pipeline allows pick and place tasks to be fully planned given a set of grasps for an object to be picked and a set of place locations where the object can be placed. The pipeline utilizes a series of manipulation planning *stages* that are configured to run in parallel. They include (a) freespace motion planning stage to plan the overall motion of the arm, (b) cartesian planning stages to plan approach and retreat motions, (c) inverse kinematics stages to compute arm configurations for pick and place and (d) perception and grasping interfaces for object recognition and grasp planning respectively.

- **Process Path Planning**: Complex processes like gluing, welding, etc. often require an end-effector to follow a prescribed path while avoiding collisions, singularities and other constraints. MoveIt! can plan such process paths using a cartesian planning routine built into MoveIt! itself.
- **Planning Visibility Paths**: MoveIt! includes the ability to process visibility constraints. Visibility planning is particularly useful when planning inspection paths, i.e. planning a path for inspecting a complex part with a camera mounted on the end-effector of an arm.
- **Tele-operation**: MoveIt! has been used for tele-operation in complex environments, e.g. in the DARPA Robotic Challenge. The ability to visualize complex environments while also planning collision free paths allows full teleoperation applications to be built with MoveIt!.

9 Conclusion

MoveIt! has rapidly emerged as the core ROS package for manipulation. In combination with ROS-Control, it provides a framework for building core functionality and full applications for any robotics task. The use of the MoveIt! setup assistant has made MoveIt! more accessible to new and intermediate users. It has allowed new robotic platforms to be easily integrated into ROS. The next goal for MoveIt! development is to extend its capabilities to provide even more out of the box capabilities and enable better integration with other types of sensing (e.g. force/tactile) sensing. We also aim to extend MoveIt! to whole-body manipulation tasks to enable more applications with humanoid robots. MoveIt! can also form the basis for collaborative robots, enabling the next generation of tasks where humans and robots work together. For more information, users are referred to the following resources for MoveIt!:

- The MoveIt! Website
- The MoveIt! Tutorials
- MoveIt! High-level Concept Documentation
- MoveIt! mailing list
- MoveIt! Source code

Acknowledgments MoveIt! is developed and maintained by a large community of users. Special mention should be made of Dave Hershberger, Dave Coleman, Michael Ferguson, Ioan Sucan and Acorn Pooley for supporting and maintaining MoveIt! and its associated components in ROS over the last few years.

References

1. S. Chitta, E.G. Jones, M. Ciocarlie, K. Hsiao, Perception, planning, and execution for mobile manipulation in unstructured environments. IEEE Robot. Autom. Mag. Special Issue on Mobile Manipulation **19**(2), 58–71 (2012)
2. S. Chitta, E.G. Jones, I. Sucan, Arm Navigation. http://wiki.ros.org/arm_navigation (2010)
3. J. Pan, S. Chitta, D. Manocha, FCL: a general purpose library for collision and proximity queries, in *IEEE International Conference on Robotics and Automation, Minneapolis*, Minnesota (2012)
4. D. Coleman, Integrating IKFast with MoveIt!: A Tutorial. http://docs.ros.org/hydro/api/moveit_ikfast/html/doc/ikfast_tutorial.html (2014)
5. S. Chitta, A. Pooley, D. Hershberger, I. Sucan, MoveIt! http://moveit.ros.org (2015)
6. The ROS Control Framework. http://wiki.ros.org/ros_control 2014

Hands-on Learning of ROS Using Common Hardware

Andreas Bihlmaier and Heinz Wörn

Abstract Enhancing the teaching of robotics with hands-on activities is clearly beneficial. Yet at the same time, resources in higher education are scarce. Apart from the lack of supervisors, there are often not enough robots available for undergraduate teaching. Robotics simulators are a viable substitute for some tasks, but often real world interaction is more engaging. In this tutorial chapter, we present a hands-on introduction to ROS, which requires only hardware that is most likely already available or costs only about 150$. Instead of starting out with theoretical or highly artificial examples, the basic idea is to work along tangible ones. Each example is supposed to have an obvious relation to whatever real robotic system the knowledge should be transfered to afterwards. At the same time, the introduction covers all important aspects of ROS from sensors, transformations, robot modeling, simulation and motion planning to actuator control. Of course, one chapter cannot cover any subsystem in depth, rather the aim is to provide a big picture of ROS in a coherent and hands-on manner with many pointers to more in-depth information. The tutorial was written for ROS Indigo running on Ubuntu Trusty (14.04). The accompanying source code repository is available at https://github.com/andreasBihlmaier/holoruch.

Keywords General introduction · Hands-on learning · Education

1 Introduction

Individual ROS packages are sometimes well documented and sometimes not. However, the bigger problem for somebody working with ROS for the first time is not the poor documentation of individual packages. Instead the actual problem

A. Bihlmaier (✉) · H. Wörn
Institute for Anthropomatics and Robotics (IAR), Intelligent Process Control
and Robotics Lab (IPR), Karlsruhe Institute of Technology (KIT),
76131 Karlsruhe, Germany
e-mail: andreas.bihlmaier@kit.edu

H. Wörn
e-mail: woern@kit.edu

© Springer International Publishing Switzerland 2016
A. Koubaa (ed.), *Robot Operating System (ROS)*, Studies in Computational
Intelligence 625, DOI 10.1007/978-3-319-26054-9_2

is to understand the big picture-to understand how all the various pieces of ROS come together.[1] Although the ROS community provides tutorials,[2] in our experience undergraduates struggle to transfer what they have learned in the sandbox tutorials to real systems. At the same time, it is difficult to start out with real ROS robots for two reasons. First, real robots are expensive, easily broken and often require significant space to work with. Therefore they are often not available in sufficient quantity for undergraduate education. Second, real robots can be dangerous to work with. This holds true especially if the goal is to provide a hands-on learning experience for the students, i.e. allow them to explore and figure out the system by themselves.

Ideally, each student would be provided with a simple robot that is safe and yet capable enough to also learn more advanced concepts. To our knowledge no such device is commercially available in the range of less than 200$. We will not suggest how one could be built, since building it for each student would be too time consuming. Instead, the goal of this tutorial is to detail a hands-on introduction to ROS on the basis of commonly available and very low-cost hardware, which does not require tinkering. The main hardware components are one or two webcams, a Microsoft Kinect or Asus Xtion and two or more low-power Dynamixel servos. The webcams may also be laptop integrated. Optionally, small embedded computers such as Raspberry Pis or BeagleBone Blacks can be utilized. We assume the reader to understand fundamental networking and operating system concepts, to be familiar with the basics of Linux including the command line and to know C++ or Python. Furthermore, some exposure to CMake is beneficial.

The remainder of the chapter is structured as follows:

- First, a brief background section on essential concepts of ROS. It covers the concepts of the ROS master, names, nodes, messages, topics, services, parameters and launch files. This section should be read on a first reading. However, its purpose is also to serve as glossary and reference for the rest of the chapter.
- Second, a common basis in terms of the host setup is created.
- Third, working with a single camera, e.g. a webcam, under ROS serves as an example to introduce the computation graph: nodes, topics and messages. In addition, rqt and tools of the image_pipeline stack are introduced.
- Fourth, a custom catkin package for filtering sensor_msgs/Image is created. Names, services, parameters and launch files are presented. Also, the definition of custom messages and services as well as the dynamic_reconfigure and vision_opencv stacks are shown.
- Fifth, we give a short introduction on how to use RGB-D cameras in ROS, such as the Microsoft Kinect or Asus Xtion. Point clouds are visualized in rviz and pointers for the interoperability between ROS and PCL are provided.
- Sixth, working with Dynamixel smart servos is explained in order to explain the basics of ros_control.

---

[1]At least this has been the experience in our lab, not only for undergraduates but also for graduate students with a solid background in robotics, who had never worked with ROS before.

[2]http://wiki.ros.org/ROS/Tutorials.

- Seventh, a simple robot with two joints and a camera at the end effector is modelled as an URDF robot description. URDF visualization tools are shown. Furthermore, tf is introduced.
- Eighth, based on the URDF model, a MoveIt! configuration for the robot is generated and motion planning is presented exemplary.
- Ninth, the URDF is extended and converted to SDF in order to simulate the robot with Gazebo.

2 Background

ROS topics are an implementation of the publish-subscribe mechanism, in which the ROS Master serves as a well-known entry point for naming and registration. Each ROS node advertises the topics it publishes or subscribes to the ROS Master. If a publication and subscription exist for the same topic, a direct connection is created between the publishing and subscribing node(s), as shown in Fig. 1. In order to have a definite vocabulary, which may also serve the reader as glossary or reference, we give a few short definitions of ROS terminology:

- **Master**[3]: Unique, well-known (ROS_MASTER_URI environment variable) entry point for naming and registration. Often referred to as roscore.
- **(Graph Resource) Name**[4]: A name of a resource (node, topic, service or parameter) within the ROS computation graph. The naming scheme is hierarchical and has many aspects in common to UNIX file system paths, e.g. they can be absolute or relative.
- **Host**: Computer within the ROS network, identified by its IP address (ROS_IP environment variable).
- **Node**[5]: Any process using the ROS client API, identified by its graph resource name.
- **Topic**[6]: A unidirectional, asynchronous, strongly typed, named communication channel as used in the publish-subscribe mechanism, identified by its graph resource name.
- **Message**[7]: A specific data structure, based on a set of built-in types,[8] used as type for topics. Messages can be arbitrarily nested, but do not offer any kind of is-a (inheritance) mechanism.

---

[3]http://wiki.ros.org/Master.

[4]http://wiki.ros.org/Names.

[5]http://wiki.ros.org/Nodes.

[6]http://wiki.ros.org/Topics.

[7]http://wiki.ros.org/Messages.

[8]http://wiki.ros.org/msg.

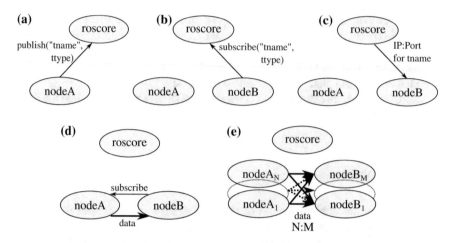

Fig. 1 Overview of the ROS topic mechanism. When a publisher (**a**) and subscriber (**b**) are registered to the same topic, the subscriber receives the network address and port of all publishers (**c**). The subscriber continues by directly contacting each publisher, which in return starts sending data directly to the subscriber (**d**). Many nodes can publish and subscribe to the same topic resulting in a $N : M$ relation (**e**). On the network layer there are $N \cdot M$ connections, one for each (publisher, subscriber) tuple. The nodes can be distributed over any number of hosts within the ROS network

- **Connection** A connection between a (publisher, subscriber) tuple carrying the data of a specific topic (cf. Fig. 1), identified by the tuple (publisher-node, topic, subscriber-node).
- **Service**[9]: A synchronous remote procedure call, identified by its graph resource name.
- **Action**[10]: A higher-level mechanism built on top of topics and services for long-lasting or preemptable tasks with intermediate feedback to the caller.
- **Parameters**[11]: "A shared, multi-variate dictionary that is accessible via network APIs." Its intended use is for slow changing data, such as initialization arguments. A specific parameter is identified by its graph resource name.
- **roslaunch**[12]: A command line tool and XML format to coherently start a set of nodes including remapping of names and setting of parameters.

Topics are well suited for streaming data, where each subscriber is supposed to get as much of the data as he can process in a given time interval and the network can deliver. The network overhead and latency is comparatively low because the connections between publisher and subscriber remain open. However, it is important to remember that messages are automatically dropped, if the subscriber queue becomes full. In contrast, services establish a new connection per call and cannot

---

[9]http://wiki.ros.org/Services.

[10]http://wiki.ros.org/actionlib.

[11]http://wiki.ros.org/ParameterServer.

[12]http://wiki.ros.org/roslaunch.

fail without notice on the caller side. The third mechanism built on top of topics and services are actions. An action uses services to initiate and finalize a (potentially) long-lasting task, thereby providing definite feedback. Between start and end of the action, continuous feedback is provided via the topics mechanism. This mapping to basic communication mechanisms is encapsulated by the actionlib.

3 ROS Environment Configuration

For the rest of this chapter, we assume a working standard installation[13] of ROS Indigo-on Ubuntu Trusty (14.04). Furthermore, everything shown in this chapter can be done on a single host. Therefore, a localhost setup is assumed, i.e. roscore and all nodes run on localhost:

```
echo 'export ROS_MASTER_URI=http://127.0.0.1:11311' >> ~/.bashrc
echo 'export ROS_IP=127.0.0.1' >> ~/.bashrc

# run "source ~/.bashrc" or open a new terminal
echo $ROS_MASTER_URI
# should: http://127.0.0.1:11311
echo $ROS_IP
# should: 127.0.0.1
```

From here on it is presumed that roscore is always running. The second part of setup requires a working catkin build system.[14] In case no catkin workspace has been initialized, this can be achieved with

```
mkdir -p ~/catkin_ws/src
cd ~/catkin_ws/src
catkin_init_workspace
```

Afterwards all packages in the catkin workspace source directory (~/catkin_ws/src) can be built with

```
cd ~/catkin_ws
catkin_make
```

The catkin_make command should run without errors for an empty workspace. In the rest of this chapter, the instruction "Install the REPOSITORY package from source" refers to cloning the repository into the ~/catkin_ws/src directory, followed by running catkin_make. Start by installing https://github.com/andreasBihlmaier/holoruch, which will be used throughout this chapter. After these preliminaries, we can start the hands-on introduction to the various aspects of ROS.

---

[13] A desktop-full installation according to http://wiki.ros.org/indigo/Installation/Ubuntu.
[14] http://wiki.ros.org/catkin.

4 Camera Sensors: Driver, Use and Calibration

The first example is a camera sensor. A USB webcam or integrated camera must be attached to the computer. If multiple cameras are connected to the computer, the first one is used here.[15] Where these settings are stored and how to change them will be shown later in this section. Use guvcview to make sure the camera is working correctly and also have a look at the "Image Controls" and "Video" tab for supported resolutions/framerates and control settings.[16]

First, install https://github.com/ktossell/camera_umd from source:

```
cd ~/catkin_ws/src ; git clone https://github.com/ktossell/
    ↪ camera_umd.git
cd ~/catkin_ws ; catkin_make
```

Second, start the camera driver node

```
roslaunch holoruch_camera webcam.launch
```

Third, use the command line utilities to check if the camera node is running as expected:

```
rosnode list
```

The output should contain /webcam/uvc_camera_webcam. We also want to check if the expected topics have been created:

```
rostopic list
```

Here, /webcam/camera_info and /webcam/image_raw is expected. If either the node or the topics do not show up, look at the output of roslaunch and compare possible errors to the explanation of the launch file in this section. Next, start rqt to visualize the image data of the webcam

```
rosrun rqt_gui rqt_gui
```

Start the "Image View" plugin through the menu: "Plugins" → "Visualization" → "Image View". Use the upper left drop-down list to select /webcam/image_raw. Now the webcam's live image should appear. Note that everything is already network transparent. If we would not be using localhost IP addresses, the camera node and rqt could be running on two different hosts without having to change anything.

After a simple data source and consumer have been setup and before we add more nodes, let's look at all of the involved ROS components. First the roslaunch file webcam.launch:

---

[15]The order is determined by the Linux kernel's Udev subsystem.

[16]Note: While it does not matter for this tutorial, it is essential for any real-world application that the webcam either has manual focus or the auto focus can be disabled. Very cheap or old webcams have the former and better new ones usually have the latter. For the second kind try v4l2-ctl -c focus_auto=0.

```
<launch>
  <node ns="/webcam"
        pkg="uvc_camera" type="uvc_camera_node" name="uvc_camera_webcam"
        output="screen">
    <param name="width" type="int" value="640" />
    <param name="height" type="int" value="480" />
    <param name="fps" type="int" value="30" />
    <param name="frame" type="string" value="wide_stereo" />

    <param name="auto_focus" type="bool" value="False" />
    <param name="focus_absolute" type="int" value="0" />
    <!-- other supported params: auto_exposure,
         exposure_absolute, brightness, power_line_frequency -->

    <param name="device" type="string" value="/dev/video0" />
    <param name="camera_info_url" type="string"
           value="file://$(find holoruch_camera)/webcam.yaml" />
  </node>
</launch>
```

The goal here is to introduce the major elements of a roslaunch file, not to detail all features, which are described in the ROS wiki.[17] Each <node> tag starts one node, in our case uvc_camera_node from the uvc_camera package we installed earlier. The name, uvc_camera_webcam can be arbitrarily chosen, a good convention is to combine the executable name with a task specific description. If output was not set to screen, the node's output would not be shown in the terminal, but sent to a log file.[18] Finally, the namespace tag, ns, allows to prefix the node's graph name, which is in analogy with pushing it down into a subdirectory of the filesystem namespace. Note that the nodes are not started in any particular order and there is no way to enforce one. The <param> tags can be either direct children of <launch> or within a <node> tag.[19] Either way they allow to set values on the parameter server from within the launch file before any of the nodes are started. In the latter case, which applies here, each <param> tag specifies a private parameter[20] for the parent <node> tag. The parameters are node specific.

In case of the uvc_camera_node, the parameters pertain to camera settings. This node uses the Linux Video4Linux2 API[21] to retrieve images from any video input device supported by the kernel. It then converts each image to the ROS image format sensor_msgs/Image and published them over a topic, thereby making them available to the whole ROS system.[22] If the launch file does not work as it is, this is most likely related to the combination of width, height and fps. Further

---

[17]http://wiki.ros.org/roslaunch/XML.

[18]See http://wiki.ros.org/roslaunch/XML/node.

[19]See http://wiki.ros.org/roslaunch/XML/param.

[20]Cf. http://wiki.ros.org/Names.

[21]http://lwn.net/Articles/203924/.

[22]See also http://wiki.ros.org/image_common.

information on the `uvc_camera` package is available on the ROS wiki[23] and in the repository's example launch files.[24]

In order to get a transformation between the 3D world and the 2D image of the camera, an instrinsic calibration of the camera is required. This functionality is available, for mono and stereo cameras, through the `cameracalibrator.py` in the `camera_calibration` package. Here one essential feature and design pattern of ROS comes into play. The `cameracalibrator.py` node does not require command line arguments for changing the relevant ROS names, such as the image topic. Instead ROS provides runtime name remapping. Any ROS graph name within a node's code can be changed by this ROS mechanism on startup of the node. In our case, the `cameracalibrator.py` code subscribes to a `"image"` topic, but we want it to subscribe to the webcam's images on `/webcam/image_raw`. The launch syntax provides the `<remap>` tag for this purpose:

```
<launch>
  <node pkg="camera_calibration" type="cameracalibrator.py"
        name="calibrator_webcam" output="screen"
        args="--size 8x6 --square 0.0255">
    <remap from="image" to="/webcam/image_raw" />
    <remap from="camera" to="/webcam" />
  </node>
</launch>
```

The same can be achieved on the command line by the `oldname:=newname` syntax:

```
rosrun camera_calibration cameracalibrator.py \
  --size 8x6 --square 0.0255 \
  image:=/webcam/image_raw camera:=/webcam
```

In both cases, the same node with the same arguments and remappings is started, albeit with a different node name. The monocular calibration tutorial[25] explains all steps to calibrate the webcam using a printed checkerboard. We continue with the assumption this calibration has been done and "commited".[26]

Commonly required image processing tasks, such as undistorting and rectification of images, are available in the `image_proc` package.[27] Due to the flexibility of the topic mechanism, we do not have to restart the camera driver, rather we just add further nodes to the ROS graph.[28] Run

```
roslaunch holoruch_camera proc_webcam.launch
```

---

[23] http://wiki.ros.org/uvc_camera.

[24] ~/catkin_ws/src/camera_umd/uvc_camera/launch/example.launch.

[25] http://wiki.ros.org/camera_calibration/Tutorials/MonocularCalibration.

[26] Check that the values of `rostopic echo -n 1 /webcam/camera_info` correspond to those printed to the terminal by `cameracalibrator.py`.

[27] http://wiki.ros.org/image_proc.

[28] If the camera driver and image processing is running on the same host, it is good practice to use nodelets instead of nodes for both-in order to reduce memory and (de)serialization overhead. However, this does not substantially change anything.

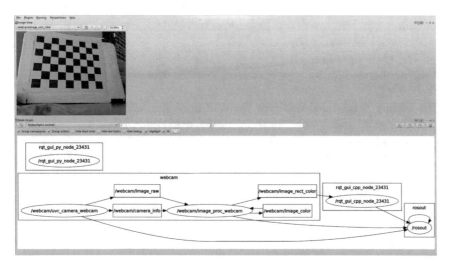

Fig. 2 The screenshot shows `rqt` with the "Image View" plugin on the *top* and the "Node Graph" on the *bottom*. The graph visualizes the simple image processing pipeline consisting of `uvc_camera` and `image_proc`

Afterwards `rostopic list` should show additional topics in the `/webcam` namespace. For a live view of the undistorted image, update the drop-down list in the "Image View" `rqt` plugin and select `/webcam/image_rect_color`.

So far the ROS graph is simple. There only exists the `uvc_camera_node`, whose `image_raw` and `camera_info` topics are subscribed by the `image_proc_webcam` node, whose `image_rect_color` topic is in turn subscribed by the `rqt` plugin. However, graphs of real systems often contain dozens of nodes with hundreds of topics. Fortunately, ROS provides introspection capabilities. That is, ROS provides mechanisms to acquire some key information about the current system state. One important information is the structure of the computation graph, i.e. which nodes are running and to which topics is each one publishing or subscribing. This can be visualized with the "Node Graph" `rqt` plugin (under "Plugins" → "Introspection"). The current ROS graph can be seen in Fig. 2. Next, we will write a custom ROS node that does custom image processing on the webcam image stream.

5 Custom Node and Messages for Image Processing with OpenCV

In this section we will create a custom catkin package for image processing on a `sensor_msgs/Image` topic.[29] Also, the definition of custom messages and services as well as the `dynamic_reconfigure` and `vision_opencv` stacks

---

[29]The goal of this section is to provide an example which is short, but at the same time very close to a real useful node. Standalone examples of how to create ROS publishers and subscribers in C++

are shown. First, create a new package, here `holoruch_custom`, and specify all dependencies that are already known[30]:

```
cd ~/catkin_ws/src
catkin_create_pkg holoruch_custom roscpp dynamic_reconfigure
cv_bridge
```

The slightly abbreviated code for the core node functionality in `holoruch_custom/src/holoruch_custom_node.cpp` is shown below

```
1   int main(int argc, char **argv) {
2     ros::init(argc, argv, "holoruch_custom");
3     ros::NodeHandle n;
4
5     sub = n.subscribe("image_raw", 1, imageCallback);
6     pub = n.advertise<sensor_msgs::Image>("image_edges", 1);
7
8     ros::spin();
9     return 0;
10  }
11
12  void imageCallback(const sensor_msgs::ImageConstPtr& msg) {
13    cv_bridge::CvImagePtr cvimg = cv_bridge::toCvCopy(msg, "bgr8");
14
15    cv::Mat img_gray;
16    cv::cvtColor(cvimg->image, img_gray, CV_BGR2GRAY);
17    cv::Canny(img_gray, img_gray, cfg.threshLow, cfg.thresHigh);
18    cv::Mat img_edges_color(cvimg->image.size(), cvimg->image.type(),
19                           cv::Scalar(cfg.edgeB, cfg.edgeG, cfg.edgeR
20  ));
21    img_edges_color.copyTo(cvimg->image, img_gray); // use img_gray as
22  mask
23
24    pub.publish(cvimg->toImageMsg());
25  }
```

After initializing the node (2), we create a subscriber to receive the images (5) and a publisher to send the modified ones (6). The remaining work will be done in the subscriber callbacks, which are handled by the ROS spinner (8). Each time a new image arrives, the `imageCallback` function is called. It converts the ROS image format to the OpenCV format (13) using the `cv_bridge` package.[31] Afterwards some OpenCV functions are applied to the data (15–20), the resulting image is converted back to ROS and published (22). In the example, we apply an edge filter

(Footnote 29 continued)
and Python are available in the ROS wiki: http://wiki.ros.org/ROS/Tutorials. Very basic standalone packages with examples can also be found at: https://github.com/andreasBihlmaier/ahb_rospy_example and https://github.com/andreasBihlmaier/ahb_roscpp_example.

[30]These steps are for illustration only, the full `holoruch_custom` and `holoruch_custom_msgs` package is already contained in the `holoruch` repository.

[31]See http://wiki.ros.org/cv_bridge. We will not go into any OpenCV details such as color encodings here. For more information see http://wiki.ros.org/cv_bridge/Tutorials/UsingCvBridgeToConvertBetweenROSImagesAndOpenCVImages and the OpenCV documentation at http://opencv.org/documentation.html.

to the image and overlay the edges on the original one. There is much more to be learned about working with images in ROS. However, since the goal of this tutorial is to give the big picture, we cannot go into further details here.[32]

Now we add the functionality to change the filter settings during runtime using `dynamic_reconfigure`. If the goal would be to change the settings only on startup (or very seldom), we would use the parameter server instead.[33] First, we need to define the parameters, which is accomplished by creating a `holoruch_custom/cfg/filter.cfg` file:

```
# imports omitted from listing
g = ParameterGenerator()
#      Name       Type    Rcfg-lvl Description             Default Min Max
g.add("edgeR", int_t, 0,          "Edge Color Red",   255,    0,   255)
g.add("edgeG", int_t, 0,          "Edge Color Green", 0,      0,   255)
# further parameters omitted from listing
exit(g.generate(PACKAGE, "holoruch_custom", "filter"))
```

Second, we need to initialize a `dynamic_reconfigure` in `main`. Therefore, we add the following in line 7 above

```
dynamic_reconfigure::Server<holoruch_custom::filterConfig> server;
server.setCallback(boost::bind(&dynamic_reconf_cb, _1, _2));
```

and a new callback function, e.g. after line 23, for reconfigure

```
void dynamic_reconf_cb(holoruch_custom::filterConfig &ncfg, uint32_t lvl
)
{ // cfg is a global variable: holoruch_custom::filterConfig cfg
  cfg = ncfg;
}
```

We refer to the `holoruch_custom` repository regarding the required additions in `CMakeLists.txt`.

After compiling the workspace with `catkin_make`, we can now run our custom node. This time we use the `ROS_NAMESPACE` environment variable to put `holoruch_custom_node`, and thereby its depend names, into `/webcam`.

```
env ROS_NAMESPACE=/webcam rosrun holoruch_custom holoruch_custom_node \
  image_raw:=image_rect_color
```

Again, the same could be achieved by remapping alone or by the roslaunch ns tag as shown in the previous section. The result is shown in Fig. 3, the filter settings can be continuously changed while the node is running by use of the "Dynamic Reconfigure" rqt plugin.

So far we have only used predefined ROS message types. Let's assume we want to publish the number of connected edge pixel components, which are detected by our node, together with the number of pixels in each one. No adequate predefined

---

[32]Good documentation and tutorials to delve into this topic can be found at http://wiki.ros. org/image_common, http://wiki.ros.org/camera_calibration and http://wiki.ros.org/cv_bridge. For more information on ROS supported sensors, see "Cameras" and "3D sensors" at http://wiki.ros. org/Sensors.

[33]See http://wiki.ros.org/roscpp/Overview/ParameterServer. for an example.

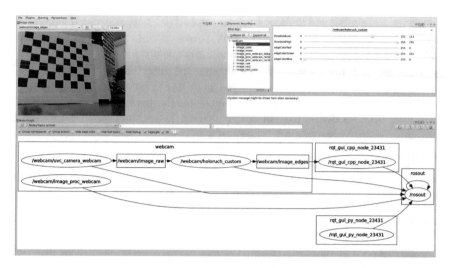

Fig. 3 A screenshot of the custom image processing node's output /webcam/image_edges in rqt. Compared to Fig. 2, the "Dynamic Reconfigure" plugin was added in the *upper right* and the "Node Graph" in the *bottom* shows the additional node holoruch_custom and its connections

message exists for this purpose, thus we will define a custom one.[34] We could put the message definitions into holoruch_custom, but it is favorable to have them in a separate package. This way other nodes that want to use the custom message only have to depend on the message package and only compile it. Thus we create it

```
catkin_create_pkg holoruch_custom_msgs \
  message_generation message_runtime std_msgs
```

Our custom message holoruch_custom_msgs/EdgePixels, is defined by the file holoruch_custom_msgs/msg/EdgePixels.msg[35]:

```
std_msgs/Header header
int32[] edge_components
```

Again, see the repository for details about message generation with CMakeLists. txt. After running catkin_make, the custom message is available in the same manner as predefined ones:

```
rosmsg show holoruch_custom_msgs/EdgePixels
```

The custom message can now be used in holoruch_custom_node.cpp to communicate information about edge components in the image. We would create a second publisher in main at line 6:

```
compPub = n.advertise<holoruch_custom_msgs::EdgePixels>("edges", 1);
```

---

[34]It is highly advisable to make sure that no matching message type already exists before defining a custom one.

[35]For available elementary data types see http://wiki.ros.org/msg.

and the code to calculate and afterwards publish the components would be inserted at line 21 in `imageCallback`. Without going into the details of catkin dependencies, please refer to the `holoruch` repository and wiki documentation,[36] it is necessary to add a dependency to `holoruch_custom_msgs` into `holoruch_custom/package.xml` and `holoruch_custom/CMakeLists.txt`. This concludes the introduction to creating a custom ROS node. Next, we look a different kind of optical sensor.

6   RGB-D Sensors and PCL

RGB-D sensors such as the Microsoft Kinect or the Asus Xtion provide two kinds of information: First, a RGB image just like a normal camera. Second, a depth image usually encoded as a grayscale image, but with each gray value measuring distance to the camera instead of light intensity. Separately, the RGB and depth image can be used as shown in the previous sections. However, the depth image can also be represented as a point cloud, i.e. an ordered collection of 3D points in the camera's coordinate system. If the RGB and depth image is registred to each other, which means the external camera calibration is known, a colored point cloud can be generated. The ROS community provides a stack for OpenNI-compatible devices,[37] which contains a launch file to bring up the drivers together with the low-level processing pipeline[38]:

```
roslaunch openni2_launch openni2.launch
```

Use dynamic reconfigure, e.g. the "Dynamic Reconfigure" plugin, to activate "depth registration" and "color_depth_synchronisation" for the `driver` node. Because by default no RGB point clouds are generated.

The most interesting new ROS subsystems that become relevant when working with RGB-D sensors, or with the `sensor_msgs/PointCloud2` message, are `rviz` and the Point Cloud Library (PCL). If no RGB-D sensor is available, but two (web)cameras, have a look at the `stereo_image_proc` stack[39], which can calculate stereo disparity images and process these to point clouds, as well. We want to stress this point once more, since it is one of the big benefits when-properly-using ROS: Message types, sent over topics, represent an abstract interface to the robot system. They abstract not only on which machine data is generated, transformed or consumed, but also how this data has been acquired. In the case at hand, once a `sensor_msgs/PointCloud2` has been created, it does not matter whether it

---

[36]http://wiki.ros.org/ROS/Tutorials/CreatingPackage.

[37]Since the end of 2014 there also exists a similar stack for the Kinect One (aka Kinect v2): https://github.com/code-iai/iai_kinect2.

[38]If `openni2` does not work, also try the older `openni` stack.

[39]In case of using two webcams, do not start a seperate driver node for each, instead use `uvc_stereo_node`. For stereo calibration refer to http://wiki.ros.org/camera_calibration/Tutorials/StereoCalibration.

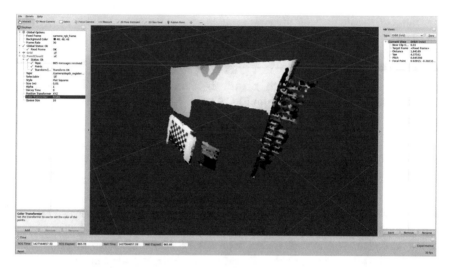

Fig. 4 Live visualization of RGB point clouds (`sensor_msgs/PointCloud2`) with `rviz`. In order to show the colored point cloud, the "Color Transformer" of the "PointCloud2" plugin must be set to RGB8

originates from a stereo camera or a RGB-D sensor or from a monocular structure from motion algorithm and so forth.

The `rviz` "PointCloud2" plugin provides a live 3D visualization of a point cloud topic. We start `rviz` using `rosrun`

```
rosrun rviz rviz
```

and use the "Add" button in order to create a new instance of a visualization plugin. Next we select the "By topic" tab and select the `depth_registered` PointCloud2 entry. The resulting display can be seen in Fig. 4. Furthermore, as will become evident till the end of this chapter, `rviz` is able to combine a multitude of sensor and robot state information into a single 3D visualization.

A custom ROS node for processing point clouds with PCL is very similar to the custom OpenCV node shown in the previous section. Put differently, it is simple and only requires a few lines of code to wrap existing algorithms-based on frameworks such as OpenCV or PCL-for ROS. Thereby benefiting from network transparency and compatibility with the large amount of software available by the ROS community. Again working with stereo cameras or RGB-D sensors is a large topic in its own right and we cannot go into further details here.[40] Having gained experience with two different optical sensor types and a lot of fundamental knowledge in working with ROS, we turn our attention to actuators.

---

[40]We refer to the excellent documentation and tutorials in the ROS wiki (http://wiki.ros.org/pcl/ Overview) as well as those for OpenCV (http://opencv.org/documentation.html) and PCL (http:// pointclouds.org/documentation/).

Fig. 5 A simple pan-tilt unit is shown, it consists of two Dynamixel AX-12A actuators together with a USB2Dynamixel adapter. Any small webcam or RGB-D camera can be attached to the pan-tilt unit, in the image a Logitech C910 is used. Together the pan-tilt unit and the camera can be used for tabletop robotics experiments. They provide a simple closed sensor actuator loop, allowing to teach advanced robotics concepts

7 Actuator Control: Dynamixel and ROS Control

The Robotis Dynamixel actuators are integrated position or torque controlled "smart servo" motors and can be daisy chained. It is sufficient to connect them on one end of the daisy chain to a computer via an USB adapter and to a 9-12 V power supply. For this chapter we use two Dynamixel AX-12A and one USB2Dynamixel adapter. Each AX-12A has a stall torque of 1.5 Nm, smooth motion is possible up to about 1/5th of the stall torque.[41] As of 2015 the overall cost of the three items is less than 150\$. The Dynamixel actuators are popular in the ROS community because-among other things-they can be used out-of-the-box with the `dynamixel_motor` stack.[42] We built a simple pan-tilt unit, which is depicted in Fig. 5. The goal here is not an introduction to `dynamixel_motor`,[43] rather we provide a hands-on example of working with actuators under ROS.

First we need to start the node, which directly accesses the Dynamixel motors via USB:

```
roslaunch holoruch_pantilt pantilt_manager.launch
```

Before proceeding make sure that motor data is received:

```
rostopic echo /motor_states/pan_tilt_port
```

Now we spawn a joint controller for each axis in the pan-tilt unit:

---

[41] http://www.robotis.com/xe/dynamixel_en.

[42] http://wiki.ros.org/dynamixel_motor.

[43] In depth tutorials are available in the ROS wiki: http://wiki.ros.org/dynamixel_controllers/ Tutorials.

```
roslaunch holoruch_pantilt pantilt_controller_spawner.launch
```

Note that the above call will terminate after spawning the joint controllers. Before moving the motors, one can set the positioning speed (default is 2.0):

```
rosservice call /pan_controller/set_speed 'speed: 0.5'
rosservice call /tilt_controller/set_speed 'speed: 0.1'
```

To check whether everything worked, try to move each axis:

```
rostopic pub -1 /pan_controller/command std_msgs/Float64 -- 0.4
rostopic pub -1 /tilt_controller/command std_msgs/Float64 -- 0.1
```

Now we could control our pan-tilt unit through a custom node that publishes to each axis' command topic. However, ROS provides higher level mechanisms to work with robots. The next step will thus be to create a model of our two axis "robot".

8 Robot Description with URDF

In ROS robots are described, in terms of their rigid parts (= links) and (moveable) axis (= joints), by the URDF format. The complete holoruch_pantilt.urdf can be found in the holoruch repository. Only the general URDF structure is described here:

```
1   <robot name="holoruch_pantilt">
2     <link name="base_link">
3       <visual>
4         <geometry>
5           <box size="0.15 0.13 0.01" />
6         </geometry>
7       </visual>
8       ...
9     </link>
10
11    <joint name="base_to_pan_joint" type="fixed">
12      <origin xyz="0 0 0.021" rpy="0 0 0"/>
13      <parent link="base_link"/>
14      <child link="pan_link"/>
15    </joint>
16
17    <link name="pan_link">
18      <visual>
19        <geometry>
20          <mesh filename="package://holoruch_pantilt_description/
21            meshes/AX-12A.stl" />
22        </geometry>
23      </visual>
24      <collision>
25        <geometry>
26          <mesh filename="package://holoruch_pantilt_description/
27            meshes/AX-12A_convex.stl" />
28        </geometry>
```

```
29        </collision>
30        <inertial>
31          <origin rpy="0 0 0" xyz="0 0 0"/>
32          <mass value="0.055"/>
33          <inertia ixx="1" ixy="0" ixz="0" iyy="1" iyz="0" izz="1"/>
34        </inertial>
35      </link>
36
37      <joint name="pan_joint" type="revolute">
38        <origin xyz="0 0 0.06" rpy="0 0 0"/>
39        <parent link="pan_link"/>
40        <child link="tilt_link"/>
41        <limit lower="-1.57079" upper="1.57079" effort="1.0" velocity=
42        "1.0" />
43        <axis xyz="0 0 1" />
44      </joint>
45
46      <link name="tilt_link"> ... </link>
47      <joint name="tilt_joint" type="revolute"> ... </joint>
48      <link name="camera_link"> ... </link>
```

Each `<link>` tag has at least a `<visual>`, `<collision>` and `<inertial>` child tag. The visual elements can be either geometric primitives (5) or mesh files (20). For the collision element the same is true with one very important exception: Collision meshes must always be *convex*. If this is not minded, collision checking might not work correctly when using the URDF in combination with MoveIt! (cf. the following section). Each `<joint>` tag has an `origin` relative to its `parent` link and connects it with a `child` link. There are several different joint types, e.g. `fixed` (11) and `revolute` (35). Furthermore, if the joint is moveable, its `axis` of motion has to be specified together with a `limit`. To understand how coordinate systems are specified in URDF please see the relevant drawings in the ROS wiki.[44]

At any time the XML is valid, the current appearance of the robot can be tested by a utility from the `urdf_tutorial` package:

```
roslaunch urdf_tutorial display.launch gui:=True model:=some_robot.urdf
```

This starts `rviz` including the required plugins together with a `joint_state_publisher`, which allows to manually set the joint positions through a GUI. Figure 6 depicts a screenshot for the pan-tilt unit.

Once a robot description has been created it can be used by the `rviz` "Robot-Model" plugin to visualize the current robot state. Furthermore, the URDF is the basis for motion planning, which will be covered next.

---

[44]http://wiki.ros.org/urdf/XML/model, http://wiki.ros.org/urdf/XML/link and http://wiki.ros.org/urdf/XML/joint.

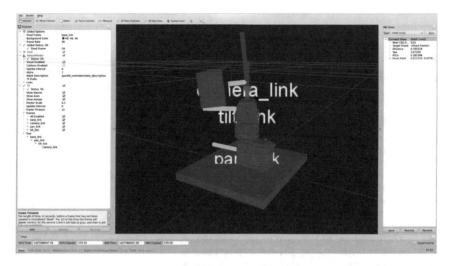

Fig. 6 rviz displaying the live state of the pan-tilt unit. The fixed frame is the base link of the pan-tilt unit (`base_link`). The "RobotModel" plugin uses the URDF robot description from the parameter server (`/pantilt_controller/robot_description`) to visualize the robot links according to the current joint position. The "TF" plugin shows all `tf` frames. In the screenshot these are only the frames of the pan-tilt joints. These joint frames are provided by the `robot_state_publisher` based on the robot description and a `sensor_msgs/JointState` topic (`joint_states`)

9   Motion Planning with MoveIt!

The MoveIt! framework[45] gives ROS users an easy and unified access to many motion planning algorithms. This does not only include collision free path planning according to meshes, but also works in combination with continuously updated obstacles. Furthermore, it provides `rviz` plugins to interactively view trajectory execution and specify motion targets with live pose visualization. In order to use MoveIt!, a lot of additional configuration is required. Fortunately, a graphical setup assistant is provided, which takes care of (almost) all required configuration:

```
roslaunch moveit_setup_assistant setup_assistant.launch
```

We will not go through the configuration process here, the completely configured `holoruch_pantilt_moveit_config`, however, is contained in the repository. For more details and background information, we refer to the online available MoveIt! setup assistant tutorials.[46] After completion of the setup, we can directly use the generated `demo.launch` to test all planning features but without actually sending the trajectory goals to the robot (cf. Fig. 7). Although having a generic motion planner can be demonstrated by means of more impressive examples than

---

[45]http://moveit.ros.org/.

[46]http://docs.ros.org/indigo/api/moveit_setup_assistant/html/doc/tutorial.html.

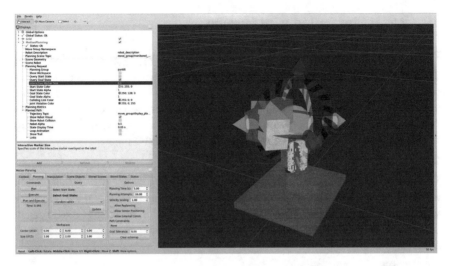

Fig. 7 The "MotionPlanning" `rviz` plugin provided by MoveIt! is shown in combination with the pan-tilt unit. Target positions can be specified by moving the interactive marker. Once the desired target is set, a trajectory from the current pose to the target can be planned by pressing the "Plan" button. If a motion plan is found, it can be directly visualized in different ways. However, this barely scratches the surface of the available "MotionPlanning" functionality

by a two axis pan-tilt unit, MoveIt! also works in this case and could be even useful in applications such as visual servoing. The last point is especially relevant when considering that no modification of code is required when switching between robots with completely different kinematics if the MoveIt! API-or at least the `actionlib` together with `trajectory_msgs/JointTrajectory`-is used instead of custom solutions.

At this point, we have seen how to work with sensor data and control actuators using ROS. In addition, it should have become clear, at least the big picture of it, how one can write custom nodes that process incoming sensor data in order to plan and finally take an action by sending commands to the actuators. The last section of this tutorial will show how the Gazebo simulator can be used together with ROS in order to work without access to the real robot.[47] Everything written so far about ROS, processing sensor data as well as moving robots, can be done without any modification with simulated sensors and simulated actuators. Nodes do not even have to be recompiled for this. Most of the time, even the same launch files can be used.

---

[47]There are also many other important uses for simulated robot instances. One of these is "Robot Unit Testing", which brings unit and regression testing for robotics to a new level, for more information see [1].

10   Robot Simulation with Gazebo

Due to historical reasons, ROS and Gazebo use different formats to describe robots.
The former, URDF, has been introduced in the previous section. The Gazebo format is called SDF and has a similar structure to URDF, i.e. there are links which
are connected by joints-however, there are also some important differences on the
conceptual level, for example how coordinate systems are represented. Thankfully
it is possible to automatically convert from URDF to SDF[48]:

```
gz sdf --print robot.urdf > model.sdf
```

We have to convert `holoruch_pantilt.urdf` in this manner and also add a
`model.config`.[49] The `model.sdf` and `model.config` file must be copied
into their own subdirectory of `~/.gazebo/models/`. Afterwards Gazebo can be
started and we can add our simulated pan-tilt unit via the "Insert" tab in the left side
menu:

```
rosrun gazebo_ros gazebo
```

After adding the model to the world, the simulated gravity will slowly pull the tilt
axis down until the camera rests on the motor. This happens for two different reasons:
First, we did not specify a `<dynamics>` tag, which contains `friction`, for our
model. Second, the SDF does not contain any sensor or model plugins.

We will add two plugins to the SDF file: First, a simulated camera sensor[50]:

```
<link name='camera_link'>
    ...
  <sensor name="cam_sensor" type="camera">
    <always_on>1</always_on>
    <visualize>0</visualize>
    <pose>
      0 0 0.116
      3.14159 3.14159 -1.5708
    </pose>
    <update_rate>30</update_rate>
    <camera>
      <horizontal_fov>1.0</horizontal_fov>
      <image>
        <width>640</width>
        <height>480</height>
        <format>R8G8B8</format>
      </image>
      <clip>
        <near>0.0100000</near>
        <far>100.000000</far>
```

---

[48]In case one starts with the SDF description, this can also be converted to an URDF using
`sdf2urdf.py model.sdf robot.urdf`, see http://wiki.ros.org/pysdf. This option might
become even more attractive with the graphical model editor in Gazebo 6, cf. http://gazebosim.org/
tutorials?tut=model_editor.

[49]Cf. http://gazebosim.org/tutorials?tut=build_robot\&cat=build_robot.

[50]http://gazebosim.org/tutorials?tut=ros_gzplugins.

```
      </clip>
    </camera>
    <plugin name="cam_sensor_ros" filename="libgazebo_ros_camera.so"
     >
      <alwaysOn>true</alwaysOn>
      <cameraName>/simcam</cameraName>
      <imageTopicName>image_raw</imageTopicName>
      <cameraInfoTopicName>camera_info</cameraInfoTopicName>
      <frameName>sim_cam</frameName>
    </plugin>
  </sensor>
</link>
```

Second, a plugin to control the joints of our simulated model[51]:

```
...
  <plugin name="joint_pos_control" filename=
  "libRobotJointPositionControlPlugin.so">
    <robotNamespace>/pantilt</robotNamespace>
    <jointsReadTopic>get_joint_positions</jointsReadTopic>
    <jointsWriteTopic>set_joint_positions</jointsWriteTopic>
  </plugin>
</model>
```

Fig. 8 The screenshot shows a simulated world containing the pan-tilt unit in Gazebo together with the live image provided by the simulated camera in `rqt`. Since the camera images are sent over a standard `sensor_msgs/Image` topic, we could run the simulated images through our common node created in an earlier section

---

[51]For ease of use, we use a custom plugin here. It must be installed from source https://github.com/andreasBihlmaier/robot_joint_position_controller_gazebo. Alternatively Gazebo also provides ROS control interfaces, however they require the robot to be externally spawned via `spawn_model`, see http://gazebosim.org/tutorials/?tut=ros_control.

Both plugins provide the usual ROS interface as would be expected from real hardware. The complete `model.sdf` is available in `holoruch_gazebo`. Figure 8 shows the simulated pan-tilt unit in Gazebo together with the simulated camera image shown in `rviz`. This concludes the last part of our hands-on introduction to ROS.

Reference

1. A. Bihlmaier, H. Wörn, Automated endoscopic camera guidance: a knowledge-based system towards robot assisted surgery, in *Proceedings for the Joint Conference of ISR 2014 (45th International Symposium on Robotics) and ROBOTIK 2014 (8th German Conference on Robotics)*, pp. 617-622 (2014)

Authors' Biography

Andreas Bihlmaier Dipl.-Inform., obtained his Diploma in computer science from the Karlsruhe Institute of Technology (KIT). He is a Ph.D. candidate working in the Transregional Collaborative Research Centre (TCRC) "Cognition-Guided Surgery" and is leader of the Cognitive Medical Technologies group in the Institute for Anthropomatics and Robotics-Intelligent Process Control and Robotics Lab (IAR-IPR) at the KIT. His research focuses on cognitive surgical robotics for minimally-invasive surgery, such as a knowledge-based endoscope guidance robot.

Heinz Wörn Prof. Dr.-Ing., studied electronic engineering at the University of Stuttgart. He did his Phd thesis on "Multi Processor Control Systems". He is an expert on robotics and automation with 18 years of industrial experience. In 1997 he became professor at the University of Karlsruhe, now the KIT, for "Complex Systems in Automation and Robotics" and also head of the Institute for Process Control and Robotics (IPR). Prof. Wörn performs research in the fields of industrial, swarm, service and medical robotics.

Threaded Applications with the roscpp API

Hunter L. Allen

Abstract We begin this tutorial by discussing the features of the Catkin build system. We then proceed by giving a thorough explanation of ROS callback functions in terms of sensor data. We utilize the Qt5 libraries to make a very simple Graphical User Interface to control the robot with on screen buttons, as well as view position information in (x, y, θ) coordinates. This GUI will use the Qt thread library as well as ROS messages to control and provide information about the state of the robot.

Keywords Tutorial · Callback functions · CMake · Concurrency · GUI · Qt5 Programming

1 Introduction

In this chapter, we develop a foundation in the practice of Graphical User Interfaces and threaded ROS implementation with the aid of the Qt5 library. In doing so, we introduce a great deal of syntax; however, we implement a more robust system. Moreover, we will see that ROS operates quite naturally in a threaded environment, thus making the code more easily adaptable for larger scale applications.

There is a large collection of Qt interfaces for ROS which have already been written. This collection is known as RQt. RQt was implemented as a replacement for rxtools. This chapter will be focused on UI implementation outside of RQt, but, due to the strong resemblance to the primary example in this chapter, its existence should certainly be pointed out [1].

The primary example in this chapter will be a User Interface for controlling a robot. The controller's source code is available online [2]. To understand this example, we must first explore aspects of the catkin build system, using the Qt libraries with ROS, as well as some fundamentals of GUI creation. The tools introduced here can be

Special thanks to Dr. Julie A. Adams.

H.L. Allen (✉)
Purdue University, West Lafayette, IN 47906, USA
e-mail: allen286@Purdue.Edu

© Springer International Publishing Switzerland 2016
A. Koubaa (ed.), *Robot Operating System (ROS)*, Studies in Computational
Intelligence 625, DOI 10.1007/978-3-319-26054-9_3

51

extended for use in research, specifically in Human-Robot Interaction. We proceed as follows:

- First, we investigate the Catkin Build system: we will discuss the contents of relevant files for our package, setting the package up for Qt and touch lightly on message generation.
- Second, we discuss different methods of implementing ROS callback functions.
- Third, we develop a simple application that listens to the position of a robot and echoes the position in (x, y, θ) to the terminal.
- Fourth, we develop the same application in a threaded environment.
- Fifth, we implement the Graphical User Interface discussed earlier.

2 ROS Environment Configuration

The following examples will require the installation of a few packages. We assume the user is working on Ubuntu 14.04 with ROS Indigo. It is also assumed that the user has set up a Catkin workspace (if this is not a valid assumption, see the following tutorial on the ROS Wiki [3]). To install the required packages, run the following commands.

```
$ sudo apt-get install ros-indigo-nav-msgs qt-sdk
$ sudo apt-get install ros-indigo-geometry-msgs
```

Finally, you may download and compile the source for this chapter's examples with the following commands.

```
$ roscd && cd ../src
$ git clone https://github.com/allenh1/ros-thread-tutorial
$ git clone https://github.com/allenh1/ros-qt-controller
$ cd ../ && catkin_make
```

At this point you should have all the necessary examples and build-dependencies for this chapter, and the code should be compiled.

3 Catkin Build System

The Catkin Build System is one of the most powerful tools at your disposal for use with ROS. Catkin was established as the primary build tool for ROS, replacing rosbuild, in ROS groovy [11].

3.1 Creating `package.xml`

In a Catkin package, there are two required files: `package.xml` and `CMake Lists.txt` (`package.xml` replaced `manifest.xml` in REP-140 [4]). The first file is responsible for identifying a directory as something Catkin needs to build as well as provide information about the order in which the packages should be built. Consider the following example from `ros_threading_example`.

```xml
<?xml version="1.0"?>
<package>
  <name>ros_threading_example</name>
  <version>0.0.1</version>
  <description>Simple threading example.</description>
  <license>BSD</license>
  <maintainer email="allen286@purdue.edu">Hunter Allen</maintainer>
  <author email="allen286@purdue.edu">Hunter Allen</author>

  <buildtool_depend>catkin</buildtool_depend>

  <build_depend>nav_msgs</build_depend>
  <build_depend>roscpp</build_depend>
  <build_depend>message_generation</build_depend>

  <run_depend>nav_msgs</run_depend>
  <run_depend>roscpp</run_depend>
  <run_depend>message_generation</run_depend>
</package>
```

(Example `package.xml` from `ros-threading-example` on GitHub [2])

In this example, we create a minimal `package.xml` file. The required xml tags include package name, version, description, license, maintainer and author. These tags are used to identify the package. The `buildtool_depend` tag identifies the package as a Catkin package. The `build_depend` tags are used to establish the *topological* building of packages: this helps Catkin to build the packages in the correct order (so that dependencies are built before the packages which depend on them). As for `run_depend`, this identifies the packages with which Catkin needs to link your package dynamically.

3.2 CMake Setup

Next we examine the file `CMakeLists.txt` from the package `ros-threading-example`.

```
cmake_minimum_required(VERSION 2.8.9)
project(ros_threading_example)
find_package(catkin REQUIRED COMPONENTS roscpp
                     message_generation
```

```
                                                        nav_msgs)
find_package(Qt5Widgets REQUIRED)
find_package(Qt5Core REQUIRED)

add_message_files(FILES Pose2D.msg)
generate_messages(DEPENDENCIES std_msgs)

catkin_package(
  CATKIN_DEPENDS roscpp message_generation
  DEPENDS system-lib
)

qt5_wrap_cpp(QT_MOC_HPP src/RobotThread.h)

include_directories(src ${catkin_INCLUDE_DIRS})
include_directories(${Qt5Widgets_INCLUDE_DIRS})

add_executable(pose_echo_threaded src/main.cpp
                            src/RobotThread.cpp
                            ${QT_RESOURCES_CPP}
                                ${QT_MOC_HPP})

add_executable(pose_echo src/PoseEcho.cc)

add_dependencies(pose_echo ${PROJECT_NAME}_generate_messages_cpp)
add_dependencies(pose_echo_threaded ${PROJECT_NAME}_generate_messages_cpp)

target_link_libraries(pose_echo_threaded ${catkin_LIBRARIES} Qt5::Widgets)
target_link_libraries(pose_echo ${catkin_LIBRARIES})

install(TARGETS pose_echo_threaded pose_echo
  RUNTIME DESTINATION ${CATKIN_PACKAGE_BIN_DESTINATION}
)
```

(Example CMakeLists.txt from ros_threading_example on GitHub [5])

In this CMakeLists.txt file, we tell Catkin to compile and link the executables with the necessary ROS libraries (as declared in package.xml) and with the necessary Qt libraries. We will discuss only the relevant commands here, and note that information regarding commands out of the scope of this text may be found on the CMake website [6].

We begin by restricting the CMake version to only those that are 2.8.9 or later (for the Qt5-specific Commands). This is handled by the command cmake_minimum_required followed by the flag VERSION and a release number.

The project command identifies the name of the project. Here, one can also specify the language or languages in which the particular program was written; however, it is of the author's opinion that this information is not relevant in the ROS world (unless you really want to be pedantic).

The find_package command is used to load settings from an external CMake package. Here it is used to set up CMake to work with Catkin and the necessary catkin packages. Successive calls are made to this command to include the Qt5 libraries in the build. The flag REQUIRED tells CMake to exit and throw an error if Catkin or any necessary ROS components are not found. The COMPONENTS flag lists the specific CMake components to require before building.

Catkin provides the macro catkin_package. This command is used to generate the package configuration and remaining CMake files. There are five arguments, all optional: INCLUDE_DIRS, LIBRARIES, CATKIN_DEPENDS, DEPENDS and CFG_EXTRAS. INCLUDE_DIRS controls the export of the inclusion paths. LIBRARIES declares the libraries that Catkin is exporting from this package. DEPENDS lists the CMake projects to incorporate that are not Catkin libraries (system-lib, for example). CFG_EXTRAS is the catch-all command for Catkin; with this command, you specify another CMake file to include in the build (for example, this would be a file named extra.cmake).

CMake also has built-in Qt support. It uses the command qt5_wrap_cpp to produce Qt wrappers.[1] These are the moc files usually generated by the make command in a qmake generated Makefile. In our context, we use this to create a qt-library from the sources command which we identify with QT_MOC_HPP. Had we created a User Interface with a tool such as QtDesigner (generating a .ui file), then we would employ the qt_wrap_ui in a similar fashion.

CMake uses include_directories to identify the directories to include when looking for header files. We use the ones with which Catkin is already familiar. It is important to note that, unless otherwise specified, CMake interprets your directory input as relative to the current directory (that is, the one with the file CMakeLists.txt). In particular, we use it to identify the source directories, the Catkin include directories and Qt5Widgets include directories.

The add_executable command identifies the binary to be built and identifies it with the sources with which to build it.

A specified target is linked with the command target_link_libraries. Such a target must have been produced by an add_executable command or an add_library command. In our example, we link both executables to the Qt5 Widgets library and the Catkin libraries.

Finally, we place the binary (binaries) and or library (libraries) in the appropriate locations with the install command. The flag TARGETS identifies the targets we are installing. There are five types of targets: ARCHIVE, LIBRARY, RUNTIME, FRAMEWORK and BUNDLE. In most ROS applications, you will exclusively use LIBRARY and RUNTIME. The DESTINATION flag identifies the end location of the specified targets.

3.3 Message Generation

Before proceeding into Callback functions, the ability to generate a custom message file with ease is a noteworthy ROS feature. Say, for example, you wanted to create a message that held the robot's position in only two coordinates. You can create a file

---

[1]This command is only found when Qt5Core is in the build environment. This command is also only available in CMake versions 2.8.9 and later.

called `Pose2D.msg` and in only a few lines implement the message. Below is the message file for the above example.

```
float64  xPose
float64  yPose
```

The above file is found a in the build path of our example under the directory `msg`. Messages are included in the CMake file with the following commands.

```
add_message_files(FILES Pose2D.msg)
generate_messages(DEPENDENCIES std_msgs)
```

Also, it is important to note that these commands must come before the `catkin_package()` command appears. Before the targets are linked in `CMakeLists.txt`, we add the message as a dependency with the following lines.

```
add_dependencies(pose_echo ${PROJECT_NAME}_generate_messages_cpp)
add_dependencies(pose_echo_threaded ${PROJECT_NAME}_generate_messages_cpp)
```

The messages are referenced in the source by including the automatically generated header file `$PROJECT_NAME/MessageName.h`. In our particular example, we include the message and redefine the type as `Pose2D` with the following lines.

```
#include <ros_threading_example/Pose2D.h>
typedef ros_threading_example::Pose2D Pose2D;
```

4   ROS Callback Functions

In this section we explore the ways a callback function can be implemented in ROS. We will be using the code examples installed in Sect. 2. We will begin with a simple example including callbacks of a non-threaded nature, then proceed to callbacks in a threaded context.

4.1   Basic Callback Functions

Nodes communicate through messages. These messages are processed in a *callback function*. In C++, callback functions are implemented through a *function pointer*.

Consider the following example from `ros-thread-tutorial`, installed in Sect. 2.

```
#include <ros/ros.h>
#include <ros/network.h>
#include <nav_msgs/Odometry.h>
```

```
#include <stdio.h>
#include <ros_threading_example/Pose2D.h>

typedef ros_threading_example::Pose2D Pose2D;

Pose2D poseMessage;
void poseCallback(const nav_msgs::Odometry& msg)
{
    float m_xPos = msg.pose.pose.position.x;
    float m_yPos = msg.pose.pose.position.y;
    float m_aPos = msg.pose.pose.orientation.w;

    std::printf("Pose: (%.4f, %.4f, %.4f)\n", m_xPos, m_yPos, m_aPos);
    poseMessage.xPose = m_xPos;
    poseMessage.yPose = m_yPos;
}

int main(int argc, char **argv)
{
  ros::init(argc, argv, "pose_listener");

  ros::NodeHandle n;

  ros::Rate rate(100);

  ros::Subscriber sub = n.subscribe("odom", 10, poseCallback);
  ros::Publisher  pub = n.advertise<Pose2D>("/pose2d", 100);

  while (ros::ok()) {
    pub.publish(poseMessage);

    ros::spinOnce();
    rate.sleep();
  }

  return 0;
}
```

We will break this example into parts for discussion. The function `main` begins by initializing the ros node named "pose_listener." A `ros::NodeHandle` is then created. This object acts as an interface for subscribing/publishing and other related activity. Such an object is essential in any ros application.

```
ros::Subscriber sub = n.subscribe("odom", 10, poseCallback);
```

The above line tells the `ros::Subscriber` object named "sub" to listen to the topic "odom" with and store the messages with a buffer size of 10. The line also tells the subscriber how the data should be processed, sending each datum to the function `poseCallback`.

```
ros::Publisher  pub = n.advertise<Pose2D>("/pose2d", 100);
```

This line tells the nodehandle to advertise a new topic called /pose2d that has a queue of size five; in other words, the topic will store at most five messages should the publisher speed fall behind the requested speed. After the publishers and subscribers have been declared, we enter a loop that publishes our custom message type. The command `ros::spinOnce()` tells the topics to refresh themselves, and we pause the loop for the amount of time we specified earlier.

Some sample output is provided below, produced from a simulated Pioneer-3AT robot moving in direction of the positive x-axis inside the Gazebo Simulator.

```
$ rosrun ros_threading_example pose_echo
Pose: (-2.0987, -0.0061, 1.0000)
Pose: (-2.0974, -0.0061, 1.0000)
Pose: (-2.0961, -0.0061, 1.0000)
Pose: (-2.0948, -0.0061, 1.0000)
Pose: (-2.0936, -0.0061, 1.0000)
Pose: (-2.0923, -0.0061, 1.0000)
Pose: (-2.0910, -0.0061, 1.0000)
```

4.2 Robots as a Thread

The implementation considered in Sect. 4.1 is not so useful when working in the context of User Interfaces. Certainly, the ROS functionality should be entirely separate from any sort of user interface, but must also be connected. Our goal is simple: we pass the necessary information to and from a continuously running ROS node to the interface. As the reader may verify, this lends very nicely to a threaded implementation, as both the interface and ROS communications are quite hindered by waiting.

To implement a thread in ROS, there are two functions we must ensure exist. These functions include the destructor `virtual ~RobotThread()` and the function that calls `ros::init(int argc, char ** argv)`, implemented here as `bool init()`. `bool init()`, in the case of Qt, will also be responsible for setting up and starting the thread on which our robot will run.

Here we provide an example of a very minimal implementation of `RobotThread` that is built upon Qt's thread library, QThread. In the following example, we will write the same program as before, but in a threaded setting. This form of implementation is not as succinct as that of the prior, but is far more robust. Such a header is also quite reusable, as it can be minimally adapted for a myriad of situations. It is important to note, however, that due to limitations in the roscpp API, multiple instances of this thread cannot be created as it is implemented here. The source for this example was installed with that of the prior example in Sect. 2, so we will only examine the file `RobotThread.cpp` (as the `RobotThread.h` file holds only the class).

```
#include "RobotThread.h"

RobotThread::RobotThread(int argc, char** pArgv)
    : m_Init_argc(argc),
      m_pInit_argv(pArgv)
{/** Constructor for the robot thread **/}
```

```
RobotThread::~RobotThread()
{
    if (ros::isStarted())
    {
        ros::shutdown();
        ros::waitForShutdown();
    }//end if

    m_pThread->wait();
}//end destructor

bool RobotThread::init()
{
    m_pThread = new QThread();
    this->moveToThread(m_pThread);

    connect(m_pThread, &QThread::started, this, &RobotThread::run);

    ros::init(m_Init_argc, m_pInit_argv, "pose_echo_threaded");

    if (!ros::master::check())
        return false;//do not if ros is not running

    ros::start();
    ros::Time::init();
    ros::NodeHandle nh;

    pose_listener = nh.subscribe("odom", 100, &RobotThread::poseCallback,
    this);
    pose2d_pub = nh.advertise<Pose2D>("/pose2d", 100);

    m_pThread->start();
    return true;
}//set up ros

void RobotThread::poseCallback(const nav_msgs::Odometry & msg)
{
    QMutex pMutex = new QMutex();
    pMutex->lock();
    m_xPos = msg.pose.pose.position.x;
    m_yPos = msg.pose.pose.position.y;
    m_aPos = msg.pose.pose.orientation.w;
    std::printf("Pose: (%.4f, %.4f, %.4f)\n", m_xPos, m_yPos, m_aPos);
    pMutex->unlock();

    delete pMutex;
}//callback method to echo the robot's position

void RobotThread::run()
{
    ros::Rate loop_rate(1000);

    QMutex * pMutex;

    while (ros::ok())
    {
        pMutex = new QMutex();
        pMutex->lock();
```

```
        poseMessage.xPose = m_xPos;
        poseMessage.yPose = m_yPos;
        pose2d_pub.publish(poseMessage);
        pMutex->unlock();

        delete pMutex;

        ros::spinOnce();
        loop_rate.sleep();
    }//run while ROS is ok.
}
```

We begin with the constructor. The function `RobotThread::RobotThread` `(int argc, char ** argv)` is responsible for creating the Robot Thread. It's function is rather limited: it simply initializes 'm_Init_argc' and 'm_pInit_argv' which are to be passed to ROS.

The destructor `RobotThread::~RobotThread()` serves as a clean exit. The function first verifies that ROS is running. If so, the function halts its ROS functionality and waits for this action to complete. Lastly, `wait()` is called to stop the thread from processing.

`RobotThread::init()` is the function responsible for setting up our thread. Older Qt implementations tend to subclass QThread. It was later decided that, instead, one should subclass QObject (the most basic of Qt constructions) and move the object onto a thread. This provides both better control and a clearer thread affinity. This process is not too difficult to accomplish, as we simply need to create a thread object, move our implementation onto said object and identify the function that will be running on the thread. We will discuss the meaning of the `connect` statement later in Sect. 5.2.

`RobotThread::init()` is also responsible for bringing up the ROS services. As in `PoseEcho`, `ros::init` starts the node named `pose_listener`. To verify that the master node is running appropriately, we make a call to `ros::master::` `check()`; if there is no master node, our function returns false, meaning that initialization of our node failed. ROS is then started, as is the timer for ROS. The `ros::subscriber` in this context is stored as a member object called `pose_listener`. This subscribes to the `/odom` topic, storing a buffer of ten messages with the callback function `RobotThread::poseCallback`. Take note of the & in front of the function in this context, in contrast to the prior example. The next line advertises topic `/pose2d` with a queue of size five.

Lastly, the function `RobotThread::run()` is the run function for our thread. We have a member object `poseMessage` which is a message of type `Pose2D`. We also have member variables m_xPos and m_yPos, which are given data by the sensor callback function and are in turn used for the data of `Pose2D`. Using the same environment as before, we obtain the following output.

```
$ rosrun ros_threading_example pose_echo_threaded
Pose: (0.0664, 0.0010, 1.0000)
Pose: (0.0677, 0.0010, 1.0000)
Pose: (0.0690, 0.0010, 1.0000)
```

```
Pose: (0.0702, 0.0010, 1.0000)
Pose: (0.0715, 0.0010, 1.0000)
Pose: (0.0728, 0.0010, 1.0000)
Pose: (0.0741, 0.0010, 1.0000)
```

5 GUI Programming with Qt5 and ROS

In this section we develop a simple user interface for controlling a robot. Our goal is to control a robot with the aid of a Graphical User Interface and simultaneously receive the robot's position. Here we implement `RobotThread` in a manner similar to that of Sect. 4.2. We will publish a velocity command (a message of type `geometry_msgs::Twist`) to the topic `/cmd_vel` containing the velocity corresponding to the button clicked by the user. The interface will also inform the user of the robot's position.

Such an implementation is fairly typical. While there are several implementations already in existence, this implementation serves as a teaching tool to demonstrate the many tools we have outlined in this chapter.

Before proceeding, please note that Sect. 5.1 deals with the construction of the interface, and is by no means ROS specific; if the reader has experience in the development of Qt5 GUI interfaces, they should feel free to move past Sect. 5.1 and delve immediately into Sect. 5.2.

5.1 An Overview of Qt5 GUI Programming

Before we consider linking the `RobotThread`, we must create a window with which to control the robot. Here we are working with the file `ControlWindow.cpp` from the package `ros_qt_controller` installed in Sect. 2. As the GUI in Fig. 1 was built without the Qt Designer tool, the source links each button manually[2]; accordingly, we will only examine sections of the constructor.

The constructor for the `ControlWindow` object, takes the arguments to pass to the Robot thread and a pointer to the parent window, which is by default NULL.

The constructor begins by creating `QPushButton` objects which will ultimately become the controls for the robot. After each button is created, we must properly set it up. Below is the setup for the right turn button.

```
ControlWindow::ControlWindow(int argc, char **argv, QWidget *parent)
    : QWidget(parent),
      m_RobotThread(argc, argv)
```

---

[2]Note that this is by no means a requirement: one may implement the UI with Qt Designer if one so chooses, as was mentioned in Sect. 3.2.

Fig. 1 Screen shot of the implemented Control Window

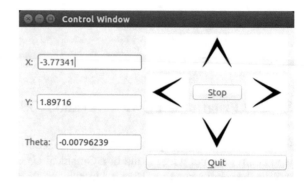

```
{
    ...
    p_rightButton = new QPushButton();
    ...
    QPalette palette = p_rightButton->palette();
    palette.setColor(QPalette::Button,QColor(255,255,255));
    p_rightButton->setAutoFillBackground(true);
    p_rightButton->setFlat(true);
    p_rightButton->setPalette(palette);
    p_rightButton->setIcon(QIcon(":/images/right.xpm"));
    p_rightButton->setIconSize(QSize(50, 50));
    ...
}
```

Here we attach the icons declared in `CMakeLists.txt` to the button. The color of the button is set to white with the `QPalette` object, and is then flattened. We then place the icon on the button with the `setIcon` function. We indicate the QResource with the : appended to the location. Lastly, we set the button to be 50 × 50 pixels. This process then repeated for the remaining buttons.

The layout of the window is managed by the `QVBoxLayout` and `QHBoxLayout` objects. These objects keep things in the same relative location when the window is resized. For our purposes, we divide the window into the left and right halves of the screen; this is done with a `QHBoxLayout` object. The left and right sides are represented by `QVBoxLayout` objects.

On the left side of the screen we place the position data. As we have labeled the places the position appears, we create horizontal layouts for each location. The below code creates the label and the box where the position data appears for the x-coordinate.

```
ControlWindow::ControlWindow(int argc, char **argv, QWidget *parent)
    : QWidget(parent),
      m_RobotThread(argc, argv)
{
    ...
    leftLayout = new QVBoxLayout();
    p_xLayout = new QHBoxLayout();
    ...
```

```
        p_xLabel = new QLabel();
        p_xLabel->setText(tr("X:"));
        p_xDisplay = new QLineEdit();
        p_xDisplay->setText(tr("0.0"));
        ...
        p_xLayout->addWidget(p_xLabel);
        p_xLayout->addWidget(p_xDisplay);
        ...
        leftLayout->addLayout(p_xLayout);
        ...
    }
```

The QLabel and QLineEdit objects are subclasses of QWidget. The widgets are constructed, then added to the layout with the function addWidget. The calls to the function tr() should be noted as well. This call is to the Qt's translation system. Qt recommends you use it for all string literals, as it allows the program to be instantly translated into several languages [7]. The p_xLayout object is then added to the left layout. Note: this must be done in the expected order; the objects are placed from left to right on the horizontal layouts and from top to bottom on the vertical layouts.

On the right side of the control window we place the controls, as well as a quit button to close the application cleanly. The below code places the buttons in their desired locations and combines the left and right layouts in the mainLayout object.

```
ControlWindow::ControlWindow(int argc, char **argv, QWidget *parent)
    : QWidget(parent),
      m_RobotThread(argc, argv)
{
    ...
    rightLayout = new QVBoxLayout();
    layout = new QHBoxLayout();
    layout2 = new QHBoxLayout();
    ...
    mainLayout = new QHBoxLayout();
    ...
    layout2->addWidget(p_rightButton);

    ...
    rightLayout->addLayout(layout2);
    ...

    mainLayout->addLayout(leftLayout);
    mainLayout->addLayout(rightLayout);
    ...
}
```

Finally, we set the layout of the window, tell the window to appear on the screen and set the title of the window to "Control Window" with the following three lines of code.

```
ControlWindow::ControlWindow(int argc, char **argv, QWidget *parent)
    : QWidget(parent),
      m_RobotThread(argc, argv)
```

```
{
    ...
    setLayout(mainLayout);

    show();

    setWindowTitle(tr("Control Window"));
    ...
}
```

At this point, we have constructed the window in Fig. 1; however, we have *only* constructed the window. The buttons must now be *connected* to the desired functionality. In the world of Qt, this is accomplished through the use of `SIGNAL` and `SLOT` functions.

5.2   Connecting the Robot to the GUI

Recall that ROS communication is performed by means of a function pointer. In a similar fashion, Qt controls when a function is called with Signals and Slots [8].

Signals (Slots) also have the ability to emit (receive) multiple pieces of data at the same time. The *signals* will be called when the data is ready, and the *slots* will do the necessary processing with whatever arguments they were supplied. We can make this even safer by sending `double`'s instead of `const double &`'s, as this would send a copy of the variable instead of a constant reference. Lastly, if there is still a possibility of synchronization issues, Qt has its own Mutex library, `QMutex` [9].

Qt manages Signals and Slots with the function `connect`. A typical call to this function is illustrated below.

```
connect(&m_closeButton, SIGNAL(clicked()), this, SLOT(close()))
```

It should also be mentioned that Qt5 supports an alternative syntax (which will soon replace the above) that is much more user friendly, as it catches incorrectly connected functions at compiler time. This syntax will not work with Qt4, but is strongly recommended for Qt5. The previous statement would be implemented as follows.

```
connect(p_quitButton,  &QPushButton::clicked,
                this,  &ControlWindow::close);
```

Here we have a member variable named `m_closeButton` of type `QPush Button`. Also, suppose we encounter this line in the constructor of the main window (`this` in the above). Then when the button is clicked, a signal is *emitted* to call the main window's `close` function.[3]

---

[3] `close()` is a function inherited from QWidget that closes the window.

A Slot (Signal) is declared as a void with the required parameters by prepending the macro Q_SLOT (Q_SIGNAL). A Signal is triggered with the macro Q_EMIT followed by the function and any required arguments.

Returning to our particular example, we modify the callback function for our odometry messages to inform the control window of the new position. The signal containing the position information is declared in RobotThread.h as follows.

```
Q_SIGNAL void newPose(double,double,double);
```

The callback function in RobotThread.cpp emits the position information when each message is received with the following function.

```
void RobotThread::poseCallback(const nav_msgs::Odometry & msg)
{
    QMutex * pMutex = new QMutex();

    pMutex->lock();
    m_xPos = msg.pose.pose.position.x;
    m_yPos = msg.pose.pose.position.y;
    m_aPos = msg.pose.pose.orientation.w;
    pMutex->unlock();

    delete pMutex;
    Q_EMIT newPose(m_xPos, m_yPos, m_aPos);
}//callback method to update the robot's position.
```

Meanwhile, in ControlWindow.h we declare the slot

```
Q_SLOT void updatePoseDisplay(double x, double y, double theta);
```

which is implemented as follows.

```
void ControlWindow::updatePoseDisplay(double x, double y, double theta)
{
    QString xPose, yPose, aPose;
    xPose.setNum(x);
    yPose.setNum(y);
    aPose.setNum(theta);

    p_xDisplay->setText(xPose);
    p_yDisplay->setText(yPose);
    p_aDisplay->setText(aPose);
}//update the display.
```

This function converts the double variables into QString variables and sets the contents of the respective displays with the converted strings.

This pair is connected in the constructor of the Control Window with the following line.

```
connect(&m_RobotThread, &RobotThread::newPose,
                this, &ControlWindow::updatePoseDisplay);
```

Next we need to change the velocity of the robot. Since the robot is a member variable of our `ControlWindow` object, we can store the linear and angular velocities to be published to the robot as member variables. The `RobotThread`'s `run` function then publishes the currently desired speed. We add the following function to `RobotThread` to modify the robot's velocity.

```
void RobotThread::SetSpeed(double speed, double angle)
{
    QMutex * pMutex = new QMutex();
    pMutex->lock();
    m_speed = speed;
    m_angle = angle;
    pMutex->unlock();

    delete pMutex;
}//set the speed of the robot.
```

This function allows the Slots for each respective arrow key to have a rather pithy implementation. The following is found in `ControlWindow.cpp`.

```
void ControlWindow::goForward() { m_RobotThread.SetSpeed(0.25, 0); }
void ControlWindow::goBackward(){ m_RobotThread.SetSpeed(-0.25, 0); }
void ControlWindow::goRight()   { m_RobotThread.SetSpeed(0, -PI / 6.0); }
void ControlWindow::goLeft()    { m_RobotThread.SetSpeed(0, PI / 6.0); }
void ControlWindow::halt()      { m_RobotThread.SetSpeed(0, 0); }
```

The final responsibilities of the constructor for `ControlWindow` is to connect these functions to their respective buttons, initialize the `RobotThread` and run the `RobotThread`. This process is depicted below.

```
ControlWindow::ControlWindow(int argc, char **argv, QWidget *parent)
    : QWidget(parent),
      m_RobotThread(argc, argv)
{
    ...
    connect(p_quitButton,   &QPushButton::clicked,
                    this,    &ControlWindow::close);
    connect(p_upButton,     &QPushButton::clicked,
                    this,    &ControlWindow::goForward);
    connect(p_leftButton,   &QPushButton::clicked,
                    this,    &ControlWindow::goLeft);
    connect(p_rightButton,  &QPushButton::clicked,
                    this,    &ControlWindow::goRight);
    connect(p_downButton,   &QPushButton::clicked,
                    this,    &ControlWindow::goBackward);
    connect(p_stopButton,   &QPushButton::clicked,
                    this,    &ControlWindow::halt);

    conenct(&m_RobotThread, &RobotThread::newPose,
                    this,    &ControlWindow::updatePoseDisplay);
    m_RobotThread.init();
}
```

5.3   Publishing the Velocity Messages

Now that the desired velocity settings have been linked to the `RobotThread`, we need to send the message to the robot itself. This implementation is almost identical to the implementation discussed Sect. 4.2. As such, we will highlight the differences here.

The initialization function is implemented as follows.

```
bool RobotThread::init()
{
    m_pThread = new QThread();
    this->moveToThread(m_pThread);

    connect(m_pThread, &QThread::started, this, &RobotThread::run);
    ros::init(m_Init_argc, m_pInit_argv, "gui_command");

    if (!ros::master::check())
        return false;//do not start without ros.

    ros::start();
    ros::Time::init();
    ros::NodeHandle nh;
    sim_velocity = nh.advertise<geometry_msgs::Twist>("/cmd_vel", 100);
    pose_listener = nh.subscribe(m_topic, 10, &RobotThread::poseCallback,
    this);

    m_pThread->start();
    return true;
}//set up the thread
```

This implementation creates the node `gui_command`. The function advertises the velocity messages on the topic "`cmd_vel`."[4] We then subscribe to the topic name stored in the variable `m_topic`, specified by the terminal arguments and defaulting to "`odom`."

We now turn our attention to the `run` function.

```
void RobotThread::run()
{
    ros::Rate loop_rate(100);
    QMutex * pMutex;
    while (ros::ok())
    {
        pMutex = new QMutex();

        geometry_msgs::Twist cmd_msg;
        pMutex->lock();
        cmd_msg.linear.x = m_speed;
        cmd_msg.angular.z = m_angle;
        pMutex->unlock();
```

---

[4]This is the standard topic for such a message; however, the reader is encouraged to find a way to initialize this with a variable topic [Hint: add a `const char *` to the constructor which defaults to "`cmd_vel`"].

```
        sim_velocity.publish(cmd_msg);
        ros::spinOnce();
        loop_rate.sleep();
        delete pMutex;
    }//main ros loop
}
```

The reader should find this function to be familiar. The message is published as before, but with a much faster loop rate (100 Hz). The message is also of a different type (`geometry_msgs::Twist`). Lastly, the variable `pMutex` prevents the `setSpeed()` function from writing over the member variables `m_speed` and `m_angle` in the middle of creating the message.

A video Demonstration of the application running can be found online [10]. The code will work on any robot that publishes on the `cmd_vel` topic. This includes most robots that support ROS. If the reader does not have a physical robot available, the same code will work in simulation.

```
$ sudo apt-get install ros-indigo-p2os-launch ros-indigo-p2os-urdf
$ roslaunch pioneer3at.gazebo.launch
```

Then, in a separate terminal, run the executable.

```
$ rosrun ros_qt_controller qt_ros_ctrl
```

The on screen robot should now be moving, and the data fields should be updating appropriately in a manner similar to that of the video demonstration.

5.4 Results

We have now demonstrated the flexibility of the Qt library, as well as the practicality of programming with threads in ROS. Solutions of this type are efficient, reusable and include modern programming libraries which are quite easily updated. Our setup is done using only CMake calls, sharing with the reader the necessary tools to connect any external CMake library with the Catkin build system.

We have discovered the power and speed of a threaded ROS application, as well as the remarkable portability of the code required to make such an application. The reader should also have some familiarity with GUI programming for ROS and should be able to adapt the code for research in many fields, particularly Human-Robot Interaction.

References

1. RQt. http://wiki.ros.org/rqt
2. Hunter Allen's Git Repository. http://github.com/allenh1/ros-qt-controller
3. Creating a Catkin Workspace. http://wiki.ros.org/catkin/Tutorials/create_a_workspace
4. R.E.P. 140. http://www.ros.org/reps/rep-0140.html
5. ROS Threading Tutorial Package. https://github.com/allenh1/ros-thread-tutorial
6. CMake Commands. http://www.cmake.org/cmake/help/v3.0/manual/cmake-commands.7.html
7. Qt Translation. http://doc.qt.io/qt-5/i18n-source-translation.html
8. Qt Online Documentation. https://doc-snapshots.qt.io
9. QMutex Class. http://doc.qt.io/qt-5.4/qmutex.html
10. Video Demonstration of Controller. https://vimeo.com/123502207
11. ROS Wiki, Catkin Build. http://wiki.ros.org/catkin

Author's Biography

Hunter Allen is currently an Undergraduate at Purdue University, double-majoring in Mathematics and Computer Science with a minor in Linguistics. Hunter spent 3 years at Vanderbilt University under Dr. Julie A. Adams in the Human-Machine Teaming Laboratory researching autonomous mobile robotics. Hunter is also the current maintainer of the P2OS ROS stack and the Gentoo ROS overlay.

Part II
Navigation, Motion and Planning

Writing Global Path Planners Plugins in ROS: A Tutorial

Maram Alajlan and Anis Koubâa

Abstract In this tutorial chapter, we demonstrate how to integrate a new planner into ROS and present their benefits. Extensive experimentations are performed to show the effectiveness of the newly integrated planners as compared to Robot Operating System (ROS) default planners. The navigation stack of the ROS open-source middleware incorporates both global and local path planners to support ROS-enabled robot navigation. Only basic algorithms are defined for the global path planner including Dijkstra, A*, and carrot planners. However, more intelligent global planners have been defined in the literature but were not integrated in ROS distributions. This tutorial was developed under Ubuntu 12.4 and for ROS Hydro version. However, it is expected to also work with Groovy (not tested). A repository of the new path planner is available at https://github.com/coins-lab/relaxed_astar. A video tutorial also available at https://www.youtube.com/playlist?list=PL8UbFU8tzwRjkxccq2zLkmTkOOYela5fu.

Keywords ROS · Global path planner · Navigation stack

M. Alajlan (✉) · A. Koubâa
Cooperative Networked Intelligent Systems (COINS) Research Group,
Riyadh, Saudi Arabia
e-mail: maram.ajlan@coins-lab.org

A. Koubâa
College of Computer and Information Sciences, Prince Sultan University,
Rafha Street, Riyadh 11586, Saudi Arabia
e-mail: akoubaa@coins-lab.org

M. Alajlan
College of Computer and Information Sciences, King Saud University,
Riyadh 11683, Saudi Arabia

A. Koubâa
CISTER/INESC-TEC, ISEP, Polytechnic Institute of Porto, Porto, Portugal

© Springer International Publishing Switzerland 2016
A. Koubaa (ed.), *Robot Operating System (ROS)*, Studies in Computational
Intelligence 625, DOI 10.1007/978-3-319-26054-9_4

1   Introduction

Mobile robot path planning is a hot research area. Indeed, the robot should have the ability to autonomously generate collision free path between any two positions in its environment. The path planning problem can be formulated as follows: given a mobile robot and a model of the environment, find the optimal path between a start position and a final position without colliding with obstacles. Designing an efficient path planning algorithm is an essential issue in mobile robot navigation since path quality influences the efficiency of the entire application. The constructed path must satisfy a set of optimization criteria including the traveled distance, the processing time, and the energy consumption.

In the literature, the path planning problem is influenced by two factors: (1) the environment, which can be static or dynamic, (2) the knowledge that the robot has about the environment; if the robot has a complete knowledge about the environment, this problem is known as global path planning. On the other hand, if the robot has an incomplete knowledge, this problem is classified as local path planning.

The navigation stack of the Robot Operating System (ROS) open-source middleware incorporates both global and local path planners to support ROS-enabled robot navigation. However, only basic algorithms are defined for the global path planner including Dijkstra, A*, and carrot planners.

In this tutorial, we present the steps for integrating a new global path planner into the ROS navigation system. The new path planner is based on a relaxed version of the A* (RA*) algorithm, and it can be found in [1]. Also, we compare the performance of the RA* with ROS default planner.

The rest of this tutorial is organized as follow.

- Section 2 introduces ROS and its navigation system.
- Section 3 introduces the relaxed A* algorithm.
- In Sect. 4, we present the steps of integrating a new global path planner into the ROS navigation system.
- Section 5 presents the ROS environment configuration.
- In Sect. 6, we conduct the experimental evaluation study to compare the performance of RA* and ROS default planner.

2   ROS

ROS (Robot Operating System) [2] has been developed by Willow Garage [3] and Stanford University as a part of STAIR [4] project, as a free and open-source robotic middleware for the large-scale development of complex robotic systems.

ROS acts as a meta-operating system for robots as it provides hardware abstraction, low-level device control, inter-processes message-passing and package management. It also provides tools and libraries for obtaining, building, writing, and running code across multiple computers. The main advantage of ROS is that it allows manipulating sensor data of the robot as a labeled abstract data stream, called topic, without having to deal with hardware drivers. This makes the programming of robots much easier for software developers as they do not have to deal with hardware drivers and interfaces. Also, ROS provides many high-level applications such as arm controllers, face tracking, mapping, localization, and path planning. This allow the researchers to focus on specific research problems rather than on implementing the many necessary, but unrelated parts of the system.

Mobile robot navigation generally requires solutions for three different problems: mapping, localization, and path planning. In ROS, the *Navigation Stack* plays such a role to integrate together all the functions necessary for autonomous navigation.

2.1 ROS Navigation Stack

In order to achieve the navigation task, the *Navigation Stack* [5] is used to integrate the mapping, localization, and path planning together. It takes in information from odometry, sensor streams, and the goal position to produce safe velocity commands and send it to the mobile base (Fig. 1). The odometry comes through `nav_msgs/Odometry` message over ROS which stores an estimate of the position and velocity of a robot in free space to determine the robot's location. The sensor information comes through either `sensor_msgs/LaserScan` or `sensor_msgs/PointCloud` messages over ROS to avoid any obstacles. The goal is sent to the navigation stack by `geometry_msgs/PoseStamped` message. The navigation stack sends the velocity commands through `geometry_msgs/Twist` message on `/cmd_vel` topic. The `Twist` message composed of two sub-messages:

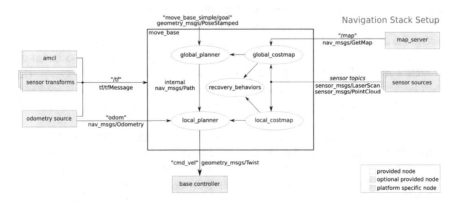

Fig. 1 ROS navigation stack [2]

```
geometry_msgs/Vector3 linear
    float64 x
    float64 y
    float64 z
geometry_msgs/Vector3 angular
    float64 x
    float64 y
    float64 z
```

linear sub-message is used for the x, y and z linear velocity components in meters per second and angular sub-message is used for the x, y and z angular velocity components in radians per second. For example the following Twist message:

```
linear: {x: 0.2, y: 0, z: 0}, angular: {x: 0, y: 0, z: 0}
```

will tell the robot to move with a speed of 0.2 m/s straight ahead. The base controller is responsible for converting Twist messages to "motor signals"which will actually move the robot's wheels [6].

The navigation stack does not require a prior static map to start with. Actually it could be initialized with or without a map. When initialized without a prior map, the robot will know about the obstacles detected by its sensors only and will be able to avoid the seen obstacles so far. For the unknown areas, the robot will generate an optimistic global path which may hit unseen obstacles. The robot will be able to re-plan its path when it receives more information by the sensors about these unknown areas. Instead, when the navigation stack initialized with a static map for the environment, the robot will be able to generate an informed plans to its goal using the map as prior obstacle information. Starting with a prior map will have significant benefits on the performance [5].

To build a map using ROS, ROS provides a wrapper for OpenSlam's Gmapping [7]. A particle filter-based mapping approach [8] is used by the gmapping package to build an occupancy grid map. Then a package named map_server could be used to save that map. The maps are stored in a pair of files: YAML file and image file. The YAML file describes the map meta-data, and names the image file. The image file encodes the occupancy data. The localization part is solved in the amcl package using an Adaptive Monte Carlo Localization [9] which is also based on particle filters. It is used to track the position of a robot against a known map. The path planning part is performed in the move_base package, and is divided into *global* and *local* planning modules which is a common strategy to deal with the complex planning problem.

The *global path planner* searches for a shortest path to the goal and the *local path planner* (also called the *controller*), incorporating current sensor readings, issues the actual commands to follow the global path while avoiding obstacles. More details about the global and local planners in ROS can be found in the next sections.

The *move_base* package also maintains two costmaps, *global_costmap* and *local_costmap* to be used with the global and local planners respectively. The costmap used to store and maintain information in the form of occupancy grid about the obstacles in the environment and where the robot should navigate. The costmap

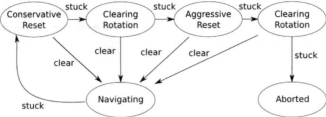

Fig. 2 Recovery behaviors [2]

initialized with prior static map if available, then it will be updated using sensor data to maintain the information about obstacles in the world. Besides that, the *move_base* may optionally perform some previously defined recovery behaviors (Fig. 2) when it fails to find a valid plan.

One reason of failure, is when the robot find itself surrounded with obstacles and cannot find a way to its goal. The recovery behaviors will be performed in some order (defined by the user), and after performing one recovery, the *move_base* will try to find a valid plan, if it succeeds, it will proceed its normal operation. Otherwise if it fails, it will perform the next recovery behavior. If it fails after performing all the recovery behaviors, the goal will be considered infeasible, and it will be aborted. The default recovery behaviors order is presented in Fig. 2 and it is in increasingly aggressive order to attempt to clear out the robot space. First recovery behavior is clearing all the obstacles outside a specific area from the robot's map. Next, an in-place rotation will be performed if possible to clear the space. Next, in case this too fails, more aggressively clearing for the map will be performed, to remove all the obstacles outside of the rectangular area in which the robot can make an in-place rotation. Next, another in-place rotation will be performed. If all this fails, the goal will be aborted.

Therefore, in each execution cycle of the *move_base*, one of three main states should be performed:

- Planning state: run the *global path planner*.
- Controlling state: run the *local path planner* and move the robot.
- Clearing state: run recovery behavior in case the robot stuck.

There are some predefined parameters in ROS navigation stack that are used to control the execution of the states, which are:

- *planner_frequency*: to determine how often the *global path planner* should be called, and is expressed in Hz. When it is set to zero, the global plan will be computed only once for each goal received.
- *controller_frequency*: to determine how often the *local path planner* or *controller* should be called, and also expressed in Hz.

For any global or local planner or recovery behavior to be used with the *move_base* it must be first adhere to some interfaces defined in *nav_core* package, which contains key interfaces for the navigation stack, then it must be added as a plugin to ROS. We developed a tutorial on how to add a new global planner as a plugin to ROS navigation stack, available at [10] and [11].

2.1.1 Global Planner

The *global path planner* in ROS operates on the *global_costmap*, which generally initialized from a prior static map, then it could be updated frequently based on the value of *update_frequency* parameter. The *global path planner* is responsible for generating a long-term plan from the start or current position to the goal position before the robot starts moving. It will be seeded with the costmap, and the start and goal positions. These start and goal positions are expressed by their x and y coordinates. A grid-based global planner that can use Dijkstra's algorithm [12] or A* algorithm to compute shortest collision free path for a robot is obtained in global_planner package. Also, ROS provide another global planner named carrot_planner, which is a simple planner that attempts to move the robot as close to its goal as possible even when that goal is in an obstacle. The current implementation of the global planner in ROS assumes a circular-shape robot. This results in generating an optimistic path for the actual robot footprint, which may be infeasible path. Besides that, the global planner ignores kinematic and acceleration constraints of the robot, so the generated path could be dynamically infeasible.

2.1.2 Local Planner

The *local path planner* or the *controller* in ROS operates on the *local_costmap*, which only uses local sensor information to build an obstacle map and dynamically updated with sensor data. It takes the generated plan from the global planner, and it will try to follow it as close as possible considering the kinematics and dynamics of the robot as well as any moving obstacles information in the *local_costmap*. ROS provides implementation of two local path planning algorithms namely the Trajectory Rollout [13] and the Dynamic Window Approach (DWA) [14] in the package *base_local_planner*. Both algorithms have the same idea to first discretely sampled the control space then to perform forward simulation, and the selection among potential commands. The two algorithm differ in how they sample the robot's control space.

After the global plan passed to the *controller*, the *controller* will produce velocity commands to send to a mobile base. For each control cycle, *the controller* will try to process part from global path (determined by the size of the *local_costmap*).

First, the *controller* will sampled the control space of the robot discretely. The number of the samples will be specified by the controller parameters *vx_samples* and *vtheta_samples* (more details about the parameters can be found in the next section). Then, the controller will perform a simulation in advance for each one of those velocity samples from the current place of the robot to foresee the situation from applying each sample for amount of time (this time will be specified in the parameter *sim_time*). Then, the controller will evaluate each resultant path from the simulation and will exclude any path having collisions with obstacles. For the evaluation, the controller will incorporates the following metrics: distance from obstacles, distance to the goal position, distance from the global plan and robot speed. Finally, the controller will send the velocity command of the highest-scoring path to the mobile base to execute it.

The "Map Grid" is used to evaluate and score the velocities. For each control cycle, the controller will create a grid around the robot (the grid size determined by the size of the *local_costmap*), and the global path will be mapped onto this area. Then each grid cell will receive a distance value. The cells containing path points and the goal will be marked with 0. Then each other grid cell will be marked with its manhattan distance from nearest zero grid by a propagation algorithm. This "Map Grid" is then used in the evaluation and scoring of the velocities. As the "Map Grid" will cover small area from global path each time, the goal position often will lie outside that area. So in that case the first path point inside the area having a consecutive point outside the area will be considered as "local goal", and the distance from that local goal will be considered when scoring trajectories for distance to goal.

3   Relaxed A*

RA* is a time linear relaxed version of A*. It is proposed to solve the path planning problem for large scale grid maps. The objective of RA* consists of finding optimal or near optimal solutions with small gaps, but at much smaller execution times than traditional A*. The core idea consists of exploiting the grid-map structure to establish an accurate approximation of the optimal path, without visiting any cell more than once.

In fact, in A* the exact cost g(n) of a node n may be computed many times; namely, it is computed for each path reaching node n from the start position. However, in the RA* algorithm g(n) is approximated by the cost of the minimum-move path from the start cell to the cell associated to node n.

In order to obtain the relaxed version RA*, some instructions of A*, that are time consuming with relatively low gain in terms of solution quality, are removed. In fact, a node is processed only once in RA*, so there is no need to use the closed set of the A* algorithm. Moreover, in order to save time and memory, we do not keep track of the previous node at each expanded node. Instead, after reaching the goal, the path can be reconstructed, from goal to start by selecting, at each step, the neighbor having the minimum g(n) value. Also, it is useless to compare the g(n) of each neighbor to the g(n) of the current node n as the first calculated g(n) is considered definite.

Finally, it is not needed to check whether the neighbor of the current node is in the open list. In fact, if its g(n) value is infinite, it means that it has not been processed yet, and hence is not in the open list. The RA* algorithm is presented in Algorithm 1.

---

input : *Grid*, *Start*, *Goal*
t Break = 1+1/(length(*Grid*)+width(*Grid*));
// Initialisation:
openSet = Start // Set of nodes to be evaluated;
for *each vertex v in Grid* **do**
  | *g_score*(*v*)= infinity;
end
g_score[*Start*] = 0;
// Estimated total cost from Start to Goal:
f_score[*Start*] = heuristic_cost(*Start*, *Goal*);
while *openSet is not empty and g_score*[*Goal*]== *infinity* **do**
  | *current* = the node in *openSet* having the lowest *f_score*;
  | remove *current* from *openSet*;
  | **for** *each free neighbor v of current* **do**
  | | **if** *g_score*(*v*) == *infinity* **then**
  | | | *g_score*[*v*] = *g_score*[*current*] + *dist_edge*(*current*, *v*);
  | | | *f_score*[*v*] = *g_score*[*v*] + *t Break* * *heuristic_cost*(*v*, *Goal*);
  | | | add neighbor to *openSet*;
  | | **end**
  | **end**
end
if *g_score*(*goal*) ! = *infinity* **then**
  | **return** reconstruct_path(*g_score*) // path will be reconstructed based on *g_score* values;
else
  | **return** failure;
end

Algorithm 1: Relaxed A*

Both terms g(n) and h(n) of the evaluation function of the RA* algorithm are not exact, then there is no guaranty to find an optimal solution.

4 Integration Steps

In this section, we present the steps of integrating a new path planner into ROS. The integration has two main steps: (1) writing the path planner class, and (2) deploying it as a plugin. Following, we describe them in details.

4.1 Writing the Path Planner Class

As mentioned before, to make a new global planner work with ROS, it must first adhere to the interfaces defined in nav_core package. A similar example can be found in the carrot_planner.h [15] as a reference. All the methods defined in nav_core::BaseGlobalPlanner class must be overridden by the new global path planner. For this, you need to create a header file, that we will call in our case, RAstar_ros.h

```
1   /** include the  libraries  you need in your planner here */
2   /** for  global  path planner interface */
3   #include <ros/ros.h>
4   #include <costmap_2d/costmap_2d_ros.h>
5   #include <costmap_2d/costmap_2d.h>
6   #include <nav_core/base_global_planner.h>
7   #include <geometry_msgs/PoseStamped.h>
8   #include <angles/angles.h>
9   #include <base_local_planner/world_model.h>
10  #include <base_local_planner/costmap_model.h>
11
12  using std::string;
13
14  #ifndef RASTAR_ROS_CPP
15  #define RASTAR_ROS_CPP
16
17  namespace RAstar_planner {
18
19  class RAstarPlannerROS : public nav_core::BaseGlobalPlanner {
20  public:
21
22    RAstarPlannerROS();
23    RAstarPlannerROS(std::string name, costmap_2d::Costmap2DROS* costmap_ros);
24
25    /** overridden classes from interface nav_core::BaseGlobalPlanner **/
26    void initialize(std::string name, costmap_2d::Costmap2DROS* costmap_ros);
27    bool makePlan(const geometry_msgs::PoseStamped& start,
28          const geometry_msgs::PoseStamped& goal,
29          std::vector<geometry_msgs::PoseStamped>& plan
30            );
31  };
32  };
33  #endif
```

Now, we will explain the different parts of the header file.

```
3   #include <ros/ros.h>
4   #include <costmap_2d/costmap_2d_ros.h>
5   #include <costmap_2d/costmap_2d.h>
6   #include <nav_core/base_global_planner.h>
7   #include <geometry_msgs/PoseStamped.h>
8   #include <angles/angles.h>
9   #include <base_local_planner/world_model.h>
10  #include <base_local_planner/costmap_model.h>
```

It is necessary to include core ROS libraries needed for path planner. The headers:

```
4   #include <costmap_2d/costmap_2d_ros.h>
5   #include <costmap_2d/costmap_2d.h>
```

are needed to use the `costmap_2d::Costmap2D` class that will be used by the path planner as input map. This map will be accessed automatically by the path planner class when defined as a plugin. There is no need to subscribe to `costmap2d` to get the cost map from ROS.

```
6   #include <nav_core/base_global_planner.h>
```

is used to import the interface *nav_core :: BaseGlobalPlanner*, which the plugin must adhere to.

```
17  namespace RAstar_planner {
18
19  class RAstarPlannerROS : public nav_core::BaseGlobalPlanner {
```

It is a good practice, although not necessary, to define namespace for your class. Here, we define the namespace as `RAstar_planner` for the class `RAstarPlanner ROS`. The namespace is used to define a full reference to the class, as `RAstar_planner::RAstarPlannerROS`. The class `RAstarPlannerROS` is then defined and inherits from the interface `nav_core::BaseGlobalPlanner`. All methods defined in `nav_core::BaseGlobalPlanner` must be overridden by the new class `RAstarPlannerROS`.

```
22  RAstarPlannerROS();
```

Is the default constructor which initializes the planner attributes with default values.

```
23  RAstarPlannerROS(std::string name, costmap_2d::Costmap2DROS* costmap_ros);
```

This constructor is used to initialize the costmap, that is the map that will be used for planning (`costmap_ros`), and the name of the planner (`name`).

```
26  void initialize(std::string name, costmap_2d::Costmap2DROS* costmap_ros);
```

Is an initialization function for the `BaseGlobalPlanner`, which initializes the costmap, that is the map that will be used for planning (`costmap_ros`), and the name of the planner (`name`).

The initialize method for RA* is implemented as follows:

```
void RAstarPlannerROS::initialize(std::string name, costmap_2d::Costmap2DROS* ↵
    costmap_ros)
{
    if (!initialized_)
    {
        costmap_ros_ = costmap_ros;
        costmap_ = costmap_ros_->getCostmap();
        ros::NodeHandle private_nh("~/" + name);

        originX = costmap_->getOriginX();
        originY = costmap_->getOriginY();
        width = costmap_->getSizeInCellsX();
        height = costmap_->getSizeInCellsY();
        resolution = costmap_->getResolution();
        mapSize = width*height;
        tBreak = 1+1/(mapSize);
        OGM = new bool [mapSize];

        for (unsigned int iy = 0; iy < height; iy++)
        {
            for (unsigned int ix = 0; ix < width; ix++)
            {
                unsigned int cost = static_cast<int>(costmap_->getCost(ix, iy));
                if (cost == 0)
                    OGM[iy*width+ix]=true;
                else
                    OGM[iy*width+ix]=false;
            }
        }
        ROS_INFO("RAstar planner initialized successfully");
        initialized_ = true;
    }
    else
        ROS_WARN("This planner has already been initialized ... doing nothing");
}
```

For the particular case of the `carrot_planner`, the initialize method is implemented as follows:

```
void CarrotPlanner::initialize(std::string name, costmap_2d::Costmap2DROS* ↵
    costmap_ros){
    if (!initialized_){
        costmap_ros_ = costmap_ros; //initialize the costmap_ros_ attribute to the ↵
            parameter.
        costmap_ = costmap_ros_->getCostmap(); //get the costmap_ from ↵
            costmap_ros_

        /* initialize other planner parameters */
        ros::NodeHandle private_nh("~/" + name);
        private_nh.param("step_size", step_size_, costmap_->getResolution());
        private_nh.param("min_dist_from_robot", min_dist_from_robot_, 0.10);
        world_model_ = new base_local_planner::CostmapModel(*costmap_);

        initialized_ = true;
    }
    else
        ROS_WARN("This planner has already been initialized ... doing nothing");
}
```

```
27   bool makePlan(const geometry_msgs::PoseStamped& start,
28        const geometry_msgs::PoseStamped& goal,
29        std::vector<geometry_msgs::PoseStamped>& plan
30            );
```

Then, the method bool `makePlan` must be overridden. The final plan will be stored in the parameter `std::vector<geometry_msgs::PoseStamped>& plan` of the method. This plan will be automatically published through the plugin as a topic. An implementation of the `makePlan` method of the `carrot_planner` can be found in [16] as a reference.

Class Implementation In what follows, we present the main issues to be considered in the implementation of a global planner as plugin. The complete source code of the RA* planner can be found in [1]. Here is a minimum code implementation of the RA* global path planner (`RAstar_ros.cpp`).

```
1    #include <pluginlib/class_list_macros.h>
2    #include "RAstar_ros.h"
3
4    //register this planner as a BaseGlobalPlanner plugin
5    PLUGINLIB_EXPORT_CLASS(RAstar_planner::RAstarPlannerROS, nav_core::↵
         BaseGlobalPlanner)
6
7    using namespace std;
8
9    namespace RAstar_planner {
10   //Default Constructor
11   RAstarPlannerROS::RAstarPlannerROS(){
12
13   }
14   RAstarPlannerROS::RAstarPlannerROS(std::string name, costmap_2d::Costmap2DROS* ↵
         costmap_ros){
15     initialize(name, costmap_ros);
16   }
17   void RAstarPlannerROS::initialize(std::string name, costmap_2d::Costmap2DROS* ↵
         costmap_ros){
18
19   }
20   bool RAstarPlannerROS::makePlan(const geometry_msgs::PoseStamped& start, const ↵
         geometry_msgs::PoseStamped& goal,
21       std::vector<geometry_msgs::PoseStamped>& plan ){
22   if (!initialized_) {
23     ROS_ERROR("The planner has not been initialized, please call  initialize () to use the ↵
           planner");
24     return false ;
25   }
26   ROS_DEBUG("Got a start: %.2f, %.2f, and a goal: %.2f, %.2f", start.pose.position.x, ↵
           start.pose.position.y,
27           goal.pose.position.x, goal.pose.position.y);
28   plan.clear();
29   if (goal.header.frame_id != costmap_ros_->getGlobalFrameID()){
30     ROS_ERROR("This planner as configured will only accept goals in the %s frame, but a ↵
           goal was sent in the %s frame.",
31             costmap_ros_->getGlobalFrameID().c_str(), goal.header.frame_id.c_str↵
               ());
32     return false ;
33   }
34   tf::Stamped < tf::Pose > goal_tf;
35   tf::Stamped < tf::Pose > start_tf;
36
37   poseStampedMsgToTF(goal, goal_tf);
38   poseStampedMsgToTF(start, start_tf);
```

```
39
40      // convert the start and goal coordinates into cells indices to be used with RA* ↩
            planner
41      float startX = start.pose.position.x;
42      float startY = start.pose.position.y;
43      float goalX = goal.pose.position.x;
44      float goalY = goal.pose.position.y;
45
46      getCorrdinate(startX, startY);
47      getCorrdinate(goalX, goalY);
48
49      int startCell;
50      int goalCell;
51
52      if (isCellInsideMap(startX, startY) && isCellInsideMap(goalX, goalY)){
53        startCell = convertToCellIndex(startX, startY);
54        goalCell = convertToCellIndex(goalX, goalY);
55      }
56      else {
57        cout << endl << "the start or goal is out of the map" << endl;
58        return false ;
59      }
60      /////////////////////////////////////////////////////////////
61      // call RA* path planner
62      if (GPP->isStartAndGoalCellsValid(OGM, startCell, goalCell)){
63        vector<int> bestPath;
64        bestPath = RAstarPlanner(startCell, goalCell); // call RA*
65
66        //if the global planner find a path
67        if ( bestPath->getPath().size()>0)
68        {
69          // convert the path cells indices into coordinates to be sent to the move base
70          for (int i = 0; i < bestPath->getPath().size(); i++){
71            float x = 0.0;
72            float y = 0.0;
73            int index = bestPath->getPath()[i];
74
75            convertToCoordinate(index, x, y);
76
77            geometry_msgs::PoseStamped pose = goal;
78            pose.pose.position.x = x;
79            pose.pose.position.y = y;
80            pose.pose.position.z = 0.0;
81            pose.pose.orientation.x = 0.0;
82            pose.pose.orientation.y = 0.0;
83            pose.pose.orientation.z = 0.0;
84            pose.pose.orientation.w = 1.0;
85
86            plan.push_back(pose);
87          }
88          // calculate path length
89          float path_length = 0.0;
90          std::vector<geometry_msgs::PoseStamped>::iterator it = plan.begin();
91          geometry_msgs::PoseStamped last_pose;
92          last_pose = *it;
93          it++;
94          for (; it!=plan.end(); ++it) {
95              path_length += hypot( (*it).pose.position.x - last_pose.pose.position.x, (*↩
                     it).pose.position.y - last_pose.pose.position.y );
96              last_pose = *it;
97          }
98          cout <<"The global path length: "<< path_length<< " meters"<<endl;
99          return true;
100       }
101       else{
102         cout << endl << "The planner failed to find a path " << endl
103              << "Please choose other goal position, " << endl;
```

```
104       return  false ;
105     }
106   }
107   else {
108     cout << "Not valid start or goal" << endl;
109     return  false ;
110   }
111 }
112 };
```

The constructors can be implemented with respect to the planner requirements and specification. There are few important things to consider:

- **Register the planner as BaseGlobalPlanner plugin**: this is done through the instruction:

```
5  PLUGINLIB_EXPORT_CLASS(RAstar_planner::RAstarPlannerROS, nav_core::↵
          BaseGlobalPlanner)
```

For this it is necessary to include the library:

```
1  #include <pluginlib/class_list_macros.h>
```

- **The implementation of the makePlan() method**: The start and goal parameters are used to get initial location and target location, respectively. For RA* path planners, the start and goal first will be converted from x and y coordinates to cell indices. Then, those indices will be passed to the RA* planner. When the planner finish its execution, it will return the computed path. Finally, the path cells will be converted to x and y coordinates, then inserted into the plan vector (plan.push_back(pose)) in the for loop. This planned path will then be sent to the move_base global planner module which will publish it through the ROS topic nav_msgs/Path, which will then be received by the local planner module.

Now that your global planner class is done, you are ready for the second step, that is creating the plugin for the global planner to integrate it in the global planner module nav_core::BaseGlobalPlanner of the move_base package.

Compilation To compile the RA* global planner library created above, it must be added (with all of its dependencies if any) to the *CMakeLists.txt*. This is the code to be added:

```
add_library(relaxed_astar_lib src/RAstar_ros.cpp)
```

Then, in a terminal run catkin_make in your catkin workspace directory to generate the binary files. This will create the library file in the lib directory ~/catkin_ws/devel/lib/librelaxed_astar_lib. Observe that "lib" is appended to the library name relaxed_astar_lib declared in the CMakeLists.txt

4.2   Writing Your Plugin

Basically, it is important to follow all the steps required to create a new plugin as explained in the plugin description page [17]. There are five steps:

Plugin Registration First, you need to register your global planner class as plugin by exporting it. In order to allow a class to be dynamically loaded, it must be marked as an exported class. This is done through the special macro PLUGINLIB_EXPORT_CLASS. This macro can be put into any source (.cpp) file that composes the plugin library, but is usually put at the end of the .cpp file for the exported class. This was already done above in RAstar_ros.cpp with the instruction

```
5   PLUGINLIB_EXPORT_CLASS(RAstar_planner::RAstarPlannerROS, nav_core::↩
        BaseGlobalPlanner)
```

This will make the class RAstar_planner::RAstarPlannerROS registered as plugin for nav_core::BaseGlobalPlanner of the move_base.

Plugin Description File The second step consists in describing the plugin in a description file. The plugin description file is an XML file that serves to store all the important information about a plugin in a machine readable format. It contains information about the library the plugin is in, the name of the plugin, the type of the plugin, etc. In our case of global planner, you need to create a new file and save it in certain location in your package and give it a name, for example relaxed_astar_planner_plugin.xml. The content of the plugin description file (relaxed_astar_planner_plugin.xml), would look like this:

```
1   <library path="lib/librelaxed_astar_lib">
2    <class name="RAstar_planner/RAstarPlannerROS"
3    type="RAstar_planner::RAstarPlannerROS"
4    base_class_type="nav_core::BaseGlobalPlanner">
5      <description>This is RA* global planner plugin by iroboapp project.</↩
           description>
6    </class>
7   </library>
```

In the first line:

```
1   <library path="lib/librelaxed_astar_lib">
```

we specify the path to the plugin library. In this case, the path is lib/librelaxed_astar_lib, where lib is a folder in the directory ~/catkin_ws/devel/ (see Compilation section above).

```
2   <class name="RAstar_planner/RAstarPlannerROS"
3   type="RAstar_planner::RAstarPlannerROS"
4   base_class_type="nav_core::BaseGlobalPlanner">
```

Here we first specify the name of the global_planner plugin that we will use later in move_base launch file as parameter that specifies the global planner to be used in `nav_core`. It is typically to use the namespace (`RAstar_planner`) followed by a slash then the name of the class (`RAstarPlannerROS`) to specify the name of plugin. If you do not specify the name, then the name will be equal to the type, which is in this case will be `RAstar_planner::RAstarPlannerROS`. It recommended to specify the name to avoid confusion.

The `type` specifies the name of the class that implements the plugin which is in our case `RAstar_planner::RAstarPlannerROS`, and the `base_class_type` specifies the name of the base class that implements the plugin which is in our case `nav_core::BaseGlobalPlanner`.

```
5   <description>This is RA* global planner plugin by iroboapp project.</description>
```

The `<description>` tag provides a brief description about the plugin. For a detailed description of plugin description files and their associated tags/attributes please see the documentation in [18].

Why Do We Need This File? We need this file in addition to the code macro to allow the ROS system to automatically discover, load, and reason about plugins. The plugin description file also holds important information, like a description of the plugin, that doesn't fit well in the macro.

Registering Plugin with ROS Package System In order for pluginlib to query all available plugins on a system across all ROS packages, each package must explicitly specify the plugins it exports and which package libraries contain those plugins. A plugin provider must point to its plugin description file in its `package.xml` inside the export tag block. Note, if you have other exports they all must go in the same export field. In our RA* global planner example, the relevant lines would look as follows:

```
1   <export>
2     <nav_core plugin="${prefix}/relaxed_astar_planner_plugin.xml" />
3   </export>
```

The ${prefix}/ will automatically determine the full path to the file `relaxed_astar_planner_plugin.xml`. For a detailed discussion of exporting a plugin, interested readers may refer to [19].

Important Note: In order for the above export command to work properly, the providing package must depend directly on the package containing the plugin interface, which is *nav_core* in the case of global planner. So, the *relaxed_astar* package must have the line below in its relaxed_astar/package.xml:

```
1   <build_depend>nav_core</build_depend>
2   <run_depend>nav_core</run_depend>
```

This will tell the compiler about the dependency on the `nav_core` package.

4.2.1 Querying ROS Package System for Available Plugins

One can query the ROS package system via `rospack` to see which plugins are available by any given package. For example:

```
1  $ rospack plugins --attrib=plugin nav_core
```

This will return all plugins exported from the `nav_core` package. Here is an example of execution:

```
1   turtlebot@turtlebot-Inspiron-N5110:~$ rospack plugins --attrib=plugin nav_core
2   rotate_recovery /opt/ros/hydro/share/rotate_recovery/rotate_plugin.xml
3   navfn /home/turtlebot/catkin_ws/src/navfn/bgp_plugin.xml
4   base_local_planner /home/turtlebot/catkin_ws/src/base_local_planner/blp_plugin.↩
        xml
5   move_slow_and_clear /opt/ros/hydro/share/move_slow_and_clear/recovery_plugin.xml
6   robot_controller /home/turtlebot/catkin_ws/src/robot_controller/↩
        global_planner_plugin.xml
7   relaxed_astar /home/turtlebot/catkin_ws/src/relaxed_astar/↩
        relaxed_astar_planner_plugin.xml
8   dwa_local_planner /opt/ros/hydro/share/dwa_local_planner/blp_plugin.xml
9   clear_costmap_recovery /opt/ros/hydro/share/clear_costmap_recovery/ccr_plugin.xml
10  carrot_planner /opt/ros/hydro/share/carrot_planner/bgp_plugin.xml
```

Observe that our plugin is now available under the package `relaxed_astar` and is specified in the file /home/turtlebot/catkin_ws/src/relaxed_astar/relaxed_astar_ planner_plugin.xml. You can also observe the other plugins already existing in `nav_core` package, including `carrot_planner/CarrotPlanner` and `navfn`, which implements the Dijkstra algorithm.

Now, your plugin is ready to use.

4.3 Running the Plugin

There are a few steps to follow to run your planner in turtlebot. First, you need to copy the package that contains your global planner (in our case `relaxed_astar`) into the catkin workspace of your Turtlebot (e.g. `catkin_ws`). Then, you need to run `catkin_make` to export your plugin to your turtlebot ROS environment.

Second, you need to make some modification to `move_base` configuration to specify the new planner to be used. For this, follow these steps:

1. In Hydro, go to this folder /opt/ros/hydro/share/turtlebot_navigation/launch/ includes

```
$ roscd turtlebot_navigation/
$ cd launch/includes/
```

2. Open the file `move_base.launch.xml` (you may need sudo to open and be able to save) and add the new planner as parameters of the global planner, as follows:

```
1    ......
2    <node pkg="move_base" type="move_base" respawn="false" name="move_base" ↵
         output="screen">
3    <param name="base_global_planner" value="RAstar_planner/RAstarPlannerROS" ↵
         />
4    ....
```

Save and close the move_base.launch.xml. Note that the name of the planner is RAstar_planner/RAstarPlannerROS the same specified in relaxed_astar_planner_plugin.xml.

Now, you are ready to use your new planner.

3. You must now bringup your turtlebot. You need to launch minimal.launch, 3dsensor.launch, amcl.launch.xml and move_base.launch. xml. Here is an example of launch file that can be used for this purpose.

```
1    <launch>
2    <include file="$(find turtlebot_bringup)/launch/minimal.launch"></include>
3
4      <include file="$(find turtlebot_bringup)/launch/3dsensor.launch">
5        <arg name="rgb_processing" value="false" />
6        <arg name="depth_registration" value="false" />
7        <arg name="depth_processing" value="false" />
8        <arg name="scan_topic" value="/scan" />
9      </include>
10
11     <arg name="map_file" default="map_folder/your_map_file.yaml"/>
12     <node name="map_server" pkg="map_server" type="map_server" args="$(arg ↵
         map_file)" />
13
14     <arg name="initial_pose_x" default="0.0"/>
15     <arg name="initial_pose_y" default="0.0"/>
16     <arg name="initial_pose_a" default="0.0"/>
17     <include file="$(find turtlebot_navigation)/launch/includes/amcl.launch.xml">
18       <arg name="initial_pose_x" value="$(arg initial_pose_x)"/>
19       <arg name="initial_pose_y" value="$(arg initial_pose_y)"/>
20       <arg name="initial_pose_a" value="$(arg initial_pose_a)"/>
21     </include>
22
23     <include file="$(find turtlebot_navigation)/launch/includes/move_base.launch.xml↵
         "/>
24
25   </launch>
```

Note that changes made in the file move_base.launch.xml will now be considered when you bring-up your turtlebot with this launch file.

4.4 Testing the Planner with RVIZ

After you bringup your turtlebot, you can launch the rviz using this command (in new terminal).

```
$ roslaunch turtlebot_rviz_launchers view_navigation.launch --screen
```

5   ROS Environment Configuration

One important step before using the planners is tuning the controller parameters as they have a big impact on the performance. The controller parameters can be categorized into several groups based on what they control such as: robot configuration, goal tolerance, forward simulation, trajectory scoring, oscillation prevention, and global plan.

The robot configuration parameters are used to specify the robot acceleration information in addition to the minimum and maximum velocities allowed to the robot. We are working with the Turtlebot robot, and we used the default parameters from *turtlebot_navigation* package. The configuration parameters are set as follow: $acc\_lim\_x = 0.5$, $acc\_lim\_theta = 1$, $max\_vel\_x = 0.3$, $min\_vel\_x = 0.1$, $max\_vel\_theta = 1$ $min\_vel\_theta = -1$ $min\_in\_place\_vel\_theta = 0.6$.

The goal tolerance parameters define how close to the goal we can get. *xy_goal_tolerance* represents the tolerance in meters in the x and y distance and should not be less than the map resolution or it will make the robot spin in place indefinitely without reaching the goal, so we set it to 0.1. *yaw_goal_tolerance* represents the tolerance in radians in yaw/rotation. Setting this tolerance very small may cause the robot to oscillate near the goal. We set this parameter very high to 6.26 as we do not care about the robot orientation.

In the forward simulation category, the main parameters are: *sim_time*, *vx_samples*, *vtheta_samples*, and *controller_frequency*. The *sim_time* represents the amount of time (in seconds) to forward-simulate trajectories, and we set it to 4.0. The *vx_samples* and *vtheta_samples* represent the number of samples to use when exploring the x velocity space and the theta velocity space respectively. They should be set depending on the processing power available, and we use the value recommended in ROS web-site for them. So, we set 8 to the *vx_samples* and 20 to *vtheta_samples*. The *controller_frequency* represents the frequency at which this controller will be called. Setting this parameter a value too high can overload the CPU. Setting it to 2 work fine with our planners.

The trajectory scoring parameters are used to evaluate the possible velocities to the local planner. The three main parameters on this category are: *pdist_scale*, *gdist_scale*, and *occdist_scale*. The *pdist_scale* represents the weight for how much the controller should stay close to the global planner path. The *gdist_scale* represents the weight for how much the controller should attempt to reach its goal by whatever path necessary. Increasing this parameter will give the local planner more freedom in choosing its path away from the global path. The *occdist_scale* represents the weight for how much the controller should attempt to avoid obstacles. Because the planners may generate paths very close to the obstacles or dynamically not feasible, we set the $pdist\_scale = 0.1$, $gdist\_scale = 0.8$ and $occdist\_scale = 0.3$ when testing our planners. Another parameter named *dwa* is used to specify whether to use the DWA when setting it to *true*, or use the Trajectory Rollout when setting it to *false*. We set it to *true* because the DWA is computationally less expensive than the Trajectory Rollout.

6 Experimental Validation

For the experimental study using ROS, we have chosen the realistic Willow Garage map (Fig. 3), with dimensions 584 * 526 cells and a resolution 0.1 m/cel. In the map, the white color represents the free area, the black color represents the obstacles, and the grey color represents unknown area.

Three performance metrics are considered to evaluate the global planners: (1) *the path length*, it represents the length of the shortest global path found by the planner, (2) *the steps*, it is the number of the steps in the generated paths. (3) *the execution time*, which represents the amount of time that the planner spend to find its best path.

To evaluate the planners, we consider two tours each with 10 random points. Figure 3 shows the points for one of the tours, where the red circle represents the start location, the blue circle is the goal location, and the green circles are the intermediate waypoints. We run each planner 30 times for each tour. Figures 4, 5, 6 and 7 show the global plans for that tour generated by the RA*, ROS-A*, and ROS-Dijkstra for grid path and gradient descent method respectively.

Tables 1 and 2 shows the path length and the execution time of the planners for the complete tour.

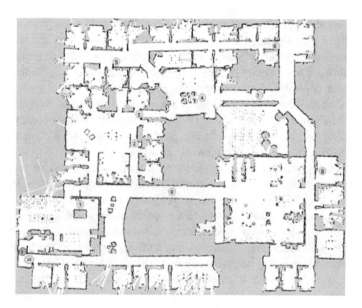

Fig. 3 Willow Garage map

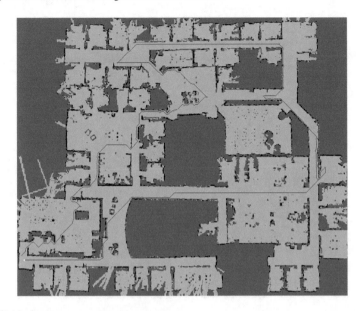

Fig. 4 RA* planner

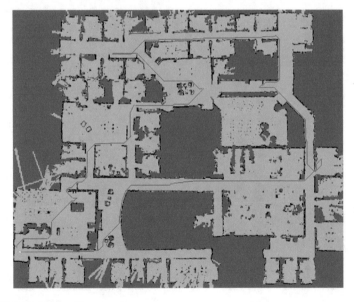

Fig. 5 ROS A* (grid path)

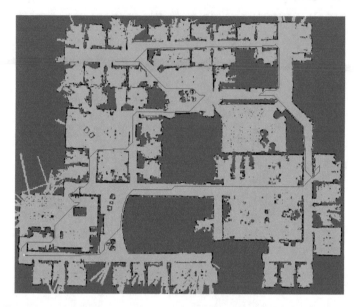

Fig. 6 ROS Dijkstra (grid path)

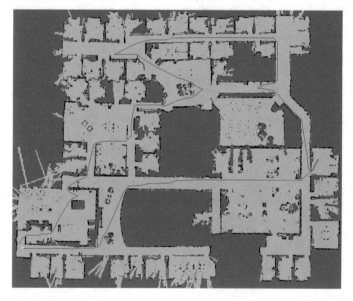

Fig. 7 ROS Dijkstra (gradient descent)

Table 1 Execution time in (microseconds) and path length in (meters) for tour 1

Planner	Execution time		Path length	Number of steps
	Total	Average		
RA* (grid path)	34.7532	3.4214 ± 0.4571	199.6486	1765
A* (grid path)	163.0028	14.4820 ± 1.1104	192.7875	1771
Dijkstra (grid path)	189.9566	19.0443 ± 1.2455	193.5490	1783
Dijkstra (gradient descent)	211.7877	19.1116 ± 1.1714	187.7220	3485

Table 2 Execution time in (microseconds) and path length in (meters) for tour 2

Planner	Execution time		Path length	Number of steps
	Total	Average		
RA* (grid path)	137.2426	13.7243 ± 0.7017	291.9332	2465
A* (grid path)	198.8332	19.8833 ± 1.2563	280.0695	2551
Dijkstra (grid path)	216.1717	21.6172 ± 1.2391	281.4142	2568
Dijkstra (gradient descent)	218.4983	21.8498 ± 1.2261	275.3235	4805

The ROS-Dijkstra with gradient descent method generates paths with more steps but little shorter in length than A* paths, this is because it has smaller granularity (more fine-grain exploration), so it has more freedom in the movements and more able to take smaller steps.

Other planners work at cell level (grid path), so they consider each cell as a single point, and they have to pass the whole cell in each step. As the resolution of the map

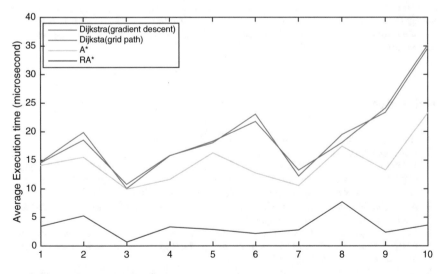

Fig. 8 Detailed execution time for tour 1

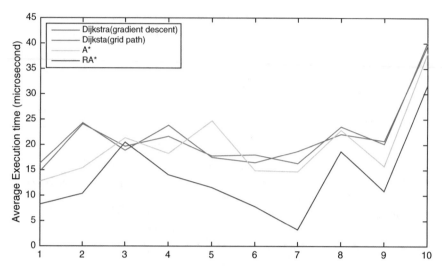

Fig. 9 Detailed execution time for tour 2

is 10 cm, so the distance in each step is at least 10 cm. Comparing the execution time, the RA*, extremely superior the other planners in all simulated cases. Using RA*, the execution time was reduced by more than 78 % from the ROS-A*. Figures 8 and 9 shows the average execution time for 30 run between each two points from the tour.

Acknowledgments This work is supported by the iroboapp project "Design and Analysis of Intelligent Algorithms for Robotic Problems and Applications" under the grant of the National Plan for Sciences, Technology and Innovation (NPSTI), managed by the Science and Technology Unit of Al-Imam Mohamed bin Saud University and by King AbdulAziz Center for Science and Technology (KACST). This work is also supported by the myBot project entitled "MyBot: A Personal Assistant Robot Case Study for Elderly People Care" under the grant from King AbdulAziz City for Science and Technology (KACST).

References

1. Relaxed A*. https://github.com/coins-lab/relaxed_astar (2014)
2. Robot Operating System (ROS). http://www.ros.org
3. K. Wyrobek, E. Berger, H. Van der Loos, J. Salisbury, Towards a personal robotics development platform: rationale and design of an intrinsically safe personal robot, in *IEEE International Conference on Robotics and Automation, ICRA 2008* (IEEE, 2008), pp. 2165–2170
4. M. Quigley, E. Berger, A.Y. Ng, STAIR: hardware and software architecture, in *AAAI, Robotics Workshop* (Vancouver, BC 2007), pp. 31–37
5. E. Marder-Eppstein, E. Berger, T. Foote, B. Gerkey, K. Konolige, The office marathon: robust navigation in an indoor office environment, in *2010 IEEE International Conference on Robotics and Automation (ICRA)*, May 2010, pp. 300–307
6. P. Goebel, *ROS By Example*, edited by. Lulu. http://www.lulu.com/shop/r-patrick-goebel/ros-by-example-hydro-volume-1/ebook/product-21393108.html (2013)

7. OpenSLAM. https://openslam.org/
8. G. Grisetti, C. Stachniss, W. Burgard, Improved techniques for grid mapping with rao-blackwellized particle filters. IEEE Trans. Robot. **23**(1), 34–46 (2007)
9. S. Thrun, D. Fox, W. Burgard, F. Dellaert, Robust Monte Carlo localization for mobile robots. Artif. Intell. **128**(1–2), 99–141 (2000)
10. Adding a global path planner as plugin in ROS. http://www.iroboapp.org/index.php?title= Adding_A_Global_Path_Planner_As_Plugin_in_ROS
11. Writing a global path planner as plugin in ROS. http://wiki.ros.org/navigation/Tutorials/ Writing%20A%20Global%20Path%20Planner%20As%20Plugin%20in%20ROS
12. E.W. Dijkstra, A note on two problems in connexion with graphs. Numerische Mathematik **1**(1), 269–271 (1959)
13. B.P. Gerkey, K. Konolige, Planning and control in unstructured terrain, in *Workshop on Path Planning on Costmaps, Proceedings of the IEEE International Conference on Robotics and Automation (ICRA)* (2008)
14. D. Fox, W. Burgard, S. Thrun, The dynamic window approach to collision avoidance. IEEE Robot. Autom. Mag. **4**(1), 23–33 (1997)
15. Carrot planner header. http://docs.ros.org/hydro/api/carrot_planner/html/carrot_planner_8h_ source.html
16. Carrot planner source. http://docs.ros.org/hydro/api/carrot_planner/html/carrot_planner_ 8cpp_source.html
17. Pluginlib package. http://wiki.ros.org/pluginlib
18. Plugin description file. http://wiki.ros.org/pluginlib/PluginDescriptionFile
19. Plugin export. http://wiki.ros.org/pluginlib/PluginExport

A Universal Grid Map Library: Implementation and Use Case for Rough Terrain Navigation

Péter Fankhauser and Marco Hutter

Abstract In this research chapter, we present our work on a universal grid map library for use as mapping framework for mobile robotics. It is designed for a wide range of applications such as online surface reconstruction and terrain interpretation for rough terrain navigation. Our software features multi-layered maps, computationally efficient repositioning of the map boundaries, and compatibility with existing ROS map message types. Data storage is based on the linear algebra library Eigen, offering a wide range of data processing algorithms. This chapter outlines how to integrate the grid map library into the reader's own applications. We explain the concepts and provide code samples to discuss various features of the software. As a use case, we present an application of the library for online elevation mapping with a legged robot. The grid map library and the robot-centric elevation mapping framework are available open-source at http://github.com/ethz-asl/grid_map and http://github.com/ethz-asl/elevation_mapping.

Keywords ROS · Grid map · Elevation mapping

1 Introduction

Mobile ground robots are traditionally designed to move on flat terrain and their mapping, planning, and control algorithms are typically developed for a two-dimensional abstraction of the environment. When it comes to navigation in rough terrain (e.g. with tracked vehicles or legged robots), the algorithms must be extended to take into account all three dimensions of the surrounding. The most popular approach is to build an elevation map of the environment, where each coordinate on the horizontal

P. Fankhauser (✉) · M. Hutter
Robotic Systems Lab, ETH Zurich, LEE J 201, Leonhardstrasse 21,
8092 Zurich, Switzerland
e-mail: pfankhauser@ethz.ch
URL: http://www.rsl.ethz.ch

M. Hutter
e-mail: mahutter@ethz.ch

© Springer International Publishing Switzerland 2016   99
A. Koubaa (ed.), *Robot Operating System (ROS)*, Studies in Computational
Intelligence 625, DOI 10.1007/978-3-319-26054-9_5

plane is associated with an elevation/height value. For simplicity, elevation maps are often stored and handled as grid maps, which can be thought of as a 2.5-dimensional representation, where each cell in the grid holds a height value.

In our recent work, we have developed a universal grid map library for use as a generic mapping framework for mobile robotics with the Robotic Operating System (ROS). The application is universal in the sense that our implementation is not restricted to any special type of input data or processing step. The library supports multiple data layers and is for example applicable to elevation, variance, color, surface normal, occupancy etc. The underlying data storage is implemented as two-dimensional circular buffer. The circular buffer implementation allows for non-destructive and computationally efficient shifting of the map position. This is for example important in applications where the map is constantly repositioned as the robot moves through the environment (e.g. robot-centric mapping [1]). Our software facilitates the handling of map data by providing several helper functions. For example, iterator functions for rectangular, circular, and polygonal regions enable convenient and memory-safe access to sub-regions of the map. All grid map data is stored as datatypes from Eigen [3], a popular C++ linear algebra library. The user can apply the available Eigen algorithms directly to the map data, which provides versatile and efficient tools for data manipulation.

A popular ROS package that also works with a grid-based map representation is the *costmap_2d* [4, 5] package. It is part of the *2D navigation stack* [6] and used for two-dimensional robot navigation. Its function is to process range measurements and to build an occupancy grid of the environment. The occupancy is typically expressed as a status such as *Occupied*, *Free*, and *Unknown*. Internally, costmaps are stored as arrays of `unsigned char` with an integer value range of 0–255. While sufficient for processing costmaps, this data format can be limiting in more general applications. To overcome these deficiencies, the grid map library presented in this work stores maps as matrices of type `float`. This allows for more precision and flexibility when working with physical types such as height, variance, surface normal vectors etc. To ensure compatibility with the existing ROS ecosystem, the grid map library provides converters to transform grid maps to *OccupancyGrid* (used by the *2D navigation stack*), *GridCells*, and *PointCloud2* message types. Converting and publishing the map data as different message types also allows to make use of existing RViz visualization plugins. Another related package is the *OctoMap* library [7] and its associated ROS interface [8]. *OctoMap* represents a map as three-dimensional structure with occupied and free voxels. Structured as octree, *OctoMap* maps can be dynamically expanded and can contain regions with different resolutions. In comparison to a 2.5-dimensional grid representation, the data structure of *OctoMap* is well suited to represent full three-dimensional structures. This is often useful when working with extended maps with multiple floors and overhanging structures or robot arm motion planning tasks. However, accessing data in the octree entails additional computational cost since a search over the nodes of the tree has to be performed [7]. Instead, the representation of grid maps allows for direct value access and simplified data management in post-processing and data interpretation steps.

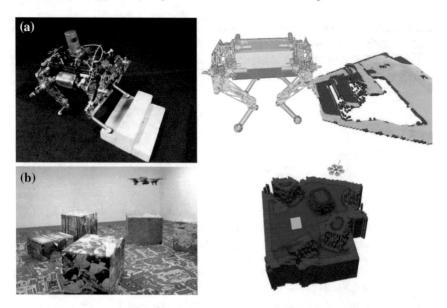

Fig. 1 The grid map library has been applied to various mapping tasks. **a** An elevation map created from an onboard Kinect depth sensor allows the quadrupedal robot StarlETH [10] to navigate in rough terrain [1]. Colors represent the uncertainty of the height estimates, *red* corresponds to less certain and *blue* to more certain estimates. **b** Autonomous landing of a multicopter is achieved by finding a safe landing spot [9]. The map is created from the depth estimation from an onboard IMU and monocular camera. Unsafe landing areas are marked in *red*, safe areas in *blue*, and the chosen landing spot in *green*

The grid map library has served as underlying framework for several applications. In [1], elevation maps are built to plan the motion of a legged robot through rough terrain (see Fig. 1a). Range measurements acquired from an onboard Kinect depth sensor and robot pose estimates are fused in a probabilistic representation of the environment. The mapping procedure is formulated from a robot-centric perspective to explicitly account for drift of the pose as the robot moves.[1] In [9], the grid map library has been used in work for autonomous landing of a Micro Aerial Vehicle (MAV) (see Fig. 1b). An elevation map is generated from estimated depth data from an onboard Inertial Measurement Unit (IMU) and a monocular camera. The map is then used to automatically find a safe landing spot for the vehicle.[2]

In this chapter, we discuss the steps required to implement our software for various applications. In the remainder of this chapter, we cover the following topics:

- First, we demonstrate how to download the grid map library and give an overview of its components. Based on a simple example node, we present the basic steps required to integrate the library.

---

[1] A video demonstration is available at http://youtu.be/I9eP8GrMyNQ.

[2] A video demonstration is available at http://youtu.be/phaBKFwfcJ4.

- Second, we describe the main functionalities of the library. We highlight the main concepts and demonstrate their usage with code samples from a tutorial node.
- Third, we discuss a use case of the software with an elevation mapping application presented in [1].

2 Overview

2.1 *Prerequisites and Installation*

In the following, we assume a functioning installation of Ubuntu 14.04 LTS (Trusty Tahr) and ROS Indigo. Installation instructions for ROS Indigo on Ubuntu are given in [11]. Although we will present the procedures for these versions, the Grid Map packages have been also tested for ROS Jade and should work with future version with no or minor adaptations. Furthermore, we presume a catkin workspace has been setup as described in [12].

Except for ROS packages that are part of the standard installation (*cmake_-modules*, *roscpp*, *sensor_msgs*, *nav_msgs* etc.), the grid map library depends only on the linear algebra library Eigen [3]. If not already available on your system, install Eigen with

```
$ sudo apt-get install libeigen3-dev
```

To use the grid map library, clone the associated packages into the /src folder of your catkin workspace with

```
$ git clone git@github.com:ethz-asl/grid_map.git
```

Finish the installation by building your catkin workspace with

```
$ catkin_make
```

To maximize performance, make sure to build in *Release* mode. You can specify the build type by setting

```
$ catkin_make -DCMAKE_BUILD_TYPE=Release
```

If desired, you can build and run the associated unit tests with

```
$ catkin_make run_tests_grid_map_core run_tests_grid_map
```

Note that our library makes use of C++11 features such as list initializations and range-based for-loops. Therefore, the CMake flag -std=c++11 is added in the CMakeLists.txt files.

2.2 Software Components

The grid map library consists of several components:

grid_map_core implements the core algorithms of the grid map library. It provides the `GridMap` class and several helper classes. This package is implemented without ROS dependencies.

grid_map is the main package for ROS dependent projects using the grid map library. It provides the interfaces to convert grid map objects to several ROS message types.

grid_map_msgs holds the ROS message and service definitions for the *GridMap* message type.

grid_map_visualization contains a node written to visualize *GridMap* messages in RViz by converting them to standard ROS message types. The visualization types and parameters are fully user-configurable through ROS parameters.

grid_map_demos contains several nodes for demonstration purposes. The *simple_demo* node is a short example on how to use the grid map library. An extended demonstration of the library's functionalities is given in the *tutorial_demo* node. Finally, the *iterators_demo* and *image_to_gridmap_demo* nodes showcase the usage of the grid map iterators and the conversion of images to grid maps, respectively.

grid_map_filters builds on the ROS filters library [13] to implement a range of filters for grid map data. The filters provide a standardized API to define a chain of filters based on runtime parameters. This allows for great flexibility when writing software to process grid maps as a sequence of configurable filters.

2.3 A Simple Example

In the following, we describe a simple example on how to use the grid map library. Use this code to verify your installation of the grid map packages and to get you started with your own usage of the library. Locate the file `grid_map_demos/src/simple_demo_node.cpp` with the following content:

```
1  #include <ros/ros.h>
2  #include <grid_map/grid_map.hpp>
3  #include <grid_map_msgs/GridMap.h>
4  #include <cmath>
5
6  using namespace grid_map;
7
8  int main(int argc, char** argv)
9  {
10   // Initialize node and publisher.
11   ros::init(argc, argv, "grid_map_simple_demo");
12   ros::NodeHandle nh("~");
```

```
13    ros::Publisher publisher =
          nh.advertise<grid_map_msgs::GridMap>("grid_map", 1, true);
14
15    // Create grid map.
16    GridMap map({"elevation"});
17    map.setFrameId("map");
18    map.setGeometry(Length(1.2, 2.0), 0.03);
19    ROS_INFO("Created map with size %f x %f m (%i x %i cells).",
20      map.getLength().x(), map.getLength().y(),
21      map.getSize()(0), map.getSize()(1));
22
23    // Work with grid map in a loop.
24    ros::Rate rate(30.0);
25    while (nh.ok()) {
26
27      // Add data to grid map.
28      ros::Time time = ros::Time::now();
29      for (GridMapIterator it(map); !it.isPastEnd(); ++it) {
30        Position position;
31        map.getPosition(*it, position);
32        map.at("elevation", *it) = -0.04 + 0.2 * std::sin(3.0 *
              time.toSec() + 5.0 * position.y()) * position.x();
33      }
34
35      // Publish grid map.
36      map.setTimestamp(time.toNSec());
37      grid_map_msgs::GridMap message;
38      GridMapRosConverter::toMessage(map, message);
39      publisher.publish(message);
40      ROS_INFO_THROTTLE(1.0, "Grid map (timestamp %f) published.",
              message.info.header.stamp.toSec());
41
42      // Wait for next cycle.
43      rate.sleep();
44    }
45    return 0;
46  }
```

In this program, we initialize a ROS node which creates a grid map, adds data, and publishes it. The code consists of several code blocks which we explain part by part.

```
10    // Initialize node and publisher.
11    ros::init(argc, argv, "grid_map_simple_demo");
12    ros::NodeHandle nh("~");
13    ros::Publisher publisher =
          nh.advertise<grid_map_msgs::GridMap>("grid_map", 1, true);
```

This part initializes a node with name *grid_map_simple_demo* (Line 11) and creates a private node handle (Line 12). A publisher of type grid_map_msgs::GridMap is created which advertises on the topic grid_map (Line 13).

```
15    // Create grid map.
16    GridMap map({"elevation"});
17    map.setFrameId("map");
```

```
18    map.setGeometry(Length(1.2, 2.0), 0.03);
19    ROS_INFO("Created map with size %f x %f m (%i x %i cells).",
20      map.getLength().x(), map.getLength().y(),
21      map.getSize()(0), map.getSize()(1));
```

We create a variable `map` of type `grid_map::GridMap` (we are setting using `namespace grid_map` on Line 6). A grid map can contain multiple map layers. Here the grid map is constructed with one layer named *elevation*.[3] The frame id is specified and the size of the map is set to 1.2 × 2.0 m (side length along the *x* and *y*-axis of the `map` frame) with a resolution of 0.03 m/cell. Optionally, the position (of the center) of the map could be set with the third argument of the `setGeometry(...)` method. The print out shows information about the generated map:

```
[INFO ][..]: Created map with size 1.200000 x 2.010000 m (40 x 67cells).
```

Notice that the requested map side length of 2.0 m has been changed to 2.01 m. This is done automatically in order to ensure consistency such that the map length is a multiple of the resolution (0.03 m/cell).

```
23    // Work with grid map in a loop.
24    ros::Rate rate(30.0);
25    while (nh.ok()) {
```

After having setup the node and the grid map, we add data to the map and publish it in a loop of 30 Hz.

```
27    // Add data to grid map.
28    ros::Time time = ros::Time::now();
29    for (GridMapIterator it(map); !it.isPastEnd(); ++it) {
30      Position position;
31      map.getPosition(*it, position);
32      map.at("elevation", *it) = -0.04 + 0.2 * std::sin(3.0 *
33          time.toSec() + 5.0 * position.y()) * position.x();
      }
```

Our goal here is to add data to the *elevation* layer of the map where the elevation for each cell is a function of the cell's position and the current time. The `GridMapIterator` allows to iterate through all cells of the grid map (Line 29). Using the *-operator on the iterator, the current cell index is retrieved. This is used to determine the position for each cell with help of the `getPosition(...)` method of the grid map (Line 31). The current time in seconds is stored in the variable `time`. Applying the temporary variables `position` and `time`, the elevation is computed and stored in the current cell of the *elevation* layer (Line 32).

```
35    // Publish grid map.
36    map.setTimestamp(time.toNSec());
37    grid_map_msgs::GridMap message;
38    GridMapRosConverter::toMessage(map, message);
```

---

[3]For simplicity, we use the list initialization feature of C++11 on Line 16.

```
39    publisher.publish(message);
40    ROS_INFO_THROTTLE(1.0, "Grid map (timestamp %f) published.",
          message.info.header.stamp.toSec());
```

We update the timestamp of the map and then use the `GridMapRosConverter` class to convert the grid map object (of type `grid_map::GridMap`) to a ROS grid map message (of type `grid_map_msgs::GridMap`, Line 38). Finally, the message is broadcasted to ROS with help of the previously defined publisher (Line 39). After building with `catkin_make`, make sure a *roscore* is active, and then run this example with

```
$ rosrun grid_map_demos simple_demo
```

This will run the node *simple_demo* and publish the generated grid maps under the topic `grid_map_simple_demo/grid_map`.

In the next step, we show the steps to visualize the grid map data. The *grid_map_visualization* package provides a simple tool to visualize a ROS grid map message in RViz in various forms. It makes use of existing RViz plugins by converting the grid map message to message formats such as *PointCloud2*, *OccupancyGrid*, *Marker* etc. We create a launch-file under `grid_map_demos/grid_map_demos/launch/simple_demo.launch` with the following content

```
1   <launch>
2     <!-- Launch the grid map simple demo node -->
3     <node pkg="grid_map_demos" type="simple_demo"
          name="grid_map_simple_demo" output="screen" />
4     <!-- Launch the grid map visualizer -->
5     <node pkg="grid_map_visualization"
          type="grid_map_visualization"
          name="grid_map_visualization" output="screen">
6       <rosparam command="load" file="$(find
            grid_map_demos)/config/simple_demo.yaml" />
7     </node>
8     <!-- Launch RViz with the demo configuration -->
9     <node name="rviz" pkg="rviz" type="rviz" args="-d $(find
          grid_map_demos)/rviz/grid_map_demo.rviz" />
10  </launch>
```

This launches *simple_demo* (Line 3), *grid_map_visualization* (Line 5), and *RViz* (Line 9). The *grid_map_visualization* node is loaded with the parameters from `grid_map_demos/config/simple_demo.yaml` with following configuration

```
1   grid_map_topic: /grid_map_simple_demo/grid_map
2   grid_map_visualizations:
3     - name: elevation_points
4       type: point_cloud
5       params:
6         layer: elevation
7     - name: elevation_grid
8       type: occupancy_grid
```

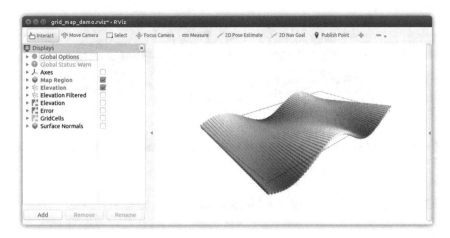

Fig. 2 Launching the `simple_demo.launch` file will run the *simple_demo* and *grid_map_visualization* node and open RViz. The generated grid map is visualized as point cloud

```
 9    params:
10      layer: elevation
11      data_min: 0.1
12      data_max: -0.18
```

This connects the *grid_map_visualization* with the *simple_demo* node via the `grid_map_topic` parameter (Line 1) and adds a *PointCloud2* (Line 3) and *OccupancyGrid* visualization (Line 7) for the *elevation* layer. Run the launch file with

```
$ roslaunch grid_map_demos simple_demo.launch
```

If everything is setup correctly, you should see the generated grid map in RViz as shown in Fig. 2.

3   Package Description

This section gives an overview of the functions of the grid map library. We describe the API and demonstrate its usage with example code from a tutorial file at `grid_map_demos/src/tutorial_demo_node.cpp`. This *tutorial_demo* extends the *simple_demo* by deteriorating the *elevation* layer with noise and outliers (holes in the map). An average filtering step is applied to reconstruct the original data and the remaining error is analyzed. You can run the *tutorial_demo* including visualization with

```
$ roslaunch grid_map_demos tutorial_demo.launch
```

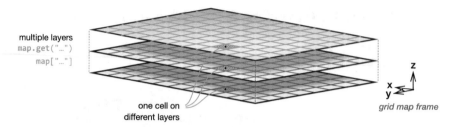

multiple layers
map.get("...")
map["..."]

one cell on
different layers

grid map frame

Fig. 3 The grid map library uses multilayered grid maps to store data for different types of information

3.1 Adding, Accessing, and Removing Layers

When working with mobile robotic mapping, we often compute additional information to interpret the data and to guide the navigation algorithms. The grid map library uses *layers* to store information for different types of data. Figure 3 illustrates the multilayered grid map concept, where for each cell data is stored on the congruent layers.

The grid map class can be initialized empty or directly with a list of layers such as

```
18  GridMap map({"elevation", "normal_x", "normal_y", "normal_z"});
```

A new layer can be added to an existing map with

```
void add(const std::string& layer, const float value) ,
```

where the argument `value` determines the value with which the new layer is populated. Alternatively, data for a new layer can be added with

```
45  map.add("noise", Matrix::Random(map.getSize()(0), map.getSize()
        (1)));
```

In case the added layer already exists in the map, it is replaced with the new data. The availability of a layer in the map can be checked with the method `exists(...)`. Access to the layer data is given by the `get(...)` methods and its short form operator `[...]`:

```
46  map.add("elevation_noisy", map.get("elevation") + map["noise"]);
```

More details on the use of the addition operator + with grid maps is given in Sect. 3.7. A layer from the map can be removed with `erase(...)`.

3.2 Setting the Geometry and Position

For consistent representation of the data, we define the geometrical properties of a grid map as illustrated in Fig. 4. The map is specified in a *grid map frame*, to which

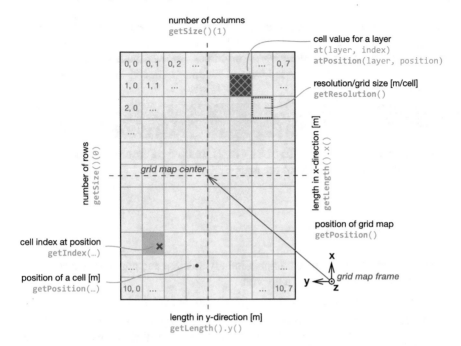

Fig. 4 The grid map is defined to be aligned with its *grid map frame*. The position of the map is specified as the center of the map relative the its frame

the map is aligned to. The map's position is defined as the center of the rectangular map in the specified frame.

The basic geometry of the map such as side lengths, resolution, and position (see Fig. 4) is set through the `setGeometry(...)` method. When working with different coordinate frames, it is important to specify the grid map's frame:

```
19   map.setFrameId("map");
20   map.setGeometry(Length(1.2, 2.0), 0.03, Position(0.0, -0.1));
```

It is important to set the geometry before adding any data as all cell data is cleared when the grid map is resized.

3.3 Accessing Cells

As shown in Fig. 4, individual cells of grid map layer can be accessed in two ways, either by direct index access with

```
float& at(const std::string& layer, const grid_map::Index& index)
```

or by referring to a position of the underlying cell with

```
float& atPosition(const std::string& layer,
                  const grid_map::Position& position).
```

Conversions between a position and the corresponding cell index and vice versa are given with the `getIndex(...)` and `getPosition(...)` methods. These conversions handle the underlying algorithmics involved with the circular buffer storage and are the recommended way of handling with cell indices/positions. An example of usage is:

```
51  if (map.isInside(randomPosition))
52    map.atPosition("elevation_noisy", randomPosition) = ...;
```

Here, the helper method `isInside(...)` is used to ensure that the requested position is within the bounds of the map.

3.4 Moving the Map

The grid map library offers a computationally efficient and non-destructive method to change the position of grid map with

void move(**const** grid_map::Position& position).

The argument `position` specifies the new absolute position of the grid map w.r.t. the grid map frame (see Sect. 3.2). This method relocates the grid map such that the grid aligns between the previous and new position (Fig. 5). Data in the overlapping region before and after the position change remains stored. Data that falls outside of the map at its new position is discarded. Cells that cover previously unknown regions are emptied (set to `nan`). The data storage is implemented as two-dimensional

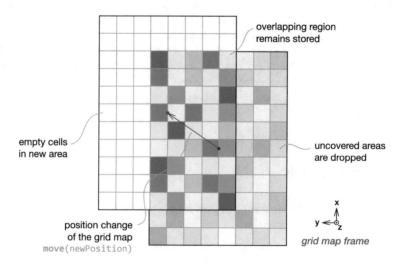

Fig. 5 The grid map library offers with the move(...) method a computationally efficient and non-destructive method to change the position of grid map. The underlying implementation of the grid map as two-dimensional circular buffer allows to move the map without allocating new memory

circular buffer, meaning that data in the overlapping region of the map is not moved in memory. This minimizes the amount of data that needs to be changed when moving the map.

3.5 Basic Layers

For certain applications it is useful to define a set of *basic layers* with `setBasic Layers(...)`. These layers are considered when checking the validity of a grid map cell. If the cell has a valid (finite) value in *all* basic layers, the cell is declared valid by the method

`bool isValid(const grid_map::Index& index) const.`

Similarly, the method `clearBasic()` will clear only the basic layers and is therefore more efficient than `clearAll()` which clears all layers. By default, the list of basic layers is empty.

3.6 Iterating over Cells

Often, one wants to iterate through a number of cells within a geometric shape within the grid map, for example to check the cells which are covered by the footprint of a robot. The grid map library offers iterators for various regions such as rectangular submaps, circular and polygonal regions, and lines (see Fig. 6). Iteration can be achieved with a `for`-loop of the form

```
for (GridMapIterator it(map); !isPastEnd(); ++it) { ... }
```

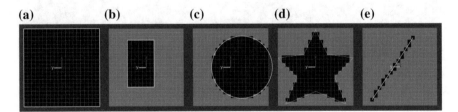

(a)   **(b)**   **(c)**   **(d)**   **(e)**

Fig. 6 Iterators are a convenient way to work with cells belonging to certain geometrical regions. **a** The `GridMapIterator` is used to iterate through all cells of the grid map. **b** The `SubmapIterator` accesses cells belonging to a rectangular region of the map. **c** The `CircleIterator` iterates through a circular region specified by a center and radius. **d** The `PolygonIterator` is defined by *n* vertices and covers all inlying cells. **e** The `LineIterator` uses Bresenham's line algorithm [14] to iterate over a line defined by a start and end point. Example code on how to use the grid map iterators is given in the *iterators_demo* node of the *grid_map_demos* package

The dereference operator * is used to retrieve the current cell index of the iterator such as in following examples:

```
35   map.at("elevation", *it) = ...;
```

```
67   Position currentPosition;
68   map.getPosition(*it, currentPosition);
```

```
76   if (!map.isValid(*circleIt, "elevation_noisy")) continue;
```

These iterators can be used at the border of the maps without concern as they internally ensure to access only cells within the map boundary.

3.7 Using Eigen Functions

Because each layer of the grid map is stored internally as an Eigen matrix, all functions provided by the Eigen library [3] can directly be applied. This is illustrated with following examples:

```
46   map.add("elevation_noisy", map.get("elevation") + map["noise"]);
```

Here, the cell values of the two layers *elevation* and *noise* are summed up arithmetically and the result is stored in the new layer *elevation_noisy*. Alternatively, the noise could have been added to the *elevation* layer directly with the += operator:

```
46   map.get("elevation") += 0.015 * Matrix::Random(map.getSize()(0),
47       map.getSize()(1));
```

A more advanced example demonstrates the simplicity that is provided by making use of the various Eigen functions:

```
91   map.add("error", (map.get("elevation_filtered") -
         map.get("elevation")).cwiseAbs());
92   unsigned int nCells = map.getSize().prod();
93   double rmse = sqrt(map["error"].array().pow(2).sum() / nCells);
```

Here, the absolute error between two layers is computed (Line 91) and used to calculate the Root Mean Squared Error (RMSE) (Line 93) as

$$e_{\text{RMS}} = \sqrt{\frac{\sum_{i=1}^{n} (f_i - c_i)^2}{n}}, \tag{1}$$

where c_i denotes the value of cell i of the original *elevation* layer, f_i the cell values in the filtered *elevation_filtered* layer, and n the number of cells of the grid map.

3.8 Creating Submaps

A copy of a part from the grid map can be generated with the method

```
GridMap getSubmap(const grid_map::Position& position,
                  const grid_map::Length& length, bool&
                  isSuccess).
```

The return value is of type `GridMap` and all described methods apply also to the retrieved submap. When retrieving a submap, a deep copy of the data is made and changing data in the original or the submap will not influence its counterpart. Working with submaps is often useful to minimize computational load and data transfer if only parts of the data is of interest for further processing. To access or manipulate the original data in a submap region without copying, refer to the `SubmapIterator` described in Sect. 3.6.

3.9 Converting from and to ROS Data Types

The grid map class can be converted from and to a ROS *GridMap* message with the methods `fromMessage(...)` and `toMessage(...)` of the `GridMapRos Converter` class:

```
97  grid_map_msgs::GridMap message;
98  GridMapRosConverter::toMessage(map, message);
```

Additionally, a grid map can be saved in and loaded from a ROS bag file with the `saveToBag(...)` and `loadFromBag(...)` methods. For compatibility and visualization purposes, a grid map can be converted also to *PointCloud2*, *OccupancyGrid*, and *GridCells* message types. For example, using the method

```
static void toPointCloud(const grid_map::GridMap& gridMap,
    const std::vector<std::string>& layers,
    const std::string& pointLayer,
    sensor_msgs::PointCloud2& pointCloud) ,
```

a grid map is converted to a *PointCloud2* message. Here, the second argument `layers` determines which layers are converted to the point cloud. The layer in the third argument `pointLayer` specifies which of these layers is used as z-value of the points (x and y are given by the cell positions). All other layers are added as additional fields in the point cloud message. These additional fields can be used in RViz to determine the color of the points. This is a convenient way to visualize characteristic of the map such as uncertainty, terrain quality, material etc.

Figure 7 shows the different ROS messages types that have been generated with the *grid_map_visualization* node for the *tutorial_demo*. The setup of the *grid_map_visualization* for this example is stored in the `grid_map_demos/config/tutorial_demo.yaml` parameters file.

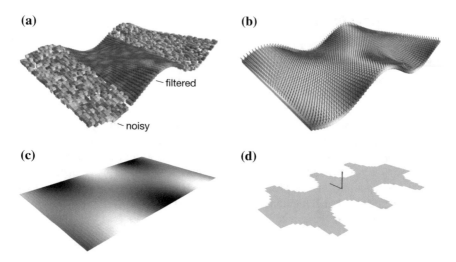

Fig. 7 Grid maps can be converted to several ROS message types for visualization in RViz. **a** Visualization as *PointCloud2* message type showing the results of the *tutorial_demo* filtering step. Colors represent the absolute error between the filtered result and the original data before noise corruption. **b** Vectors (here surface normals) can be visualized with help of the *Marker* message type. **c** Representation of a single layer (here the *elevation* layer) of the grid map as *OccupancyGrid*. **d** The same layer is shown as *GridCells* type where cells within a threshold are visualized

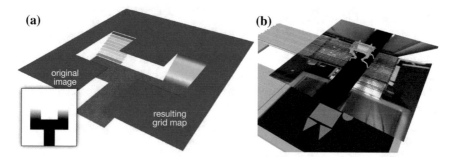

Fig. 8 Image data can be used to populate a grid map. **a** A terrain is drawn in an image editing software and used to generate a height field grid map layer. **b** A color layer in the grid map is updated by the projected video stream of two wide-angle cameras in the front and back of the robot. Example code on how to add data from images to a grid map is given in the *image_to_gridmap_demo* node of the *grid_map_demos* package

3.10 Adding Data from Images

The grid map library allows to load data from images. In a first step, the geometry of the grid map can be initialized to the size of the image with the method `GridMapRosConverter::initializeFromImage(...)`. Two different methods are provided to add data from an image as layer in a grid map. To add data as scalar values (typically from a grayscale image), the method `addLayerFrom` `Image()` can be used. This requires to specify the lower and upper values for the corresponding black and white pixels of the image. Figure 8a shows an example, where an image editing software was used to draw a terrain. To add data from an image as color information, the `addColorLayerFromImage()` is available. This can be used for example to add color information from a camera to the grid map as shown in Fig. 8b.

4   Use Case: Elevation Mapping

Based on the grid map library, we have developed a ROS package for local terrain mapping with an autonomous robot. This section introduces the technical background of the mapping process, discusses the implementation and usage of the package, and presents results from real world experiments.

4.1   Background

Enabling robots to navigate in previously unseen, rough terrain, requires to reconstruct the terrain as the robot moves through the environment. In this work, we assume an existing robot pose estimation and an onboard range measurement sensor. For many autonomous robots, the pose estimation is prone to drift. In case of a drifting pose estimate, stitching fresh scans with previous data leads to inconsistent maps. Addressing this issue, we formulate a probabilistic elevation mapping process from a robot-centric perspective. In this approach, the generated map is an estimate of the shape of the terrain from a local perspective. We illustrate the robot-centric elevation mapping procedure in Fig. 9. Our mapping method consists of three main steps:

1. **Measurement Update:**   New measurements from the range sensor are fused with existing data in the map. By modeling the three-dimensional noise distribution of the range sensor, we can propagate the depth measurement errors to the corresponding elevation uncertainty. By means of a Kalman filter, new measurements are fused with existing data in the map.
2. **Map Update:**   For a robot-centric formulation, the elevation map needs to be updated when the robot moves. We propagate changes of pose covariance matrix to the spatial uncertainty of the cells in the grid map. This reflects errors of the pose estimate (such as drift) in the elevation map. At this step, the uncertainty for each cell is treated separately to minimize computational load.
3. **Map Fusion:**   When map data is required for further processing in the planning step, an estimation of the cell heights is computed. This requires to infer the elevation and variance for each cell from its surrounding cells.

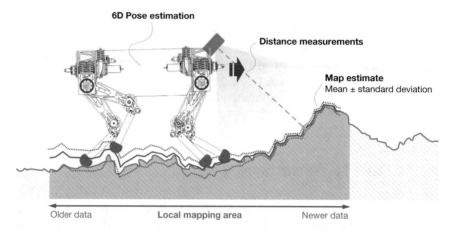

Fig. 9 The elevation mapping mapping procedure is formulated from a robot-centric perspective.

Applying our mapping method, the terrain is reconstructed under consideration of the range sensor errors and the robots pose uncertainty. Each cell in the grid map holds information about the height estimate and a corresponding variance. The region ahead of the robot has typically the highest precision as it is constantly updated with new measurements from the forward-looking distance sensor. Regions which are out of the sensor's field of view (below or behind the robot) have decreased certainty due to drift of the robot's relative pose estimation. Applying a probabilistic approach, motion/trajectory planning algorithms can take advantage of the available certainty estimate of the map data.

4.2 Implementation

We have implemented the robot-centric elevation mapping process as a ROS package.[4] The node subscribes to *PointCloud2* depth measurements and a robot pose estimate of type *PoseWithCovarianceStamped*. For the depth measurements, we provide *sensor processors* which generate the measurement variances based on a noise model of the device. We support several range measurement devices such as Kinect-type sensors [2], laser range sensors, and stereo cameras.

Building a local map around the robot, the map's position is updated continuously to cover the area around the robot. This can be achieved with minimal computational overhead by using the move(...) method provided by the grid map library (see Sect. 3.4).

---

[4] Available at: http://github.com/ethz-asl/elevation_mapping.

Fig. 10 The quadrupedal robot StarlETH [10] is walking over obstacles by using the *elevation mapping* node to map the environment around the its current position. A Kinect-type range sensor is used in combination with the robot pose estimation to generate a probabilistic reconstruction of the terrain

The elevation mapping node publishes the generated map as *GridMap* message (containing layers for elevation, variance etc.) on two topics. The unfused elevation map (considering only processing steps 1 and 2 from Sect. 4.1) is continuously published under the topic `elevation_map_raw` and can be used to monitor the mapping process. The fused elevation map (the result of processing step 3 Sect. 4.1) is computed and published under `elevation_map` if the ROS service `trigger_fusion` is called. Partial (fused) maps can be requested over the `get_submap` service call. The node implements the map update and fusion steps as multi-threaded processes to enable continuous measurement updates.

Using the *grid_map_visualization* node, the elevation maps can be visualized in RViz in several ways. It is convenient to show the map as point cloud and color the points with data from different layers. For example, in Fig. 1 the points are colored with the variance data of the map and in Fig. 10 with colors from an RGB camera. Using a separate node for visualization is a good way to split the computational load to different systems and limit the data transfer from the robot to an operator computer.

4.3 Results

We have implemented the elevation mapping software on the quadrupedal robot StarlETH [10] (see Fig. 10). A downward-facing PrimeSense Carmine 1.09 structured light sensor is attached in the front of the robot as distance sensor. The state estimation is based on stochastic fusion of kinematic and inertial measurements [15], where the position and the yaw-angle are in general unobservable and are therefore subject to drift. We generate an elevation map with a size of 2.5×2.5 m with a resolution of 1 cm/cell and update it with range measurements and pose estimates at 20 Hz on an Intel Core i3, 2.60 GHz onboard computer. The mapping software can process the data at sufficient speed such that the map can be used for real-time navigation. As a result of the high resolution and update rate of the range sensor, the map holds

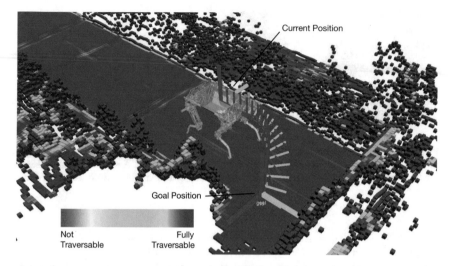

Fig. 11 The traversability of the terrain is judged based on the acquired elevation map. The traversability estimation takes factors such as slope, step size, and roughness of the terrain into consideration. A collision free path for the pose of the robot is found using an RRT-based [16] planner

dense information and contains only very few cells with no height information. This is an important prerequisite for collision checking and foothold selection algorithms. Thanks to the probabilistic fusion of the data, the real terrain structure is accurately captured and outliers and spurious data is suppress effectively.

In another setup, we have also used the elevation map data to find a traversable path through the environment. Figure 11 depicts the robot planning a collision free path in the map of a corridor. A rotating Hokuyo laser range sensor in the front of the robot is used the generate the elevation map with a size of 6×6m with a resolution of 3 cm/cell. In this work, the elevation information is processed to estimate the traversability of the terrain. Multiple custom grid map filters (see Sect. 2.2) are used to interpret the data based on the slope, step height, and roughness of the terrain. An RRT-based [16] planner finds a traversable path to the goal pose for the horizontal translation x and y and yaw-rotation ψ. At each expansion of the search tree, the traversability of all the cells contained in the footprint of the robot is checked with help of the `PolygonIterator` (see Sect. 3.6).

5 Summary and Conclusion

In this chapter, we introduced a grid map library developed for mapping applications with mobile robots. Based on a simple example, the first steps to integrate the library in existing or new frameworks were presented. We highlighted some of the features of the software and illustrated their usage with code samples from a tutorial example. Finally, an application of the grid map library for elevation mapping was shown, whereby the background, usage, and results were discussed.

In our current work, we use the grid map library (in combination with the elevation mapping package) in applications such as multi-robot mapping, foothold quality evaluation, collision checking for motion planning, and terrain property prediction through color information. We hope our software will be useful to other researcher working on rough terrain mapping and navigation with mobile robots.

Acknowledgments This work was supported in part by the Swiss National Science Foundation (SNF) through project 200021_149427/1 and the National Centre of Competence in Research Robotics.

References

1. P. Fankhauser, M. Bloesch, C. Gehring, M. Hutter, R. Siegwart, Robot-centric elevation mapping with uncertainty estimates, in *International Conference on Climbing and Walking Robots (CLAWAR)* (2014)
2. P. Fankhauser, M. Bloesch, D. Rodriguez, R. Kaestner, M. Hutter, R. Seigwart, Kinect v2 for mobile robot navigation: evaluation and modeling, in *International Conference on Advanced Robotics (ICAR)* (2015)
3. G. Guennebaud, B. Jacob, Eigen 3, http://eigen.tuxfamily.org. Accessed Aug 2015
4. E. Marder-Eppstein, D.V. Lu, D. Hershberger. costmap_2d, http://wiki.ros.org/costmap_2d. Accessed Aug 2015
5. D.V. Lu, D. Hershberger, W.D. Smart, Layered costmaps for context-sensitive navigation, in *IEEE International Conference on Intelligent Robots and Systems (IROS)* (2014)
6. D.V. Lu, M. Ferguson, navigation, http://wiki.ros.org/navigation. Accessed Aug 2015
7. A. Hornung, K.M. Wurm, M. Bennewitz, C. Stachniss, W. Burgard, OctoMap: an efficient probabilistic 3D mapping framework based on octrees. Auton. Robot. **34**(3), 189–206 (2013)
8. K.M. Wurm, A. Hornung. octomap, http://wiki.ros.org/octomap. Accessed Aug 2015
9. C. Forster, M. Faessler, F. Fontana, M. Werlberger, D. Scaramuzza, Continuous on-board monocular-vision–based elevation mapping applied to autonomous landing of micro aerial vehicles, in *IEEE International Conference on Robotics and Automation (ICRA)* (2015)
10. M. Hutter, C. Gehring, M. Hoepflinger, M. Bloesch, R. Siegwart, Towards combining speed, efficiency, versatility and robustness in an autonomous quadruped. IEEE Trans. Robot. **30**(6), 1427–1440 (2014)
11. ROS.org. Ubuntu install of ROS Indigo, http://wiki.ros.org/indigo/Installation/Ubuntu. Accessed Aug 2015
12. ROS.org. Installing and Configuring Your ROS Environment, http://wiki.ros.org/ROS/Tutorials/InstallingandConfiguringROSEnvironment. Accessed Aug 2015
13. ROS.org. filters, http://wiki.ros.org/filters. Accessed Aug 2015
14. J.E. Bresenham, Algorithm for computer control of a digital plotter. IBM Syst. J. **4**(1), 25–30 (1965)
15. M. Bloesch, C. Gehring, P. Fankhauser, M. Hutter, M.A. Hoepflinger, R. Siegwart, State estimation for legged robots on unstable and slippery terrain, in *IEEE/RSJ International Conference on Intelligent Robots and Systems (IROS)* (2013)
16. LaValle, S.M. Rapidly-Exploring Random Trees: A New Tool for Path Planning (1998)

Authors' Biography

Péter Fankhauser is a Ph.D. student at the Robotic Systems Lab at ETH Zurich since 2013. He received his MSc degree in Robotics, Systems and Control from ETH Zurich in 2012. His research interests are in mapping and motion planning for legged robots.

Marco Hutter is assistant professor for Robotic Systems at ETH Zurich and Branco Weiss Fellow. He received his M.Sc. in mechanical engineering (2009) and his doctoral degree in robotics (2013) from ETH Zurich. In his work, Marco focuses on design, actuation, and control of legged robotic systems that can work in challenging environments.

ROS Navigation: Concepts and Tutorial

Rodrigo Longhi Guimarães, André Schneider de Oliveira, João Alberto Fabro, Thiago Becker and Vinícius Amilgar Brenner

Abstract This tutorial chapter aims to teach the main theoretical concepts and explain the use of ROS Navigation Stack. This is a powerful toolbox to path planning and Simultaneous Localization And Mapping (SLAM) but its application is not trivial due to lack of comprehension of the related concepts. This chapter will present the theory inside this stack and explain in an easy way how to perform SLAM in any robot. Step by step guides, example codes explained (line by line) and also real robot testing will be available. We will present the requisites and the how-to's that will make the readers able to set the odometry, establish reference frames and its transformations, configure perception sensors, tune the navigation controllers and plan the path on their own virtual or real robots.

Keywords ROS · Navigation Stack · Tutorial · Real robots · Transformations

1 Introduction

Simultaneous Localization And Mapping (SLAM) is an important algorithm that allows the robot to acknowledge the obstacles around it and localize itself. When combined with some other methods, such as path planning, it is possible to allow robots to navigate unknown or partially known environments. ROS has a package

R.L. Guimarães (✉) · A.S. de Oliveira · J.A. Fabro · T. Becker · V.A. Brenner
LASER - Advanced Laboratory of Embedded Systems and Robotics, Federal University
of Technology - Paraná, Av. Sete de Setembro, Curitiba 3165, Brazil
e-mail: rguimaraes@alunos.utfpr.edu.br

A.S. de Oliveira
e-mail: andreoliveira@utfpr.edu.br

J.A. Fabro
e-mail: fabro@dainf.ct.utfpr.edu.br

T. Becker
e-mail: beckerthiago@gmail.com

V.A. Brenner
e-mail: brenner@alunos.utfpr.edu.br

© Springer International Publishing Switzerland 2016   121
A. Koubaa (ed.), *Robot Operating System (ROS)*, Studies in Computational
Intelligence 625, DOI 10.1007/978-3-319-26054-9_6

that performs SLAM and path planning along with other functionalities for navigation, named Navigation Stack. However, some details of its application are hidden, considering that the programmer has some expertise. The unclear explanation and the many subjective aspects within the package can lead the user to fail using the technique or, at least, consume extra effort.

This chapter aims to present the theory inside ROS Navigation Stack and explain in an easy way how to perform autonomous navigation in any robot. It will also explain how to use a virtual environment to use the Navigation Stack on virtual robots. These robots are designed to publish and subscribe the same information in real and virtual environments, where all sensors and actuators of real world are functional in virtual environment.

We begin the chapter explaining what the Navigation Stack is with some simple and straight-forward definitions and examples where the functionalities of the package are explained in conjunction with reminders of some basic concepts of ROS. Moreover, a global view of a system running the Navigation Stack at its full potential is shown, where the core subset of this system is explained bit by bit. In addition, most of the indirectly related components of the system, like multiple machine communication, will be indicated and briefly explained, with some tips about its implementation.

The chapter will be structured in four main topics. Firstly, an introduction of mobile robot navigation and a discussion about SLAM will be presented, along with some discussion about transformations.

In the second section, ROS Environment, the main purpose is to let the reader know everything he needs to configure his own virtual or real robot. Firstly, the limitations of the Navigation Stack are listed together with the expected hardware and software that the reader should have to follow the tutorial. Secondly, an explanation about odometry and kinematics is given, focusing on topics like precision of the odometry and the impact of the lack of it in the construction of costmaps for navigation (do not worry, you will see that in detail later). Perception sensors are also discussed. The differences and advantages of each of the main kinds of perception sensors, like depth sensors, light detection and ranging (LIDAR) sensors are also emphasized. In addition, reference frames and its transformations are discussed, showing how to achieve the correct merge of the sensors information. Still on the second section, a dense subsection about the Navigation Stack is presented. It is in respect to the configuration of the Navigation Stack to work with as many robots as possible, trying to organize the tutorial in a way that both reach high level of detail and generalization, ensuring that the reader is apt to perceive a way to have his own robot navigate itself. An in-depth discussion of map generation and map's occupancy is performed. To do that, the tutorial is structured in a step-by-step way in which all the navigation configuration files will be analyzed, examining the parameters and setting them to example values that correspond to pioneer (ARIA based) robots (e.g. Pioneer 3-AT, as seen on Fig. 1). The explanation of the whole process in addition to the demonstration of the effects of the parameter changes should be enough to clear up any doubts the reader might have.

Fig. 1 Side by side are two
pioneer ARIA based robots:
Pioneer 3-AT and Pioneer
LX. These robots, owned by
the Advanced Laboratory of
Embedded Systems and
Robotics of UTFPR, can use
the Navigation Stack

In the third section we will discuss the experiments in real and virtual environments
to prove the accuracy of the SLAM method. All steps to use a virtual environment
where the reader should be able to test his own configuration for the Navigation
Stack are demonstrated. Lastly, some experiments are run in virtual and real robots,
to illustrate some more capabilities of the package. In this section we will also take
a glance over rviz usage.

Lastly, a brief biography of the authors will be presented, showing why this team
is able to write the tutorial chapter here presented.

2   Background

ROS has a set of resources that are useful so a robot is able to navigate through a
known, partially known or unknown environment, in other words, the robot is capable
of planning and tracking a path while it deviates from obstacles that appear on its
path throughout the course. These resources are found on the Navigation Stack.

One of the many resources needed for completing this task and that is present
on the Navigation Stack are the SLAM systems (also called localization systems),
that allow a robot to locate itself, whether there is a static map available or SLAM
is required. amcl is a tool that allows the robot to locate itself in an environment
through a static map, a previously created map. The entire area in which the robot
could navigate would have to be mapped in a metrically correct way to use static maps,
a big disadvantage of this resource. Depending on the environment that surrounds
the robot, these static maps are capable of increasing or decreasing the confidence
of the localization systems. To bypass the lack of flexibility of static maps, two
other localization systems are offered by ROS to work with the Navigation Stack:
Gmapping and hector_mapping.

Both Gmapping and hector_mapping are implementations of SLAM, a technique that consists on mapping an environment at the same time that the robot is moving, in other words, while the robot navigates through an environment, it gathers information from the environment through its sensors and generates a map. This way you have a mobile base able not only to generate a map of an unknown environment as well as updating the existent map, thus enabling the use of the device in more generic environments, not immune to changes.

The difference between Gmapping and hector_mapping is that the first one takes in account the odometry information to generate and update the map and the robot's pose. However, the robot needs to have proprioceptive sensors, which makes the usage of it hard for some robots (e.g. flying robots). The odometry information is interesting because they are able to aid on the generation of more precise maps, since understanding the robot kinematics we can estimate its pose.

Kinematics is influenced, basically, by the way that the devices that guarantee the robot's movement are assembled. Some examples of mechanic features that influence the kinematics are: the wheel type, the number of wheels, the wheel's positioning and the angle at which they are disposed. A more in-depth explanation of kinematics is done in Introduction to Mobile Robots [1].

However, as much useful as the odometry information can be, it is not immune to faults. The faults are caused by the lack of precision on the capture of data, friction, slip, drift and other factors. These accumulated factors can lead to inconsistent data and prejudice the maps formation, that tend to be distorted under these circumstances.

Other indispensable data to generate a map are the sensors' distance readings, for the reason that they are responsible in detecting the external world and, this way, serve as reference to the robot. Nonetheless, the data gathered by the sensors must be adjusted before being used by the device. These adjustments are needed because the sensors measure the environment in relation to themselves, not in relation to the robot, in other words, a geometric conversion is needed. To make this conversion simpler, ROS offers the `tf` tool, which makes it possible to adjust the sensors positions in relation to the robot and, this way, suit the measures to the robot's navigation.

3 ROS Environment

Before we begin setting up the environment in which we will work, it is very important to be aware of the limitations of the Navigation Stack, so we are able to adapt our hardware. There are four main limitations in respect to the hardware:

- The stack was built aiming to address only differential drive and holonomic robots, although it is possible to use some features with another types of robots, which will not be covered here.
- The Navigation Stack assumes that the robot receives a twist type message [2] with X, Y and Theta velocities and is able to control the mobile base to achieve these velocities. If your robot is not able to do so, you can adapt your hardware or

just create a ROS node that converts the twist message provided by the Navigation Stack to the type of message that best suits your needs.

- The environment information is gathered from a LaserScan message type topic. If you have a planar laser, such as a Hokuyo URG or a sick Laser, it should be very easy to get them to publish their data, all you need to do is install the Hokuyo_node, Sicktoolbox or similar packages, depending on your sensor. Moreover, it is possible to use other sensors, as long as you can convert their data to the LaserScan type. In this chapter, we will use converted data from Microsoft Kinect's depth sensor.
- The Navigation Stack will perform better with square or circular robots, whereas it is possible to use it with arbitrary shapes and sizes. Unique sizes and shapes may cause the robot to have some issues in restricted spaces. In this chapter we will be using a custom footprint that is an approximation of the robot.

If your system complies with all the requirements, it is time to move to the software requirements, which are very few and easy to get.

- You should have ROS Hydro or Indigo to get all the features, such as layered costmaps, that are a Hydro+ feature. From here, we assume you are using a Ubuntu 12.04 with ROS Hydro.
- You should have the Navigation Stack installed. In the full desktop version of ROS it is bundled, but depending on your installation it might be not included. Do not worry with that for now, it is just good to know that if some node is not launching it may be because the package you are trying to use is not installed.
- As stated on the hardware requirements, you might need some software to get your system navigation ready:
  - If your robot is not able to receive a twist message and control its velocity as demanded, one possible solution is to use a custom ROS node to transform the data to a more suitable mode.
  - You need to have drivers able to read the sensor data and publish it in a ROS topic. If you're using a sensor different from a planar laser sensor, such as a depth sensor, you'll most likely also need to change the data type to LaserScan through a ROS node.

Now that you know all you need for navigation, it is time to begin getting those things. In this tutorial we'll be using Pioneer 3-AT and Pioneer LX and each of them will have some particularities in the configuration that will help us to generalize the settings as much as possible. We'll be using Microsoft's Kinect depth sensor in the Pioneer 3-AT and the Sick S300 laser rangefinder that comes built-in in the Pioneer LX. Most of this chapter is structured in a "by example" approach from now on and if you like it, you should check out the ROS By Example—Volume 1 [3] book. In this book, there are a lot of useful codes for robots and a navigation section that complements very well what we discuss in this chapter. You also might find useful to check their example codes [4], where you can find a lot of useful codes, such as rviz configuration files and navigation examples.

Important Note: no matter what robot you're using, you will probably need to allow the access to the USB ports that have sensors or to the ports that communicate to your robots. For example, if you have the robot connected in ttyUSB0 port, you can give the access permission by issuing the following command:

```
1  sudo chmod 777 -R /dev/ttyUSB0
```

This configuration will be valid only for the current terminal, so it is recommended to add this instruction to the ~/.bashrc file. Edit this file by writing the following lines at the bottom of the file:

```
1  if [ -e /dev/ttyUSB0 ]; then
2  sudo chmod 777 -R /dev/ttyUSB0
3  echo "Robot found on USB0!"
4  fi
```

This command will check if there is something in the /dev/ttyUSB0 port and, if the condition holds, it will change the permissions on the port and will print to the screen the message "Robot found on USB0!". It is noteworthy that, as said above, this command is assuming that the robot is on /dev/ttyUSB0. Also, this script will be executed every time a new terminal is opened and since a sudo command is issued a password will be prompted for non-root users.

3.1 Configuring the Kinect Sensor

The Kinect is a multiple sensor equipment, equipped with a depth sensor, an RGB camera and microphones. First of all, we need to get those features to work by installing the drivers, which can be found in the ros package Openni_camera [5]. You can get Openni_camera source code from its git repository, available on the wiki page, or install the stack with a linux terminal command using apt-get (REC-OMMENDED). We will also need a ROS publisher to use the sensor data and that publisher is found in the Openni_launch [6] package, that can be obtained the exactly same ways. To do so, open your terminal window and type:

```
1  $ sudo apt-get install ros-<rosdistro>-openni-camera
2  $ sudo apt-get install ros-<rosdistro>-openni-launch
```

Remember to change "<rosdistro>" to your ros version (i.e. 'hydro' or 'indigo'). With all the software installed, it's time to power on the hardware.

Depending on your application plugging Kinect's USB Cable to the computer you're using and the AC Adapter to the wall socket can be enough, however needing a wall socket to use your robot is not a great idea. The AC Adapter converts the AC voltage (e.g. 100 V @ 60 Hz) to the DC voltage needed to power up the Kinect (12 V), therefore the solution is exchanging the AC adapter for a 12 V battery. The procedure for doing this is explained briefly in the following topics:

- Cut the AC adapter off, preferably near the end of the cable.
- Strip a small end to each of the two wires (white and brown) inside of the cable.
- Connect the brown wire to the positive(+) side of the 12 V battery and the white wire to the negative(−). You can do this connection by soldering [7] or using connectors, such as crimp [8] or clamp [9] wire connectors.

Be aware that the USB connection is enough to blink the green LED in front of the Kinect and it does not indicate that the external 12 V voltage is there. You can also learn a little more about this procedure by reading the "Adding a Kinect to an iRobot Create/Roomba" wiki page [10].

Now that we have the software and the hardware prepared, some testing is required. With the Kinect powered on, execute the following command on the Ubuntu terminal:

```
1 $ roslaunch openni_launch openni.launch
```

The software will come up and a lot of processes will be started along with the creation of some topics. If your hardware is not found for some reason, you may see the message "No devices connected...waiting for devices to be connected". If this happens, please verify your hardware connections (USB and power). If that does not solve it, you may try to remove some modules from the Linux Kernel that may be the cause of the problems and try again. The commands for removing the modules are:

```
1 $ sudo modprobe -r gspca_kinect
2 $ sudo modprobe -r gspca_main
```

Once the message of no devices connected disappears, you can check some of the data supplied by the Kinect in another Terminal window (you may open multiple tabs of the terminal in standard Ubuntu by pressing CTRL+SHIFT+T) by using one or more of these commands:

```
1 $ rosrun image_view disparity_view image:=/camera/depth/disparity
2 $ rosrun image_view image_view image:=/camera/rgb/image_color
3 $ rosrun image_view image_view image:=/camera/rgb/image_mono
```

The first one will show a disparity image, while the second and third commands will show the RGB camera image in color and in black and white respectively.

Lastly, we need to convert the depth image data to a LaserScan message. That is needed because Gmapping, the localization system we are using, accepts as input a single LaserScan message. Fortunately, we have yet another packages to do this for us, the Depthimage_to_laserscan package and the ira_laser_tools package. You just need one of the two packages, and, although the final result is a LaserScan message for both, the data presented will be probably different. The Depthimage_to_laserscan package is easier to get and use, and as we have done with the other packages, we can check out the wiki page [11] and get the source code from the git repository or we can simply get the package with apt-get:

```
1 $ sudo apt-get install ros-<rosversion>-depthimage-to-laserscan
```

This package will take a horizontal line of the PointCloud and use it to produce the LaserScan. As you can imagine, it is possible that some obstacles aren't present at the selected height, the LaserScan message will not represent the PointCloud data so well and gmapping will struggle to localize the robot in the costmap generated by the Navigation Stack using the PointCloud directly.

The other option, `ira_laser_tools` package, produces a more representative result, but it is harder to use and consumes a lot more processing power. In this package you have two nodes: `laserscan_virtualizer` and `laserscan_multi_merger`. The first one, can be configured to convert multiple lines of your Point-Cloud into different LaserScan messages. The second one, the merger node, receives multiple LaserScan messages and merges them in a single LaserScan message. You can check the article found on [12] to find out more about the package.

To install this package, you have to first clone the git repository found on [13] to your catkin workspace source folder and then compile it. This is done with the following commands:

```
1 $ cd /home/user/catkin_ws/src
2 $ git clone https://github.com/iralabdisco/ira_laser_tools.git
3 $ cd /home/user/catkin_ws
4 $ catkin_make
```

Now that the nodes are compiled, you have to configure the launch files to suit your needs. Go to the launch folder, inside the `ira_laser_tools` package folder, and open `laserscan_virtualizer.launch` with your favorite editor. You should see something like this:

```
1 <!--
2 FROM: http://wiki.ros.org/tf#static_transform_publisher
3
4 <<static_transform_publisher x y z yaw pitch roll
       frame_id child_frame_id period_in_ms>>
5 Publish a static coordinate transform to tf using an x/
       y/z offset and yaw/pitch/roll. The period, in
       milliseconds, specifies how often to send a
       transform. 100ms (10hz) is a good value.
6 == OR ==
7 <<static_transform_publisher x y z qx qy qz qw frame_id
       child_frame_id period_in_ms>>
8 Publish a static coordinate transform to tf using an x/
       y/z offset and quaternion. The period, in
       milliseconds, specifies how often to send a
       transform. 100ms (10hz) is a good value.
9
10 -->
11
```

```
12 <launch>
13
14        <!-- DEFINE HERE THE STATIC TRANFORMS, FROM
             BASE_FRAME (COMMON REFERENCE FRAME) TO THE
             VIRTUAL LASER FRAMES-->
15        <!-- WARNING: the virtual laser frame(s) *must*
             match the virtual laser name(s) listed in
             param: output_laser_scan -->
16        <node pkg="tf" type="static_transform_publisher"
              name="ira_static_broadcaster1" args="0 0 0
              0 0.3 0 laser_frame scan1 1000" />
17        <node pkg="tf" type="static_transform_publisher"
              name="ira_static_broadcaster2" args="0 0 0
              0 0.0 0 laser_frame scan2 1000" />
18        <node pkg="tf" type="static_transform_publisher"
              name="ira_static_broadcaster3" args="0 0 0
              0.0 0.0 0.3 laser_frame scan3 1000" />
19        <node pkg="tf" type="static_transform_publisher"
              name="ira_static_broadcaster4" args="0 0 0
              0.0 0.0 -0.3 laser_frame scan4 1000" />
20
21        <node pkg="ira_laser_tools" name="
             laserscan_virtualizer" type="
             laserscan_virtualizer" output="screen">
22            <param name="cloud_topic" value="/
                 cloud_in"/> <!-- INPUT POINT CLOUD --
                 >
23            <param name="base_frame" value="/
                 laser_frame"/> <!-- REFERENCE FRAME
                 WHICH LASER(s) ARE RELATED-->
24            <param name="output_laser_topic" value ="
                 /scan" /> <!-- VIRTUAL LASER OUTPUT
                 TOPIC, LEAVE VALUE EMPTY TO PUBLISH
                 ON THE VIRTUAL LASER NAMES (param:
                 output_laser_scan) -->
25            <param name="virtual_laser_scan" value ="
                 scan1 scan2 scan3 scan4" /> <!-- LIST
                 OF THE VIRTUAL LASER SCANS. YOU MUST
                 PROVIDE THE STATIC TRANSFORMS TO TF,
                 SEE ABOVE -->
26        </node>
27 </launch>
```

This launch file is responsible for converting some parts of the PointCloud in multiple LaserScans. As you can notice from the file, it is taking four lines from the Kinect PointCloud: two horizontal lines, one at the kinect level and another 30 cm above the kinect level; two oblique lines, each of them with 0.3 rad of rotation to one side in relation to the Kinect and in the same horizontal level. You can have more information about the usage in the comments in the code itself and on this article [12], but here are the things you will most likely need to modify to correctly use the package:

- **The number of scans you want**: You have to find a number good enough to summarize a lot of information without using too much CPU. In our case, we tested with 8 horizontal lines, from 25 cm below the Kinect to 90 cm above of it. We used horizontal lines because that way the ground would not be represented in the LaserScan message. To do this, simply duplicate one of the node launching lines, changing the name of the node and the name of the scan to an unique name. For each line, you should configure the transform coordinates to whatever you see fit, although we recommend the horizontal lines approach. Do not forget to put the name of the new scan on the `virtual_laser_scan` parameter.
- **The tf frame of the Kinect**: Change `laser_frame` to the name of the frame of your Kinect. In our case, `camera_link`.
- **Input PointCloud**: Change this line to the name of the topic that contains the PointCloud. `Openni_launch` publishes it, by default, to `/camera/depth/points`.
- **The base_frame parameter**: Selects the frame to which the LaserScans will be related. It is possible to use the same frame as the Kinect.
- **The output_laser_topic parameter**: Selects the output topic for the LaserScan messages. We left it blank, so each of the LaserScans would go to the topic with their respective name.

Lastly, you have to configure the merger node. Start by opening up the `laserscan_multi_merger.launch` file, on the same folder of the virtualizer launch file. You should see this:

```
<!--
DESCRITPION
-->


<launch>
        <node pkg="ira_laser_tools" name="
            laserscan_multi_merger" type="
            laserscan_multi_merger" output="screen">
                <param name="destination_frame" value="/
                    cart_frame"/>
                <param name="cloud_destination_topic"
                    value="/merged_cloud"/>
```

```
10              <param name="scan_destination_topic"
                    value="/scan_multi"/>
11              <param name="laserscan_topics" value ="
                    scandx scansx" /> <!-- LIST OF THE
                    LASER SCAN TOPICS TO SUBSCRIBE -->
12          </node>
13 </launch>
```

This launch file is responsible for merging multiple LaserScans into a single one. As you can notice from the file, it is taking two LaserScan topics (`scandx` and `scansx`) and merging. The three things you will most likely modify are:

- **The destination_frame parameter**: Chooses the frame to which the merged LaserScan is related. Again, you can simply use the Kinect frame.
- **Cloud** : Change `laser_frame` to the name of the frame of your Kinect. In our case, `camera_link`.
- **The scan_destination_topic**: You will probably want gmapping to read this information, so, unless you have some special reason not to do that, use the `/scan` topic here.
- **The laserscan_topics paramter**: List all the topics that contain the LaserScan messages you want to merge. If you are using eight, you will most likely use scan1 to scan8 here.

Important Notes: When using this package, there are some important things you have to notice. First, it won't work correctly if you launch the merger before the virtualizer, and since ROS launchers do not guarantee the order, you may use a shell script to do that for you. Second, if the transformations take too long to be published, you may end up with some issues in the merger. To correct that, we modified the 149th line of the code, changing the ros::Duration parameter from 1 to 3(setting this parameter too high will make the LaserScan publication less frequent). The line will then look like that:

```
1 tfListener_.waitForTransform(scan->header.frame_id.c_str(), destination_frame.c_str(),
  scan->header.stamp, ros::Duration(3));
```

With this last installation, you should have all the software you need for Kinect utilization with navigation purposes, although there is a lot of other software you can use with it. We would like to point out two of these packages that can be very valuable at your own projects:

- **kinect_aux** [14]: this package allows to use some more features of the Kinect, such as the accelerometer, tilt, and LED. It can be used along with the openni_camera package and it is also installed with a simple apt-get command.
- **Natural Interaction—openni_tracker** [15]: One of the most valuable packages for using with the Kinect, this package is able to do skeleton tracking functionalities and opens a huge number of possibilities. It is kind of tough to install and the

process can lead to problems sometimes, so we really recommend you to do a full system backup before trying to get it to work. First of all, install the openni_tracker package with an apt-get, as stated on the wiki page. After that, you have to get these recommended versions of the files listed below.

– NITE-Bin-Linux-x86-v1.5.2.23.tar.zip
– OpenNI-Bin-Dev-Linux-x86-v1.5.7.10.tar.bz2
– SensorKinect093-Bin-Linux-x86-v5.1.2.1.tar.bz2

The official openni website is no longer in the air, but you can get the files on Openni.ru [16] or on [17], where I made them available. The first two files (NITE and Openni) can be installed following the cyphy_people_mapping [18] tutorial and the last file should be installed by:

• Unpacking the file

```
1 $ tar -jxvf SensorKinect093-Bin-Linux-x86-v5.1.2.1.tar.bz2
```

• Changing the permission of the setup script to allow executing.

```
1 $ sudo chmod a+x install.sh
```

• Installing.

```
1 $ sudo ./install.sh
```

Now that we have all software installed we should pack it all together in a single launch file, to make things more independent and do not need to start a lot of packages manually when using the Kinect. Here is an example of a complete launch file for starting the Kinect and all the packages that make its data navigation-ready:

```
<launch>
        <include file="$(find openni_launch)/launch/
            openni.launch" />
        <node respawn="true" pkg="
            depthimage_to_laserscan" type="
            depthimage_to_laserscan" name="laserscan">
                <remap from="image" to="/camera/depth/
                    image" />
        </node>
</launch>
```

As you can see and you might be already used to because of the previous chapters, the ROS launch file is running one node and importing another launch file: openni.launch, imported from the openni_launch package. Analyzing the code in a little more depth:

- The first and sixth lines are the launch tag, that delimits the content of the launch file;
- The second line includes the openni.launch file from the `openni_launch` package, responsible for loading the drivers of the Kinect, getting the data, and publishing it to ROS topics;
- The third line starts the package `depthimage_to_laserscan` with the "laserscan" name. It also sets the respawn parameter to true, in case of failures. This package is responsible for getting a depth image from a ROS topic, converting it to a LaserScan message and publishing it to another topic;
- The fourth line is a parameter for the `depthimage_to_laserscan`. By default, the package gets the depth image from the `/image` topic, but the `openni_launch` publishes it in the `/camera/depth/image` topic, and that is what we are saying to the package.

There is still the transform (tf) missing, but we will discuss that later, because the configuration is very similar to all sensors.

3.2 Sick S300 Laser Sensor

The pioneer LX comes bundled with a Sick S300 laser sensor and we'll describe here how to get its data, since the process should be very similar to other laser rangefinder sensors. The package that supports this laser model is sicks300, that is currently only supported by the fuerte and groovy versions of ROS. We are using a fuerte installation of ROS in this robot, so it's no problem for us, but it must be adapted if you wish to use it with hydro or indigo. For our luck, it was adapted and it is available at STRANDS git repository [19]. The procedure for getting it to work is:

- Cloning the repository, by using the command git clone https://github.com/bohle nder/sicks300.git (the URL will change depending on your version);
- Compile the files. For rosbuild versions, use rosmake sicks300 sicks300 and for catkinized versions, use catkin_make.
- Run the files by using the command rosrun sicks300 sicks300_driver. It should work with the default parameters, but, if it doesn't, check if you have configured the baud rate parameter correctly (the default is 500000 and for the pioneer LX, for example, it is 230400).

The procedure should be very similar for other laser sensors and the most used packages, sicktoolbox_wrapper, hokyuo_node and urg_node, are very well documented on the ROS wiki. Its noteworthy that there is another option for reading pioneer LX laser sensor data: cob_sick_s300 package.

3.3 Transformations

As explained on the background section, the transforms are a necessity for the Navigation Stack to understand where the sensors are located in relation to the center of the robot (base_link). It is possible to understand a little more of transformations by examining the Fig. 2.

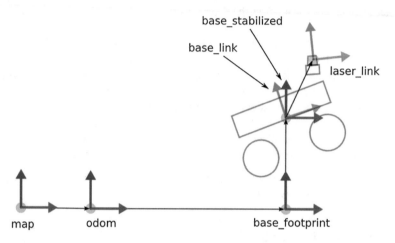

Fig. 2 Figure containing the most commonly used tf frames [20]

As you can see on the Fig. 2, we have several frames. A brief explanation of each of the topics is done below:

- **map** is the global reference frame and the robot pose should not drift much over time in relation to it [21].
- **odom** drifts and can cause discrete jumps when new sensor information is available.
- **base_link** is attached to the robot center.
- **base_footprint** is very straightforward: it is the base_link projection on the ground (zero height). Therefore, its transform is published in relation to the base_link.
- **base_stabilized** is the center position of the robot, not computing the roll and pitch angles. Therefore, its transform is also published in relation to the base_link.
- **laser_link** is the center position of the laser sensor, and its transform is published in relation to the base_link.

There is a clear hierarchy on these frames: map is the parent of odom and odom is the parent of base_link. The transform between odom and base_link has to be computed over the odometry sensor data and published. The transformation between base_link and map is computed by the localization system and the other transformation, between the map and the odom frames, uses this information to be computed.

We will publish a transform between the Pioneer 3-AT base_link and the Kinect sensor laser_link(the center of the Kinect sensor) as example, given that this procedure is equal for any type of sensor. The first thing in order to use transforms is getting the distances you need. You will have to measure three-dimensional coordinates in respect to the center of the robot and will also need to get the angles between the robot pointing direction and the sensor pointing direction. Our Kinect sensor is aligned with the robot pointing direction on the yaw and its z coordinate is also aligned with the robot center (and that's the usual case for most of the robots), therefore we have to measure only the x and y distances between the sensor and the robot centers. The x distance for our robot is 35 cm (0.35 m) and our height distance (y) is 10 cm (0.1 m). There are two standard ways to publish the angle values on transforms: using quaternion or yaw/pitch/roll. The first one, using quaternion, will respect the following order:

```
1  static_transform_publisher x y z qx qy qz qw frame_id child_frame_id period_in_ms
```

Where qx, qy, qz and qw are the versors in the quaternion representation of orientations and rotations. To understand the quaternion representation better, refer to Modeling and Control of Robot Manipulators [22]. The second way of publishing angles, using yaw/roll/pitch and the one we will be using, is published in the following order:

```
1  static_transform_publisher x y z yaw pitch roll frame_id child_frame_id period_in_ms
```

The common parameters for both quaternion and yaw/row/pitch representations are:

- x, y and z are the offset representation, in meters, for the three-dimensional distance between the two objects;
- frame_id and child_frame_id are the unique names that will bound the transformations to the object to which they relate. In our case, frame_id is the base_link of the robot and child_frame_id is the laser_link of the sensor.
- period_in_ms is the time between two publications of the tf. It is possible to calculate the publishing frequency by calculating the reciprocal of the period.

In our example, for the Pioneer 3-AT and the Kinect, we have to include, in the XML launcher (just write this line down at a new launch file, we will indicate further in the text where to use it), the following code to launch the tf node:

```
1  <node pkg="tf" type="static_transform_publisher" name="Pioneer3AT_laserscan_tf"
   args="0.1 0 0.35 0 pi/2 pi/2 base_link camera_link 100" />
```

If you have some doubts on how to verify the angles you measured, you can use rviz to check them, by including the laser_scan topic and verifying the rotation of the obtained data. If you do not know how to do that yet, check the tests section of the chapter, which includes the rviz configuration.

3.4 Creating a Package

At this point, you probably already have some XML files for launching nodes containing your sensors initialization and for initializing some packages related to your platform (robot powering up, odometry reading, etc.). To organize these files and make your launchers easy to find with ros commands, create a package containing all your launchers. To do that, go to your src folder, inside your catkin workspace, and create a package with the following commands (commands valid for catkinized versions of ROS):

```
1 $ cd /home/user/catkin_ws/src
2 $ catkin_create_pkg packageName std_msgs rospy roscpp move_base_msgs
```

These two commands are sufficient for creating a folder containing all the files that will make ROS find the package by its name. Copy all your launch files to the new folder, namesake to your package, and then compile the package, by going to your workspace folder and issuing the compile command:

```
1 $ cd /home/user/catkin_ws
2 $ catkin_make
```

That's all you need to do. From now on, remember to put your launch and config files inside this folder. You may also create some folder, such as launch and config, provided that you specify these sub-paths when using find for including launchers in other launchers.

3.5   The Navigation Stack—System Overview

Finally, all the pre-requisites for using the Navigation Stack are met. Thus, it is time to begin studying the core concepts of the Navigation Stack, as well as their usage. Since the overview of the Navigation Stack concepts was already done in the Background section, we can jump straight to the system overview, which is done in Fig. 3. The items will be analyzed in the following sections block by block.

You can see on Fig. 3 that there are three types of nodes: provided nodes, optional provided nodes and platform specific nodes.

- The nodes inside the box, provided nodes, are the core of the Navigation Stack, and are responsible, mainly, by managing the costmaps and for the path planning functionalities.
- The optional provided nodes, amcl and map_server, are related to static map functions, as will be explained later, and since using a static map is optional, using these nodes is also optional.
- The platform specific nodes are the nodes related to your robot, such as sensor reading nodes and base controller nodes.

In addition, we have the localization systems, not shown in Fig. 3. If the odometry source was perfect and no errors were present on the odometry data, we would not need to have localization systems. However, in real applications that is not the case, and we have to account other types of data, such as IMU data, so we can correct odometry errors. The localization systems discussed on this chapter are:

amcl, gmapping and hector_mapping. Below, a table is available to relate a node to a keyword, so you can understand better the relationship of the nodes of the Navigation Stack.

Localization	Environment inter-action	Static maps	Trajectory planning	Mapping
amcl	sensor_sources	amcl	global_planner	local_costmap
gmapping	base_controller	map_server	local_planner	global_costmap
hector_mapping	odometry_source		recovery_behaviors	

3.5.1 Amcl and Map_server

The first two blocks that we can focus on are the optional ones, responsible for the static map usage: amcl and map_server. map_server contains two nodes: map_server and map_saver. The first one, namesake to the package, as the name indicates, is a ROS node that provides static map data as a ROS Service, while the second one, map_saver, saves a dynamically generated map to a file. amcl does not manage the maps, it is actually a localization system that runs on a known map. It uses the base_footprint or base_link transformation to the map to work, therefore it needs a static map and it will only work after a map is created. This localization system is based on the Monte Carlo localization approach: it randomly distributes particles in a known map, representing the possible robot locations, and then uses a particle filter to determine the actual robot pose. To know more about this process, refer to Probabilistic Robotics [24].

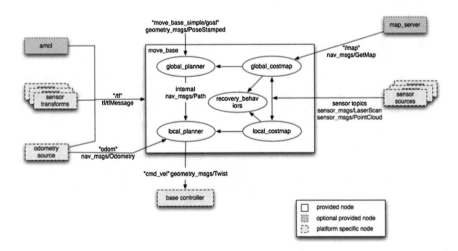

Fig. 3 Overview of a typical system running the Navigation Stack [23]

3.5.2   Gmapping

gmapping, as well as amcl, is a localization system, but unlike amcl, it runs on an unknown environment, performing Simultaneous Localization and Mapping (SLAM). It creates a 2D occupancy grid map using the robot pose and the laser data (or converted data, i.e. Kinect data). It works over the odom to map transformation, therefore it does not need the map nor IMU information, needing only the odometry.

3.5.3   Hector_mapping

As said in the background section, hector_mapping can be used instead of using gmapping. It uses the base_link to map transformation and it does not need the odometry nor the static map: it just uses the laser scan and the IMU information to localize itself.

3.5.4   Sensors and Controller

These blocks of the system overview are in respect to the hardware-software interaction and, as indicated, are platform specific nodes. The odometry source and the base controller blocks are specific to the robot you are using, since the first one is usually published using the wheel encoders data and the second one is the responsible for taking the velocity data from the cmd_vel topic and assuring that the robot reproduces these velocities. It is noteworthy that the whole system will not work if the sensor transforms are not available, since they will be used to calculate the position of the sensor readings on the environment.

3.5.5   Local and Global Costmaps

The local and global 2D costmaps are the topics containing the information that represents the projection of the obstacles in a 2D plane (the floor), as well as a security inflation radius, an area around the obstacles that guarantee that the robot will not collide with any objects, no matter what is its orientation. These projections are associated to a cost, and the robot objective is to achieve the navigation goal by creating a path with the least possible cost. While the global costmap represents the whole environment (or a huge portion of it), the local costmap is, in general, a scrolling window that moves in the global costmap in relation to the robot current position.

3.5.6 Local and Global Planners

The local and global planners do not work the same way. The global planner takes the current robot position and the goal and traces the trajectory of lower cost in respect to the global costmap. However, the local planner has a more interesting task: it works over the local costmap, and, since the local costmap is smaller, it usually has more definition, and therefore is able to detect more obstacles than the global costmap. Thus, the local planner is responsible for creating a trajectory rollout over the global trajectory, that is able to return to the original trajectory with the fewer cost while it deviates from newly inserted obstacles or obstacles that the global costmap definition was not able to detect. Just to make it clear, move_base is a package that contains the local and global planners and is responsible for linking them to achieve the navigation goal.

3.6 The Navigation Stack—Getting It to Work

At last, it is time to write the code for the full launcher. It is easier to do this with an already written code, such as the one that follows:

```
<launch>
        <master auto="start"/>


        <!-- PLATFORM SPECIFIC -->
        <node pkg="p2os_driver" type="p2os_driver" name=
            "p2os_driver" >
                <param name="port" value="/dev/ttyUSB0" /
                    >
                <param name="pulse" value="1.0" />"
        </node>


        <node pkg="rostopic" type="rostopic" name="
            enable_robot" args="pub /cmd_motor_state
            p2os_driver/MotorState 1" respawn="true">
        </node>


        <!-- TRANSFORMS -->
        <node pkg="tf" type="static_transform_publisher"
            name="Pioneer3AT_laserscan_tf" args="0.1 0
            0.35 0 pi/2 pi/2 base_link camera_link 100"
            />


        <!-- SENSORS CONFIGURATION -->
```

```
 17         <arg name="kinect_camera_name" default="camera"
             />
 18             <param name="/$(arg kinect_camera_name)/
                 driver/data_skip" value="1" />
 19             <param name="/$(arg kinect_camera_name)/
                 driver/image_mode" value="5" />
 20             <param name="/$(arg kinect_camera_name)/
                 driver/depth_mode" value="5" />
 21
 22     <include file="$(find course_p3at_navigation)/
           myKinect.launch" />
 23
 24     <!-- NAVIGATION -->
 25
 26     <node pkg="gmapping" type="slam_gmapping"
           respawn="false" name="slam_gmapping" output=
           "screen">
 27             <param name="map_update_interval" value="
                 2.0"/>
 28             <param name="maxUrange" value="6.0"/>
 29             <param name="iterations" value="5"/>
 30             <param name="linearUpdate" value="0.25"/>
 31             <param name="angularUpdate" value="0.262"
                 />
 32             <param name="temporalUpdate" value="-1.0"
                 />
 33             <param name="particles" value="300"/>
 34             <param name="xmin" value="-50.0"/>
 35             <param name="ymin" value="-50.0"/>
 36             <param name="xmax" value="50.0"/>
 37             <param name="ymax" value="50.0"/>
 38             <param name="base_frame" value="base_link
                 "/>
 39             <param name="minimumScore" value="200.0"/
                 >
 40             <param name="srr" value="0.01"/>
 41             <param name="str" value="0.01"/>
 42             <param name="srt" value="0.02"/>
 43             <param name="stt" value="0.02"/>
 44     </node>
 45
 46     <node pkg="move_base" type="move_base" respawn="
           false" name="move_base" output="screen">
 47             <rosparam file="$(find
                 course_p3at_navigation)/
```

```
                         sg_costmap_common_params_p3at.yaml"
                         command="load" ns="global_costmap" />
48    <rosparam file="$(find
                         course_p3at_navigation)/
                         sg_costmap_common_params_p3at.yaml"
                         command="load" ns="local_costmap" />
49    <rosparam file="$(find
                         course_p3at_navigation)/
                         sg_local_costmap_params.yaml" command
                         ="load" />
50    <rosparam file="$(find
                         course_p3at_navigation)/
                         sg_global_costmap_params.yaml"
                         command="load" />
51    <rosparam file="$(find
                         course_p3at_navigation)/
                         becker_base_local_planner_params.yaml
                         " command="load" />
52    <param name="base_global_planner" type="
                         string" value=" navfn/NavfnROS" />
53    <param name="controller_frequency" type="
                         double" value="6.0" />
54        </node>
55
56
57 </launch>
```

As you can see the code is divided by commentaries in four sections as enumerated below:

3.6.1   PLATFORM SPECIFIC

Before explaining this section it is useful to clarify something: these PLATFORM SPECIFIC code is related to the platform specific nodes and it is not, by any means, the only platform specific sections of the navigation configuration, since almost all parameters presented on this chapter can be modified depending on your needs. This first section of the code is relative to the nodes you have to run so your robot is able to read the information in the /cmd_vel topic and translate this information into the respective real velocities to the robot's wheels. In the case here represented, for the Pioneer 3-AT, two nodes are run: p2os_driver [25] and an instance of the rostopic [26] node. The first one, p2os_driver, is a ROS node specific for some Pioneer robots, including the Pioneer 3-AT, able to control the motors in accordance to the information it receives from the ros topics /cmd_vel and /cmd_motor_state.

/cmd_vel has the velocities information and /cmd_motor_state tells the package if the motors are enabled. That is the reason the rostopic node should be run: it publishes a true value to the /cmd_motor_state topic so the p2os_driver knows that the motor should be enabled. p2os_driver also publishes some useful data, like sonar sensor data, transformations, battery_state and more.

3.6.2 TRANSFORMS

As discussed in the transforms section, you should create a transform between the robot's base_link and the sensor laser_link. Here is the place we recommend it to be put, although it can be launched in a separate launch file or in any particular order in this launch file. Any other transforms you may have should be put here too.

3.6.3 SENSORS CONFIGURATION

This section of the code was left to initialize all the nodes regarding the sensors powering up and configuration. The first four lines of this section contain some configurations of the Kinect sensor:

- (1) The first line changes the camera name to kinect_camera_name;
- (2) The second lines sets it to drop 1 frame of the Kinect for each valid one, outputting at approximately 15 Hz instead of 30 Hz;
- (3) The third and fourth lines are in respect to the resolution, where we have selected 320×240 QVGA 30 Hz for the image and 30 Hz QVGA for the depth image. The available options for image mode are:

  - **2**: (640×480 VGA 30 Hz)
  - **5**: (320×240 QVGA 30 Hz)
  - **8**: (160×120 QQVGA 30 Hz)

  And for the depth image mode:

  - **2**: (VGA 30 Hz)
  - **5**: (QVGA 30 Hz)
  - **8**: (QQVGA 30 Hz)

It is noteworthy that these parameters configurations for the Kinect, although recommended, are optional, since the default values will work. Besides the parameter configuration, there is also the line including the Kinect launcher that we wrote at the "Configuring the Kinect Sensor" section, which powers the sensor up and gets its data converted to laser data. If you're using any other kind of sensor, like a laser sensor, you should have your own launcher include here. Finally, if you odometry sensors aren't configured yet(in our case, the p2os_driver is responsible for this) you may do this here.

3.6.4   NAVIGATION

Understanding which nodes you should run, why and what each of them does is the main focus of this chapter. To get our Pioneer 3-AT navigating, we are using two navigation nodes (gmapping and move_base) and a lot of parameters. As explained before, gmapping is a localization system, while move_base is a package that contains the local and global planners and is responsible for linking them to achieve the navigation goal. Therefore, let us start explaining the gmapping launcher, the localization system we are using, since we have odometry available and our map is unknown (we will not use any static maps). For that, each parameter will be analyzed at once, as follows:

- *map_update_interval*: time (in seconds) between two updates of the map. Ideally, the update would be instantaneous, however, it would cost too much for the CPU to do that. Therefore, we use a interval, for which the default is 5 s.
- *maxUrange*: the maximum range for which the laser issues valid data. Data farther from this distance will be discarded.
- *iterations*: the number of iterations of the scanmatcher.
- *linearUpdate, angularUpdate and temporalUpdate*: thresholds for a scan request. temporalUpdate asks for a new scan whenever the time passed since the last scan exceeds the time indicated in the parameter, while linearUpdate and angularUpdate ask for scan when the robot translates or rotates (respectively) the amount specified in the parameters.
- *particles*: sets the number of particles used in the filter.
- *xmin, ymin, xmax and ymax*: these four coordinates form, together, the map size.
- *base_frame*: indicates the frame that corresponds to the mobile base in the transform tree.
- *minimumScore*: its a threshold value for considering the outcome of the scan matching good. Here we set the parameter to 200, but you should test some values between 0 and 300 depending on your configuration. Keep in mind that scores go from about −3000 to more than 600.
- *srr, srt, str and stt*: these parameters relate to the odometry errors. You should test your configuration and measure the four possible errors respectively: translation as a function of translation, translation as a function of rotation, rotation as a function of translation and rotation as a function of rotation. Here, r goes for *rho* (translation) and t goes for *theta* (rotation). The easiest way to do this is by sending goals to your robot and measuring the difference between the expected movement and the actual movement.

As to move_base, it bases its path planning techniques on the current location and the navigation goal. In the node launcher code, we have the usual syntax to launch a node, followed by a list of seven parameters, five of which are rosparams. The params are two:

- *base_global_planner* is a parameter for selecting the plugin (dynamically loadable classes). The plugin we use is the default for 1.1+ series, so we put this statement

here just to ensure we've selected the correct one. As you will see on the test
section, we changed this parameter to use the dwa_local_planner, since it works
better on our configuration.

- **controller_ frequency** is a parameter that fixes the rate (in Hz) at which the control
  loop will run and velocity commands will be issued.

The rosparams, in turn, are files that contain more parameters for the move_base,
and which are done this way to keep the files organized and easy to read. Thus, we
will take advantage of this fact and analyze the parameter files separately. First, we
begin looking at the costmap_common_params file, the one that contains parameters
that apply for both the local and global costmaps:

```
 1  obstacle_range: 5.0
 2  raytrace_range: 6.0
 3
 4  max_obstacle_height: 1.0
 5  min_obstacle_height: 0.05
 6
 7  footprint: [ [0.3302, -0.0508], [0.254, -0.0508],
        [0.254, -0.254], [-0.254, -0.254], [-0.254, 0.254],
        [0.254, 0.254], [0.254, 0.0508], [0.3302, 0.0508] ]
 8  #robot_radius: 0.35
 9  inflation_radius: 0.35
10  footprint_padding: 0
11
12  transform_tolerance: 1.0
13  map_type: costmap
14  cost_scaling_factor: 100
15
16
17  observation_sources: laser_scan_sensor
18  laser_scan_sensor: {sensor_frame: camera_link,
        data_type: LaserScan, topic: scan, marking: true,
        clearing: true}
19
20  #observation_sources: pointcloud_sensor
21  #pointcloud_sensor: {sensor_frame: camera_link,
        data_type: PointCloud2, topic: /camera/depth/points,
         marking: true, clearing: true}
```

As you may know, the sharp(#) represents a commented line or value, and does
not affect the results. This way, let us present the meaning of each of the params used
in the costmap common parameters file:

- **obstacle_range and raytrace_range**: obstacle_range relates to the maximum distance (in meters) that will be considered when taking the obstacle data and putting

it to the costmap, while raytrace_range is the maximum distance (also in meters) that will be considered when taking the free space around the robot and putting it to the costmap.

- *max_obstacle_height and min_obstacle_height*: these parameters set the area that will consider the sensor data as valid data. The most common is setting the min height near the ground height and the max height slightly greater than the robot's height.

- *robot_radius and inflation_radius*: when you're considering your robot as circular, you can just set the robot_radius parameter to the radius(in meters) of your robot and you get a circular footprint. Although, even if you don't have a circular robot, it is important to set the inflation_radius to the "maximum radius" of your robot, so the costmap creates a inflation around obstacles and the robot doesn't collide, no matter what is it direction when getting close to obstacles.

- *footprint and footprint_padding*: when you want a most precise representation of your robot, you have to comment the robot_radius parameter and create a custom footprint, as we did, considering [0, 0] as the center of your robot. footprint_padding is summed at each of the footprint points, both at the x and y coordinates, and we do not use it here, so we set it to zero.

- *transform_tolerance*: sets the maximum latency accepted so the system knows that no link in the transform tree is missing. This parameter must be set in an interval that allows certain tolerable delays in the transform publication and detects missing transforms, so the Navigation Stack stops in case of flaws in the system.

- *map_type*: just here to enforce we are using a costmap.

- *cost_scaling_factor*: this parameter sets the scaling factor that applies over the inflation. This parameter can be adjusted so the robot has a more aggressive or conservative behavior near obstacles.

$$e^{-cost\_scaling\_factor \times (distance\_from\_obstacle - inscribed\_radius) \times (costmap\_2d::INSCRIBED\_INFLATED\_OBSTACLE-1)}$$

- *observation_sources*: This last parameter is responsible for choosing the source of the sensor data. We can both use here point_cloud, as we're using for the Kinect, or laser_scan, as the commented lines suggest and as may be used for a Hokuyo or sick laser sensor. Along with the laser type, it is very important to set the correct name of the subscribed topic, so the Navigation Stack takes the sensor data from the correct location. The marking and clearing values are self-explanatory, since they set if the sensor data from the observation source is allowed to mark and clear the costmap. This parameter raises yet some **very important** discussion:

  – When selecting which observation source you are going to use, you have to consider that gmapping is using a LaserScan message as its observation source. If you are using a Kinect, you can choose the PointCloud message as your observation source here, and, although it will represent a lot more obstacles than the LaserScan would, that can bring a lot of problems to gmapping, that will struggle to get the current robot position.

Now that we have set all the costmap common parameters, we must set the parameter specific to the local and global costmaps. We will analyze them together, since most of the parameters are very similar. First, take a look at the files. For the global costmap we have:

```
global_costmap:
  global_frame: /map
  robot_base_frame: base_link
  update_frequency: 1.0
  publish_frequency: 1.0 #0
  static_map: false
  width: 50 #3.4
  height: 50 #3.4
  origin_x: -25 #-1.20 is the actual position; -0.95 is
      the old one, for the frond wheel at the marker
  origin_y: -25 #-1.91
  resolution: 0.1
```

And for the local:

```
local_costmap:
  global_frame: /odom
  robot_base_frame: base_link
  update_frequency: 5.0
  publish_frequency: 10.0
  static_map: false
  rolling_window: true
  width: 3.0
  height: 3.0
  resolution: 0.025
```

As you can see, both of them start with a tag specifying the costmap to which they relate. Then, we have the following common parameters:

- **global_frame**: indicates the frame for the costmap to operate in. They are set to different values because gmapping publishes the transform from odom to map, the global reference frame. If you set both of them to /odom you will be using the odometry data exclusively.
- **robot_base_frame**: indicates the transformation frame of the robot's base_link.
- **update_frequency and publish_frequency**: The frequency (in Hz) for map update and for publication of the display data.
- **static_map**: indicates whether the system uses or not a static map.
- **width and height**: width and height of the map, in meters.
- **resolution**: resolution of the map in meters per cell. This parameter is usually higher in smaller maps (local).

Aside from these common parameters, there's the definition of the map size along with the choosing between rolling window map or not. For the global map, we adopted the fixed map (there is no need to set rolling_windows to false, since it is the default), therefore we need to declare the x and y initial positions of the robots in respect to the map window. For the local_costmap, we use a rolling window map and the only parameter we have to set is the rolling_window to true.

Lastly, we have the base_local_planner parameters file. The base_local_planner treats the velocity data according to its parameters so the base_controller receives coherent data. Thus, the base_local_planner parameters are platform_specific. Take a look at the configuration for the Pioneer 3-AT:

```
TrajectoryPlannerROS:
   max_vel_x: 0.5
   min_vel_x: 0.1
   max_rotational_vel: 0.5
   max_vel_theta: 0.5
   min_vel_theta: -0.5
   min_in_place_rotational_vel: 0.5
   min_in_place_vel_theta: 0.5
   escape_vel: -0.1


   acc_lim_th: 0.5
   acc_lim_x: 0.5
   acc_lim_y: 0.5

   holonomic_robot: false
```

Again, we should analyze the most important parameters separately.

- *min_vel_x and max_vel_x*: The minimum and maximum velocities (in meters/second) allowed when sending data to the mobile base. The minimum velocity should be great enough to overcome friction. The maximum velocity adjust is good for limiting the robot's velocity in narrow environments.
- *max_rotational_vel and min_in_place_rotational_vel*: limits for the rotational velocities, the difference is that rotational_vel is the maximum rotation velocity when the mobile base is also moving forward or backward, while in_place _rotational_vel is the minimum rotation vel so the robot can overcome friction and turn without having to move forward or backward.
- *min_vel_theta and max_vel_theta*: the minimum and maximum rotational velocities (in radians/second).
- *min_in_place_vel_theta*: alike min_in_place_rotational_vel, but in radians per second.
- *escape_vel*: this speed delimits the driving speed during escapes (in meters per second). Its noteworthy that this value should be negative for the robot to reverse.
- *acc_lim_x, acc_lim_y and acc_lim_theta*: accelerations limits. They are the x,y and rotational acceleration limits respectively, wherein the first two are in meters per squared second and the last is radians per squared second.

- ***holomic_robot***: this is a boolean responsible to choose between holonomic and non-holonomic robots, so the base_local_planner can issue velocity commands as expected.

Finally, we have a basic set up, contemplating all the usual parameters that you have to configure and some more. There is a small chance that some parameter is missing for your configuration, therefore it is a good idea to do a quick check in the base_local_planner [27] and costmap_2d [28] wiki pages.

The way that we have presented does not use layers, although ROS Hydro+ supports this feature. Porting these files to this new approach of costmaps is not a hard task, and that is what we will cover now.

3.7 Layered Costmaps

For this approach, we use the launchers and the configuration files from the previous package. First, we create a package named p3at_layer_navigation, as stated on the creating a package section. Then, we copy all files from the previous package but the package.xml and CMakelists.txt files to the folder of the newly created package. For the base_local_planner, nothing should be modified, since the planning will not be affected in any way when exploding the maps in layers. The common costmaps file is the one that will be affected the most, and here is one example costmap_common_params.yaml file that illustrates this:

```
robot_base_frame: base_link

transform_tolerance: 1.0

robot_radius: 0.35

footprint: [ [0.3302, -0.0508], [0.254, -0.0508],
    [0.254, -0.254], [-0.254, -0.254], [-0.254, 0.254],
    [0.254, 0.254], [0.254, 0.0508], [0.3302, 0.0508] ]

inflater:
  robot_radius: 0.35
  inflation_radius: 0.35

obstacles:
  observation_sources: pointcloud_sensor
  pointcloud_sensor:
    data_type: PointCloud2
    topic: camera/depth/points
    min_obstacle_height: 0.2
    max_obstacle_height: 2.0
    marking: true
    clearing: true
```

```
22   z_voxels: 8
23   z_resolution: 0.25
24   max_obstacle_height: 2.0
```

As you can see in the file, the parameters do not change much, the difference is that they are organized in a different way: there are some parameters that are common for all costmaps and there are some parameters that are common between layers. In this example, we create two layers: a inflater layer, that considers a circular robot with 35 cm of radius, and, therefore, an inflation radius of 35 cm so it doesn't collide with anything; a obstacles layer, that takes the pointcloud data (if you are using a laser, please change that here) and passes this data to the costmap.

The two other files have a slight modification: you should specify the layers they are using by using the plugins mnemonic, as shown for the global_costmap configuration file:

```
 1   global_frame: map
 2
 3   robot_base_frame: base_link
 4   update_frequency: 1.0
 5   publish_frequency: 1.0
 6   static_map: false
 7   width: 50
 8   height: 50
 9   origin_x: -25
10   origin_y: -25
11   resolution: 0.1
12
13   plugins:
14    - {name: obstacles, type: "costmap_2d::VoxelLayer"}
15    - {name: inflater, type: "costmap_2d::InflationLayer"}
```

The local_costmap should have the same plugins statement at the end. Moreover, you can add any extra layers you want. The structure of the topics will change a little bit, since the footprint is now inside the layers and the costmaps are divided in multiple topics. You can get to know a little more about this organization in the Using rviz section.

4   Starting with a Test

Before we begin the testing, we must find a way to visualize the navigation in action. That can be done through the software rviz, that allows us, amongst other things, to visualize the sensor data in a comprehensive way and to check the planned paths as they are generated. The execution of rviz is often in a computer different from the one that operates on the robot, so you may use multiple machines that share the same ROS topics and communicate.

4.1   Using Rviz

To run rviz, simply issue the following command:

```
1 $ rosrun rviz rviz
```

The interface of rviz depends on your version, but the operation should be very similar. It is way easier to configure rviz with the Navigation Stack up and running, although it is possible to do so without it. In this tutorial we will only cover the configuration steps when the Navigation Stack launcher is already implemented, so make sure you have launched all the nodes needed for navigation and just then launched rviz. After launching rviz, you should add the topics you wish to display. First, as an example, add the PointCloud 2 from the Kinect, as shown in Fig. 4.

As you can see in Fig. 4, there are four steps for adding a new topic when Navigation Stack is already running:

- (1) Click on the button "add" at the bottom left-hand side of the screen;
- (2) Choose the tab "By topic" on the windows that appears. This is only possible when the topics are available, so if you don't have the Navigation Stack running

Fig. 4 Steps for adding new information for rviz to display

you will have to choose the info in the tab "By display type" and manually insert the topic names and types.

- (3) Select the topic and its message type on the central frame of the window. In this example, we are selecting the PointCloud2 data that the Kinect provides on the /camera/depth_registered/points topic.
- (4) Write a meaningful display name in the textbox, so you don't forget what the data is representing in the future.
- (5) Confirm the addition by pressing "Ok".

The process is equal for all kinds of topics, so a list of the most common topics (note: depending if you changed some topic names, some things on the list may differ) should be enough to understand and add all the topics you need.

Name	Topic	Message type
ROBOT FOOT-PRINT	/local_costmap/robot_footprint	geometry_msgs/PolygonStamped
LOCAL COSTMAP	/move_base/local_costmap/costmap	nav_msgs/GridCells
OBSTACLES LAYER	/local_costmap/obstacles	nav_msgs/GridCells
INFLATED OBSTA-CLES LAYER	/local_costmap/inflated_obstacles	nav_msgs/GridCells
STATIC MAP	/map	nav_msgs/GetMap or nav_msgs/OccupancyGrid
GLOBAL PLAN	/move_base/TrajectoryPlannerROS/global_plan	nav_msgs/Path
LOCAL PLAN	/move_base/TrajectoryPlannerROS/local_plan	nav_msgs/Path
2D NAV GOAL	/move_base_simple/goal	geometry_msgs/PoseStamped
PLANNER PLAN	/move_base/NavfnROS/plan	nav_msgs/Path
LASER SCAN	/scan	sensor_msgs/LaserScan
KINECT POINT-CLOUD	/camera/depth_registered/points	sensor_msgs/PointCloud2

It is interesting to know a little more about topics that you haven't heard about, because every topic listed here is very valuable at checking the navigation functionalities at some point. Therefore, lets do a brief explanation at each of the topics:

- **ROBOT FOOTPRINT:** These message is the displayed polygon that represents the footprint of the robot. Here we are taking the footprint from the local_costmap, but it is possible to use the footprint from the global_costmap and it is also possible to take the footprint from a layer, for example, the footprint may be available at the /move_base/global_costmap/obstacle_layer_footprint/footprint_stamped topic.
- **LOCAL COSTMAP:** If you're not using a layered approach, your local_costmap in its whole will be displayed in this topic.
- **OBSTACLES LAYER:** One of the main layers when you're using a layered costmap, containing the detected obstacles.

- **INFLATED OBSTACLES LAYER:** One of the main layers when you're using a layered costmap, containing areas around detected obstacles that prevent the robot from crashing with the obstacles.
- **STATIC MAP:** When using a pre-built static map it will be made available at this topic by the map_server.
- **GLOBAL PLAN:** This topic contains the portion of the global plan that the local plan is considering at the moment.
- **LOCAL PLAN:** Display the real trajectory that the robot is doing at the moment, the one that will imply in commands to the mobile base through the /cmd_vel topic.
- **2D NAV GOAL:** Topic that receives navigation goals for the robot to achieve. If you want to see the goal that the robot is currently trying to achieve you should use the /move_base/current_goal topic.
- **PLANNER PLAN:** Contains the complete global plan.
- **LASER SCAN:** Contains the laser_scan data. Depending on your configuration this topic can be a real reading from your laser sensor or it can be a converted value from another type of sensor.
- **KINECT POINTCLOUD:** This topic, shown in the example, is a cloud of points, as the name suggests, that forms, in space, the depthimage captured by the Kinect. If you are using a laser sensor, this topic will not be available.

These are the most used topics, however you may have a lot more depending on your setup and in what you want to see. Besides, just the local_costmap and the most used layers of it were presented, but you may want to see the global costmap and its layers, in addition to another layers that you may use. Explore the topics you have running with the rviz and you may find more useful info.

4.2 Multiple Machines Communication

Multiple Machines Communication is a rather complex topic and it is possible that you have to use more than one computer at the same time when navigating with a robot. Usually, in navigation scenarios, two computers are used: one is mounted on the mobile base and is responsible for getting sensor data and passing velocities commands to the robot, while the other is responsible for heavy processing and monitoring.

Although a ROS network can be very complex, if you are using just two machines as specified above, you will probably be able to communicate them with the following steps. If that is not your case, please refer to the Multiple Machines Communication ROS wiki page [29]. To get the machines working together you must specify names for both machines in the /etc/hosts file on your system. The usual hosts file is similar to: You have to choose a name for both the machines (here we chose robot and masterpc) and add to both the /etc/hosts files the entries for them. The entry on the /etc/hosts must have the IP of the machine in the wireless lan they share and the name

```
1 IPAddress Hostname
2 127.0.0.1 localhost
3 192.168.1.101 robot
4 192.168.1.100 masterpc
```

you picked (two new entries per file, one for itself and other for the other PC). After that, you should set the ROS_MASTER_URI variables. In the master machine, it should be:

```
1 $ export ROS_MASTER_URI=http://localhost:11311
```

In the other machine, you should type:

```
1 $ export ROS_MASTER_URI=http://mastermachine:11311
```

Test your configuration this way, and if the configurations work, add the export lines to the end of the ~/.bashrc file of both computers, so every time a terminal window is opened these commands are issued.

4.3 Real Tests on Pioneer 3-AT

Finally, it is time to see the robot navigating. Launch your navigation file and then run rviz. If you made all the correct configuration for the Navigation Stack and for rviz you should be able to see your robot footprint, pose and the costmaps. Try selecting the 2D Nav goal at the top of the screen in rviz, click and hold at some point on the map and then choose the direction, so the robot knows where to go and in what position it should stop. A global path would be generated, as well as a local, and you should see them indicated by green lines. You can see an example of the robot navigating on Figs. 5 and 6.

As you can see in the pictures, a global plan is drawn from the start point to the finish point and a local plan is being drawn along the way, trying to follow the global path without crashing. As for the costmaps, the yellow section indicates where the obstacle is, and is an infinite cost area. The radius around the obstacle, in our case almost entirely blue, is the inflation radius, where the cost is exponentially decreasing from the obstacle to the border of the inflation radius. Depending on your inflation_radius and cost_scaling_factor parameters, this coloring can be different.

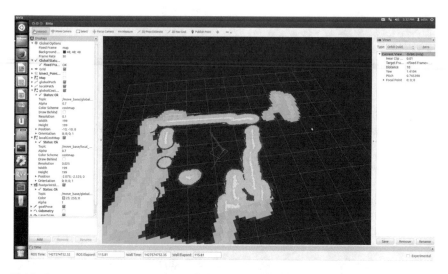

Fig. 5 Example costmaps for the Pioneer 3-AT navigating in a room

Fig. 6 Photograph of the robot navigating, at the same moment the print screen of Fig. 5 was taken

Conducting the tests we have found that the navigation does not work so great with the default planner. Therefore, we tested the dwa_local_planner and the results were way better. The dwa_local_planner is, in general, better for slow robots, and to use it you should modify your move_base launcher as follows:

```
<node pkg="move_base" type="move_base" respawn="false"
    name="move_base" output="screen" clear_params="true"
    >
            <param name="controller_frequency" value="
                10.0" />
            <param name="controller_patience" value="15.0
                " />
            <param name="planner_frequency" value="2.0" /
                >
            <param name="clearing_rotation_allowed" value
                ="false" />
            <rosparam file="$(find course_p3at_navigation
                )/sg_costmap_common_params_p3at.yaml"
                command="load" ns="global_costmap" />
        <rosparam file="$(find course_p3at_navigation)/
            sg_costmap_common_params_p3at.yaml" command="
            load" ns="local_costmap" />
        <rosparam file="$(find course_p3at_navigation)/
            sg_local_costmap_params.yaml" command="load"
            />
        <rosparam file="$(find course_p3at_navigation)/
            sg_global_costmap_params.yaml" command="load"
             />
            <param name="base_local_planner" value="
                dwa_local_planner/DWAPlannerROS" />
            <rosparam file="$(find course_p3at_navigation
                )/dwa_planner_ros.yaml" command="load" />
</node>
```

As you can see, a new param file is added, and this file is very similar to the base_local_planner parameters. You can find the original version of the file, taken from the scitos_2d_navigation github [30] on their URL or you can see our modified version here:

```
DWAPlannerROS:
  acc_lim_x: 1.0
  acc_lim_y: 0.0

  acc_lim_th: 2.0

  min_vel_x: -0.55
  min_vel_y: 0.0
  max_vel_y: 0.0
  max_rot_vel: 1.0
  min_rot_vel: 0.4
```

```
12
13  yaw_goal_tolerance: 0.1
14
15  xy_goal_tolerance: 0.3
16
17  latch_xy_goal_tolerance: true
18
19  sim_time: 1.7
20
21  path_distance_bias: 5.0
22  goal_distance_bias: 9.0
23  occdist_scale: 0.01
24
25  oscillation_reset_dist: 0.05
26
27  prune_plan: true
28
29  holonomic_robot: false
```

If you are having some troubles, you may verify the view_frames functionality of the tf ROS package. It is very useful for spotting problems on your transformation tree and it also helps to learn better some of the concepts about transformations that we have presented. To generate a "frames.pdf" file containing the current tf tree, insert the following lines on the shell:

```
1  $ rosrun tf view_frames
```

As a result, we got the tf tree seen on Fig. 7, where you can clearly see the hierarchy between the frames and understand a little better the relation between them. Each transform on the tf tree (represented by the arrows) has some related data, such as the average rate and the node that is broadcasting the transform. If you don't have a tree structure like that, where some frame is not linked to the tree or the map, odom and base_link are not presented on this order, please check your tf configuration. Please note that this is not the only possible configuration, as you can have base_footprint when you are using hector_mapping, for example. Another thing that you may have noticed is that we do not have only one frame for the Kinect: the openni_launcher node publishes 4 transformations between 5 distinct frames, and the only transformation that we have manually launched, in this case, is the transformation between the base_link and the camera_link.

After seeing the system was working as expected, we tried to do some more significant tests. First we arranged our test room in two ways: a single wood object on the middle of the room and the robot at one side; a corridor formed by some objects. You can see the two scenarios and the results of the navigation in rviz in Figs. 8 and 9.

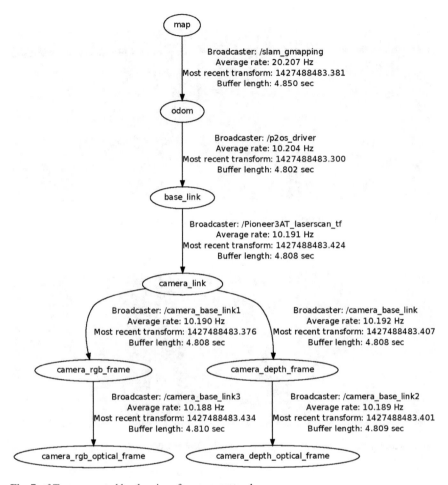

Fig. 7 tf Tree generated by the view_frames command

To do these experiments, a initial and final position on the room were chosen for each one. After that the distance and the difference in angle were measured, and then we published the goal with these values. We have done that with a simple node, very similar to the node you can find here [31]. You can also publish a goal by issuing the following command (**CHANGE THE VALUES TO SUIT YOUR NEEDS**):

```
1 rostopic pub /move_base_simple/goal geometry_msgs/PoseStamped ' header:  frame_id:
  "/base_link", pose:  position:  x: 0.2, y: 0 , orientation:  x: 0, y: 0, z: 0, w: 1  '
```

It is possible to see on Fig. 8 that the robot would get to the goal by following a straight line if no object was present, but it successfully avoids the obstacle and gets to the exact point it intended to. It is also possible to see on the images that it doesn't just avoid the obstacle, but a region around it, the inflation radius, since there is a low

(a)   (b)   (c)

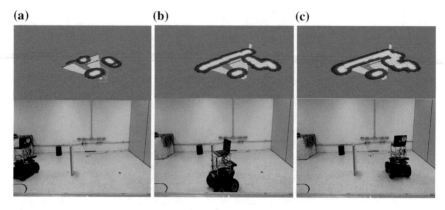

Fig. 8 **a** Robot moving right after it receives a goal. **b** Robot avoiding the obstacle. **c** Robot getting to the goal

(a)   (b)   (c)

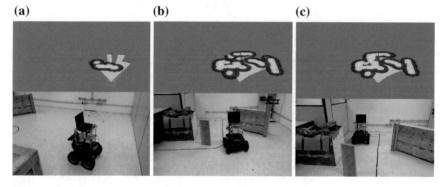

Fig. 9 **a** Robot moving right after it receives a goal. **b** Robot entering the corridor. **c** Robot getting to the goal

cost area right around of it. When it gets to the goal, it rotates until it gets the right angle. In the way, you can see that the local and global CostMaps do not perfectly match and start to drift, but the gmapping transformation is able to adjust that with very little errors.

On Fig. 9 the task of the robot is harder: it has to avoid some obstacles and get to a point that is after a corridor. At first, it tries to avoid the obstacles by going around them, but it sees the wall. After that, it decides to enter the inflation radius, a high cost area, but trying to stay as close to the border as possible. At the end, it successfully gets to the goal and adjusts itself to the right direction.

In `rviz` it is possible to see some details about the navigation:

- Obstacles on the gmapping map and on the CostMaps should overlap each other perfectly, and that is almost the case in most of the situations;
- Obstacles on local and global CostMap should also overlap each other, and gmapping does a good job in correcting the odometry errors to get that;

- The inflation radius is avoided by the robot, that usually translates right at its border to get to the goal. This behavior can be different depending on the cost factors you chose at the planner parameters;
- The global path is not perfectly followed by the local path in almost every situation, because the parameters to calculate the smallest cost of them are usually different. The frequency of update and cost parameters should alter this behavior as needed.

4.3.1 Conclusion

Navigating through unknown environments is a very hard task, however we had great results doing it while using the Navigation Stack. This set of tools helped us to achieve great results, allowing the robot to create a dynamically generated map and achieve goals without crashing. We hope that this tutorial chapter, along with some background reading of the references we have presented, is enough for you to also get your robot navigating.

References

1. I.R.S.D. Siegwart, *Introduction to Autonomous Mobile Robots*, Intelligent Robotics and Autonomous Agent series (The MIT Press, Cambridge, 2011)
2. ROS.org, Ros package geometry_msgs (2015), http://wiki.ros.org/geometry_msgs. Accessed 03 Jan 2015
3. R.P. Goebel, Ros by example, vol. 1 (2014), http://www.lulu.com/shop/r-patrick-goebel/ros-by-example-hydro-volume-1/ebook/product-21393108.html. Accessed 13 June 2015
4. R.P. Goebel, Ros metapackage rbx1 (2014), https://github.com/pirobot/rbx1/tree/hydro-devel. Accessed 13 June 2015
5. ROS.org, Ros package openni_camera (2015), http://wiki.ros.org/openni_camera Accessed 03 Jan 2015
6. ROS.org, Ros package openni_launch (2015), http://wiki.ros.org/openni_launch. Accessed 03 Jan 2015
7. AnthonyJ350, How to solder / soldering basics tutorial (2011), https://www.youtube.com/watch?v=BxASFu19bLU. Accessed 03 Jan 2015
8. mjlorton, Tutorial: How to crimp connectors, strip wire and use heat shrink. (2013), https://www.youtube.com/watch?v=kjSGCSwNuAg. Accessed 03 Jan 2015
9. E.C. Wireconnector, E-clamp quick wire connector (2013). https://www.youtube.com/watch?v=GI8lRSQQbJk. Accessed 03 Jan 2015
10. ROS.org, Adding a kinect to an irobot create/roomba (2015), http://wiki.ros.org/kinect/Tutorials/Adding. Accessed 03 Jan 2015
11. ROS.org, Ros package depthimage_laserscan (2015), http://wiki.ros.org/depthimage_to_laserscan. Accessed 03 Jan 2015
12. S.F.A.S D.G. Ballardini, A.L. Fontana, ira_tools: a ros laserscan manipulation toolbox (2014), arXiv:1411.1086. Accessed 12 Jan 2015
13. S.F.A.S.D.G. Ballardini, A.L. Fontana, Ros package ira_tools (2014), https://github.com/iralabdisco/ira_laser_tools. Accessed 12 Jan 2015
14. ROS.org, Ros package kinect_aux (2015), http://wiki.ros.org/kinect_aux. Accessed 03 Jan 2015

15. ROS.org, Ros package openni_tracker (2015), http://wiki.ros.org/openni_tracker. Accessed 03 Jan 2015
16. O.T.S. for 3D Sensing, Openni sdk history (2014), http://www.openni.ru/openni-sdk/openni-sdk-history-2/index.html. Accessed 12 June 2015
17. R.L. Guimaraes, Complementary files for using more kinect functions (2015), https://github.com/rosnavigation/KINECT_FILES. Accessed 12 June 2015
18. ROS.org, Ros package cyphy_mapping (2015), http://wiki.ros.org/cyphy_people_mapping. Accessed 03 Jan 2015
19. STRANDS, Ros package sicks300 (2014), https://github.com/strands-project-releases/sicks300/tree/release/hydro/sicks300. Accessed 03 Jan 2015
20. ROS.org, How to set up hector_slam for your robot (2015), http://wiki.ros.org/hector_slam/Tutorials/SettingUpForYourRobot. Accessed 03 Jan 2015
21. W. Meeussen, Coordinate frames for mobile platforms (2015), http://www.ros.org/reps/rep-0105.html. Accessed 03 Jan 2015
22. B. Sciavicco, L. Sicoliano, *Modelling and Control of Robot Manipulators*, Advanced Textbooks in Control and Signal Processing (CreateSpace Independent Publishing Platform, Charleston, 2000)
23. ROS.org, Navigation tutorials: Robotsetup (2015), http://wiki.ros.org/navigation/Tutorials/RobotSetup. Accessed 03 Jan 2015
24. W.F.D. Thrun, S. Buggard, *Probabilistic Robotics*, Intelligent Robotics and Autonomous Agent Series (MIT Press, Cambridge, 2005)
25. ROS.org, Ros package p2os_driver (2015), http://wiki.ros.org/p2os_driver. Accessed 03 Jan 2015
26. ROS.org, Ros package rostopic (2015), http://wiki.ros.org/rostopic. Accessed 03 Jan 2015
27. ROS.org, Ros package base_planner (2015), http://wiki.ros.org/base_local_planner. Accessed 03 Jan 2015
28. ROS.org, Ros package costmap_2d/layered (2015), http://wiki.ros.org/costmap_2d/layered. Accessed 03 Jan 2015
29. ROS.org, Running ros across multiple machines (2015), http://wiki.ros.org/ROS/Tutorials/MultipleMachines. Accessed 19 June 2015
30. N. Bore, Ros scitos_navigation package github (2013), https://github.com/nilsbore/scitos_2d_navigation. Accessed 20 June 2015
31. R. Yehoshua, Sendgoals.cpp (2013), http://u.cs.biu.ac.il/~yehoshr1/89-685/Fall2013/demos/lesson7/SendGoals.cpp. Accessed 20 June 2015

Localization and Navigation of a Climbing Robot Inside a LPG Spherical Tank Based on Dual-LIDAR Scanning of Weld Beads

Ricardo S. da Veiga, Andre Schneider de Oliveira,
Lucia Valeria Ramos de Arruda and Flavio Neves Junior

Abstract Mobile robot localization is a classical problem in robotics and many solutions are discussed. This problem becomes more challenging in environments with few and/or none landmarks and poor illumination conditions. This article presents a novel solution to improve robot localization inside a LPG spherical tank by robot motion of detected weld beads. No external light source and no easily detectable landmarks are required. The weld beads are detected by filtering and processing techniques applied to raw signals from the LIDAR (Light Detection And Ranging) sensors. A specific classification technique—SVM (Support Vector Machine)—is used to sort data between noises and weld beads. Odometry is determined according to robot motion in relation with the weld beads. The data fusion of this odometry with another measurements is performed through Extended Kalman Filter (EKF) to improve the robot localization. Lastly, this improved position is used as input to the autonomous navigation system, allowing the robot to travel through the entire surface to be inspected.

Keywords Mobile robots · Localization · LIDAR · No landmarks · Weld beads · Data fusion · Autonomous navigation

R.S. da Veiga (✉) · A.S. de Oliveira · L.V.R. de Arruda · F. Neves Junior
Automation and Advanced Control System Laboratory (LASCA), Federal University
of Technology - Parana (UTFPR), Av. Sete de Setembro, 3165 - Reboucas CEP,
Curitiba, Parana 80230-901, Brazil
e-mail: ricardo_veiga@mail.com

A.S. de Oliveira
e-mail: andreoliveira@utfpr.edu.br

L.V.R. de Arruda
e-mail: lvrarruda@utfpr.edu.br

F. Neves Junior
e-mail: neves@utfpr.edu.br

© Springer International Publishing Switzerland 2016
A. Koubaa (ed.), *Robot Operating System (ROS)*, Studies in Computational
Intelligence 625, DOI 10.1007/978-3-319-26054-9_7

1   Introduction

This work presents a ROS-based robot designed to inspect metallic spherical tanks. Spherical tanks are used to store Liquefied Petroleum Gas (LPG) and they are also known as Horton Spheres in honor of Horace Ebenezer Horton, founder of CB&I—Chicago Bridge and Iron Company, which built the very first spherical tank in 1923 in Port Arthur, Texas. Often inspections and verifications of weld beads are necessary during the sphere life cycle to prevent failures and leaks, as shown on Fig. 1—this was a spherical tank explosion after an earthquake in Chiba city, Japan, march 11, 2011, which can give an idea of how tragic one explosion in this equipments can be.

Inspection usually applies long and expensive techniques, as seen in [1], many times exposing the technicians to unhealthy and often hazardous environments. Mobile robots can be used to reduce or avoid these risks. Inspection robots have the task of navigating through the entire inner and outer surfaces of the storage tank, searching for structural failures in the steel plates or in the weld beads.

Finding the inspection robot accurate location inside the sphere might be a harsh task, but it is essential to a successful inspection job. When the inspection system detects any surface problem or disturbance, its exact position must be stored to a later deeper and more detailed inspection. The navigation module of an inspection robot must perform several turns around the sphere's inner surface to cover the entire tank and the full path must extend for many meters. A small localization error can deviate the robot a lot from its original path, introducing a large error in inspection fault mapping. Several localization techniques may be merged to increase robot localization precision, e.g. [2–4].

Fig. 1   Major failure and explosion of a LPG tank

A simple solution to localize the robot could be visual feedback from cameras. Some papers suggest this approach, e.g., [5–7]. However, the LPG internal environment is almost totally dark and powerful lights must be avoided due to the possibility that heat generated by the lamps may create sparks inside a highly explosive atmosphere, therefore this technique should be discarded.

However, robot navigation inside LPG spherical tanks are also not achievable only via wheel odometry because many uncertainties are introduced in trajectory tracking, making the inspection job to fail. Other sources of odometry should be considered to improve localization. The Inertial Measurement Unit (IMU) is an important tool to increase the accuracy of robot localization, as reported on some works e.g. [8, 9]. This unit is essential to climbing robots because the gravitational force may act as a disturbance and it cannot be neglected as discussed in [10], especially when the robot is traveling at the equatorial zones of the spheres. Aiming a better precision, the fusion of different odometry sources is typically performed often with the widely known Extended Kalman Filter (EKF) like in [11].

Data fusion between the wheel odometry and IMU improves meaningfully the robot localization. However, these are 'blind' sources related only with robot's parameters and without environment data (proprioceptive sensors). Both methods compute odometry based on current sensor readings and previous robot pose. This is also know as dead-reckoning, where the errors are accumulative [12].

LIDAR (Light Detection And Ranging) sensors are also commonly used to environment mapping, but they can be used like an odometry source as well. Many articles discuss this approach, e.g. [13–15]. This type of sensor does a pseudo-visual mapping due its output data that is a one-dimensional array of distances with n-points. LIDAR odometry is related to distance between the robot and its environment's landmarks, like walls, obstacles, etc. [16–18]. However, this approach cannot be used in spherical tanks because the distance (over inside and outside surfaces) are invariable and there are no other easily detectable landmarks.

This work presents a novel approach to localize a climbing robot, especially inside LPG spherical tanks, based on LIDAR sensors. Signals from two LIDAR sensors placed horizontally facing downwards are processed by several techniques. The goal is to make possible the detection of small structures like weld beads, thus increasing the accuracy in determining the inspection robot pose. ROS modules are used to allow easy navigation in this environment and all source code is available at https:// github.com/AIR-LASCA/air1.git.

2 AIR—Autonomous Inspection Robot

The robot presented in this work is more detailed in [10]. The same base was used with more sensors allowing its localization and autonomous navigation. The robot's basic concept is shown on Fig. 2.

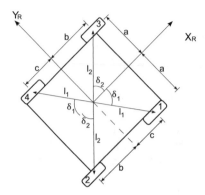

Fig. 2 Robot parameters and footprint

Fig. 3 Robot on a real tank

This robot have four unaligned wheels to overpass obstacles in tank surface, like weld beads and possible structural failures without adherence losses. It was tested outside on a real tank, as shows Fig. 3.

Adherence to the tank walls are performed by magnetic wheels specially designed to LPG tank surface, ensuring that all magnetic force is applied towards the tank surface, even when these wheels are over some obstacle. One experiment with this kind of wheel is shown on Fig. 4.

More informations about this project, as the mechanical parameters and specifications can be found at [10, 19].

The robot uses an active adhesion control system that can estimate how much each wheel is loosing its adhesion power, using a Artificial Neural Network (ANN). This controller workflow is shown on Fig. 5. Based on the IMU, using the magnetic

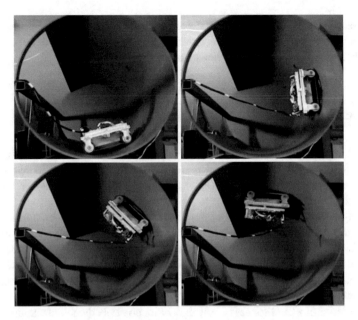

Fig. 4 Magnetic Wheels experiments and testing [10]

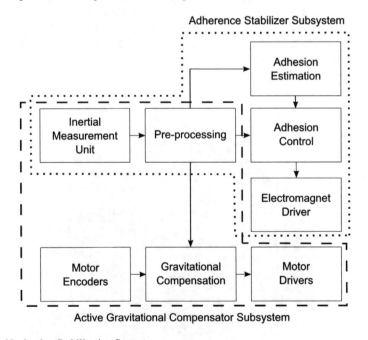

Fig. 5 Navigation Stabilization System

field measurements, this system can detect and accurately determine if there is any adhesion loss. Furthermore, it also controls a security electromagnet to prevent the robot from falling when it detects an adhesion loss.

3 Localization Problems in Spherical Tanks

Storage tanks, like the one presented in Fig. 6, are metal structures used in industries to store liquids and gases under high pressures. Since they are spherical and therefore have no corners that would be the weakest region of the structure, they are one good solution to this storage problem.

During autonomous navigation, the robot must cover the entire surface of the tank so the inspection module can detect problems and inconsistencies properly, both in the weld beads and in the whole metallic surface of the plates. Accurate robot localization in this environment is crucial to a good inspection job because when a surface fault is detected, its exact position must be stored to further inspection and maintenance.

The interior of a LPG sphere is fully dark, with no easily detectable landmarks and no different faces—from every viewpoint on the inner side of the sphere structure, the robot will always see the same semi-sphere. The only possible reference inside the sphere are the weld beads that connect the sphere plates, and these are hard to detect because they are just a couple millimeters high. Figure 6 shows a sphere being assembled, and the weld beads are digitally highlighted.

Fig. 6 Assembly of a LPG Sphere tank with digitally highlighted weld beads

4 Weld Beads Scanning with LIDAR Sensors

In this paper, the weld bead scanning through LIDAR sensor is evaluated in a part of a real LPG tank plate. This plate is made with two sphere assembly pieces linked by a weld bead, as seen in Fig. 7.

The LIDAR sensor fires a 2D laser beam around itself with 240° of scan area and 0.36° of angular resolution. It detects objects within a semicircle with 4000 mm radius and it returns a one-dimensional matrix containing approximately 683 scan points, called laser beam. The LIDAR sensor measurements are the distances around itself and now it is necessary to linearize distances without angular displacement as shown in Fig. 8.

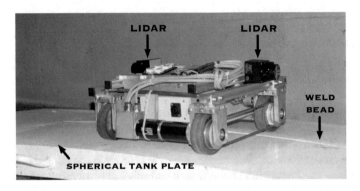

Fig. 7 Robot sitting on a real section plate of a LPG sphere tank with weld bead

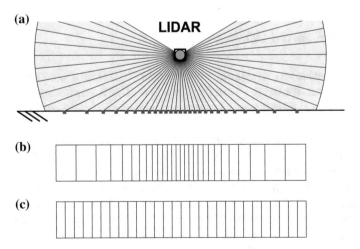

Fig. 8 Entire linearization process. **a** LIDAR sensor detecting a at surface, gray dots are the points where the laser beam contacts the surface. Each slice represents a $(1/p)$ of the total reading data. **b** Data packets transcribed to linear scale, still having linearization errors. **c** Data packets transcribed to linear scale without linearization errors

This linearization is performed with raw data through simple trigonometry rules, based on laser height from the ground (17.1334 cm), as

$$x' = tan\left(\left(\frac{sa}{p}\right) * \left(x - \frac{sa}{2}\right)\right) * h, \tag{1}$$

where, x is the measured point, x' is the processed point, sa is the scan area angle, p is the number of scan points and h represents the height of laser with respect to the robot's base.

The entire linearization process to the 240° spectrum is illustrated in Fig. 8.

This linearized signal resulting from the LIDAR spread beam is the core element to detect the weld beads.

5 Localization Inside LPG Spheres

As discussed above, the main goal of this work is improve robot localization inside a pitch-black spherical tank. In the previous sections, it was proved that LIDAR sensor accurately detects weld beads. Now, this sensor will be applied to improve robot's odometry. From now on, the experiments are performed in a virtual spherical tank, which is designed with same physical characteristics of a real one. Dynamic effects like gravity, magnetic adhesion, friction and others are also considered in the virtual environment. The interaction between tank and robot aims to mimic the real world as closely as possible.

5.1 LIDAR Based Odometry

Localization inside spherical tanks is made based on motion of weld beads. A dual-LIDAR approach is used to odometry where two LIDAR sensors are used, each of one robot's side and both parallel to robot, as shown previously on Fig. 7, on page xxx.

LIDAR odometry comprises various steps. The first is the detection of any motion on dual-LIDAR signals. The idea is simple: Since the sensor can detect a weld bead then this weld bead can be tracked. If the bead moves 10 cm in one direction, this means that the robot has moved 10 cm in the opposite direction. Notice that this technique does not need any knowledge a priori about the position neither the displacement of the weld beads on the sphere or its diameter. In the case when no bead is detected, that is, when the trajectory is collinear with the beads, no calculations are made and the output of the whole system is cloned from the Wheel Odometry.

Figure 9 shows a schematic drawing of the robot while scanning a given section of a planar structure. Two small arrows point the weld beads detected by the LIDAR sensor.

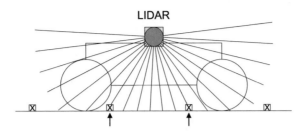

Fig. 9 Schematic of the weld beads being detected by the LIDAR sensor. Small arrows point the beads

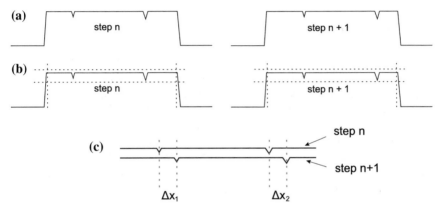

Fig. 10 Example of LIDAR odometry computation. **a** Processed signal coming from the LIDAR sensors. **b** Computation of interest areas by lower and upper limits. **c** Motion estimation

The scanned surface signal, after the processing described in Sect. 4, looks like Fig. 10a. Peak regions, represented by point-down triangles, are the weld bead detections, called as interest points. The motion of interest points represents the motion of weld beads related to the robot reference and consequently robot motion related to the odometry frame.

When any interest point motion occurs, the calculation of lower and upper limits is made to cut nonessential signal parts through a band-pass filter, as illustrated in Fig. 10b. Areas around the interest points are called as interest areas, these represents the LIDAR detection of the weld beads in its whole range. The edge points of the interest areas are used to synchronize between dual-LIDAR signals.

The difference between current data and previous data are made with delta computation and it represents the robot's motion in two LIDAR reading steps, as shown in Fig. 10c. The last step is the delta signal filtering to avoid any noises. Mean filters and low-pass filters are applied and results are integrated to obtain the position value.

Global motion estimation is performed by data fusion of these processed LIDAR data with the other odometry sources through Extended Kalman Filter (EKF), as seen in Fig. 11. Since the working procedure of the EKF filter is widely known, it won't

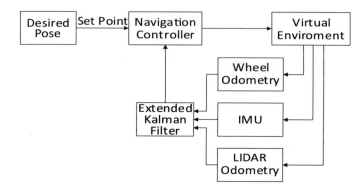

Fig. 11 Fusion of different odometry sources

be discussed here. One source is wheel odometry, that contributes mainly in linear displacement but at long distances it presents serious disturbances, especially in cases when the robot is located sideways, traveling near the equatorial zones of the sphere, where Gravity force acts as a massive disturbance. Other source is the 3DOF IMU that ensures a correction on linear displacement and angular displacement estimation. However, both methods are strongly disturbed by dynamic influences, like gravity and friction. The dual-LIDAR odometry corrects these disturbances and improves the robot's localization.

6 Experimental Results

Experimental tests are subdivided in two environments, each one with many runs. In the first test, it is applied a linear trajectory on planar environment to evaluate the proposed odometry with classical approaches. Second test is inside the virtual spherical tank aiming to evaluate the dynamic influences of gravity. Table 1 identify the used methods.

Table 1 Used odometry methods

Abbreviation	Meaning
Des	Desired pose
Sim	Virtual odometry calculated by virtual environment without errors
W	Wheels odometry based only in wheels encoders
W+I	EKF fusion of Inertial Measurement Unit (IMU) odometry and wheels odometry
W+I+L	EKF fusion of IMU odometry, wheels odometry and LIDAR odometry

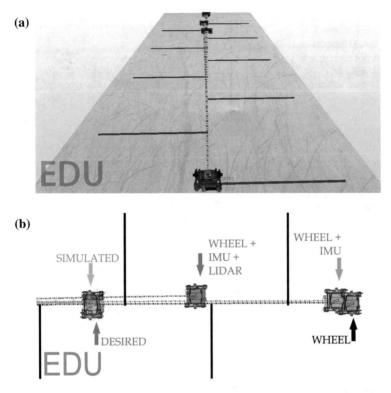

Fig. 12 Comparison between odometry methods in planar environment. **a** Planar environment. **b** Final pose of trajectory in planar environment

6.1 Experiments on Planar Environment

The initial experiment is performed on planar environment and the robot must track a linear path without any degree of angular motion (Fig. 12a). The trajectory tracking is run four times, each one with a different method of pose feedback. The first approach uses odometry directly from the virtual environment, without errors, labeled Simulated—no calculations, errors or estimations employed. The deviance from the original desired track that can be noticed on Fig. 12b comes from the robot's dynamic model, and it is considered normal. The second method performs the odometry using only wheel encoders. The fusion between wheel odometry and IMU is the third experiment. Finally, the trajectory tracking is performed via fusion of wheel odometry, IMU odometry and LIDAR odometry. Figure 12b shows a more detailed picture of the day point, as Fig. 12a brings the entire scenery of the first experiment.

The distance covered by four methods is presented in Fig. 13. Numerically, the fusion between IMU and wheel odometry has 1.75 m of error. However, the addition of the LIDAR odometry in EKF fusion reduces this error in 46 % (0.8 m of error). Figure 14 shows the normalized error of these methods.

Fig. 13 Trajectory tracking in planar environment. Follow Table 1 information about the abbreviations on the graph legends

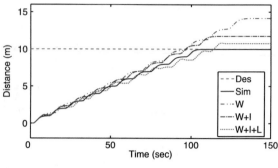

Fig. 14 Normalized errors in planar environment

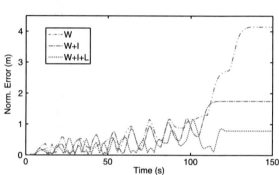

6.2 Experiments in LPG Spherical Tank

This experiment evaluates trajectory tracking inside spherical tanks. It is desired that the robot perform two turns around the tank with gravity acting sideways (Fig. 15). The same four feedback methods are tested.

The distance covered by four methods inside LPG tank is presented in Fig. 16, while Fig. 17 shows that the best performance is with the use of the LIDAR based odometry. To a path with two turns and 88 meters of distance, this method reduces normalized error of 24 meters (wheels + IMU) to 7.5 meters, meaning a reduction of about 68 %.

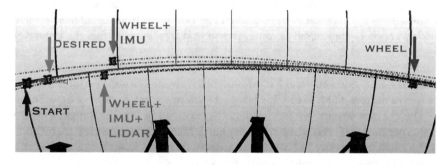

Fig. 15 Comparison between odometry methods in LPG spherical tank

Fig. 16 Trajectory tracking in spherical tank

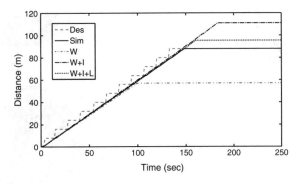

Fig. 17 Normalized errors in spherical tank

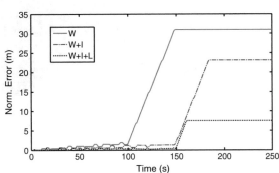

7 Navigation

In order to the robot accomplish its main navigating goal through a certain area on the tank surface, a navigation system is necessary, so it can avoid obstacles while traveling to its goals. The ROS navigation stack fits this needs perfectly, with some hardware requirements, such as a 2-D laser sensor, a moving base and a few data about the robot's geometry and dynamic behavior—like acceleration and top speed limits—it is possible to move the robot around without many problems.

To configure the AIR robot to this needs, some minor changes were made to its initial form. First, another LIDAR sensor was placed on the top-front position. A 3D camera from a Kinect sensor was also embedded on the robot for monitoring purposes, since it is not necessary to the Navigation Stack. This version of the AIR robot is shown on Fig. 18. The complete diagram showing the control loop is shown on Fig. 19. Each of this modules will be detailed on the forthcoming paragraphs.

Analyzing from the top part in Fig. 19, the first module called Switch reads a file with the goals—points on the trajectory that the robot must follow—and send these goals to one of two controllers, named Auxiliary and Navigation. This is done to circumvent the fact that ROS Navigation stack does not handle well a goal where this robot in particular must make a turn on its own z axis (sometimes called an *yaw* axis turn). So, depending on the current goal settings, the Switch module chooses the appropriate controller and sends it the goal.

Fig. 18 Diagonal view.
LIDAR sensor and 3D
camera are mounted on top
of the robot

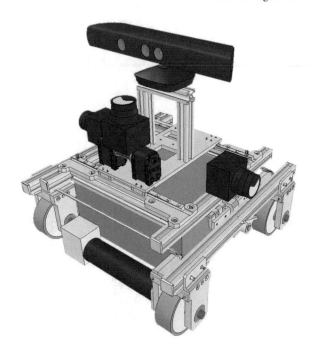

The Auxiliary controller will read the current odometry (*nav_msgs/Odometry*) and send *geometry_msgs/Twist* messages to the base controller. If the Navigation controller is the chosen one, it will also read the map from another block, called gmapping, and produce messages of the same type as Auxiliary controller to the base controller. The Navigation uses data from the map and from the 2D laser scan to assist the routing process by building 2D costmaps that tells the robot where are the obstacles, and where can it go without harm. Map construction and navigation at the same time will be explained on the next section.

Two costmaps are built to plan the trajectory, one local and one global. The local one takes care of the current destination obstacles, and the global costmap builds a larger route, in a smaller detail. The higher the cost of a given cell in the map, the harder it is to the robot move across this cell. Obstacles are marked as a impossible-to-cross cell, with the highest value. Figure 20 shows an example of both global and local costmaps, in gray lines. The pentagonal shape represents the robot.

The next block on the flowchart is the base controller, which works as an inverse kinematics calculation module, reading the desired speed to the robot as a whole and calculating each joint speed. This are then inputted on the virtual environment, where the main parts of the simulation takes place. Figure 21 shows a screenshot of this virtual environment (Fig. 21a) and a real spherical tank—both tanks have the same physical characteristics.

From the virtual environment, three modules are attached, IMU—that gathers data from the accelerometer and the gyroscope inside the virtual robot and publishes this data, the Wheel Odometry, connected to the encoders in each motor,

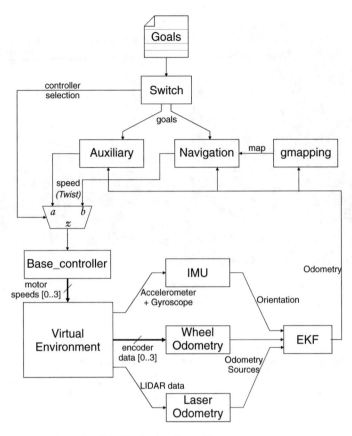

Fig. 19 Complete schematic of the Navigation Control Loop

Fig. 20 Example map. The gray borders over the black lines form the global costmap. Inside the dotted square, the local costmap takes place

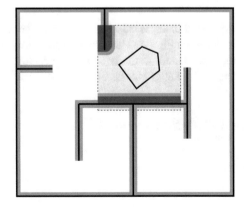

(a)   **(b)**

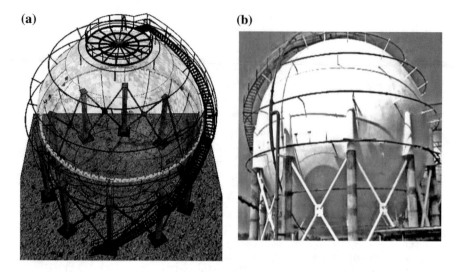

Fig. 21 Comparison between the two spherical tanks, virtual and real. **a** Spherical tank used on the virtual environment. **b** A real sphere, with the same dimensions of the virtual one

performing calculations somewhat similar to a forward kinematics, and keeping track of the robot's odometry, but it does not count slips on the wheels and this errors are cumulative—this is known as Dead Reckoning—, and lastly the LIDAR odometry, that has been covered in detail in the previous section.

7.1 Wheel Odometry

Wheel odometry is calculated based on the wheel's encoders. A very simple encoder is shown on Fig. 22. This disc-like structure is attached to the motor and for each full turn on the motor, the two beams are tampered sixteen times, making the encoder generate sixteen pulses. The encoder attached to the real motors has a resolution

Fig. 22 Simplified encoder schematic

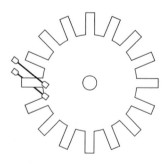

way higher, having 13.500 pulses for each wheel turn, given the gear ratio from the motor to the wheel itself and publish its informations on topic */to_odom*. Given the diameter of each wheel and the number of pulses per turn, it is possible to calculate how much the robot have moved in a given time.

This is known as Dead Reckoning because it does not take into account slips and drifts of the wheel on the environment. If the wheel slips and the robot gets stuck in a given place, the odometry will still count as if the robot was moving.

The odometry main code is the following (based on Ubuntu 14.04 and ROS Indigo):

```
#include "ros/ros.h"
#include "nav_msgs/Odometry.h"
#include "geometry_msgs/Twist.h"

nav_msgs::Odometry MyOdom;
ros::Time T;
ros::Publisher pub;

void to_odom_callback(const geometry_msgs::Twist::ConstPtr& msg){
ros::Duration difft = ros::Time::now() - T;
MyOdom.header.stamp = ros::Time::now();
MyOdom.header.frame_id = "/odom";
MyOdom.child_frame_id = "/base_link";

MyOdom.pose.pose.position.x = MyOdom.pose.pose.position.x + msg->linear.x * difft.toSec() / M_PI;
MyOdom.pose.pose.position.y = MyOdom.pose.pose.position.y + msg->linear.y * difft.toSec() / M_PI;
MyOdom.pose.pose.orientation.z = MyOdom.pose.pose.orientation.z + msg->angular.z * difft.toSec() / M_PI;

MyOdom.twist.twist.linear.x = msg->linear.x;
MyOdom.twist.twist.linear.y = msg->linear.y;
MyOdom.twist.twist.angular.z = msg->angular.z;
pub.publish(MyOdom);
T = ros::Time::now();
}

int main (int argc, char** argv) {

ros::init(argc, argv, "Odometry");
ros::NodeHandle n;
ros::Subscriber sub = n.subscribe("/to_odom", 1000, to_odom_callback);
pub = n.advertise<nav_msgs::Odometry>("/odom", 1000);

ros::Rate loop_rate(10);
T = ros::Time::now();

while(ros::ok()) {
ros::spinOnce();
loop_rate.sleep();
}
return 0;
}
```

7.2 EKF

The Extended Kalman Filter fuses the three sources of information about the robot movements. Its working principle is simple, conceptually speaking, and can be

Fig. 23 Simple diagram of
EKF working principle

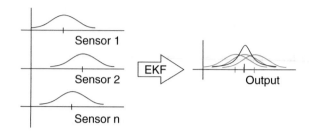

resumed like it is shown on Fig. 23. It operates as a probabilistic filter, and to each
sensor input there is a covariance attached. Based on this covariance, the EKF calculates how good is a given data source, so it can assign a weight to this source in the
fusion. Then, it fuses together all the data sources, taking into account their weights.
The result, in this case, is a Odometry message with a better estimation of where the
robot actually is. Figure 23 shows a simple diagram of a fusion between three data
sources, outputting a result with a better estimation, and a lower covariance attached
to it.

To this specific project, the EKF launch file is listed below.

```
<launch>
        <node pkg="robot_pose_ekf" type="robot_pose_ekf" name="robot_pose_ekf">
                <param name="output_frame" value="odom_combined"/>
                <param name="freq" value="10.0"/>
                <param name="sensor_timeout" value="1.0"/>
                <param name="odom_used" value="true"/>
                <param name="imu_used" value="true"/>
                <param name="vo_used" value="true"/>
                <param name="debug" value="true"/>
                <param name="self_diagnose" value="true"/>
        </node>
</launch>
```

7.3 Simultaneous Localization and Mapping—SLAM

Maps are valuable items when exploring new areas. The same concept of a human
map is applied here to the robot mapping system. When the robot is exploring a brand
new area and if the map is not available, one must be built on-the-fly, at the same time
when the navigation is running. This is done by the Gmapping block and it is called
Simultaneous Localization and Mapping—SLAM. The idea is, then again, simple in
concept. An empty-cell grid is populated by data coming from a 2D LIDAR sensor,
as shown on Fig. 24.

When the sensor detects an obstacle, the module traces where that obstacle is,
given the localization of the sensor itself and the distance and position of the detected
obstacle, and marks the obstacle in the map. In some cases, the empty space between
the robot and the obstacle is also marked on the map as free space, so it differs from

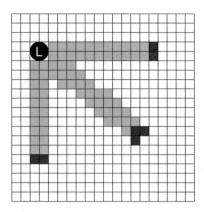

Fig. 24 Slam map building, based on the localization and obstacle detection by the LIDAR sensor, marked as "L". Light-gray areas are marked as free space, white areas are unknown space, dark-gray spots are obstacles

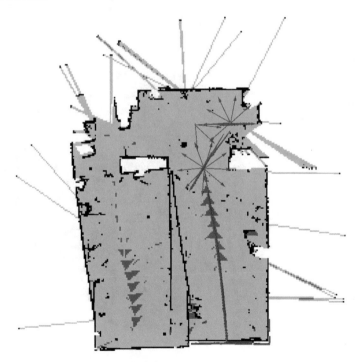

Fig. 25 SLAM example, built by the AIR robot

the unknown space (empty cells). A more realistic example of a SLAM built map is shown on Fig. 25, that represents the floor plan of the room where the AIR robot is being built and tested.

Fig. 26 Navigation
example, in a simplified
environment. As it is
possible to see, the robot
succeeded to arrive at it's
destination, avoiding
obstacles

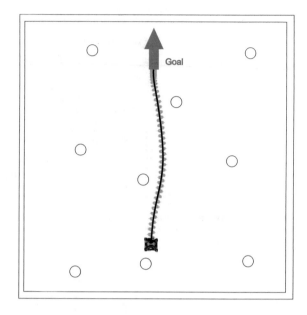

Fig. 27 2D path and
environment visualization,
using RVIZ software

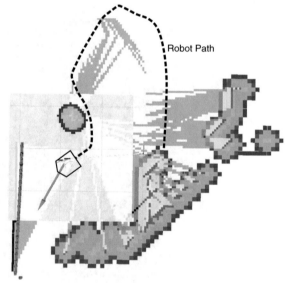

Using the SLAM package along with the Navigation one, it is possible to send
a specific goal to the robot and it will fetch this goal avoiding obstacles on its path.
This can be seen on a simplistic example, on Fig. 26 and on the virtual environment,
on Figs. 27 and 28.

Fig. 28 3D view of the same path, using VREP simulation software

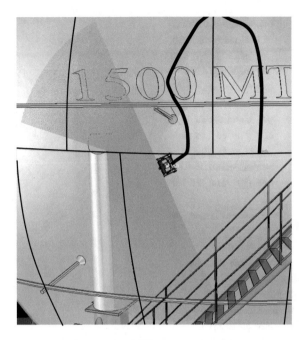

The launch file and configuration files for this mapping system are listed as follows.

- main launch file.

```
<launch>
        <node pkg="move_base" type="move_base" respawn="false" name="move_base" output="screen">
                <rosparam file="costmap_common_params.yaml" command="load" ns="global_costmap" />
                <rosparam file="costmap_common_params.yaml" command="load" ns="local_costmap" />
                <rosparam file="local_costmap_params.yaml" command="load" />
                <rosparam file="global_costmap_params.yaml" command="load" />
                <rosparam file="base_local_planner_params.yaml" command="load" />
        </node>
</launch>
```

– base_local_planner_params.yaml.

```
TrajectoryPlannerROS:
      acc_lim_x: 0.025
      acc_lim_theta: 0.01
      max_vel_x: 0.20
      min_vel_x: 0.01
      max_rotational_vel: 0.2
      min_in_place_rotational_vel: 0.18
```

```
holonomic_robot: false
xy_goal_tolerance: 0.250
yaw_goal_tolerance: 0.20
sim_granularity: 0.025
vx_samples: 10
vtheta_samples: 20
goal_distance_bias: 10
path_distance_bias: 22
occdist_scale: 0.01
heading_lookahead: 0.325
oscillation_reset_dist: 0.08
```

– costmap_common_params.yaml

```
obstacle_range: 3.0
raytrace_range: 4.0
inflation_radius: 0.3
transform_tolerance: 1
footprint: footprint_padding: 0.03
controller_patience: 2.0
observation_sources: laser_scan_sensor
laser_scan_sensor: {
sensor_frame: /Hokuyo_ROS,
data_type: LaserScan,
topic: /vrep/front_scan,
marking: true,
clearing: true}
```

– global_costmap_params.yaml

```
global_costmap:
    global_frame: /map
    robot_base_frame: base_footprint
    update_frequency: 5
    publish_frequency: 1
    static_map: false
    origin_x: -10
    origin_y: -10
    width: 20.0
    height: 20.0
    resolution: 0.1
```

– local_costmap_params.yaml
```
local_costmap:
    global_frame: odom
```

```
robot_base_frame: base_footprint
update_frequency: 10.0
publish_frequency: 2.0
static_map: false
rolling_window: true
width: 3.0
height: 3.0
resolution: 0.05
origin_x: -1.50
origin_y: -1.50
```

8   Conclusions

This paper presented a localization method to a climbing robot to be used in inspection of liquefied petroleum gas storage tanks. LPG spheres are fully dark, with no easily detectable landmarks and no different faces. The unique reference inside sphere are weld beads and these are hard to detect because they are only a few millimeters high.

An odometry based on dual-LIDAR sensor is proposed in order to improve the classical odometry calculation through inertial measurement unit and wheel encoders. The proposed localization estimation used two LIDAR sensors placed each robot side to identify the weld beads. Sensor accuracy is expanded with application of some filters and a Support Vector Machine which classify data coming from the LIDAR signals as weld beads and noises. Odometry is determined according to motion of weld beads in relation of robot.

Data fusion between wheels odometry, IMU and LIDAR odometry is made through extended Kalman filter to improve localization estimation. The result is a better accuracy of robot localization through minimal landmarks and without light influence.

9   Compliance with Ethical Standards

Conflict of Interest: The authors declare that they have no conflict of interest.

Acknowledgments This project was partially funded by Brazil's National Counsel of Technological and Scientific Development (CNPq), Coordination for the Improvement of Higher Level People (CAPES) and the National Agency of Petroleum, Natural Gas and Biofuels (ANP) together with the Financier of Studies and Projects (FINEP) and Brazil's Ministry of Science and Technology (MCT) through the ANPs Human Resources Program for the Petroleum and Gas Sector - PRH-ANP/MCT PRH10-UTFPR.

References

1. W. Guan, Y. Tao, H. Cheng, C. Ma, and P. Guo, Present status of inspection technology and standards for large-sized in-service vertical storage tanks, in *ASME 2011 Pressure Vessels and Piping Conference* (American Society of Mechanical Engineers, 2011), pp. 749–754
2. C. Pfitzner, W. Antal, P. Hess, S. May, C. Merkl, P. Koch, R. Koch, M. Wagner, 3d multi-sensor data fusion for object localization in industrial applications, in *ISR/Robotik 2014; Proceedings of 41st International Symposium on Robotics*, (June 2014), pp. 1–6
3. A. Corominas Murtra, J. Tur, Imu and cable encoder data fusion for in-pipe mobile robot localization, in *Technologies for Practical Robot Applications (TePRA), 2013 IEEE International Conference on* (April 2013), pp. 1–6
4. N. Ganganath, H. Leung, Mobile robot localization using odometry and kinect sensor, in *Emerging Signal Processing Applications (ESPA), 2012 IEEE International Conference on*, (Jan 2012), pp. 91–94
5. T. Whelan, H. Johannsson, M. Kaess, J. Leonard, J. McDonald, Robust real-time visual odometry for dense rgb-d mapping, in *Robotics and Automation (ICRA), 2013 IEEE International Conference on* (May 2013), pp. 5724–5731
6. C. Siagian, C.-K. Chang, and L. Itti, Mobile robot navigation system in outdoor pedestrian environment using vision-based road recognition, in *2013 IEEE International Conference on Robotics and Automation (ICRA)* (May 2013), pp. 564–571
7. H. Azartash, N. Banai, T.Q. Nguyen, An integrated stereo visual odometry for robotic navigation. Robot. Auton. Syst. **62**(4), 414–421 (2014)
8. V. Malyavej, W. Kumkeaw, M. Aorpimai, Indoor robot localization by rssi/imu sensor fusion, in *Electrical Engineering/Electronics, Computer, Telecommunications and Information Technology (ECTI-CON), 2013 10th International Conference on* (May 2013), pp. 1–6
9. K. Saadeddin, M. Abdel-Hafez, M. Jarrah, Estimating vehicle state by gps/imu fusion with vehicle dynamics. J. Intell. Robot. Syst. **74**(1–2), 147–172 (2014)
10. R.V. Espinoza, A.S. de Oliveira, L.V.R. de Arruda, F.N. Junior, Navigations stabilization system of a magnetic adherence-based climbing robot. J. Intell. Robot. Syst. **78**, 65–81 (2015)
11. J. Simanek, M. Reinstein, V. Kubelka, "Evaluation of the ekf-based estimation architectures for data fusion in mobile robots. IEEE/ASME Trans. Mechatron. **99**, 1–6 (2014)
12. J. Knuth, P. Barooah, Error growth in position estimation from noisy relative pose measurements. Robot. Auton. Syst. **61**(3), 229–244 (2013)
13. S. Hu, C. Chen, A. Zhang, W. Sun, L. Zhu, A small and lightweight autonomous laser mapping system without gps. J. Field Robot. **30**(5), 784–802 (2013)
14. S.A. Hiremath, G.W. van der Heijden, F.K. van Evert, A. Stein, C.J. ter Braak, Laser range finder model for autonomous navigation of a robot in a maize field using a particle filter. Comput. Electron. Agric. **100**, 41–50 (2014)
15. H. Surmann, A. Nuchter, J. Hertzberg, An autonomous mobile robot with a 3d laser range finder for 3d exploration and digitalization of indoor environments. Robot. Auton. Syst. **45**(3–4), 181–198 (2003)
16. H. Dong, T. Barfoot, Lighting-invariant visual odometry using lidar intensity imagery and pose interpolation, in *Field and Service Robotics* eds. by K. Yoshida and S. Tadokoro, Springer Tracts in Advanced Robotics, vol. 92 (Springer, Berlin, 2014), pp. 327–342
17. G. Fu, P. Corradi, A. Menciassi, P. Dario, An integrated triangulation laser scanner for obstacle detection of miniature mobile robots in indoor environment. IEEE/ASME Trans. Mechatron. **16**, 778–783 (2011)
18. H. Kloeden, D. Schwarz, E.M. Biebl, R.H. Rasshofer, Vehicle localization using cooperative rf-based landmarks, in *Intelligent Vehicles Symposium (IV), 2011 IEEE* (IEEE, 2011), pp. 387–392
19. R.V. Espinoza, J.P.B. Nadas, A.S. de Oliveira, L.V.R. de Arruda, N.F, Adhesion force control and active gravitational compensation for autonomous inspection in lpg storage spheres, in *Robotics Symposium and Latin American Robotics Symposium (SBR-LARS), 2012 Brazilian* (IEEE, 2012), pp. 232–238

Part III
Service and Experimental Robots

People Detection, Tracking and Visualization Using ROS on a Mobile Service Robot

Timm Linder and Kai O. Arras

Abstract In this case study chapter, we discuss the implementation and deployment of a ROS-based, multi-modal people detection and tracking framework on a custom-built mobile service robot during the EU FP7 project SPENCER. The mildly humanized robot platform is equipped with five computers and an array of RGB-D, stereo and 2D laser range sensors. After describing the robot platform, we illustrate our real-time perception pipeline starting from ROS-based people detection modules for RGB-D and 2D laser data, via nodes for aggregating detections from multiple sensors, up to person and group tracking. For each stage of the pipeline, we provide sample code online. We also present a set of highly configurable, custom RViz plugins for visualizing detected and tracked persons and groups. Due to the flexible and modular structure of our pipeline, all of our components can easily be reused in custom setups. Finally, we outline how to generate test data using a pedestrian simulator and Gazebo. We conclude with quantitative results from our experiments and lessons that we learned during the project. To our knowledge, the presented framework is the functionally most complete one that is currently available for ROS as open-source software.

Keywords People detection · People tracking · Group tracking · Perception · Service robot · Mobile robot · Sensors · Visualization

T. Linder (✉) · K.O. Arras
Social Robotics Laboratory, University of Freiburg,
Georges-Köhler-Allee 074, 79110 Freiburg im Breisgau, Germany
e-mail: linder@cs.uni-freiburg.de
URL: http://github.com/spencer-project

K.O. Arras
e-mail: arras@cs.uni-freiburg.de

© Springer International Publishing Switzerland 2016
A. Koubaa (ed.), *Robot Operating System (ROS)*, Studies in Computational
Intelligence 625, DOI 10.1007/978-3-319-26054-9_8

1 Introduction

In this chapter, we discuss our experiences with the implementation and deployment of a ROS-based people detection and tracking framework on a complex, custom mobile service robot platform equipped with a large array of sensors.

Contributions of the Book Chapter

One key contribution of this chapter is a set of reusable message definitions for a people and group detection and tracking pipeline. In our research, we demonstrated that these definitions can successfully be applied across different sensor modalities and real-time detection and tracking algorithms, for which we provide exemplary implementations. Based upon these message definitions, we also provide a reusable and highly configurable set of visualization plugins for the standard ROS visualization tool, RViz. To our knowledge, the resulting people detection and tracking framework is the functionally most complete one that is publicly available for ROS at present.

The code for the detection and tracking modules, as well as our message definitions and visualization plugins, can be found online[1] in a Git repository, mostly under a BSD license. The content of this repository needs to be cloned into a ROS workspace and built using `catkin_make`. An up-to-date list of dependencies can be found in the accompanying Readme file. Our components have been tested on ROS Hydro and Indigo on 64-bit Ubuntu systems.

2 Background of the SPENCER Project

The aim of the EU FP7 research project SPENCER is to develop algorithms for service robots that can guide groups of people through highly dynamic and crowded pedestrian environments, such as airports or shopping malls, while behaving in a socially compliant manner by e. g. not crossing in between families or couples. Possible situations that such a robot could encounter are visualized in Fig. 1. To this end, robust and computationally efficient components for the perception of humans in the robot's surroundings need to be developed.

2.1 Robot Hardware and Sensory Setup

As none of the commercially available robot platforms offered the computational and sensory capabilities required for the research in SPENCER, a custom mobile robot was developed by an industrial partner within the project. The differential-drive robot

---

[1] https://github.com/spencer-project/spencer_people_tracking.

Fig. 1 Typical situations encountered by a mobile service robot in crowded pedestrian environments, such as shopping malls or airports. To be able to behave in a socially compliant way while driving, by for instance not crossing through a group, the robot needs to gain a precise understanding of the persons in its environment

Fig. 2 *Left* Picture and renderings of the SPENCER robot platform. *Right* The custom-built mobile data capture platform, including a custom odometry solution using quadrature encoders in dynamo housings on both main wheels, connected to a microcontroller (*bottom right picture*). Both platforms contain numerous ROS-based software components distributed over multiple computers

platform built for SPENCER, shown in Fig. 2 (left), is around 2 m tall and equipped with two onboard Intel Core i7-3520M machines for planning and navigation, an i3-2120 machine for interaction, two i7-4700MQ gaming laptops with nVidia GeForce GTX 765M for perception and one embedded PowerPC system for low-level motion control. All systems communicate over a gigabit ethernet connection and, except for the embedded system, run ROS Indigo on Ubuntu 14.04.

The robot is equipped with two front- and two rear-looking RGB-D sensors, a front-facing stereo camera system and a pair of 2D laser scanners offering, when combined, a 360° coverage. For human-robot interaction, a touchscreen and a boarding pass reader have been integrated. Custom ROS wrappers for communication with the embedded system and the robot's head joints were developed as part of the project.

Mobile Data Capture Platform To obtain groundtruth data for the design, training and testing of new algorithms, we also built a mobile data capture platform (Fig. 2, right) consisting of an off-the-shelf bicycle trailer equipped with sensors in a similar configuration as on the actual robot. Wheel odometry for localization is obtained using a custom microcontroller solution that communicates with a ROS node on a laptop. For data recording using *rosbag*, we used 3 high-end laptops with fast SSDs. Overall, using this platform, we captured over 1.4 TB of data at a major European airport to train our perception components.

Fig. 3 People and group tracking pipeline developed during the SPENCER project

2.2 People Tracking Pipeline

Figure 3 shows the real-time people and group detection and tracking pipeline developed in the context of the SPENCER project.

Starting with sensory data such as 2D laser scans, RGB-D point clouds or stereo or monocular camera images, we first detect people using detectors devised specifically for the particular sensor modality (Sect. 3). The resulting person detections are then fed into a person tracker (Sect. 4), which integrates the information over time and attempts to maintain a consistent ID for a given person for as long as possible. A simple approach for tracking entire groups of persons by estimating their spatial and social relations is presented in Sect. 5.

All of this becomes more complex if information from multiple detectors operating on different sensor modalities, such as 2D laser and RGB-D, shall be combined to make tracking more robust and cover a larger field of view (Sect. 6). Finally, using the powerful RViz visualization tool and custom plugins developed by us as well as a custom SVG exporter script, the outputs of the tracking pipeline can be visualized (Sect. 7). In Sect. 8, we briefly outline how to generate test data for experiments using a pedestrian simulator. Finally, we show qualitative results and discuss runtime performance in Sect. 9.

The entire communication between different stages of our pipeline occurs via ROS messages defined in the corresponding sections, which we explain in detail to encourage reuse of our components in custom setups. The architecture allows for easy interchangeability of individual components in all stages of the pipeline.

3 People Detection

In this section, we present ROS message definitions and exemplary implementations for detecting people in RGB-D and 2D laser range data. We start by a short summary of existing research in this field.

While still relatively expensive, 2D laser range finders offer data at high frequencies (up to 100 Hz) and cover a large field of view (usually around 180°). Due to the sparseness, the sensor data is very cheap to process, but does not offer any appearance-based cues that can be used for people detection or re-identification. Early works based on 2D laser range data detect people using ad-hoc classifiers that find local minima in the scan [10, 27]. A learning approach is taken by Arras et al. [2], where a classifier for 2D point clouds is trained by boosting a set of geometric and statistical features.

In RGB-D, affordable sensors are gaining popularity in many indoor close-range sensing scenarios since they do not require expensive disparity map calculations like stereo cameras and usually behave more robustly in low lighting conditions. Spinello and Arras [28] proposed a probabilistically fused HOD (histogram of oriented depths) and HOG (histogram of oriented gradients) classifier. Munaro et al. [21] present a person detector in RGB-D that uses a height map-based ROI extraction mechanism and linear SVM classification using HOG features, which has been integrated into the Point Cloud Library (PCL). Jafari et al. [12] at close range also use a depth-based ROI extraction mechanism and evaluate a normalized depth template on the depth image at locations where the height map shows local maxima corresponding to heads of people. For larger distances, a GPU-accelerated version of HOG is used on the RGB image.

3.1 ROS Message Definitions

In the following, we present our message definitions for person detection. Due to the multi-language support of ROS, the resulting messages can be processed by any kind of ROS node, regardless if implemented in C++, Python, Lua or Lisp. We want to emphasize that our message definitions are intentionally kept as simple and generic as possible, to allow reuse over a wide range of possible sensor modalities and detection methods. Therefore, for instance, we do not include image bounding boxes or visual appearance information which would be specific to vision-based approaches and not exist e.g. in 2D laser range data.

Our definition of a detected person is similar to a geometry_msgs/PoseArray, but additionally, for each detection we also specify a unique detection ID, a confidence score, a covariance matrix and the sensor modality. The detection ID allows the detection to be matched against e.g. the corresponding image bounding box, published under the same detection ID on a separate ROS topic. The covariance matrix expresses the detector's uncertainty in the position (and orientation, if known). It could, for instance, be a function of the detected person's distance to the sensor, and is therefore not necessarily constant. The confidence score can be used to control track initialization and to compute track scores. Finally, information about the modality can be used by a tracking algorithm to, for example, assign a higher importance to visually confirmed targets.

A spencer_tracking_msgs/**DetectedPerson** thus has the following attributes:

- *detection_id* [uint64]: Unique identifier of the detected person, monotonically increasing over time. To ensure uniqueness, different detector instances, should use distinct ID ranges, by e.g. having the first detector issue only IDs that end in 1, the second detector IDs that end in 2, and so on.
- *confidence* [float64]: A value between 0.0 and 1.0 describing the confidence that the detection is a true positive.

Fig. 4 *Left* Area visible to the front laser scanner (in *grey*), candidate laser scan segments (with numbers that identify the segments), and laser-based person detections (*orange boxes*). *Right* Projection of the laser-based person detections into a color image of the scene. Segment 72 on the left border is a false negative classification. As the 2D laser scanner has a very large horizontal field of view (190°), the camera image does not cover the entire set of laser scan segments (e.g. segment 24 on the *right*)

- *pose* [geometry_msgs/PoseWithCovariance]: Position and orientation of the detection in metric 3D space, along with its uncertainty (expressed as a 6 × 6 covariance matrix). For unknown components, e.g. position on the z axis or orientation, the corresponding elements should be set to a large value.[2] The pose is relative to the coordinate frame specified in the DetectedPersons message (see below).
- *modality* [string]: A textual identifier for the modality or detection method used by the detector (e.g. RGB-D, 2D laser). Common string constants are pre-defined in the message file.

A **DetectedPersons** message aggregates all detections found by a detector in the same detection cycle, and contains the following attributes:

- *header* [std_msgs/Header]: Timestamp and coordinate frame ID for these detections. The timestamp should be copied from the header of the sensor_msgs/LaserScan, Image or PointCloud2 message that the detector is operating on. Similarly, the coordinate frame can be a local sensor frame as long as a corresponding transformation into the tracking frame (usually "odom") exists in the TF hierarchy.
- *detections* [array of DetectedPerson]: The array of persons that were detected by the detector in the current time step.

3.2 Person Detection in 2D Laser Data

Boosted Laser Segment Classifier For people detection in 2D laser range data (Fig. 4), we use a re-implementation of [2] using an Ada-Boost implementation from the OpenCV library. The classifier has been trained on a large, manually annotated

---

[2]We usually use a value of 10^5 to indicate this. If set to infinity, the covariance matrix becomes non-invertible, causing issues later on during tracking.

data set consisting of 9535 frames captured using our mobile data capture platform in a pedestrian zone, with the laser scanner mounted at about 75 cm height. We used an angular scan resolution of 0.25° and annotated detections up to a range of 20 m.

Prior to classification, the laser scan is segmented by a separate ROS node using either jump distance or agglomerative hierarchical clustering with a distance threshold of 0.4 m. The combined laser-based segmentation and detection system is launched via the command-line

```
roslaunch srl_laser_detectors
    adaboost_detector_with_segmentation.launch
```

and expects sensor_msgs/LaserScan messages on the /laser topic and publishes the resulting detections at /detected_persons. The names of these topics can be reconfigured via parameters passed to the launch file. A short example dataset for testing, and instructions on how to play back this recorded data, can be found online in our README file.

Leg Detector We also provide a wrapper to make the existing leg_detector ROS package compatible with our message definitions. This package needs to be downloaded and built separately.[3] The underlying algorithm uses a subset of the 2D features also included in our implementation of [2], but first tracks both legs separately and in our experience works best if the laser sensor is mounted very close to the ground, below 0.5 m height.

3.3 Person Detection in RGB-D

Upper-Body Detector We modified the publicly available close-range upper-body detector by [12] (Fig. 5, left) to also output DetectedPersons messages. It operates purely on depth images and is launched via:

```
roslaunch rwth_upper_body_detector upper_body_detector.launch
```

The input and output topics can be configured via the camera_namespace and detected_persons parameters. It is assumed that a ground plane estimate is published on the topic specified by the ground_plane parameter, which can e.g. be achieved using the ground_plane_fixed.launch file provided in the rwth_ground_plane package.

Further RGB-D Detectors In the pcl_people_detector package, we also integrated the RGB-D person detector by [21] from the Point Cloud Library (PCL), which extracts regions of interest from a depth-based height map and then applies a linear HOG classifier. We extended the code to output markers for visualization of

---

[3]http://wiki.ros.org/leg_detector.

Fig. 5 Different RGB-D detectors which we integrated into our framework. *Left* Upper-body detector from [12] which slides a normalized depth template over the depth image. *Middle* RGB-D detector from PCL which first extracts regions of interest, visualized here by boxes, from the point cloud [21]. *Right* Combo-HOD (histogram of oriented depths) detector [28] (closed-source)

ROIs (Fig. 5, middle) and output DetectedPersons messages. Configurable parameters are documented in a launch file that can be started via:

```
roslaunch pcl_people_detector start.launch
```

Likewise, as a proof of concept, a closed-source implementation of Combo-HOD (histogram of oriented depths) [28] was integrated (Fig. 5, right).

3.4 Person Detection in Monocular Vision

groundHOG We also integrated the GPU-accelerated medium- to far-range ground-HOG detector from [12] into our framework. It is configured in a similar way as the upper-body detector described above, and also requires a groundplane estimate to narrow down the search space for possible person detections. On a computer with a working CUDA installation, it can be launched via:

```
roslaunch rwth_ground_hog ground_hog_with_GP.launch
```

As can be seen from these examples, it is very easy to integrate new detectors into our people tracking framework. This makes it possible to easily swap out detectors in experiments, as well as to draw upon a common visualization toolkit for displaying the detections along with raw sensor data in 3D space (Sect. 7).

4 People Tracking

The output of a person detector just represents a single snapshot in time and may be subject to false alarms and missed detections. To gain a more long-term understanding of the scene, to filter out spurious misdetections and to extract trajectory information and velocity profiles, the detected persons need to be associated over time, a process called *people tracking*. The goal of a people tracking system is to maintain a persistent identifier for the same person over time, as long as the person remains visible in the scene and while bridging short moments of occlusion.

Different multi-target data association and tracking algorithms have been studied in the context of person tracking. Simpler algorithms such as the Nearest-Neighbor Standard Filter (NNSF), Global Nearest Neighbor (GNN) [4, 22] or Nearest-Neighbor Joint Probabilistic Data Association (NNJPDA) [3] make hard data association decisions, whereas other variants including the Joint Probabilistic Data Association Filter (JPDAF) [27] use soft assignments. All of these methods are single-hypothesis algorithms that at the end of each tracking cycle, only keep the most likely data association hypothesis in memory. In contrast, multi-hypothesis tracking approaches (MHT) [1, 8, 12, 25] use a hypothesis tree which allows to correct wrong initial decisions at a later point in time. This, however, comes at the price of higher computational complexity and complex implementation.

4.1 ROS Message Definitions

As in the detection case, we tried to keep the message definitions for people tracking as generic as possible such that they can be re-used across a large variety of different people tracking algorithms. In the following sections, we illustrate concrete implementations that we have experimented with during the project, all of which yield the following output.

A **TrackedPerson**, according to our definition, possesses the following attributes:

- *track_id* [uint64]: An identifier of the tracked person, unique over time.
- *is_matched* [bool]: False if no matching detection was found close to the track's predicted position. If false for too long, the track usually gets deleted.
- *is_occluded* [bool]: True if the person is physically occluded by another person or obstacle. False if the tracking algorithm cannot determine this.
- *detection_id* [uint64]: If is_matched is true, the unique ID of the detection associated with the track in the current tracking cycle. Otherwise undefined.
- *pose* [geometry_msgs/PoseWithCovariance]: The position and orientation of the tracked person in metric 3D space, along with its uncertainty (expressed as a 6×6 covariance matrix). The pose is relative to the coordinate frame specified in the TrackedPersons message (see below).
- *twist* [geometry_msgs/TwistWithCovariance]: The linear and possibly angular velocity of the track, along with their uncertainties.
- *age* [duration]: The time span for which this track has existed in the system.

The **TrackedPersons** message aggregates all TrackedPerson instances that are currently being tracked:

- *header* [std_msgs/Header]: The coordinate frame is usually a locally fixed frame that does *not* move with the robot, such as "odom". The timestamp is copied from the incoming DetectedPersons message; this is important to synchronize DetectedPersons and TrackedPersons messages to be able to look up details about a DetectedPerson identified by its detection_id.
- *tracks* [array of TrackedPerson]: All persons that are being tracked in the current tracking cycle, including occluded and unmatched tracks.

4.2   People Tracking Algorithms Used in Our Experiments

During the SPENCER project, we have conducted experiments with different person tracking algorithms of varying complexity. Next to a computationally cheap nearest-neighbor algorithm which we will describe in the following section, we converted an existing hypothesis-oriented multi-hypothesis tracker from our past work [1, 17] into a ROS package[4] to exploit the visualization capabilities and interchangeability of detectors of our framework. Further experiments were conducted using a vision-based track-oriented multi-hypothesis tracker[5] [12].

A detailed study of which person tracking algorithm to choose under which circumstances is subject of our ongoing research. We believe that in general, the choice of models (e.g. for motion prediction, occlusion handling, inclusion of visual appearance) has a greater influence on overall tracking performance than the tracking algorithm itself. Here, various methods have been proposed in the past, e.g. the use of a motion model influenced by social forces [19], adapting occlusion probabilities and motion prediction within groups [17], improved occlusion models [16], or an online-boosted RGB-D classifier that learns to tell apart tracks from background or other tracked persons [18, 22].

4.3   Example: Nearest-Neighbor Tracker

As an illustrative example, we show how a person tracking system can be implemented using a very simple nearest-neighbor standard filter, for which the full code is provided online. Combined with the extensions discussed in the following section, the system can in most situations already track persons more robustly than other publicly available ROS-based people tracking implementations that we are aware of (e.g. in wg-perception/people).

The nearest-neighbor data association method greedily associates a detection with the closest (existing) track, if that track's predicted position is closer than a certain

---

[4]Not (yet) publicly available due to open questions on licensing.

[5]ROS version to be made available in the rwth_pedestrian_tracking package.

gating threshold. While not as robust as more advanced methods like multi-hypothesis tracking (MHT), it is very simple to implement and real time-capable even with a high number of tracks (up to 40–80). In our experience, such single-hypothesis methods deliver satisfactory results if there are not too many closely spaced persons in the scene and the detector has a relatively low false alarm rate and high recall, which can also be achieved by fusing information from multiple detectors in different modalities (Sect. 6).

A new tracking cycle starts once a new, and potentially empty, set of Detected-Persons is received by the tracker from the detector. The resulting actions that are performed are summarized in the following ROS callback:

People Tracking Workflow

```
1  void newDetectedPersonsCallback(DetectedPersons& detections)
2  {
3    transformIntoFixedFrame(detections);
4    predictTrackStates();
5    predictMeasurements();
6    Pairings pairings = performDataAssociation(m_tracks, detections);
7    updateTrackStates(pairings);
8    initNewTracks(detections, pairings);
9    deleteObsoleteTracks();
10 }
```

First, in line 3, all detections are transformed into the locally fixed "odom" coordinate frame, by looking up the robot's current position in the world as reported by odometry using the tf library. The states of all existing tracked persons—internally represented as a 4D vector $(x, y, \dot{x}, \dot{y})$ of position and velocity—are predicted into the future (line 4), e.g. using a Kalman Filter, and converted into measurement predictions by dropping their velocity components (line 5). Then, an association matrix is constructed which contains the distance between predicted and real measurement position for each possible pairing of detections and tracks. This usually involves a gating step in which incompatible pairings are discarded beforehand. The nearest-neighbor standard filter then starts searching for the pairing with minimum distance until all tracks and/or detections have been associated. In line 7, the states of all matched tracks are updated with the position of the associated detection, or in the case of an unmatched track, the state prediction is used as the new state. Finally, new tracks are initialized from detections that could not be associated with any track (line 8). Tracks that have not been associated with a detection for too long (i.e. occluded) are deleted.

4.4 Improving Robustness of Tracking

For more robust tracking of maneuvering persons, we extended this tracking method with an interacting multiple models (IMM) approach that dynamically mixes four different motion models depending on the evolution of a tracked person's trajectory and the prediction error: (1) A constant velocity (CV) model with low process noise, (2) a CV model with high process noise, (3) a Brownian motion model, (4) a coordinated-turn model. In lab experiments, this combination helped to reduce the number of track losses caused by abruptly stopping or turning persons, without us having to manually tune process noise levels, especially in smaller or very crowded environments where motions are less linear.

Our implementation also optionally subscribes to a second DetectedPersons ROS topic, on which detections of a high-recall, low-confidence detector (e.g. a primitive laser blob detector) can be published. After regular data association, we perform another round of data association on these detections only with tracks that have not yet been associated with a detection (from the high-precision detector). This simple modification can improve tracking performance significantly if the main detector has a low sensitivity.

Finally, an implementation of a track initialization logic as described in [6] helps to prevent false track initializations in case of spurious misdetections.

Launching the People Tracker

The nearest-neighbor tracker implementation comes with a launch file configured with parameters that yielded good results in our test cases. We recommend the reader to copy this launch file and use it as a starting point for own experiments:

```
roslaunch srl_nearest_neighbor_tracker nnt.launch
```

4.5   Tracking Metrics

For performance evaluations, we wrapped two publicly available Python implementations of the CLEAR MOT [5] and OSPA [26] multi-target tracking metrics into ROS nodes that follow our message conventions. These metrics are helpful for tuning parameters of the tracking system, as well as evaluating completely new tracking algorithms. They assume that annotated groundtruth tracks are given in the form of time-synchronized TrackedPersons messages, and can be found in the `spencer_tracking_metrics` package.

5   Group Tracking

For socially compliant navigation, which is the main research objective in the SPENCER project, tracking just individual persons is not sufficient. If the task of the robot is to guide a group of persons through a shopping mall or an airport, or to avoid crossing through groups of people, knowledge of groups in the surroundings is important. In the latter case, we regard as a group a closely spaced formation of standing or walking persons. In the former case, a more long-term definition of *group* might be necessary, in which a person can, for instance, briefly leave the formation to avoid opposing traffic before re-joining the group.

Group tracking has been studied before in the computer vision community on photos and movies [7, 9, 29]. Methods have also been developed for group tracking using overhead cameras in the context of video surveillance [14, 23, 24, 30]. For people tracking on mobile robots in a first-person perspective, a multi-model multi-hypothesis tracker for groups has been proposed by Lau et al. [13], which was extended by Luber and Arras [17] and Linder and Arras [15].

In the following, we outline a few basic principles for online group detection and tracking, and describe sample code that we provide online.

5.1 Social Relation Estimation

Before detecting groups, we first estimate the pairwise social relations of all tracked persons by feeding the output of the people tracker into a social relation estimation module. This module, based upon the work described in [15], builds a social network graph as shown in Fig. 6 (left), where the edge weights encode the likelihood of a positive social relation between a pair of persons. To estimate these likelihoods, we rely on *coherent motion indicators*, which are motion-related features that were found to indicate group affiliation between people in large-scale crowd behavior analysis experiments [20]. They consist of relative spatial distance, difference in velocity and difference in orientation of a given pair of tracks. Using a probabilistic SVM classifier trained on a large dataset annotated with group affiliations, these features are mapped to a social relation probability that indicates the strength of the relation. This method allows for an easy integration of additional social cues (such as persons' eye gaze, body pose) in the future. An implementation and a trained SVM model are provided in the `spencer_social_relations` package. Figure 7 shows the

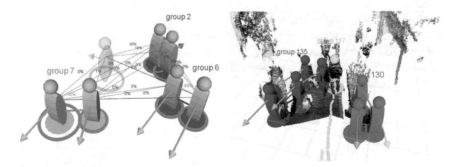

Fig. 6 *Left* Example of a social network graph, where the edge weights encode the likelihood of a positive social relation between persons. Colored circles below each person indicate resulting group affiliations and *grey* covariance ellipses visualize position uncertainty. *Right* Three different groups that are being tracked in RGB-D

Fig. 7 Message definitions for the social network graph. The strength of a social relation is specified as a real number between 0.0 and 1.0. The type string can be used to distinguish between different types of relations, e.g. spatial, family or romantic

corresponding SocialRelation and SocialRelations ROS messages that constitute the resulting social network graph.

5.2 Group Detection and Tracking

To detect groups, we perform graph-cutting in the social network graph by removing all edges below a fixed edge weight threshold. The remaining connected components of the graph then become the resulting groups.

A simple group tracking algorithm can now be implemented by treating groups as separate entities, and tracking their centroids. This data association might fail if the group splits apart very suddenly, causing a large shift in the resulting new centroids. Alternatively, it is possible to track just the composition of a group, in terms of its individual group members, and then keep a memory of previously seen group configurations. This works as long as at least one person in the group is always visible. A simple Python-based implementation of this approach can be found in the `spencer_group_tracking` package. The output is in the form of Tracked-Groups messages, as outlined in Fig. 8.

In ongoing research shown in Fig. 6 (right), we use a more complex approach which interleaves a group-level data association layer with regular person-level data association in a multi-model multi-hypothesis tracker [15, 17]. By explicitly modelling group formation in terms of merge and split events, groups can be tracked more robustly. Internally, a prototype of this system has already been integrated with our visualization and detection framework and yielded good experimental results. While currently not publicly available, this system uses exactly the same message interfaces as the provided Python code.

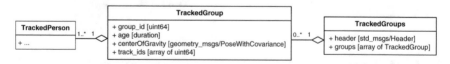

Fig. 8 Message definition for a single tracked group, a collection of all tracked groups in one cycle, and their relation towards tracked persons

6 Multi-Modal Tracking

In this section, we describe how we can fuse the output of multiple detectors, potentially operating in different sensor modalities, to make people tracking more robust and extend the robot's field of view. A broad overview of different strategies for fusing multi-modal detections for general tracking purposes is given in [4]. Here, we mainly focus on fusion at the detection level, as track-to-track fusion involves associating two or more trajectories with each other, which is computationally more complex and not straightforward. Our framework allows for two different fusion strategies at the detection level, namely detection-to-detection and detection-to-track fusion.

6.1 ROS Message Definitions

To keep a memory of which detections have been fused together, we define a CompositeDetectedPerson message (Fig. 9). Retaining this information is essential for components at later stages in the perception pipeline, like a human attribute classifier that takes as an input the regions of interest found by a vision-based person detector. If these ROIs likewise bear a corresponding detection ID, they can later on be associated with the correct tracked person such that the extracted information can be smoothed over time.

In order to provide all of our fusion components with a streamlined, homogenous interface, we initially convert all DetectedPersons messages into CompositeDetectedPersons via a converter node, even if there is only a single DetectedPerson involved. This allows for easy chaining of the components.

6.2 Strategies for Fusion at the Detection Level

Detection-to-Detection Fusion Fusing detections to detections has got the advantage that the resulting composite detections can be fed into any existing (unimodal) tracking algorithm, without requiring special provisions to cope with information

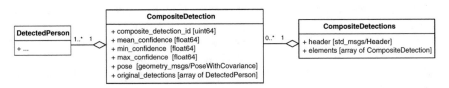

Fig. 9 Message definitions for composite detections. Retaining the information which original detections have been combined into a composite detection or a track is important for perception components at later stages in the pipeline

from multiple detectors. This approach can also be helpful if the tracking algorithm itself is computationally very complex and scales badly with the number of detections. For detection-to-detection fusion, we have implemented a series of nodelets[6] which can be used to flexibly compose a fusion pipeline by means of roslaunch XML files:

- *Converter nodelets* for converting DetectedPersons messages into CompositeDetectedPersons messages, and vice versa.
- *Aggregator nodelets* which simply concatenate two sets of CompositeDetectedPersons. Useful for combining detections from sensors with non-overlapping fields of view, in which case no data association is required.
- *Fusion nodelets* that perform data association between two sets of CompositeDetectedPersons, e.g. using a nearest-neighbor data association technique.

Detection-to-Track Fusion Although presently not realized, our framework also allows for detection-to-track fusion. As the tracker's knowledge of a track's previous trajectory can help with the association of the incoming detections, this approach may provide better results in case of very crowded environments. In this case, the fusion stage needs to be implemented within the tracker itself. It then becomes the tracker's responsibility to publish a corresponding CompositeDetectedPersons message to let other perception components know which original detections have been fused.

6.3 Post-Processing Filters

Often, higher-level reasoning components are not interested in all tracks that are output by the people tracking system. Therefore, we provide a series of post-processing filters for TrackedPersons messages such that the output, for instance, only includes visually confirmed tracks, non-static persons, or the n persons closest to the robot (useful for human-robot interaction). These are implemented in `spencer_tracking_utils` and can be set up in a chain if needed.

6.4 Multi-Modal People Tracking Setup on the Robot

On the robot platform, we run the front laser detector, the front upper-body RGB-D detector, the front HOG detector, and the fusion-related nodes on a single laptop. Likewise, the detectors for the rear-oriented sensors are executed on the second laptop, as well as the people tracking system itself. The lower RGB-D sensor on each side is tilted downwards and currently only used for close-range collision avoidance. Since all components, including the laser-based segmentation as a pre-processing

---

[6]A powerful mechanism to dynamically combine multiple algorithms into a single process with zero-copy communication cost. See http://wiki.ros.org/nodelet.

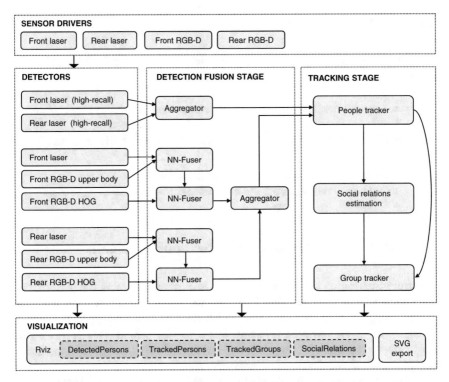

Fig. 10 A possible multi-modal people and group detection and tracking architecture that can be configured using our framework. Arrows represent ROS message flows. This setup was used during experiments on the SPENCER robot platform. Rounded rectangles represent reusable components that can be run as separate ROS nodes, and therefore easily be distributed across different computers. Connections between nodes are configured entirely in roslaunch XML without having to modify source code

step, are implemented as individual ROS nodes, they are separate system processes and therefore automatically make full use of modern multi-core CPUs.

Figure 10 shows a configuration which we used during experiments with our robot platform. Due to the flexible configuration in roslaunch XML files, different setups can easily be tested without modifying any source code. Of course, if the used people tracker implementation directly supports multiple detection sources, most of the detection-to-detection fusion stage can be left out (up to a possible aggregation of front and rear detections). We also ran experiments with an integrated person- and group tracker [15] that combines all of the steps in the tracking stage in order to feed back information from group-level tracking into person tracking, while still publishing TrackedPersons, SocialRelations and TrackedGroups messages. In all these cases, the remaining components including detectors, visualization, post-processing filters and evaluation metrics can still be used as-is.

6.5   Exemplary Launch File

As an inspiration for own experiments, we provide a launch file for the multi-modal
people tracking setup that we used on our robot:

```
roslaunch spencer_people_tracking_launch tracking_on_robot.launch
```

Alternatively, we provide a launch file that just uses a single RGB-D sensor:

```
roslaunch spencer_people_tracking_launch
    tracking_single_rgbd_sensor.launch height_above_ground:=1.6
```

7   Visualizing the Outputs of the Perception Pipeline

Powerful visualization tools can be of great help when analyzing the performance
of a complex perception pipeline. The standard visualization tool for sensor data
which is shipped with ROS is RViz, which uses the feature-rich OGRE graphics engine as its visualization backend. The core idea behind RViz is that it provides different visualization plugins *(displays)* for different types of ROS messages,
e.g. sensor_msgs/LaserScan or nav_msgs/Odometry.
In general, RViz provides two ways of implementing custom visualizations:

- *Markers and marker arrays*, using existing displays provided by RViz and message
  types from the `visualization_msgs` package. They allow to easily display
  primitives such as lines, triangles, texts, cubes or static 3D meshes. The pose,
  dimensions and color of the shape are specified within the message itself, published
  by a ROS node in any supported programming language.
- *RViz plugins*, written in C++, can benefit from the entire set of capabilities offered
  by the OGRE graphics engine. As each display comes with a property panel based
  on the UI framework Qt, they can easily be customized within the RViz user
  interface. RViz automatically takes care of persisting any settings when the user
  saves the configuration, which allows to load entire visualization setups via a single
  mouse-click.

In our experience, markers allow to quickly implement component-specific visualizations without requiring much developer effort. On the other hand, RViz plugins
are great for more complex visualizations that are reused often. They offer a better
end-user experience (due to the ability to change settings on-the-fly inside the RViz
GUI) at the cost of larger implementation effort.

7.1   Custom RViz Visualization Plugins

During the course of the SPENCER project, we were often in need of visualizing the outputs of our people tracking system and changing visualization settings on-the-fly, for instance during live robot experiments and for demonstration videos and publications. Therefore, in the `spencer_tracking_rviz_plugin` package, we developed a series of custom RViz plugins for displaying detected persons, tracked persons, social relations, tracked groups and human attributes.[7] They are automatically discovered by RViz and include features such as:

- Different visual styles: 3D bounding box, cylinder, animated human mesh
- Coloring: 6 different color palettes
- Display of velocity arrows
- Visualization of the 99 % covariance ellipse for position uncertainty
- Display of track IDs, status (matched, occluded), associated detection IDs
- Configurable reduction of opacity when a track is occluded
- Track history (trajectory) display as dots or lines
- Configurable font sizes and line widths

In the Results section in Fig. 14, we show a number of different visualizations generated using our RViz plugins. For the first row of that figure, we additionally used the existing RViz *Camera* display to project 3D shapes into a camera image. Examples of the social relations and tracked groups display are shown in Fig. 6, left and right. As opposed to other RViz displays, these displays subscribe to two topics at the same time (e. g. social relations and tracked persons).

7.2   URDF Model for Robot Visualization

To visualize the robot in RViz, we built a URDF (Unified Robot Description Format) file for the SPENCER robot, which defines the different joints of the platform, including fixed joints at the base of the robot and the sensor mounting points. The `robot_state_publisher` and `joint_state_publisher` packages use the URDF to publish a hierarchy of transforms (TF tree) for transforming between local sensor and robot coordinate frames as shown in Fig. 11 (right).

To create the visual robot model shown in Fig. 11 (left), we used a CAD model provided by the robot manufacturer and first removed any unnecessary components inside the robot's shell that are not visible from the outside using FreeCAD. We then exported a single Collada (DAE) file for each movable part and simplified the resulting mesh to reduce its polygon count. This can be done using software such as MeshLab or 3DS Max. We then manually aligned the visual parts of the model against the TF links specified in the URDF, by cross-checking with RViz and the CAD file.

---

[7]Such as gender and age group, not discussed here in detail.

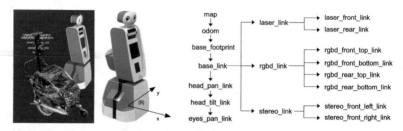

Fig. 11 *Left* Robot models (based upon URDF descriptions) of the SPENCER robot platform and the mobile data capture platform. *Middle* Robot coordinate frame. *Right* Simplified version of the TF hierarchy used on the SPENCER robot

For the mobile data capture platform, we used a point cloud registration software and an RGB-D sensor that we rotated around the platform at different azimuths and elevations to generate a visual representation. We built a transform hierarchy similar to that of the actual robot, to ensure that the same perception algorithms can be run on recorded and live data without modifications.

7.3 ROS-based SVG Exporters

For the visualization of detected and tracked persons and groups, we additionally provide a Python script in the `srl_tracking_exporter` package for exporting scalable vector graphics (SVGs). This is useful to analyze person trajectories from a 2D top-down view of the scene, which can be animated to display the evolution of tracks over time. In contrast to videos recorded from Rviz, they support free zooming to permit the analysis of smaller details (e.g. the cause of track identifier switches, by looking at the positions of individual detections). The resulting SVGs (Fig. 12) are best viewed in a web browser such as Chrome.

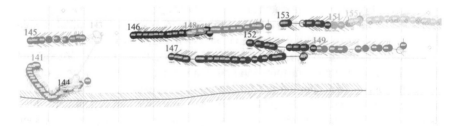

Fig. 12 Scalable and optionally animated vector graphic showing a 2D *top*-down view of a scene with several person tracks and their track IDs. Detections are shown as small diamonds. Arrowheads symbolize velocity, with the *bottom grey* track showing robot odometry. The user can zoom into the SVG to read detection IDs and timestamps

8 Integration with 3rd-Party Simulation Tools

Often, experimenting with real, recorded data can be cumbersome due to the need for manual groundtruth annotations and because it can be difficult to capture or provoke specific situations without scripting the behavior of the agents in the scene. For example, testing socially compliant motion planning in combination with a real people tracking system (including occlusions) proves to be difficult from pre-recorded datasets as the sensor position is decided at the time of recording. Testing on the real robot, however, can bring up its own challenges, where lack of open space, limited physical access to the robot, hardware issues or discharged batteries are just a few obstacles to overcome.

8.1 Integration with the Robot Simulator Gazebo

For this reason, we integrated our framework with the robot simulator Gazebo to generate simulated sensor data of moving persons from a moving robot in static environments (Fig. 13, left). We simulate 2D laser scan data with a noise model matching that of the real sensor on the basis of datasheet specifications.

As simulating the environment becomes computationally very complex with more than a dozen persons in the scene, we disable the Gazebo physics engine and instead manually update the agents' and robot's position at 25 Hz using the Gazebo ROS API. Person shapes are approximated by static 3D meshes that are processed by a GPU-accelerated raytracing algorithm in Gazebo to simulate laser scans. The resulting sensor_msgs/LaserScan is published on ROS and fed to our laser-based person detector, and RViz for visualization (Fig. 13, right).

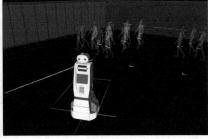

Fig. 13 *Left* Positions of simulated pedestrians are constantly sent to the Gazebo simulator via ROS, which then generates simulated laser scans and again publishes them via ROS. These simulated laser scans can then be fed into the people tracking pipeline for synthetic experiments. *Right* Moving pedestrians in a ROS adaptation of PedSim, the pedestrian simulator, visualized in RViz. The pedestrian behavior is modelled using a social force model from behavioral sciences

8.2 Integration with the Pedestrian Simulator PedSim

Simulating realistic pedestrian motion behaviors is a research topic of its own. Here, we use the publicly available PedSim library,[8] which simulates pedestrian crowd behaviors using a social force model from behavioral sciences [11]. We wrapped the library into a ROS node and added an RViz-based visualization of persons and obstacles using visualization markers (visible in Fig. 13, right). The positions of the simulated pedestrians are published as TrackedPersons messages, which can serve as a groundtruth to evaluate tracking performance (see Sect. 4.5). A separate converter node then sends these positions to Gazebo, where they are used to position 3D meshes as described in the previous section.

9 Results

In this section, we show qualitative results of our tracking system, discuss runtime performance and summarize important lessons that we learned during the project. As research in SPENCER is still going on at the time of this writing, we are currently preparing a detailed quantitative study of our tracking system, including a comparison of different multi-modal tracking approaches, for a publication at the end of the project.

9.1 Qualitative Results

In Fig. 14, we show illustrative results of our tracking system running on data recorded with our mobile data capture platform (Fig. 2) during a crowded situation at an airport, shortly after passengers have disembarked from an airplane. At the detection stage, we observe that the different detectors (2D laser, RGB-D upper-body, groundHOG in RGB) complement each other well and significantly increase the field of view around the robot in which persons can be tracked. In group guidance scenarios, it can make sense to filter out tracks in a post-processing step which have not been visually confirmed (shown in different color in the third row of Fig. 14), as especially the laser-based detector can sometimes trigger false alarms when objects appear like humans in the laser scan.

Multiple videos which show our tracking system in action on both the real robot platform and pre-recorded datasets can be found on our YouTube channel.[9]

---

[8]http://pedsim.silmaril.org/.

[9]http://www.youtube.com/spencereuproject.

Fig. 14 Multi-modal people detection and tracking from a mobile robot in a crowded airport environment. *First row* Color images of the upper rear and front RGB-D sensors, with projected laser points and bounding boxes of tracked persons. *Second row* Detections of 2D laser (*orange boxes*), upper-body detector (*cyan*) and groundHOG (yellow) on top of front and rear RGB-D and 2D laser point clouds. Fused detections are shown as *grey* cylinders. The area visible to the laser scanners is shaded in *grey*. *Third row* Resulting person tracks. As opposed to blue persons which are only tracked in laser, red tracks have been confirmed a number of times by one of the visual detectors. *Last row* Tracked groups that were detected via coherent motion indicator features

9.2 Runtime Performance

Our system runs in real-time at 20–25 Hz using the configuration described in Sect. 6.4 on two Intel Core i7-4700MQ gaming laptops with 8 GB RAM and nVidia GeForce GTX 765M graphics.[10] These laptops are dedicated to perception tasks on the robot. We avoid expensive streaming of RGB-D images over network by executing the RGB-D detectors directly on the laptop where the corresponding sensor is connected via USB. The computationally most expensive components are the person detectors, each requiring about one CPU core, with the people tracker itself requiring only 10–15 % of a single CPU core when 20–30 tracks are simultaneously being tracked.

9.3 Lessons Learned

Finally, we want to note down a few points that we learned about ROS and our hardware during the course of our project:

Parallelization: A complex perception pipeline such as the one developed in SPENCER requires a high amount of computational power. We learned that ROS can be a great help here due to its node-based modular structure, which easily allows to split up complex processing tasks into separate processes that can be run on different CPU cores or even different computers, without requiring extensive reconfiguration.

Collaboration: We think it is important to devise mock components early on in the project to enable other collaborators to test their dependent components. Due to the multi-language support in ROS, these can be simple Python scripts that e.g. publish fake detections.

Transform issues: We often had to deal with transform-related issues, which is natural due to the high number of sensor coordinate frames involved. Here, we found the tools `tf_monitor`, `tf_echo`, `view_frames` of the `tf` package as well as the TF display in RViz to be of great help. Especially the first-mentioned tool allows to identify latency issues that can arise when the clocks of the computers on the robot are not properly synchronized, which will cause transform lookups to fail. The `roswtf` tool lists unconnected ROS topic subscriptions, which helps when dealing with over 300 concurrently active topics.

Hardware: In terms of hardware, instead of using gaming laptops onboard a robot, in the future we would rather use traditional computers with extra GPU and USB PCIe cards to circumvent the USB bus sharing issues which we repeatedly encountered with the first-generation RGB-D sensors, since most laptops nowadays only expose

---

[10]The GPU is only used by the groundHOG person detector. For a single RGB-D sensor and 2D laser scanner, a single laptop is sufficient.

a single USB bus. It might also be worthwhile to consider using many small-form-factor PCs with integrated GPUs, as opposed to few high-end computers. Due to the modular structure of ROS, individual nodes can easily be spread across different computers.

10 Conclusion

In this chapter, we presented a multi-modal, ROS-based people detection and tracking framework developed for and tested on a large mobile service robot. We believe that due to its modular structure with clearly defined interfaces between components, our framework can easily be re-used on other robot platforms equipped with similar sensor configurations. Also, as we found out in our own research, standardized message interfaces allow to easily replace individual components of the pipeline, such as the core tracking algorithm, by a different implementation for comparative purposes. When doing so, the user can leverage our wide infrastructure of existing visualization, evaluation and simulation tools to quickly set up a usable tracking system. To our knowledge, the presented people tracking framework is the functionally most complete one that is currently available as open-source ROS software.

In future work, we want to extend our tracking framework with person re-identification capabilities, the lack of which is currently the most apparent bottleneck when dealing with lengthy occlusions of tracked persons.

Acknowledgments This work has been supported by the EC under contract number FP7-ICT-600877 (SPENCER). The authors would like to thank Fabian Girrbach for implementing several important extensions to the tracking system; Stefan Breuers from the Computer Vision Group at RWTH Aachen for MHT experiments and integrating the RGB-D and monocular vision detectors into the framework; and Christian Dondrup from the Lincoln Centre for Autonomous Systems Research for the original ROS integration of these components.

References

1. K.O. Arras, S. Grzonka, M. Luber, W. Burgard, Efficient people tracking in laser range data using a multi-hypothesis leg-tracker with adaptive occlusion probabilities, in *Proceedings of the IEEE International Conference on Robotics and Automation (ICRA)* (Pasadena, California, USA, 2008)
2. K.O. Arras, O.M. Mozos, W. Burgard, Using boosted features for the detection of people in 2D range data, in *Proceedings of the IEEE International Conference on Robotics and Automation (ICRA)* (Rome, Italy, 2007)
3. Y. Bar-Shalom, *University of California, L.A.U.E.: Multitarget-Multisensor Tracking: Applications and Advances* (No. 3 in Radar Library, Artech House, 2000)
4. Y. Bar-Shalom, P. Willett, X. Tian, *Tracking and Data Fusion: A Handbook of Algorithms* (YBS Publishing, Storrs, 2011)
5. K. Bernardin, R. Stiefelhagen, Evaluating multiple object tracking performance: the CLEAR MOT metrics. J. Image Video Process. **2008**, 1:1–1:10 (2008)

6. S. Blackman, R. Popoli, *Design and Analysis of Modern Tracking Systems* (Radar Library, Artech House, 1999)
7. W. Choi, S. Savarese, A unified framework for multi-target tracking and collective activity recognition, in *Proceedings of the European Conference on Computer Vision (ECCV)* (Florence, Italy, 2012), pp. 215–230
8. I. Cox, S. Hingorani, An efficient implementation of Reid's multiple hypothesis tracking algorithm and its evaluation for the purpose of visual tracking. IEEE Trans. Pattern Anal. Mach. Intell. (PAMI) **18**(2), 138–150 (1996)
9. L. Ding, A. Yilmaz, Inferring social relations from visual concepts, in *IEEE International Conference on Computer Vision (ICCV)* (Barcelona, Spain, 2011)
10. A. Fod, A. Howard, M.J. Matari, Laser-based people tracking, in *Proceedings of the IEEE International Conference on Robotics and Automation (ICRA)* (Washington, DC, 2002)
11. D. Helbing, P. Molnar, A social force model for pedestrian dynamics. Phys. Rev. E **51** (1995)
12. O.H. Jafari, D. Mitzel, B. Leibe, Real-time RGB-D based people detection and tracking for mobile robots and head-worn cameras, in *Proceedings of the IEEE International Conference on Robotics and Automation (ICRA)* (Hong Kong, China, 2014)
13. B. Lau, K.O. Arras, W. Burgard, Multi-model hypothesis group tracking and group size estimation. Int. J. Soc. Robot. **2**(1), 19–30 (2010)
14. L. Leal-Taixé, G. Pons-Moll, B. Rosenhahn, Everybody needs somebody: modeling social and grouping behavior on a linear programming multiple people tracker, in *ICCV Workshop on Modeling, Simulation and Visual Analysis of Large Crowds* (2011)
15. T. Linder, K.O. Arras, Multi-model hypothesis tracking of groups of people in RGB-D data, in *IEEE International Conference on Information Fusion (FUSION'14)* (Salamanca, Spain, 2014)
16. Luber, M., Tipaldi, G.D., Arras, K.O.: Better models for people tracking, in *Proceedings of the IEEE International Conference on Robotics and Automation (ICRA)* (Shanghai, 2011)
17. M. Luber, K.O. Arras, Multi-hypothesis social grouping and tracking for mobile robots, in *Robotics: Science and Systems (RSS'13)* (Berlin, Germany, 2013)
18. M. Luber, L. Spinello, K.O. Arras, People tracking in RGB-D data with online-boosted target models, in *Proceedings of theInternational Conference on Intelligent Robots and Systems (IROS)* (San Francisco, USA, 2011)
19. M. Luber, J.A. Stork, G.D. Tipaldi, K.O. Arras, People tracking with human motion predictions from social forces, in *Proceedings of the IEEE International Conference on Robotics and Automation (ICRA)* (Anchorage, USA, 2010)
20. M. Moussaïd, N. Perozo, S. Garnier, D. Helbing, G. Theraulaz, The walking behaviour of pedestrian social groups and its impact on crowd dynamics. PLoS ONE **5**(4) (2010)
21. M. Munaro, F. Basso, E. Menegatti, Tracking people within groups with RGB-D data, in Proceedings of the International Conference on Intelligent Robots and Systems (IROS) (2012)
22. M. Munaro, E. Menegatti, Fast RGB-D people tracking for service robots. Auton. Robot. **37**(3), 227–242 (2014)
23. S. Pellegrini, A. Ess, van Gool, L. Improving data association by joint modeling of pedestrian trajectories and groupings, in *Proceedings of the European Conference on Computer Vision (ECCV)*, (Heraklion, Greece, 2010)
24. Z. Qin, C.R. Shelton, Improving multi-target tracking via social grouping, in *Proceedings of the IEEE Conference on Computer Vision and Pattern Recognition (CVPR)*, (Providence, RI, USA, 2012)
25. D.B. Reid, An algorithm for tracking multiple targets. Trans. Autom. Control. **24**(6) (1979)
26. D. Schuhmacher, B.T. Vo, B.N. Vo, A consistent metric for performance evaluation of multi-object filters. Signal Process. IEEE Trans. **56**(8), 3447–3457 (2008)
27. D. Schulz, W. Burgard, D. Fox, A.B. Cremers, People tracking with mobile robots using sample-based joint probabilistic data association filters. Int. J. Robot. Res. **22**(2), 99–116 (2003)
28. L. Spinello, K.O. Arras, People detection in RGB-D data, in *Proceedings of the International Conference on Intelligent Robots and Systems (IROS)* (San Francisco, USA, 2011)

29. G. Wang, A. Gallagher, J. Luo, D. Forsyth, Seeing people in social context: recognizing people and social relationships, in *Proceedings of the European Conference on Computer Vision (ECCV)* (Heraklion, Greece, 2010)
30. T. Yu, S.N. Lim, K.A. Patwardhan, N. Krahnstoever, Monitoring, recognizing and discovering social networks, in *Proceedings of the IEEE Conference on Computer Vision and Pattern Recognition (CVPR)* (Miami, FL, USA, 2009)

Authors' Biography

Timm Linder is a PhD candidate at the Social Robotics Laboratory of the University of Freiburg, Germany. His research in the SPENCER project focusses on people detection and tracking in 2D laser range and RGB-D data, as well as human attribute classification in RGB-D. He is an experienced C++ and Python developer and has been a ROS user for over two years. He contributed to the ROS visualization tool RViz and played a major role in the deployment and integration of the software stack on the SPENCER robot platform.

Kai O. Arras is an assistant professor at the Social Robotics Laboratory of the University of Freiburg, and coordinator of the EU FP7 project SPENCER. After his diploma degree in Electrical Engineering at ETH Zürich, Switzerland and a PhD at EPFL, he was a post-doctoral research associate at KTH Stockholm, Sweden and at the Autonomous Intelligent Systems Group of the University of Freiburg. He also spent time as a senior research scientist at Evolution Robotics, Inc. and later became a DFG Junior Research Group Leader at the University of Freiburg.

A ROS-Based System for an Autonomous Service Robot

Viktor Seib, Raphael Memmesheimer and Dietrich Paulus

Abstract The Active Vision Group (AGAS) has gained plenty of experience in robotics over the past years. This contribution focuses on the area of service robotics. We present several important components that are crucial for a service robot system: mapping and navigation, object recognition, speech synthesis and speech recognition. A detailed tutorial on each of these packages is given in the presented chapter. All of the presented components are published on our ROS package repository: http:// wiki.ros.org/agas-ros-pkg.

Keywords Service robots · SLAM · Navigation · Object recognition · Human-robot interaction · Speech recognition · Robot face

1 Introduction

Since 2003 the Active Vision Group (AGAS) is developing robotic systems. The developed software was tested on different robots in the RoboCup[1] competitions. Our robot *Robbie* participated in the RoboCup Rescue league where it became world champion in autonomy (*Best in Class Autonomy Award*) in 2007 and 2008. It also won the *Best in Class Autonomy Award* in the RoboCup German Open in the years 2007–2011. Our robot *Lisa* depicted in Fig. 1 is the successor of Robbie and is participating in the RoboCup@Home league. Lisa's software is based on Robbie's, but was adopted and extended to the tasks of a service robot in domestic

---

[1]RoboCup website: http://www.robocup.org

V. Seib (✉) · R. Memmesheimer · D. Paulus
Active Vision Group (AGAS), University of Koblenz-Landau, Universitätsstr. 1,
56070 Koblenz, Germany
e-mail: vseib@uni-koblenz.de
URL: http://agas.uni-koblenz.de

R. Memmesheimer
e-mail: raphael@uni-koblenz.de

D. Paulus
e-mail: paulus@uni-koblenz.de

Fig. 1 The current setup of
our robot Lisa. A laser range
finder in the lower part is
used for mapping and
navigation. On top, an
RGB-D camera, a high
resolution camera and a
microphone are mounted on
a pan-tilt unit. Additionally,
Lisa possesses an 6-DOF
robotic arm for object
manipulation and a screen
displaying different facial
expressions to interact with
the user

environments. Lisa has already won four times the 3rd place and in 2015 the 2nd
place of the RoboCup German Open and is frequently participating in the finals of
the RoboCup@Home World Championship. Further, in 2015 Lisa won the 1st place
in the RoboCup@Home World Championship in Hefei, China. Additionally, Lisa
has won the Innovation Award (2010) and the Technical Challenge Award (2012) of
the RoboCup@Home league. Inspired by this success, we want to share our software
with the ROS community and give a detailed instruction on how to use this software
with other robots. Further information about our research group and robotic projects
are available online[2] and information about Lisa at our project website.[3]

After a common environment setup, presented in Sec. 2, we will introduce the
following ROS components in this chapter:

- First, we introduce a graphical user interface, `homer_gui` in Sect. 3.
- Second, we present nodes for mapping and navigation in Sect. 4.
- This will be followed by a description of our object recognition in Sect. 5.
- Finally, we demonstrate components for human robot interaction such as speech
  recognition, speech synthesis and a rendered robot face in Sect. 6

With the presented software, a ROS enabled robot will be able to autonomously create a map and navigate in a dynamic environment. Further, it will be able
to recognize objects. Finally, the robot will be able to receive and react to speech
commands and reply using natural language. The speech recognition is implemented
by interfacing with an Android device that is connected to the robot. A robot face
with different face expressions and synchronized lip movements can be presented on
an onboard screen of the robot.

---

[2] Active Vision Group: http://agas.uni-koblenz.de.

[3] Team homer@UniKoblenz: http://homer.uni-koblenz.de.

The presented software can be downloaded from our ROS software repository.[4] All example code in this chapter is also available online. Further, we provide accompanying videos that can be accessed via our YouTube channel.[5]

2 ROS Environment Configuration

All contributed packages have been tested with Ubuntu 14.04 LTS (Trusty Tahr) and ROS Indigo. For convenience, the script `install_homer_dependencies.sh` (in the software repository) installs all needed dependencies. However, also the step-by-step instruction in each section can be used.

There are common steps to all presented packages that need to be taken to configure the ROS environment properly. To configure all paths correctly, it is necessary to define an environment variable `HOMER_DIR`. It should contain the path of the folder that contains the *catkin workspace* on your hard disk. For example, on our robot the path is `/home/homer/ros`. To set the environment variable `HOMER_DIR` to point to this path, type:

```
export HOMER_DIR=/home/homer/ros
```

Please replace the path by the one that you use. To test whether the variable was set correctly, type:

```
echo $HOMER_DIR
```

This command displays the path the variable points to.

Note that you need to set the path in every terminal. To avoid this, the path can be defined in your session configuration file `.bashrc` and will be automatically set for each terminal that is opened. In order to do this, please type:

```
echo"export HOMER_DIR=/home/homer/ros">> ~/.bashrc
```

After setting the path, make sure that you did not change the directory structure after downloading the software from our repository. However, if you changed it, please adjust the paths written out after `$ENV{HOMER_DIR}` in the *CMakeLists.txt* file of the corresponding packages.

3 Graphical User Interface

Although a graphical user interface (GUI) is not necessary for a robot, it provides a lot of comfort in monitoring or controlling the state of the system. ROS already

---

[4]AGAS ROS packages: http://wiki.ros.org/agas-ros-pkg.

[5]HomerUniKoblenz on YouTube: http://www.youtube.com/c/homerUniKoblenz.

has `rviz`, a powerful tool for visualization of sensor data that can be extended with own plugins. Thus, we do not aim at creating our own visualization tool. Rather, in our package `homer_gui` we include `rviz`'s visualization as a plugin. This means that it is able to visualize everything that `rviz` does and can be adapted using the same configuration files. However, our GUI provides more than pure data visualization. It offers a convenient way of creating a map, defining navigation points and monitoring the navigation. Also, it enables the user to train and test new objects for object recognition in a convenient way and to monitor and define commands for human robot interaction. Additionally, basic state machine-like behavior and task execution can be defined in the GUI with only a few mouse clicks. The `homer_gui` is intended to run directly on the on-board computer of the robot. However, it can also be used to control the robot's state and behavior remotely. Please note that the version of the `homer_gui` presented here is a reduced version, encompassing only functionality that we provide additional packages for. As we continue releasing more and more components from our framework, more tabs and widgets will be included in the `homer_gui`.

3.1 Background

The `homer_gui` package is a Qt-based application. It serves as the central command, monitoring and visualization interface for our robot. The main concept is to define several task specific tabs that provide a detailed control of all task specific components. As an example, in the currently provided `homer_gui` package we added the *Map* and *Object Recognition* tabs. These two tabs are closely linked to our mapping and navigation (see Sect. 4) and to our object recognition (see Sect. 5). Their functionality will be described in the corresponding sections, together with the related packages. The third tab in `homer_gui` is the main tab with the title *Robot Control*. Its purpose is the visualization of the sensor data and the robot's belief about the world. Further, some frequently used functionality of the robot can be accessed through buttons in this tab.

3.2 ROS Environment Configuration

Please make sure to take the configuration steps described in Sect. 2 before proceeding. In order to use the package `homer_gui` you have to install the following libraries:

- Qt: http://qt.nokia.com
- Rviz: http://wiki.ros.org/rviz (usually comes with ROS)
- PCL: http://pointclouds.org/ (usually comes with ROS)

You can install the Qt libraries by opening a terminal and typing (all in one line):

```
sudo apt-get install libqt4-core libqt4-dev libqt4-gui
```

3.3   Quick Start and Example

To start the `homer_gui`, type

```
roslaunch homer_gui homer_gui.launch
```

A window, similar to Fig. 2 will appear. By itself, the GUI is not very exciting. Its full potential unfolds when used on a robot platform or e.g. with the packages described in Sects. 4 and 5. If started on a robot platform, the 3D view of the *Robot Control* tab will, among others, display the laser data and the robot model (Fig. 2).

3.4   Package Description

Since the *Object Recognition* and *Map* tabs are described in detail in Sects. 4 and 5, this description will focus on the *Robot Control* tab. The main functionality of this tab is the 3D view on the left part of the tab. We included the `rviz` plugin into the 3D view widget. On startup, the `homer_gui` will load the file `homer_display.rviz` contained in the `config` folder of the package. Every `rviz` display configured in that file will also be available in our GUI. However, it can only be reconfigured using `rviz`.

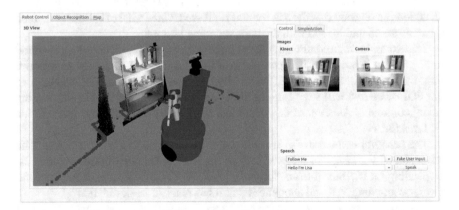

Fig. 2 Graphical user interface `homer_gui`. On the *left* the well known 3D view of `rviz` is included as plugin. On the *right* images of two connected cameras are displayed

The right side of the *Robot Control* tab contains two subtabs, *Control* and *SimpleAction*. The *Control* has two image widgets on top. The left one subscribes to the topic /camera/rgb/image_raw, the default image topic of the widespread RGB-D cameras (e.g. Kinect). The right one can be used to visualize other images as needed. It subscribes to the topic /image and in the same manner as the left widget, expects a message of type sensor_msgs/Image. The *Control* tab further contains two text line widgets with corresponding buttons. The first one is used to simulate recognized speech. When the button *Fake User Input* is pushed, the text in the corresponding text widget is published on the topic /recognized_speech as a std_msgs/String message. The second button is used to let the robot speak. Pushing the button *Speak* publishes a std_msgs/String message on the topic /robot_face/speak. If the robot_face described in Sect. 6 is running, the speech will be synthesized.

The second subtab, *SimpleAction*, is used to spontaneously create basic state machine-like behavior. So far, three simple actions can be used: *Drive to*, *Speak* and *Hear*. After selecting one of the actions on the left side of the tab, a parameter needs to be specified.

The *Speak* action takes a string that will be spoken by the robot (using the robot_face package described in Sect. 6) as parameter. The *Drive to* action takes a point of interest (POI) name as string as its parameter. For now it is sufficiently to define a POI as a "named navigation goal". On execution, the robot will navigate to this POI in the map. For details on how to create POIs, please refer to Sect. 4. The *Hear* action is used for speech recognition and is the only action that takes multiple parameters, separated by commas (without spaces). If only one parameter is specified, the action will block until speech is recognized that matches the given parameter string. However, with multiple parameters a case distinction is achieved. Depending on the recognized words the robot can perform different behavior.

By using the *Add* button, an Action is added to the list of actions. Added actions, except *Hear*, can be moved inside the list. Clicking *Start* will start the execution until all actions are executed or the *Stop* button is pressed. Once a list of actions is defined it can be saved and loaded with the corresponding buttons.

Two examples for a command sequence defined using simple actions are presented in Fig. 3. The behavior of the robot defined in the upper image will be as follows: first, the robot will say "starting". After finishing the speech action, it navigates to the POI *office* and will say "I am in the office". After that the robot waits for the "exit" command. As soon as this command is received, the robot navigates to the POI *outside*.

The behavior defined in the lower image is an example for case distinction. The robot starts by saying "I am ready". After that it waits for one of three commands. Depending on whether it hears "office", "kitchen" or "living room, it navigates to the corresponding POI and announces that it has reached its destination.

Thus, using the *Simple Action* tab basic robotic behavior can be achieved without programming. In the following sections further functionality of the homer_gui and the corresponding packages will be introduced.

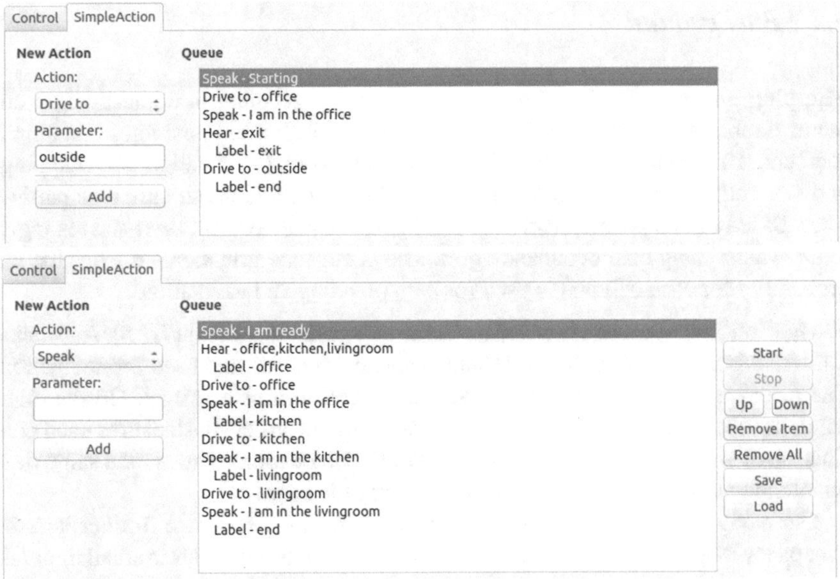

Fig. 3 Two examples of task sequences defined using the simple actions

4 Mapping and Navigation

The approach used in our software follows the algorithms described in [16] that were developed in our research group. To generate a map, the algorithm uses the robot's odometry data and the sensor readings of a laser range finder. The map building process uses a particle filter to match the current scan onto the occupancy grid. For navigation a path to the target is found using the A*-algorithm on the occupancy grid. The found path is post processed to allow for smooth trajectories. The robot then follows waypoints along the planned path.

The main benefit of our mapping and navigation compared to well established packages such as `hector_mapping`[6] and `amcl`[7] is its close integration with the graphical user interface `homer_gui`. It allows the user to define and manage different named navigation targets (points of interest). Further, maps can be edited by adding or removing obstacles in different map layers. Despite the points of interest and map layers, the grid maps created with our application are completely compatible with the ROS standard. The additional information is stored in a separate *yaml*-file that is only processed by the `homer_gui`, while the actual map data is not changed.

A video showing an example usage of the `homer_gui` with our mapping and navigation is available online.[8]

---

[6]Hector Mapping: http://wiki.ros.org/hector_mapping.

[7]AMCL: http://wiki.ros.org/amcl.

[8]Example video for mapping and navigation: http://youtu.be/rH4pNq3nlds.

4.1 Background

Mapping and Localization For autonomous task execution in domestic environments a robot needs to know the environment and its current position in that environment. This problem is called SLAM (Simultaneous Localization and Mapping) and is usually solved by a probabilistic Bayes formulation, in our case by a particle filter. Based on the position estimate from odometry the current laser scan is registered against the global occupancy grid. The occupancy grid stores the occupation probability for each cell and is used for path planning and navigation.

Navigation This approach is based on Zelinsky's path transform [17, 18]. A distance transform is combined with a weighted obstacle transform and used for navigation. The distance transform holds the distance per cell to a given target. On the other hand, the obstacle transform holds the distance to the closest obstacle for each cell. This enables the calculation of short paths to target locations while at the same time maintaining a required safety distance to nearby obstacles.

Our navigation system merges the current laser range scan as a frontier into the occupancy map. A calculated path is then checked against obstacles in small intervals during navigation, which can be done at very little computational expense. If an object blocks the path for a given interval, the path is re-calculated. This approach allows the robot to efficiently navigate in highly dynamic environments.

4.2 ROS Environment Configuration

Please make sure to take the configuration steps described in Sect. 2 before proceeding. Further, the package homer_gui will be used in the examples presented here. Please also follow the configurations steps described in Sect. 3. The mapping and navigation packages require the *Eigen* library. To install this dependency, please type the following command in a terminal:

```
sudo apt-get install libeigen3-dev
```

4.3 Quick Start and Example

Before proceeding, do not forget to execute the catkin_make command in your ROS workspace.

Mapping To be able to use the mapping, your robot needs to publish odometry data on the topic /odom and laser range data on the topic /scan. Please refer to the ReadMe file[9] if your robot is publishing encoder ticks instead of odometry data. With

---

[9] A ReadMe is available online in the software repository.

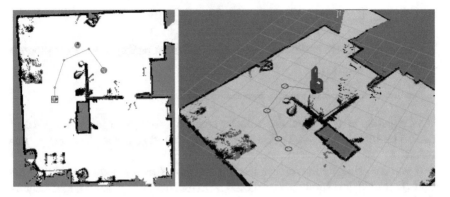

Fig. 4 2D and 3D view of a map and a planned path (*blue line*). *Red dots* indicate the current laser scan, while *orange circles* in the 2D map represent named navigation targets (points of interest)

the odometry and laser range data we are able to construct 2D occupancy grid maps using the homer_mapping package. However, the position of the laser range finder relative to the robot needs to be given. This is achieved by creating a transformation from /base_link to /laser for instance by adding:

```
<node pkg="tf"type="static_transform_publisher"
name="base_link_to_laser_broadcaster"
args="0.0 0.0 0.1 0.0 0.0 0.0 1.0 /base_link /laser 100"/>
```

to your launch file. Please substitute the frame name of the laser range finder (here: /laser) by the frame frame_id of the /scan messages from your laser range finder. For further information on transformations and frames please check out the tf package documentation.[10]

With these preparations completed the mapping can be started with the following command:

```
roslaunch homer_mapping homer_mapping.launch
```

The mapping node constructs the map incrementally and refines the current position of the robot using a particle filter SLAM approach. The map is published on the /map topic and the robot's pose on the topic /pose. The map can be inspected with rviz or our interface presented in Sect. 3 by typing

```
roslaunch homer_gui homer_gui.launch
```

and selecting the *Map* tab. An example visualization is shown in Fig. 4.

Navigation Prerequisite for navigation is the map and the robot's pose, both are computed by the homer_mapping package:

---

[10]tf package documentation: http://wiki.ros.org/tf.

```
roslaunch homer_mapping homer_mapping.launch
```

For navigation, please additionally start the navigation package with the following command:

```
roslaunch homer_navigation homer_navigation.launch
```

The navigation node will publish velocity commands of type `geometry_msgs/Twist` on topic `/cmd_vel`. Please make sure that your robot subscribes to this topic.

If you are using `rviz` you can use the *2D Pose Estimate* button to localize the robot (if needed) and subsequently the *2D Nav Goal* button to define a target location for navigation and start navigating.

Navigation can also be started via command line. The equivalent command to the *2D Nav Goal* button is

```
rostopic pub /move_base_simple/goal geometry_msgs/PoseStamped
```

After typing this command, press the *TAB*-button twice and fill in the desired target location. Alternatively, you can use the following command to start navigation

```
rostopic pub /homer_navigation/start_navigation
             homer_mapnav_msgs/StartNavigation
```

The latter option has additional parameters that allow you to define a distance to the target or to ignore the given orientation. In the next section, named target location, the so called points of interest (POIs), will be introduced. The following command can be used to navigate to a previously defined POI (in this example to the POI *kitchen*):

```
rostopic pub /homer_navigation/navigate_to_POI  homer_mapnav_msgs/
        NavigateToPOI
               "poi_name: 'kitchen'
                distance_to_target: 0.0
                skip_final_turn: false"
```

In all cases the navigation can be canceled by the command

```
rostopic pub /homer_navigation/stop_navigation std_msgs/Empty"{}"
```

As soon as the target location is reached, a message of type `std_msgs/Empty` is published on the topic `/homer_navigation/target_reached`. In case the navigation goal could not be reached, a message of type `homer_mapnav_msgs/TargetUnreachable` is published on the topic `/homer_navigation/target_unreachable` which holds the reason why the target could not be reached.

Please note that the full strength of our mapping and navigation packages comes from its close integration with the `homer_gui` package, which will be explained in the following section.

4.4 Using the `homer_gui` for Mapping and Navigation

This section describes the integration of the mapping and navigation with the `homer_gui` and the involved topics and messages. Please type the following commands to start all components needed for the examples in this section:

```
roslaunch homer_mapping homer_mapping.launch roslaunch
homer_navigation homer_navigation.launch
roslaunch homer_gui homer_gui.launch
```

A window showing the `homer_gui` should open. Please select the *Map* tab where you should see the current map and the robot's pose estimate similar to Fig. 5. Optionally, a previously created map can be loaded with the *Open...* button. After loading a map, usually the robot's pose needs to be changed. This is achieved by *Ctrl + Left-Click* in the map view. To change the robot's orientation please use *Ctrl + RightClick*, clicking on the point the robot should look at. This corresponds to the functionality encapsulated in the *2D Pose Estimate* button in `rviz`, however, by splitting the definition of the robot's location and orientation a more precise pose estimate can be given than if both have to be set at once.

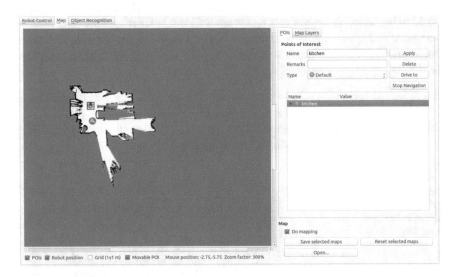

Fig. 5 The mapping tab in the `homer_gui`. On the *left side* the occupancy gridmap is shown. The current laser measurements are aligned to the map and drawn in *red*. The robot's pose estimate is denoted by the *blue circle* with an *arrow*. *Orange circles* are points of interest

However, the functionality of the *2D Nav Goal* button in `rviz` is handled completely differently in our system. In `rviz`, the *2D Nav Goal* button is used to define a navigation goal, a target orientation *and* start navigating towards this goal at once. In the `homer_gui` a click on the map adds a point of interest (POI) to the map view. After adding a POI symbol a name needs to be assigned on the right side of the tab (Fig. 5). A close-up of this dialog is shown in Fig. 6. A click on *Apply* informs the `homer_map_manager` that a new POI is available. The target orientation of a POI is denoted by a small arrow inside the POI symbol. To change its orientation a POI must be selected. A *Ctrl + LeftClick* on a location in the map defines the direction the robot should face after reaching the target. The specified orientation needs to be confirmed with the *Apply* button. An arbitrary number of POIs can be added to the map specifying all required locations for navigation. To actually navigate to a POI, it needs to be selected and the *Drive to* button pressed (Fig. 6). The *Stop navigation* button allows to abort navigation and stop the robot. Please note that the *Drive to* button uses the same topic and message as the *2D Nav Goal* button in `rviz`. However, the location in `rviz` is determined from a click on the map whereas here the location is taken from the predefined POI.

The `homer_gui` further allows to use different map layers. This is used e.g. if the robot is equipped with only one laser range finder on a fixed height that is not capable of detecting all obstacles in the area. The map layers are shown in the *Map Layers* subtab of the *Map* tab (Fig. 7). The *SLAM Map* layer is the normal occupancy grid for navigation. The *laser data* layer only shows the current laser scan, while the *Kinect Map* layer is currently not used, but will contain sensor readings from an RGB-D camera in the future. All selected layers are merged into the *SLAM Map* and used for navigation and obstacle avoidance. The *Masking Map* layer is used to add additional obstacles that the robot was not able to perceive automatically. To modify the *Masking Map*, please select it and press *Shift + LeftClick* into the map view. A red square will appear whereby its corners can be dragged to give it the desired shape and location. A click on *apply* will finally add the modification to the layer. There are different types of modifications that can be added:

block: An area that the robot will consider as an obstacle that will be avoided in any case (e.g. stairs or other not perceived obstacles).

obstacle: Similar to *block*, however this type can be used to allow manipulation in this area while navigation will still be prohibited.

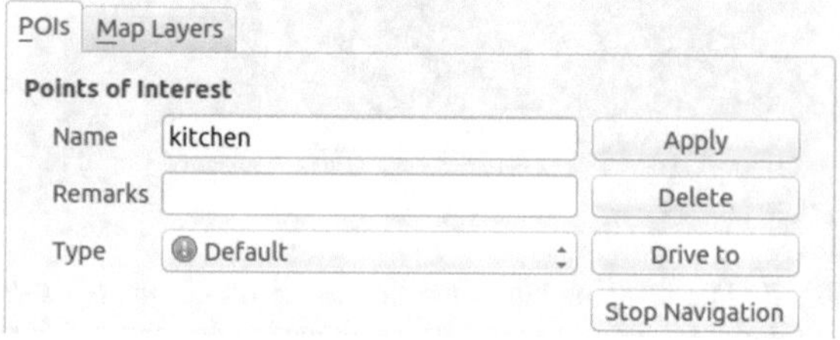

Fig. 6 The interface for adding and editing POIs, as well as to start and abort navigation

Fig. 7 The map layers
interface to manually modify
the existing map in different
ways

free: This area will be considered as free regardless of the sensor readings or the
stored map data. This is useful if navigating with a fixed map where an object has
moved or a closed door is now open.

sensitive: This type enables you to define high sensitive areas where obstacles are
faster included into the map. This is useful for known dynamic parts of the map,
for instance door areas.

These modifications can be added to any map layer. However, we advise to only
modify the *Masking Map* manually since other layers are continuously updated by
sensor readings and thus might alter the manual modifications.

4.5 Package Description and Code Examples

The graph in Fig. 8 shows the two nodes that are encapsulated in the homer_
mapping package with the most important topics. The homer_mapping node
is in charge of the mapping and the associated algorithms. The node homer_map_
manager handles the loading and storage of maps, as well as manages the POIs
and map layers. The most important topics are explained in the following:

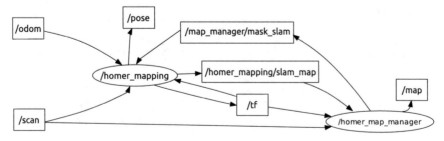

Fig. 8 ROS graph of the nodes inside the homer_mapping package with the most important
topics

Mapping Publishers

/pose (geometry_msgs/PoseStamped): The current pose estimate in the
/map frame as calculated by the particle filter.

/homer_mapping/slam_map (nav_msgs/OccupancyGrid): The current map as output from the SLAM algorithm (without any map layers). The
default publishing frequency is 2 Hz.

/map (nav_msgs/OccupancyGrid): The current map enhanced with map
layers.

Mapping Subscribers

/odom (nav_msgs/Odometry): Current robot odometry values, needed by the
particle filter for the SLAM algorithm.

/scan (sensor_msgs/LaserScan): Current laser measurements, needed by
the particle filter for the SLAM algorithm.

Mapping Services

/homer_map_manager/save_map (homer_mapnav_msgs/SaveMap):
Saves the current occupancy map with all POIs and layers.

/homer_map_manager/load_map (homer_mapnav_msgs/LoadMap):
Loads a map from a specified file path.

The graph in Fig. 9 shows the homer_navigation node with its most important topics. As the navigation relies on the node homer_mapping and homer_map
_manager, these nodes need to be running for the robot to be able to navigate. The
homer_navigation node handles the path planning algorithm and sends control
commands to the robot. The most important topics are explained in the following.
For a better overview not all of these topics are shown in Fig. 9.

Navigation Publishers

/cmd_vel (geometry_msgs/Twist): Velocity commands calculated by the
navigation node are published on this topic.

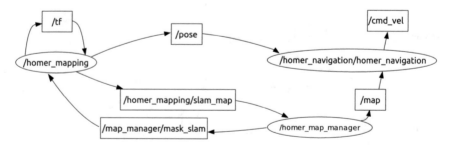

Fig. 9 ROS graph of the nodes inside the homer_navigation package with the most important
topics

/homer_navigation/target_reached (std_msgs/Empty): When the robot reaches a navigation goal a message will be send on this topic.

/homer_navigation/target_unreachable (homer_mapnav_msgs /Target Unreachable): The system is notified on this topic if a target is not reachable and the navigation is canceled.

Navigation Subscribers

/map (nav_msgs/OccupancyGrid): The current map with all defined map layers merged is used for path calculation and obstacle avoidance.

/pose (geometry_msgs/PoseStamped): The current pose of the robot for path planning.

/scan (nav_msgs/LaserScan): The current laser scan is used for obstacle avoidance.

/homer_navigation/start_navigation (homer_mapnav_msgs /Start Navigation): Plans a path and starts navigating to a target location.

/homer_navigation/navigate_to_POI (homer_mapnav_msgs/ NavigateToPOI): Plans a path and starts navigating to a given POI.

/homer_navigation/stop_navigation (homer_mapnav_msgs/ StopNavigation): Stops the current navigation task.

To use the mapping and navigation in your own application, be sure to include the homer_mapnav_msgs as *build* and *run* dependency in your package's package.xml file. Further, homer_mapnav_msgs has to be added to the find_package section and as a dependency to the catkin_package section in the CMakeLists.txt file. As with the command line examples from the *Quick Start* section we provide an example for navigating to a location and to a predefined POI by its name. An example header file for navigation could look like this:

```
1   #include <ros/ros.h>
2   #include <tf/tf.h>
3   #include <homer_mapnav_msgs/NavigateToPOI.h>
4   #include <homer_mapnav_msgs/StartNavigation.h>
5
6   class NavigationExample {
7     public:
8       NavigationExample(ros::NodeHandle nh);
9       virtual ~NavigationExample(){};
10
11      void driveToPOI(std::string name, float distance_to_target = 0.03);
12      void driveToPosition(float x, float y, float z,
13                 float orientation_in_rad = 0.0,
14                 float distance_to_target = 0.03)
15    private:
16      void targetReachedCallback(
17              const std_msgs::Empty::ConstPtr& msg);
18      void targetUnreachableCallback(
19              const homer_mapnav_msgs::TargetUnreachable::ConstPtr& msg);
20      ros::Publisher navigate_to_poi_pub_;
21      ros::Publisher start_navigation_pub_;
```

```
22      ros::Subscriber target_reached_sub_;
23      ros::Subscriber target_unreachable_sub_;
24    };
```

In your class implementation you can use the following code snippets. First, include the example header and initialize all subscribers and publishers in the constructor:

```
1     #include"NavigationExample.h"
2
3     NavigationExample::NavigationExample(ros::NodeHandle nh)
4     {
5       navigate_to_poi_pub_ = nh->advertise<homer_mapnav_msgs::NavigateToPOI>
6         ("/homer_navigation/navigate_to_POI", 1);
7       start_navigation_pub_ = nh->advertise<homer_mapnav_msgs::
8         StartNavigation>("/homer_navigation/start_navigation", 1);
9
10      target_reached_sub_ = nh->subscribe<std_msgs::Empty>
11        ("/homer_navigation/target_reached", 1,
12        &NavigationExample::targetReachedCallback, this);
13
14      target_unreachable_sub_ = nh->subscribe<homer_mapnav_msgs::
15        TargetUnreachable>("/homer_navigation/target_unreachable", 1,
16        &NavigationExample::targetUnreachableCallback, this);
17    }
```

The following function is used to navigate to a predefined POI.

```
1     void NavigationExample::driveToPOI(std::string name,
2           float distance_to_target)
3     {
4         ROS_INFO_STREAM("DRIVING TO "+ name);
5         homer_mapnav_msgs::NavigateToPOI msg;
6         msg.poi_name = name;
7         msg.distance_to_target = distance_to_target;
8         navigate_to_poi_pub_.publish(msg);
9     }
```

Further, the following function allows to navigate to arbitrary coordinates in the /map frame.

```
1     void NavigationExample::driveToPosition(float x, float y, float z,
2                         float orientation_in_rad,
3                         float distance_to_target)
4     {
5       homer_mapnav_msgs::StartNavigation start_msg;
6       start_msg.goal.position.x = x;
7       start_msg.goal.position.y = y;
8       start_msg.goal.position.z = z;
9       start_msg.goal.orientation =
10      tf::createQuaternionMsgFromYaw(orientation_in_rad);
11      start_msg.distance_to_target = distance_to_target;
```

```
12    m_start_navigation_pub.publish(start_msg);
13  }
```

Finally, the following two callbacks provide a feedback on whether the target location could be reached or not:

```
1   void NavigationExample::targetReachedCallback(
2         const std_msgs::Empty::ConstPtr& msg)
3   {
4       ROS_INFO_STREAM("Reached the goal location");
5   }
6   void NavigationExample::targetUnreachableCallback(
7         const homer_mapnav_msgs::TargetUnreachable::ConstPtr& msg)
8   {
9       ROS_WARN_STREAM("Target unreachable");
10  }
```

5 Object Recognition

The ability to recognize objects is crucial for robots and their perception of the environment. Our object recognition algorithm is based on SURF feature clustering in Hough-space. This approach is described in detail in [14]. We applied it in the Technical Challenge of the RoboCup 2012 where we won the 1st place. Additionally, with an extended version of the algorithm we won the 2nd place in the Object Perception Challenge of the RoCKIn competition in 2014.

A video showing an example usage of the homer_gui with our object recognition is available online.[11]

5.1 Background

Our object recognition approach is based on 2D camera images and SURF features [2]. The image processing pipeline for the training procedure is shown in Fig. 10. In order to train the object recognition classifier an image of the background and of the object has to be captured. These two images are used to calculate a difference image to segment the object. From segmented object we extract a number of SURF features f. A feature is a tuple $f = (x, y, \sigma, \theta, \delta)$ containing the position (x, y), scale σ and orientation θ of the feature in the image, as well as a descriptor δ. Thus, the features are invariant towards scaling, position in the image and in-plane rotations. Further images with a different object view need to be acquired and added to the object model in the database to capture the object's appearance from all sides.

---

[11]Example video for object recognition: https://youtu.be/dptgFpu7doI.

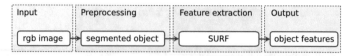

Fig. 10   Image processing pipeline for the training phase

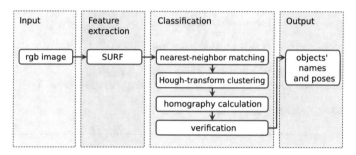

Fig. 11   Image processing pipeline for the recognition phase

The image processing pipeline for the recognition step is shown in Fig. 11. Since no information about the background is available during the object recognition phase, SURF features are extracted from the whole input image. The obtained features are then matched against the features stored in the object database. For fast nearest-neighbor and distance computation in the high dimensional descriptor space we use an approximate nearest neighbor approach [10]. Since simple distance thresholds do not perform well in high dimensional space, Lowe introduced the distance ratio [7], which is used here to sort out ambiguous matches. The result of feature matching is a set of matches between features extracted from training images and the scene image. This set may still contain outliers, i.e. matches between learned features and the scene's background.

The feature matches are clustered in a four dimensional Hough-space representing the 2D position of the object's centroid in the image (x, y), it's scale σ and it's rotation θ. The goal is to find a consistent object pose in order to eliminate false feature correspondences. Each feature correspondence is a hypothesis for an object pose and is added to the corresponding bin in the accumulator. As suggested in [5, 7], to reduce discretization errors, each hypothesis is added to the two closest bins in each dimension, thus resulting in 16 accumulator entries per feature correspondence. As a result of the Hough-transform clustering all features with consistent poses are sorted into bins, while most outliers are removed since they do not form maxima in Hough-space.

In the next step, bins representing maxima in Hough-space are inspected. A perspective transformation is calculated between the features of a bin and the corresponding points in the database under the assumption that all features lie on a 2D plane. As most outliers were removed by discarding minima in Hough-space, a consistent transformation is obtained here. Random Sample Consensus (RANSAC) is used to identify the best homography for the set of correspondences. The

(a)

(b)

Fig. 12 The input image for object recognition as acquired by our robot during the Technical Challenge of the RoboCup (**a**) and the output image depicting the recognition results (**b**). During training and recognition an image resolution of 8 MP was used

homography with most point correspondences is considered to be the correct object pose. Finally, the object presence in the scene is verified by comparing the number of matched features of the object with the number of all features in the object area. For further details on the presented algorithm, please refer to [14].

We applied the described algorithm in the Technical Challenge of the RoboCup @Home World Championship 2012 in Mexico-City where we won the 1st place. The cropped input image as well as the recognition result of the Technical Challenge are shown in Fig. 12.

5.2 ROS Environment Configuration

Please make sure to take the configuration steps described in Sect. 2 before proceeding. The following *Quick Start and Example* section will use the `homer_gui` package described in Sect. 3. Please make sure to follow the steps described there for the *ROS Environment Configuration*, as well.

The nodes for object recognition are contained in the `or_nodes` package. This package depends on the following packages that are also provided in our software repository:

- `or_libs` This package contains all algorithms used for object recognition.
- `or_msgs` All messages used by the `or_nodes` package are contained here.
- `robbie_architecture` Core libraries from our framework that other packages depend on are enclosed here.

All libraries needed to run these packages are installed with ROS automatically. However, you will need a camera to follow the presented examples. Although our approach uses RGB data only, in our experience the most widespread and ROS supported camera is a Kinect-like RGB-D camera. For the camera in our example we use OpenNI that you can install by typing (all in one line):

```
sudo apt-get install ros-indigo-openni-camera
 ros-indigo-openni-launch
```

You can use any other camera, of course.

5.3   Quick Start and Example

Before proceeding, please make sure that you completed the *ROS Environment Configuration* for the object recognition packages. Further, do not forget to execute the `catkin_make` command in your ROS workspace.

To recognize objects you need to connect a camera to your computer. In our example we use a Microsoft Kinect. After connecting the device, start the camera driver by typing

```
roslaunch openni_launch openni.launch
```

This should advertise several topics, including `/camera/rgb/image_rect_color`. This topic will be assumed as the default topic that our application uses to receive camera images from. You can specify a different topic by changing the parameter `/OrNodes/sInputImageTopic` inside the `params_kinect.yaml` file. This file is located in the `config` folder inside the `or_nodes` package.

You can start the object recognition by typing the following command:

```
roslaunch or_nodes or_nodes.launch
```

This will start the necessary nodes, as well as load the default object `lisa`. Please start now the `homer_gui` described in Sect. 3 by typing

```
roslaunch homer_gui homer_gui.launch
```

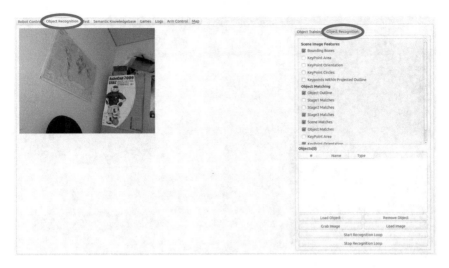

Fig. 13 Graphical user interface homer_gui with the current image of the Kinect in the upper left image widget. The *red ellipses* indicate the selected *Object Recognition* tabs

As soon as the main window appears, select the *Object Recognition* tab on the upper part of the window. In the tab you are now in please again select the *Object Recognition* on the right part of the screen. If everything works as expected, you should see the camera image of the Kinect in the top left image widget of the *Object Recognition* tab (see Fig. 13). Since a default object is already loaded, we can immediately test the recognition. In order to do this, click on the *Start Recognition Loop* button in the lower right part of the window. Now the object recognition is performed on a loop using the incoming camera images of the Kinect. Please go back to page 2 where an image of our robot Lisa is shown. If you point the kinect at that image, you should see a recognition result similar to Fig. 14. As soon as you are done testing, press the *Stop Recognition Loop* button. Please note that for obvious reasons we could not test this example with the final version of the book and used a printed image of the robot instead.

5.4 Using the homer_gui for Object Learning and Recognition

This section describes the integration of the object recognition and training pipelines with the homer_gui. Please connect an RGB-D camera (or any camera you like) with your computer and type the following commands to start all nodes that are used in the following use case

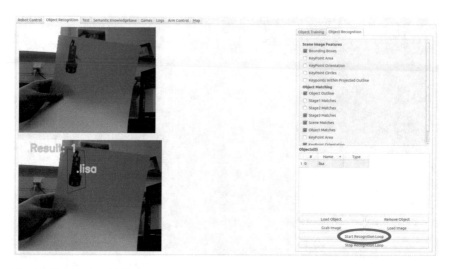

Fig. 14 Graphical user interface `homer_gui` displaying the recognition result of our robot Lisa in the *bottom left* image widget. The *red ellipse* indicates the button to start the recognition

```
roslaunch or_nodes or_nodes.launch
roslaunch homer_gui homer_gui.launch
roslaunch openni_launch openni.launch
```

Replace the last command by the launch-file of your camera driver.

To train a new object, click on *Object Recognition* tab in the `homer_gui` and in the appearing window on the tab *Object Training*. As in the quick start example, you should see the current camera image in the top left image widget. To obtain better results, point the camera at a static background then click *Grab Background*. The next incoming image on the image source topic (default: `/camera/rgb/image_rect_color`) will be processed as background image. The background image will appear in the bottom left widget (received over the topic `/or/obj_learn_primary`) and a grayscale version of the image will appear in the top right widget (received over the topic `/or/debug_image_gray`). Now place an object that you want to train in front of the camera and click on *Grab Foreground*. Again, the next incoming image on the image source topic will be processed. The result is sent to the `homer_gui` and displayed. The bottom left widget shows the outline of the object, while the top right widget displays the extracted object mask (Fig. 15).

In some cases, the initial object mask is far from being optimal. Therefore, several thresholds need to be adjusted in the GUI using the three sliders and one checkbox on the right side. Each change will be immediately displayed in the colored image with the outline and the mask image. Note that the initially displayed state of the sliders and the checkbox must not necessarily reflect their real states as the GUI might have been started before the object learning node. It is therefore advised to change each of the sliders and re-activate the checkbox and see whether the segmentation improves. The

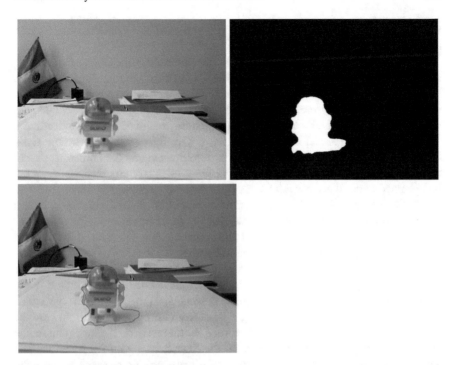

Fig. 15 The image widget part of the *Object Recognition* tab is shown. The *top left* image shows the current camera image. The *top right* image shows the object mask obtained by subtracting the background and foreground images. The *bottom* image shows the outline of the object according to the mask. Only features inside this outline will be saved for this object view

semantics of the sliders and the checkbox are as follows. If the checkbox is checked, only the largest separated segment of the difference between the background and the foreground will be used. The threshold determining the separation between the foreground and background is set with the *Background Deletion Threshold* slider. Around the segmented mask, a morphological opening is performed. The radius for this operation is determined with the *Mask Open Radius* slider. Finally, an additional border can be added to the mask by adjusting the *Additional Border* slider. The obtained mask might contain a small portion of the table that was segmented due to the shadow of the object. Usually, this has no negative effect on the recognition result, as long as the table plane is homogeneous without many features.

To save the obtained object view, click on the *AddImage* button. Optionally, you can specify a name for the image in the text field next to the button. An image can be removed again by selecting the image in the list on the right and clicking the button *RemoveImage*. By clicking the *Reset* button, all images saved so far will be removed. Several views of the object (about 8–12) should be acquired and saved by repeating the described procedure. After obtaining enough object views, the object file has to be saved by specifying a name and clicking the *Save* button (top right corner of the window). The resulting object file will be the specified object name with the file extension *.objprop*. The file will be placed in the package `or_nodes` inside the folder `objectProperties`.

Besides learning new objects by the described procedure, the *Object Training* tab offers some more functionalities. Instead of grabbing the background and foreground images, images can also be loaded from the hard disk using the buttons *Load Background* and *Load Foreground*, respectively. Further, with the *Load Object* button, a previously saved object file can be loaded and more images added or removed.

To recognize a learned object, select the *Object Recognition* tab on the right. Please note that the `homer_gui` is still in development and the numerous checkboxes in upper right part the *Object Recognition* subtab have no functionality so far. It is planned to use these checkboxes in future to display additional debug information in the recognition result image widget. Please click on the *Load Object* button to select an object for recognition. Note that only loaded objects will be used for recognition. To remove a loaded object, the corresponding object has to be selected and subsequently the *Remove Object* button pushed.

With the desired objects loaded, the recognition can be started by pressing the *Grab Single Image*, the *Load Image* or the *Start Recognition Loop* buttons. While the *Load Image* button is used to recognize objects on an image from the hard disk, the other two buttons use the incoming images from the camera sensor. In case the recognition loop was started, recognition runs continuously on the incoming image stream and can be stopped by a click on the *Stop Recognition Loop* button. The recognition result is sent as a `or_msgs/OrMatchResult` message on the topic `or/match_result`. Additionally, a `std_msgs/Image` message is published on the topic `/or/obj_learn_primary` displaying the recognition results (Fig. 16).

5.5   Package Description and Code Examples

The graph in Fig. 17 shows the two nodes contained in the `or_nodes` package with the most important topics. The `obj_learn` node is used for object learning and the

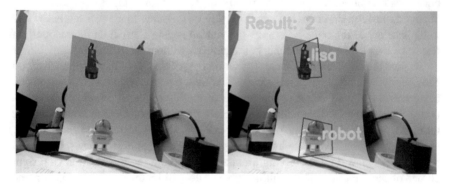

Fig. 16 Recognition example of the default object `lisa` and the learned object `robot`. The input image is shown on the *left*, the recognition results and bounding boxes are shown on the *right*

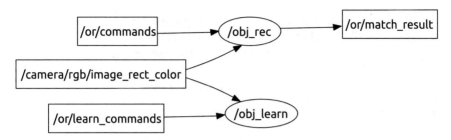

Fig. 17 ROS graph of the nodes inside the `or_nodes` package with the most important topics

node `obj_rec` for object recognition. The most important topics are explained in the following:

Publishers

`/or/match_result`   (or_msgs/OrMatchResult): Recognition results are published on this topic.

`/or/debug_image_gray`   (std_msgs/Image): Debug images and extracted object mask images are published on this topic.

`/or/obj_learn_primary_color`   (std_msgs/Image): Topic with images showing the object outline during learning and the recognition results.

Subscribers

`/camera/rgb/image_rect_color`   (std_msgs/Image): Input image topic for object recognition and learning.

`/or/learn_commands`   (or_msgs/OrLearnCommand): Topic for commands regarding the object learning procedure.

`/or/commands`   (or_msgs/OrCommand): Topic for commands regarding object recognition.

To use the object recognition in your own application, be sure to include the `or_msgs` as *build* and *run* dependency in your package's `package.xml` file. Further, `or_msgs` has to be added to the `find_package` section and as a dependency to the `catkin_package` section in the `CMakeLists.txt` file. The easiest way to load objects for recognition is to include the following line in your launch file:

```
<rosparam param="or_objects">lisa,robot</rosparam>
```

This line sets the parameter `or_objects` on the ROS parameter server and assigns it the object names. The object names must be separated by a comma without spaces. In this example the object files `lisa.objprop` and `robot.objprop` will be loaded automatically on startup. The object files have to be located inside the *object-Properties* folder in the `or_nodes` package. Note that this parameter has to be set before launching the object recognition launch file. In case this parameter is not set, default objects from the legacy (deprecated) configuration file are loaded (see below). An example header for object recognition could look like this:

```
#include <ros/ros.h>
#include <or_msgs/OrCommand.h>
#include <or_nodes/src/Modules/ORControlModule.h>
#include <or_msgs/OrMatchResult.h>
#include <or_msgs/MatchResult.h>
#include <or_msgs/RecognizeImage.h>

class ObjectsExample {
  public:
    ObjectsExample(ros::NodeHandle nh);
    virtual ~ObjectsExample(){};
  private:
    void recognizeObjects();
    void orResultsCallback(const or_msgs::OrMatchResult::ConstPtr& msg);
    void recognizeImagePart(const sensor_msgs/Image &img_msg);
    ros::Publisher or_command_pub_;
    ros::Subscriber or_result_sub_;
    ros::ServiceClient or_client_;
};
```

In your class implementation you can use the following code snippets. First, include the example header and initialize all subscribers and publishers in the constructor:

```
#include"ObjectsExample.h"

ObjectsExample::ObjectsExample(ros::NodeHandle nh) {
    or_command_pub_ = nh.advertise<or_msgs::OrCommand>
        ("/or/commands", 10);
    or_result_sub_ = nh.subscribe("/or/match_result", 1,
        &ObjectsExample::or_result_callback, this );
    or_client_ = nh.serviceClient<or_msgs::RecognizeImage>
        ("/or/recognize_object_image");
}
```

The following function calls the recognition on the input image topic:

```
void ObjectExample::recognizeObjects()
{
    or_msgs::OrCommand msg;
    msg.command = ORControlModule::GrabSingleImage;
    or_command_pub_.publish(msg);
}
```

The recognition result is obtained in this callback:

```
void ObjectExample::orResultsCallback(const
    or_msgs::OrMatchResult::ConstPtr& msg)
{
  for(MatchResult mr : msg->match_results)
  {
    ROS_INFO_STREAM("Recognized object "<< mr.object_name);
```

```
7        // mr.b_box is the bounding box of this object
8    }
9  }
```

In some application you might not want to use the whole input image for recognition, but only a certain region of interest. This is especially the case when you have segmented an object using geometrical information. The extracted RGB image part of the segmented object can then be put into an `sensor_msgs/Image` message and a service call used for recognition. The following code snippet shows an example:

```
1   void ObjectExample::recognizeImagePart(const sensor_msgs/Image &img_msg)
2   {
3     or_msgs::RecognizeImage srv;
4     srv.request.image = img_msg;
5     if(or_client_.call(srv))
6     {
7       // this vector contains all recognized objects inside img_msg
8       std::vector<std::string> result = srv.response.names;
9     }
10    else ROS_ERROR_STREAM("Failed to call service!");
11  }
```

Finally, a word about configuration files of the object recognition. These packages contain a mixture of `.xml` and `.yaml` files for configuration inside the `config` folder of the `or_nodes` package. The `.xml` files are the deprecated configuration files from the pre-ROS era of our software framework. On the other hand, the `.yaml` files are the well known configuration files of ROS. The most important files are the `params_asus.yaml` and `params_kinect.yaml`. Both define same parameters, but differ in the default topic for the image input. Since the Kinect sensor was used in the examples above, we continue with it. The default content of the `params_kinect.yaml` is:

```
/OrNodes/sConfigFile:          /config/custom.xml
/OrNodes/sProfile:             drinks
/OrNodes/sInputImageTopic:     /camera/rgb/image_rect_color
/OrNodes/debugOutput:          /true
/OrNodes/debugOutputPath:      /default
```

The first value is the *package relative* path to the pre-ROS custom configuration file and the second line defines the profile to be used inside the custom configuration file. There are two `.xml` files: `default.xml` and `custom.xml`. The file `default.xml` is automatically loaded on startup. It contains all parameters necessary for object learning and object recognition, encapsulated in the profile `default`. Normally, you don't need to change values in this file. To specify custom values, please use the file `custom.xml` that is the default custom file in the `.yaml` file. Currently, the profile `drinks` is defined in `custom.xml` that overrides the default value of `ObjectRecognition/sLoadObject`. This is the parameter specifying the object files that should be loaded on startup in case the parameter `or_objects` is missing on the ROS parameter server. When loading the configuration files, a merged.xml is generated by the config loader that contains all available profiles from `default.xml` and `custom.xml`. The third line in the `.yaml` file is the default image topic that the object recognition is using. The last two lines define the debug output. However, they are not fully integrated, yet.

6 Human Robot Interaction

Human-Robot interaction is another basic component needed by service robots. We provide a ROS package coupled with an Android app for speech recognition and synthesis. It integrates with any modern Android device and enables the robot to use the speech recognition and synthesis of the Android system. Android provides speech recognition and synthesis for many languages and also works completely offline, once the dictionaries have been downloaded.

As a second component of the presented interaction system we present our robot face as introduced and evaluated in [13]. Apart from a neutral face expression it can show 6 other emotions. The robot face captures the speech output of the robot and moves the lips synchronously to the spoken words.

A video showing an example animated face with synthesized speech is available online.[12]

6.1 Background

Speech Recognition Spoken commands belong to the most user-friendly interaction with robots. There are grammar based speech recognition systems that are optimized by understanding a subset of commands defined in a grammar. On the other hand, grammar free systems exist which are potentially able to understand all commands without prior training. In this section we present a grammar free system that integrates with the Android speech API.

Robot Face Different talking heads were developed in the last years for research in the field of human-robot interaction [3, 4, 8, 12, 15]. All of these heads were constructed in hardware, posing a challenge in designing and building these heads. A strong advantage, however, is the possibility to place cameras inside the head's eyes. This allows for intuitive interaction in a way that a person can show an object to the robot by holding it in front of the robot's head. Although this is not possible with a face completely designed in software, we chose this approach to create our animated robot face. In our opinion the high number of advantages of an animated head outweights its drawbacks. Apart from a screen, which many robots already have,

---

[12]Example video for the robot face: http://youtu.be/jgcztp_jAQE.

there is no specific hardware that needs to be added to the robot. Moreover, it is highly customizable and can be adjusted to everyone's individual needs. Several animated robot heads were developed in the recent years that likewise possess the ability to e.g. move their lips synchronously [1, 6, 11]. In contrast to our approach, these animated heads were designed with the goal of modeling a realistic and human-like appearance. However, we focus on human robot interaction and have intentionally created an abstracted robot face with a cartoon-like appearance.

Human-like or even photo-realistic faces are employed to convey realism and authenticity to the interacting person. On the other hand the purpose of stylized cartoon faces is to invoke empathy and emotions. So far, most robots lack humanoid features and stature, thus a realistic human face is not appropriate to interact with it. We therefore modeled an abstract cartoon face exhibiting only the most important facial features to express emotions: eyes, eyebrows and a mouth. Further, by choosing a cartoon face we minimize the risk of falling into the *uncanny valley* [9].

The lips of the robot face move synchronously to the synthesized speech. For speech synthesis the text-to-speech system *Festival* is used in the `robot_face` node. However, it can be replaced by the presented the Android speech synthesis (see below). We achieve synchronization by mapping visemes to phonemes of the synthesized text. Visemes are visually distinguishable shapes of the mouth and lips that are necessary to produce certain sounds. Phonemes are groups of similar sounds that feel alike for the speaker. Usually, only a few visemes are sufficient to achieve a realistic animation of the lips.

Additionally to the synchronized lips, the robot face is capable of expressing 6 different emotions and a neutral face. The different emotions are depicted in Fig. 18. To animate arbitrary text with the desired face expression we need dynamically generated animations. Apart from visemes we defined shape keys containing several different configurations for the eyes and eyebrows to display different face expressions. Animations are created by interpolating between the different shape keys. For animation we use Ogre3d as graphics engine, Qt as window manager and Blender for creating the meshes for the face.

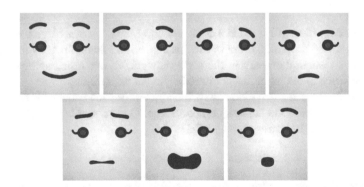

Fig. 18 Animated face of our service robot Lisa. The depicted face expressions are: happy, neutral, sad, angry, disgusted, frightened and surprised (from *left* to *right*)

6.2 ROS Environment Configuration

Speech Recognition As the communication between the Android app and the ROS node is solved by sockets you have to ensure that the ROS machine and the Android device are in the same network. An easy solution for doing this is to use your Android smartphone as wifi hotspot and connect the ROS machine to it. Additionally you have to ensure that both applications operate on the same port. You can change the port in the `recognition.launch` file in the `android_speech_pkg`. This port needs to be the same as entered in the `homer_speech` app.

```
<!-- Bind to all addresses -->
<param name="~host"value=""/>
<!-- Port -->
<param name="~port"value="8051"/>
```

Robot Face In order to use this package you have to install the following libraries:

- Festival: http://www.cstr.ed.ac.uk/projects/festival
- Ogre3d: http://www.ogre3d.org
- Qt: http://qt.nokia.com
- OpenCV: http://opencv.org (usually comes with ROS)

If you want to create your own robot face you will additionally need Blender (http://www.blender.org/). You can install all of these libraries and voices for speech synthesis by opening a terminal and typing (all in one line):

```
sudo apt-get install libqt4-core libqt4-dev libqt4-gui festival
festival-dev festlex-cmu festlex-poslex festlex-oald mbrola
mbrola-us1 libogre-1.8-dev libesd0-dev libestools2.1-dev libpulse-dev
```

6.3 Quick Start and Example

Speech Recognition We provide a compiled app `homer_speech.apk` (Fig. 19) for your Android device. In order to use this app, ensure that both, the robot's computer and the smartphone are in the same network. In our tests, we attached the Android device to the robot in a comfortable height for speaking.

To use the *speech recognition*, start the app and set the robot's IP and port in the edit boxes inside the `homer_speech` app. On the robot, run the following command to start the corresponding ROS node:

```
roslaunch android_speech_pkg recognition.launch
```

The *Recognize* button initiates the recognition process and either stops listening after recognizing silence or when the *Stop Recognition* button is pressed. So

Fig. 19 The
homer_speech app. The
Speak and *Recognize* buttons
provide speech synthesis and
recognition, respectively.
The two edit boxes in the
lower part of the screen are
used to configure the
connection for speech
recognition

far, the app allows to switch between *German* and *English US* speech recognition. The recognized speech will be send as a std_msgs/String on the /recognized_speech topic. You can test the recognition by pressing the *Recognize* button and typing the command

```
rostopic echo /recognized_speech
```

on the robot's computer.

To use the *speech synthesis*, the host parameter in the synthesis.launch file needs to be adjusted to the Android device's IP. The device's IP is shown in the homer_speech app in the bottom line (Fig. 19). On the robot, run the following command for speech synthesis:

```
roslaunch android_speech_pkg synthesis.launch
```

You can test your configuration either by pressing the *Speak* button or with the following command:

```
rostopic pub /speak std_msgs/String"data: 'Hello World'"
```

Robot Face After setting up the ROS environment and executing the catkin_make command in your ROS workspace you can start the robot_face by typing the following command:

Fig. 20 The female and male version of the `robot_face` showing in white the synthesized speech and in red the recognized speech

```
roslaunch robot_face robot_face.launch
```

A window showing the robot face appears similar to the one shown in Fig. 20 on the left. Note that with this launch file the speech will be synthesized by *festival*. Thus, the `robot_face` can be used without any Android device. In order to use the speech synthesis of the presented app, please use the following command:

```
roslaunch robot_face robot_face_android_synthesis.launch
```

Type the following command to let the robot speak:

```
rostopic pub /robot_face/speak std_msgs/String
"Hello, how are you?"
```

You will hear a voice and the face moves its lips accordingly.

If the speech recognition node is running, you can answer and see the recognized speech displayed below the face. Without speech recognition, you can fake the displayed message by typing the command:

```
rostopic pub /recognized_speech std_msgs/String
"Thank you, I am fine!"
```

Table 1 Available emoticons and the corresponding face expressions

Face expression	Neutral	Happy	Sad	Angry	Surprised	Frightened	Disgusted
Emoticon	.	:)	:(	>:	:o	:&	:!

To activate different face expressions, just include the corresponding emoticon in the string that you sent to the robot. For instance, to show a happy face expression, you can type:

```
rostopic pub /robot_face/speak std_msgs/String
"This is the happy face :)"
```

Table 1 shows all available face expressions and the intended emotions. Please refer to the user study in [13] on how these face expressions are perceived by humans. Apart from showing face expressions and synchronized lip movements, the `robot_face` can also show images and video streams instead of the face. Please refer to the following section to use this functionality. If you want to create your own robot face for this package, please refer to the ReadMe file.[13]

6.4 Package Description and Code Examples

Speech Recognition The speech recognition component is divided into an Android app and a ROS node that converts the recognized text into a ROS conform message. The recognition is able to understand different languages that can be changed in the android system settings. An active internet connection improves the recognition results, but is not mandatory. Currently, the speech recognition needs to be started by pressing the *Recognize* button. If needed a topic for starting the recognition and sending a start recognition request can be easily integrated.

Robot Face The graph in Fig. 21 shows the two nodes contained in the `robot_face` package with all topics. The creation of phonetic features including speech synthesis is handled by the `FestivalSynthesizer` node, while the `RobotFace` node creates the corresponding animations. When using an Android device, festival still creates the phonetic features for animation. However, the Google speech API takes care of the actual speech synthesis.

The topics are explained in the following:

Publishers

`/robot_face/talking_finished` (std_msgs/String): Published once the `robot_face` has completed the speech synthesis and stopped speaking. This is used to synchronize the animation and can be used in your application to know when the speech is finished.

---

[13] A ReadMe is available online in the software repository.

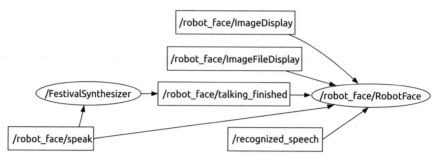

Fig. 21 Nodes and topics involved in the robot_face package

Subscribers

/robot_face/speak (std_msgs/String): Text to be spoken and displayed
below the face.

/recognized_speech (std_msgs/String): Text recognized by the speech recognition. It will be displayed below the face.

/robot_face/ImageDisplay (robot_face/ImageDisplay): Used to display
an std_msgs/Image message instead of the face for a specified time.

/robot_face/ImageFileDisplay (robot_face/ImageFileDisplay): Used to
display an image from hard disk instead of the face for a specified time.

To use all features of the robot_face in your application, please include the
robot_face as *build* and *run* dependency in your package's *package.xml* file.
Further, robot_face has do be added to the find_package section and as
a dependency to the catkin_package section in the CMakeLists.txt file.
Note, that this is only needed if you want to use the robot_face to display images
instead of the face, since custom messages are used for this functionality. An example
header to use the robot_face could look like this:

```
1   #include <ros/ros.h>
2   #include <std_msgs/String.h>
3   #include <sensor_msgs/Image.h>
4   #include <robot_face/ImageDisplay.h>
5
6   class FaceExample {
7     public:
8       FaceExample(ros::NodeHandle nh);
9       virtual ~FaceExample(){};
10    private:
11      void speak(std::string text);
12      void talkingFinishedCallback(const std_msgs::String::ConstPtr& msg);
13      void imageCallback(const sensor_msgs::Image::ConstPtr& msg);
14      void send_image(std::string& path_to_image);
15      ros::Publisher speak_pub_;
16      ros::Publisher image_pub_;
17      ros::Subscriber talking_finished_sub_;
18      ros::Subscriber image_sub_;
19  };
```

In your class implementation you can use the following code snippets. First, include the example header and initialize all subscribers and publishers in the constructor:

```
#include"FaceExample.h"

FaceExample::FaceExample(ros::NodeHandle nh)
{
  speak_pub_ = nh.advertise<std_msgs::String>("robot_face/speak", 1);
  image_pub_ = nh.advertise<robot_face::ImageDisplay>
      ("/robot_face/ImageDisplay", 1);
  talking_finished_sub_ = nh.subscribe("robot_face/talking_finished",1,
      &FaceExample::talking_finished_callback, this);
  image_sub_ = nh.subscribe("/camera/rgb/image_color",1,
      &FaceExample::image_callback, this);
}
```

Use the following function to send speech commands to the face:

```
void FaceExample::speak(std::string text)
{
  std_msgs::String msg;
  msg.data = text;
  speak_pub_.publish(msg);
}
```

As soon as the face finishes talking, the following callback will be called:

```
void FaceExample::talkingFinishedCallback(
    const std_msgs::String::ConstPtr& msg)
{
  ROS_INFO_STREAM("Finished speaking");
}
```

Apart from displaying text, the robot_face can be used to display images instead of the face. The following code snippet demonstrates how to display a sensor_msgs/Image message from an RGB-D camera:

```
void FaceExample::imageCallback(const sensor_msgs::Image::ConstPtr& msg)
{
  robot_face::ImageDisplay img_msg;
  img_msg.time = 0; // display time [s]
  img_msg.Image = *msg;
  image_pub_.publish(img_msg);
}
```

A display time of 0 seconds means that the image will be displayed until another message is received. To use this code, make sure that a camera is connected to your computer and the corresponding driver is loaded, e.g. by typing

```
roslaunch openni_launch openni.launch
```

To display an image from the hard disk use the following function:

```
void FaceExample::send_image(std::string& path_to_image)
{
  robot_face::ImageFileDisplay img_msg;
  img_msg.time = 0; // display time [s]
  img_msg.filename = path_to_image;
  image_pub_.publish(img_msg);
}
```

7 Conclusion

In this chapter we presented several components that are crucial for autonomous service robots and components that enhance human robot interaction. The presented mapping, navigation and object recognition are closely integrated into a graphical user interface. We believe that this close integration into a common graphical interface facilitates the development, control and monitoring of autonomous robots. Additionally, the speech recognition and animated robot face provide important interfaces for human robot interaction. We hope that this extensive tutorial together with the presented software components and videos are a valuable contribution to the ROS community.

Acknowledgments The presented software was developed in the Active Vision Group by research associates and students of the practical courses with the robots "Lisa" and "Robbie". The authors would like to thank Dr. Johannes Pellenz, David Gossow, Susanne Thierfelder, Julian Giesen and Malte Knauf for their contributions to the software. Further, the authors would like to thank Baharak Rezvan for her assistance in testing the setup procedure of the described packages.

References

1. I. Albrecht, J. Haber, K. Kahler, M. Schroder, H.P. Seidel, May i talk to you?:-)-facial animation from text. In: *Proceedings 10th Pacific Conference on Computer Graphics and Applications, 2002.* (IEEE, 2002) pp. 77–86
2. H. Bay, T. Tuytelaars, L.J. Van Gool. SURF: Speeded up robust features. ECCV, pp. 404–417, 2006
3. C. Breazeal, Toward sociable robots. Robot. Auton. Syst. **42**(3), 167–175 (2003)
4. C. Breazeal, B. Scassellati, How to build robots that make friends and influence people. In: *Proceedings IEEE/RSJ International Conference on Intelligent Robots and Systems, 1999. IROS'99*, vol. 2, (IEEE, 1999), pp. 858–863
5. W. Eric, L. Grimson, D.P. Huttenlocher, On the sensitivity of the hough transform for object recognition. IEEE Trans. Pattern Anal. Mach. Intell. PAMI **12**(3), 255–274 (1990)
6. J. Gustafson, M. Lundeberg, J. Liljencrants, Experiences from the development of august—a multi- modal spoken dialogue system. In: *ESCA Workshop on Interactive Dialogue in Multi-Modal Systems (IDS-99)* (1999), pp. 61–64
7. David G. Lowe, Distinctive image features from scale-invariant keypoints. Int. J. Comput. Vis. **60**(2), 91–110 (2004)

8. I. Lütkebohle, F. Hegel, S. Schulz, M. Hackel, B. Wrede, S. Wachsmuth, G. Sagerer, The biele-
    feld anthropomorphic robot head flobi. In: *2010 IEEE International Conference on Robotics
    and Automation*, (IEEE, Anchorage, Alaska, 2010) 5
9. M. Mori, K.F. MacDorman, N. Kageki, The uncanny valley [from the field]. Robot. Autom.
    Mag. IEEE **19**(2), 98–100 (2012)
10. M. Muja, Flann, fast library for approximate nearest neighbors (2009). http://mloss.org/
    software/view/143/
11. A. Niswar, E.P. Ong, H.T. Nguyen, Z. Huang, Real-time 3d talking head from a synthetic viseme
    dataset. In: *Proceedings of the 8th International Conference on Virtual Reality Continuum and
    its Applications in Industry*, (ACM, 2009), pp. 29–33
12. J. Ruiz-del-Solar, M. Mascaró, M. Correa, F. Bernuy, R. Riquelme, R. Verschae, Analyzing
    the human-robot interaction abilities of a general-purpose social robot in different naturalis-
    tic environments, *RoboCup Symposium 2009*, Lecture Notes in Computer Science (Springer,
    Berlin, 2009), pp. 308–319
13. V. Seib, J. Giesen, D. Grüntjens, D. Paulus, Enhancing human-robot interaction by a robot face
    with facial expressions and synchronized lip movements, ed. by V. Skala. *21st International
    Conference in Central Europe on Computer Graphics, Visualization and Computer Vision*
    (2013)
14. Viktor Seib, Michael Kusenbach, Susanne Thierfelder, and Dietrich Paulus. Object recognition
    using hough-transform clustering of surf features. In: *Workshops on Electronical and Computer
    Engineering Subfields*, (Scientific Cooperations Publications, 2014), pp. 169–176
15. S. Sosnowski, A. Bittermann, K. Kuhnlenz, M. Buss, Design and evaluation of emotion-display
    eddie. In: *IEEE/RSJ International Conference on Intelligent Robots and Systems, 2006* (IEEE,
    2006), pp. 3113–3118
16. S. Wirth, J. Pellenz, Exploration transform: A stable exploring algorithm for robots in rescue
    environments. In: *IEEE International Workshop on Safety, Security and Rescue Robotics, 2007.
    SSRR 2007*, (2007), pp. 1–5
17. A. Zelinsky, Robot navigation with learning. Aust. Comput. J. **20**(2), 85–93 (1988). 5
18. A. Zelinsky, Environment Exploration and Path Planning Algorithms for a Mobile Robot using
    Sonar, Ph.D. thesis, Wollongong University, Australia, 1991

Author's Biography

Dipl.-Inform. Viktor Seib studied Computer Sciences at the
University of Koblenz-Landau where he received his Diploma
in 2010. In 2011 he became a PhD student at the Active Vision
Group (AGAS) at the University of Koblenz-Landau, supervised
by Prof. Dr.-Ing. Dietrich Paulus. His main research areas are
object recognition and service robotics. He has attended multiple RoboCup and RoCKIn competitions and is currently team
leader and scientific superviser of the RoboCup@Home team
homer@UniKoblenz.

B.Sc. Raphael Memmesheimer finished his Bachelor in Computational Visualistics and is currently in his master studies focusing on robotics and computer vision. His special interests are visual odometry, visual SLAM and domestic service robots. As research assistant he is working in the Active Vision Group (AGAS) and has successfully attended multiple RoboCup and RoCKIn competitions. Currently he is technical chief designer of the homer@UniKoblenz RoboCup@Home team and states that "Robots are the real art"!

Prof. Dr.-Ing. Dietrich Paulus obtained a Bachelor degree in Computer Science from University of Western Ontario, London, Ontario, Canada, followed by a diploma (Dipl.-Inform.) in Computer Science and a PhD (Dr.-Ing.) from Friedrich-Alexander University Erlangen-Nuremberg, Germany. He worked as a senior researcher at the chair for pattern recognition (Prof. Dr. H. Niemann) at Erlangen University from 1991–2002. He obtained his habilitation in Erlangen in 2001. Since 2001 he is at the institute for computational visualistics at the University Koblenz-Landau, Germany where he became a full professor in 2002. From 2004-2008 he was the dean of the department of computer science at the University Koblenz-Landau. Since 2012 he is head of the computing center in Koblenz. His primary research interest are active computer vision, object recognition, color image processing, medical image processing, vision-based autonomous systems, and software engineering for computer vision. He has published over 150 articles on these topics and he is the author of three textbooks. He is member of Gesellschaft für Informatik (GI) and IEEE.

Lisa is the robot of team homer@UniKoblenz. She first participated in the RoboCup GermanOpen and the RoboCup World Championship in 2009. Since then, her design was continuously changed and improved to better adapt to the requirements of service robot tasks. Lisa is developed by PhD, master and bachelor students in practical courses of the Active Vision Group. Lisa and team homer@UniKoblenz won the 1st Place in the RoboCup@Home World Championship 2015 in Hefei, China.

Robotnik—Professional Service Robotics Applications with ROS

Roberto Guzman, Roman Navarro, Marc Beneto and Daniel Carbonell

Abstract This chapter summarizes our most relevant experiences in the use of ROS in the deployment of Real-World professional service robotics applications: a mobile robot for CBRN intervention missions, a tunnel inspection and surveillance robot, an upper body torso robot, an indoor healthcare logistic transport robot and a robot for precision viticulture. The chapter describes the mentioned projects and how ROS has been used in them. It focuses on the application development, on the ROS modules used and the ROS tools and components applied, and on the lessons learnt in the development process.

Keywords Professional service robotics applications with ROS · Civil protection · Agriculture robotics · Logistic robotics · Service robotics · Autonomous robots · Mobile robots · Mobile manipulators · AGVS · UGVS

1 Contributions of the Book Chapter

ROS has become a standard for the development of advanced robot systems. There is no other platform that integrates a comparable community of developers. This chapter describes a number of professional service robotics applications developed in ROS. ROS was designed for the development of large service robots [1] as the ones presented in this chapter.

R. Guzman (✉) · R. Navarro · M. Beneto · D. Carbonell
Robotnik Automation, SLL, Ciutat de Barcelona, 1H,
P.I. Fte. Del Jarro - Paterna Valencia, Spain
e-mail: rguzman@robotnik.es
URL: http://www.robotnik.eu

R. Navarro
e-mail: rnavarro@robotnik.es

M. Beneto
e-mail: mbeneto@robotnik.es

D. Carbonell
e-mail: dcarbonell@robotnik.es

© Springer International Publishing Switzerland 2016
A. Koubaa (ed.), *Robot Operating System (ROS)*, Studies in Computational
Intelligence 625, DOI 10.1007/978-3-319-26054-9_10

While it is relatively easy to find R&D projects using it, the number of publicly documented Real-World applications and in particular in product development and commercialization is still relatively low as can be deducted from the last editions of the ROSCon (ROS Conference) [2] programmes.

This chapter presents as main contribution the description of five real products that use ROS, detailing the principal challenges found from the point of view of a ROS developer.

2 RESCUER: Robot for CBRN Intervention

Intervention in dangerous environments requires the use of robots to monitor and interact remotely without putting human life at risk. There are several robots in the CBRN market but few of them based on ROS. RESCUER [3] is a robot proto-type developed for the Spanish military emergency unit (UME) to be used in CBRN (Chemical, Biological, Radiological and Nuclear) environments. It's a mobile manipulator able to operate in extreme environments. The robot has two available kinematic configurations, wheels or "flippers", that can be set before every mission, depending on the type of terrain.

This part includes a brief description of the system and work done, the different components and packages used and developed, the developed HMI and the problems found in some points, like the communication/teleoperation in low latency networks.

Due to the nature of the project and associated NDA (Non Distribution Agreement) with the end user, the packages of this robot are not available.

2.1 Brief Description of the System

RESCUER was designed as an intervention robot to be used in the following set of missions:

- Detection and alert of aggressive agent concentrations
- Identification of aggressive agents
- Recognition and sampling (take field samples of liquid, solid and mud) for further lab analysis
- Assistance in industrial accidents or terrorist action scenarios
- Environment protection and safeguarding
- Human rescue

RESCUER is a modular robot for intervention missions. It has IP67 enclosure class and has been designed with the focus on system decontamination. As a difference with many robots in the market, RESCUER is able to operate in CBRN environments and it is easy to decontaminate after the mission by means of standard and fast procedures. The following pictures show the robot in the two main kinematic configurations and including the arm, sensor set, winding unit and CBRN sensor unit (Fig. 1).

Fig. 1 Rescuer platform with flippers (*left*) and wheel (*right*)

Robot Configuration

RESCUER implements an innovative modular system that allows the following kinematic configurations:

- Skid steering with 4 flippers (8 axes): high mobility—medium speed (0.75 m/s), the robot is able to surpass obstacles, gaps and climb ramps and stairs. With the combined motion of the flippers (all axes can rotate indefinitely) an incomparable mobility is obtained.
- Skid steering with 4 wheels (4 axes): high speed (5 m/s)—medium mobility. The robot does not have the mobility performance of previous configuration, but it still has all-terrain capability and a very high speed which can be critical in reconnaissance or rescue missions (Fig. 2).

Fig. 2 Rescuer platform with flippers (*right*) and wheels (*left*)

The wheels and flippers, the arm, the winding unit and the sensor unit, implement a unique electro-mechanical interface. I.e. the connector acts not only as an electrical interface, but also provides the mechanical connection between the device and the platform.

The robot chassis contains the batteries, motors, the robot controller (an embedded PC using Ubuntu 12.04 and ROS), etc.

Robot Arm

RESCUER mounts a special IP67 version of the Terabot arm (Oceaneering). The arm has 5 DOF and is provided with internal clutches that protect the joints of torque peaks in collisions. The arm mounts an electric two-finger gripper with interchangeable fingers and able to use a set of tools (drill machine, wire cutter, etc.).

Robot Sensors

The set of sensors can be classified in the following groups:

- Navigation, motion and internal state: Inertial measurement unit, wheel encoders, flipper absolute encoders, DGPS (Differential Global Positioning System), outdoor laser range-finder, internal temperature.
- Platform cameras: front and rear IR (Infra-Red) Cameras (plus focus leds), 360°-PTZ (Pan-Tilt-Zoom) camera.
- Arm cameras and distance sensor: thermal camera, gripper camera, wrist camera, and laser telemeter.
- Bidirectional audio: IP67 audio (microphone and amplified loudspeaker).
- CBRN Sensor Suite: Radiation sensor, gas sensor, CWA/TIC (Chemical Warfare Agents/Toxic Industrial Chemicals) sensor, biological war-agents detector and pressure temperature and humidity sensor.

Communications

The RESCUER communication system permits the interconnection of the robot with the operator HMI and the control base station. RESCUER implements a multi-protocol communication system. The used interfaces are:

- Fiber Optic Ethernet Cable: 100 Mbps, winding unit able to release cable and recover it autonomously. The winding unit mounted a standard length of 100 m but is easily extensible.
- WiFi: 100 Mbps, distance with high gain antennas up to 75 m with LOS (Line Of Sight).
- Radio (900 MHz): 120 Kbps, distances up to 50 Km with NLOS (Near Line Of Sight).

The use of alternative communication interfaces in the mentioned scenario provide robustness as the system is configured as fallback, i.e. it is programmed to select the highest bandwidth channel from the available.

HMI

RESCUER HMI is arranged in two main cases. The main HMI case includes the PC and mounts a dual tactile array screen configuration. The second case includes the batteries, communication equipment and antennas. The top screen displays the robot state (arm, flippers, robot attitude), cameras, telemeter, laser, etc. while the bottom screen displays the CBRN sensor data. The top screen allows the configuration of the cameras, and the selection of the operational mode. Depending on the operational mode, the HMI joysticks control the robot arm, the platform motion or flipper motion or pan-tilt-zoom camera motion (Fig. 3).

The RESCUER HMI was implemented using rqt [4], rqt is a Qt-based framework for GUI development for ROS. The rqt_robot_dashboard infrastructure was used for building robot plugins. In order to control the RESCUER robot a number of rqt plugins were developed to control the mobile platform, robot arm, input/output (front focus, rear focus, arm focus, winding unit), control of the pan-tilt-zoom camera, display the different specific CBRN sensors and cameras, or display robot general information (battery, control mode arm/base, etc.). A set of packages (non rqt) were also developed to bring up the whole HMI, monitor the network, control the robot components (base platform, arm and gripper, input/output, ptz camera) via the HMI joysticks, use interactive markers attached to base_link to show the robot state (robot state, position and battery), or visualize GPS (Global Positioning System) data in order to track the robot route via external GIS (Geographic Information System) tool (e.g. GoogleEarth [5]).

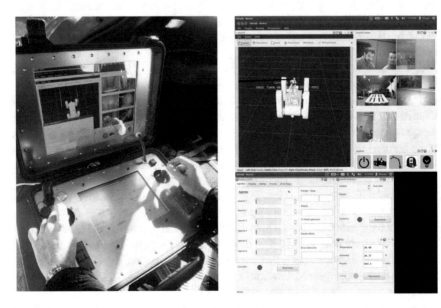

Fig. 3 Rescuer HMI dual tactile screen control box (*left*). HMI screen capture during testing (*right*)

The use of rqt with the tactile array was optimal. The main difficulties were in the configuration of a dual tactile array with different resolutions due to the lack of configuration examples for this configuration. After many tests, the right system configuration was found and worked.

One pitfall of rqt for this kind of interfaces is the lack of an option to set the rqt layout fixed to disable the accidental edition, since you can drag and drop any component and set the size by hand. This issue could not be solved and remains present in the interface.

The ROS version used for this project (in the robot platform and in the HMI) was Groovy, running on a Intel PCs with Ubuntu 12.04.

2.2 Challenges

2.2.1 Communications

Real world intervention missions impose complex communication requirements. There is a number of missions where the robot will not have line-of-sight conditions nor short distances. In particular, missions to be carried on inside buildings, subways, parkings, etc., cannot be executed relying only on standard Wi-Fi networks. In order to cope with the different limitations in terms of coverage, range and bandwidth, a multi-protocol architecture was designed.

During the project there were important hardware challenges, e.g. a fiber optic winding unit with the specific mission requirements was not available in the market so it had to be designed from scratch. The number of devices and architectures tested would be also of interest but are out of the scope of this text. The final selected architecture is described in the following picture (Fig. 4).

As explained the system includes six cameras, a set of CBRN sensors, the internal robot sensors, and bidirectional audio. The data flow is principally from the robot to the HMI, while the amount of data sent from the HMI to the robot is limited to task commands (arm tasks, platform configuration tasks), control commands (platform teleoperation) and voice stream (the robot in this case is used to send voice commands to the humans in the mission area). The following table shows an estimation of the data bandwidth requirements of the system cameras (Table 1).[1]

Due to the nature of the intervention mission, the available bandwidth can change at any time without prior notice. E.g. if the robot enters in a building and it is communicating via Wi-Fi at some point the effect of the walls and distance may reduce the bandwidth to some Kbps. These bandwidth changes will lead to packet loss. The ROS version used in this project (Groovy), supports TCP/IP-based and UDP-based message transport. However, during the software development stage the UDP-based message transport available in ROS was limited to C++ libraries. As the

---

[1] Front, Rear, Arm.

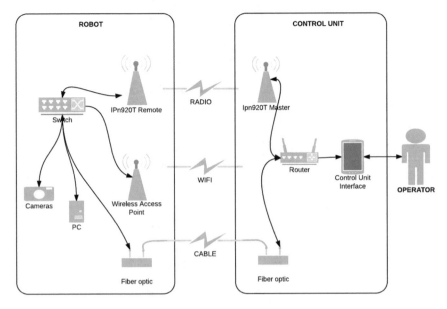

Fig. 4 Rescuer multi-protocol communication hardware architecture

HMI was developed using rviz and rqt and both tools depend on python, it was not possible to use UDP-based message transport within the ROS framework.

The first implementations of the architecture used therefore TCP for all the components. As a consequence, when the communication bandwidth was too low and after some time the retransmission of TCP frames caused a major fault in ROS being necessary a system restart.

To overcome this problem, the video and audio streams were finally transported via UDP, i.e. data was received and displayed by ROS nodes, but the data was not transported using the ROS middleware. From the presented bandwidth table, it is clear that the major part is used for the cameras and audio (up to 10 Mbps depending on configuration), while only a minor part is needed for the sensors and control (usually less than 50 Kbps).

A second level of adaptation to the available bandwidth occurs via the robot HMI. The system starts with a low bandwidth configuration, and the operator may ask to increase it by activating more cameras, or increasing frame rate or definition. If at some point, the requested bandwidth is not available, this condition is immediately marked in the screens, and the operator can compare the data throughput requested with the system monitor network bandwidth measured, thus being able to disconnect some cameras and adjust frame rates. The described communication architecture proved to be reliable and functional, failing only in those cases where the communication channel was completely lost for long periods.

Table 1 Camera bandwidth and storage requirements

Camera	F, R, A	F, R, A	F, R, A	F, R, A	PTZ	PTZ	FLIR	FLIR
Compression	H.264	H.264	H.264	H.264	H.264	H.264	MPEG-2	MPEG-2
Resolution [px]	1280 × 1024	1280 × 1024	704 × 480	704 × 480	704 × 576	704 × 576	320 × 240	320 × 240
Video quality	Maximal	Standard	Maximal	Standard	Maximal	Standard	Maximal	Standard
1 frame [Kb]	13.9	5.35	4.29	1.64	5.14	1.97	4.2	1.6
Acquisition frequency [fps]	8	8	8	8	8	8	1	1
Bandwidth	1.74 Mbps	669 Kbps	535 Kbps	205 Kbps	600 Kbps	246 Kbps	66 Kbps	25 Kbps
2 h storage	16.85 Gb	6.48 Gb	5.18 Gb	1.99 Gb	5.6 Gb	2.3 Gb	626 Mb	237 Mb

2.2.2   Robot Control

As the mobile robot platform can be switched from 4-axis wheels to 8-axis flippers, the system had to be launched with two different launch configurations. The one with the flipper urdf and controller, that launched also the absolute encoders of the flippers and the one with the wheel urdf and controller that did not launch the absolute encoders node. As the two configurations used different hardware, it was possible to detect in which one the robot had been switched on and configure the launch files accordingly automatically. The configuration change was detected by launching the node to read the absolute encoders (of the 4 flippers). If it is possible to establish the connection with the absolute encoders, the node sets an environment variable and sets a parameter to "flippers" in the parameter server, if it is not possible to establish the connection the environment variable is set to false and the ROS parameter is set to "wheels".

As we did not find a standard procedure for the flipper position control, the way the flippers were controlled via remote operation was developed experimentally. The final implementation permitted to move each flipper independently and to pair the front and back to receive the same position or velocity reference. Excluding some specific operations (e.g. gap cross passing or high obstacle overpassing), the most easy and intuitive way to move the flippers was using the front pair and rear pair as separated joystick inputs.

In any case the continuous monitoring of the flipper absolute position and robot state was of major concern for two main reasons:

- The range of states that the platform can achieve by controlling the flipper position is high, e.g. it is possible to lift the robot making it rest on the flipper tips or make it sit on the two rear ones, etc. This provides very high mobility but the control has to be done carefully as a wrong decision can compromise the platform stability. Controlling the platform from the viewpoint of the operator is easy, however, this becomes very complex from the viewpoint of the robot (by using the cameras). For this reason, the platform inclination and exact position of each flipper is displayed on the HMI.
- The kinematics permitted the potential collision of the front and rear flipper, and also the collision between the arm and the flippers. The flipper rotation motors are powerful and provide a very high torque able to damage the arm or other flipper in case of collisions. A collision of the arm against the flipper should not cause damage as the arm joints are protected by clutches. However, it is possible to trap the arm between the flipper and the chassis when the arm is accessing soil on the robot sides, in this case the collision of the flipper against the arm may break it.

In summary, potential collisions between them and between the flippers and the robot arm have to be avoided. The way to deal with this was implementing a SelfCollisionCheck function (calls planning_scene::checkSelfCollision) in the

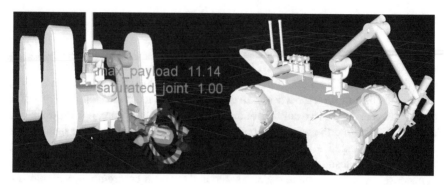

Fig. 5 Control of the end effector via MoveIt Interactive Markers (*left*), tcp positions marked (*right*)

rescuer_robot_control package. This function computed a collision distance between the robot elements on each iteration, once a predefined threshold is surpassed, only new motions that increase the collision distance are permitted.

The arm control and trajectory generator is based on MoveIt! libraries. MoveIt! is state of the art software for mobile manipulation, incorporating the latest advances in motion planning, manipulation, 3D perception, kinematics, control and navigation [6]. The arm controller implements a JointTrajectoryAction server. It is based on MoveIt! components (moveit::planning_interface::MoveGroup, moveit::core::Robot Model, moveit::core::RobotState, etc.) to allow the Cartesian/Euler, joint-by-joint control and check self collisions.

A requirement of the Military Emergency Unit was related with the TCP (Tool Center Point) selection. The arm can mount a set of tools and the TCP can also be used to control the arm cameras. We did not find an easy way to dynamically change the TCP as the input to MoveIt! is the .srdf file generated by the setup assistant. The way we solved this issue was creating a set of TCPs in the robot urdf file, relating them with each camera and tool and creating a set of MoveIt groups with the arm plus each of the predefined TCPs, thus allowing the TCP selection as a group selection (Fig. 5).

Another major issue was related with the solution of the 5-DOF Terabot Arm inverse kinematics. At the time the robot was developed the KDL library [7] was not able to solve it. All our attempts to create a working plugin using ikfast failed also, so we finally ended modifying the KDL plugin and adapting the solution for 5-DOF.

The operator can move the arm by moving the end effector via interactive markers in the interface, or with the joysticks (joint-by-joint and cartesian/euler control). This teleoperation task is assisted by the robot state feedback of rviz, the telemeter data (distance to grasp) visualized in rviz and the cameras, especially the arm and gripper cameras. In addition, the operator had a set of predefined arm locations and tasks: pick&place of tool i, move arm to pre-grasp, pick & place radiation gauge, fold arm, etc., that automate part of the teleoperation work.

3 R-INSPECT: Mobile Robot for Tunnel Inspection

In order to bring security and reliability to the new high voltage power line between France and Spain, Robotnik has developed a fleet of 4 robotic trains intended to supervise the state of all the tunnel sections.

The line has a length of 64.5 km, 33.5 in France and 31 in Spain, crossing the Pyrenees thought a 8.5 km tunnel in the middle part of the route. This is an ambitious European project developed by corporation formed by the public electricity companies of each country.

The objectives of the interconnection between both countries are the optimization of the electrical power stations daily production, increase the opportunities to operate with renewable energies and the improvement of supply conditions.

The main tasks of the vehicles are: perform autonomous missions to scan and record all the data obtained by their sensors, in order to post-process and detect any problem the tunnel; provide tele-operation capabilities to the maintenance workers; provide manual transport to the maintenance workers and firefighters.

Due to the nature of the project and associated NDA agreements with the end user, the packages of this robot are not available.

3.1 Brief Description of the System

The vehicles are distributed between France and Spain (two on every entry of the tunnel). One of the two vehicles is reserved for rescue/emergency operations.

A normal use case would be:

- The operator selects the vehicle and initializes an autonomous scan of entire the tunnel.
- The vehicle performs the mission and saves all the data.
- The vehicle returns to its garage and post-process all the recorded data, generating a report, available for the allowed users.
- If there were any errors, the operators of each part of the tunnel would have to coordinate with the other part to access it by using the tele-operated or manual mode of the vehicle (Fig. 6).

Each vehicle is controlled by a PC-based controller running ROS. It is intended to control and read from the following components:

- Motor controllers
- Input/Outputs devices
- Batteries components
- Lasers (safety and measurement)
- Cameras
- Other sensors

Fig. 6 R-INSPECT Autonomous Inspection Robot inside the tunnel

- Humidity/temperature sensor
- Oxygen sensor
- Nitrogen monoxide and dioxide sensors
- Carbonic monoxide and dioxide sensors
- Methane sensor

The ROS version used is Hydro running on Ubuntu 12.04. The controller has an i7 processor with 16 GB of memory and at least 4 TB of hard disk.

3.2 Challenges

3.2.1 Simulation

One of the most important project challenges was the creation of a real and reliable model of the vehicles and the environment, that allows the development and tests of most of the parts the software system.

The vehicles and a reduced and simplified part of the tunnel was modelled and simulated with Gazebo [8]. This was difficult at the beginning because the trains and tunnel models had to be modeled and got running properly. However, these simulations allowed to begin to work in the global behaviour of the whole system, to try different fleet management systems and collision detection algorithms in a safe environment, even without having visited the facilities and even though during the manufacture of the robots. Robot system simulation is a necessary development

Fig. 7 Gazebo simulation of the tunnel infrastructure, robot and sensor systems

tool almost in every project. Simulation allows to prototype, develop and test the whole system independently and reducing significantly the integration time in the real environment (Fig. 7).

To model the traction of the vehicle two *effort_controllers/JointEffortController* controllers were used, one for the frontal wheels and another for the rear ones. The rails were modelled as depicted in the Fig. 8.

All the sensors and cameras were modelled by using Gazebo standard plugins: libgazebo_ros_camera.so for the cameras, libgazebo_ros_openni_kinect.so for the 3D lasers, libgazebo_ros_laser.so for the 2D lasers. All the cameras except one are located in the frontal part; the 3d sensors are in both parts; the 2d lasers are mounted one in the horizontal plane for safety purposes and one in the vertical plane to scan the environment when the robot is moving.

3.2.2 Point Cloud Library

Although Point Cloud Library(PCL) [9] is currently a stand-alone library, it was developed by Willow Garage as part of the ROS core in precedent distributions, that's why it's easy to include in new ROS projects and there are a lot of packages minded to convert from and to both format types. PCL was used in this project for two main purposes:

Fig. 8 Vehicle's wheel in contact with the rail in Gazebo

- Safety: Two 2D lasers are mounted in the bottom of the robot for the basic safety checking with an approximate range of 40 m. However, there are a lot of areas that are outside of this plane. In order to avoid this problem, a 3D camera is mounted on the top of the robot. This camera gets a complete 3D representation of the 10 posterior meters of the tunnel. An algorithm was developed using PCL to filter a projection of the surface of the train through this point cloud, if there are some point inside this volume, is considered that there is an obstacle and the train will brake according to the distance to it.
- Inspection: In order to get a 3D representation of the tunnel to work with the PCL libraries, two other 2D measurement lasers were mounted in upward direction. With this measurements, the robot pose and the use of the corresponding methods of PCL, it is possible to generate a 3D representation of the tunnel. Afterwards, it is possible to use other methods and filters of the library to get the interesting parts of the cloud (for example, the cable sections) and perform a further analysis with this simplified cloud (Fig. 9).

3.2.3 Multimaster

Another important challenge is the coordination between the different robots. The robots communicate with the control servers via WiFi. The tunnel has an installation of WiFi nodes in a roaming configuration that allows the continuous communication between the vehicles and the servers. To coordinate and communicate with the robots, the following solutions were considered:

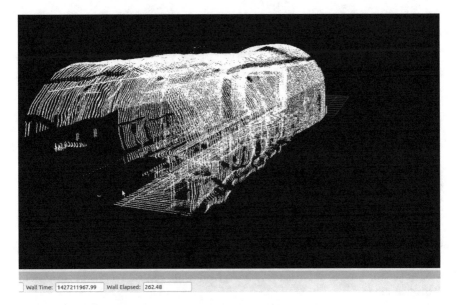

Wall Time: 1427211967.99   Wall Elapsed: 262.48

Fig. 9 PointCloud representation of a tunnel's section

- Manage all the system/network with a single roscore node, depending all the robots on this core.
- Create two cores, one to manage the robots in each part.
- Create one master per train and per control room, creating each of this a private network, and communicate them through a common network using the package multimaster_fkie [10], this package offers a set of nodes to establish and manage a multimaster network.

The third option was considered the best one; with this method each train is independent from the others, allowing to keep the complete system running independently even if one of the trains loses the communication with the control room. Furthermore, deploying this kind of network is pretty easy using this package and it requires only a minimal configuration (by default, all the topics are shared between the different masters, but it can be changed to reduce the bandwidth charge).

Once the system is deployed, the communication between topics and services of the different networks are completely transparent, and it seems like all the system is running in a huge common network, but with the advantage that one master could die without fatal consequences for the other masters (Fig. 10).

3.2.4 User Interface

A complex system like this one needs to be accessed in the simplest possible way. The user interface has to provide four main functionalities:

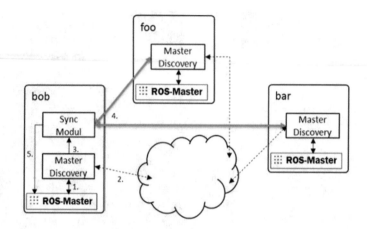

Fig. 10 Example diagram of the multimaster architecture

- Robot teleoperation: as the names stands, it allows to control every robot in remote-manual mode.
- Autonomous missions: it allows to configure and send a mission to the robots. There are three types of missions: "Tunnel scan", "Go to", "Rescue mode" and "Charge batteries".
- Monitoring of the system: it checks the state of the system or other robots.
- Reports: it allows the access to the previously generated reports.

The selected technology to implement the HMI is web. Using a web page based user interface has some advantages over other systems:

- Easy integration with ROS, by using the package rosbridge_suite [11], rosbridge provides a JSON API to ROS functionality for non-ROS programs.
- Integration with other technologies like MySQL, Django and Bootstrap.
- As the system has to be multilingual (Spanish/French), the use of a webpage allows to create different versions of the web page easily.
- Possibility to access the interface from different devices.

The webpage UI was developed using the framework Bootstrap [12] as the front-end framework and Django [13] as the application/base framework of the website. Bootstrap is a popular HTML, CSS, and JS framework for developing responsive, mobile first projects on the web. Django is a high-level Python Web framework that encourages rapid development and clean, pragmatic design. Django was selected because it provides a powerful and simple way to create a database-driven website, in addition to the simplicity to integrate ROS and Django thanks to Python and the roslibjs library provided by the rosbridge_suite. Furthermore, the template web structure of Django facilitated the multilingual support of the web. Regarding the design, Bootstrap is an excellent tool for the creation of responsive webpages. These webpages can be shown in devices with different resolutions, providing a clear and nice design.

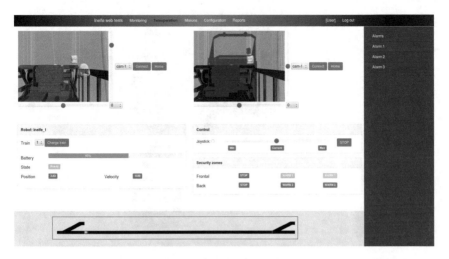

Fig. 11 Example of the screen to monitorize the vehicles, connected to Gazebo simulation

There are several users (with different permissions) that can access to the application. MySQL databases were chosen to save all the user information, to save some information related to the missions and reports and to save some kind of log information (Fig. 11).

4 CROM: Upper Body Torso Robot

This part describes the experience in the development of CROM, an upper body robot conceived and intended for research and development of Human Robot Interaction(HRI) in manufacturing environments (Fig. 12).

Working in a safe HRI manufacturing environment has become an important field in robotics and some robots are being developed to fill this market. CROM is a robot created from the integration of advanced components in the market, like Barrett and Schunk devices.

This section presents a description of the work and the different components and packages used and developed.

The main topics involved in this development were: URDF / Xacro [14], Gazebo [8], MoveIt [6], robotwebtools [15].

The simulation packages for the CROM torso can be found in:

https://github.com/RobotnikAutomation/crom_sim
https://github.com/RobotnikAutomation/crom_common

The ROS version used is Groovy and Hydro running on Ubuntu 12.04. The controller has an i7 processor with 8 GB.

Fig. 12 CROM Torso

4.1 Brief Description of the System

The Robot is composed by the following components:

- Two LWA4.P Schunk arms.
- One WSG-50 Schunk gripper with an attached camera.
- One BH8-282 hand from Barrett Technology with an attached camera.
- A torso made of aluminum.
- A head made with a module PW70 from Schunk for the movements, two cameras in a stereo vision configuration and a Kinect.
- A floor mount chassis. This section has a Schunk PRL120 module to permit one degree of freedom in the torso waist.

The following picture shows the different sections of the robot configuration (Fig. 13).

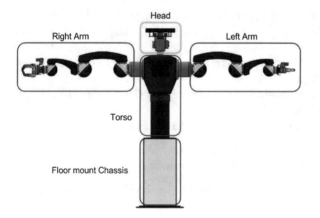

Weight	105 kg
Payload	6 kg
Total DOF	23
Arm DOF	6
Hand DOF	7
Gripper DOF	1
Head DOF	2
Waist DOF	1
Communication	Wifi / ethernet

Fig. 13 CROM robot main parts

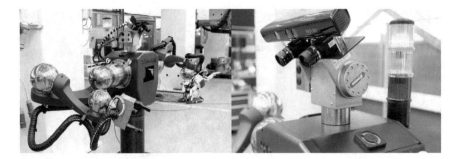

Fig. 14 CROM torso in operation

The Right arm is formed by one Schunk LWA4.P arm, one BH8-282 from Barrett Technology and one 1.3 MP camera from Point-Grey. The Left arm is formed by one Schunk LWA4.P arm, one WSG-50 gripper from (Weiss Robotics) and one 1.3 MP Point-Grey camera. The head is formed by one PW-70 Schunk pan-tilt module, one Kinect RGBD camera and two 2.0 MP Cameras from Point-Grey.

The Torso is formed by an aluminium chassis that joints the two arms, the head and the floor mounting chassis, and one cylinder made of plastic that has inside a PRL+ 120 module by Schunk. All the control equipments of the torso are mounted inside of it.

The floor mount chassis is formed by an aluminium structure. This structure forms a cabinet that contains the safety electrical components and the power supplies (Fig. 14).

4.2 Main Topics Covered

4.2.1 URDF/Xacro

Sources: crom_description

For the proper work and simulation of the torso, the robot's model description is needed. Every robot part is distributed in several urdf files and assembled in a main crom.urdf.xacro file. It is important to keep the parts as independent as possible in order to improve the modularity and maintainability of the robot.

```
$ roslaunch crom_description crom_rviz.launch
```

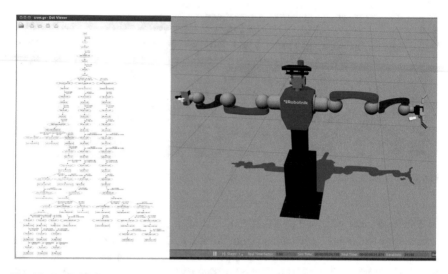

Fig. 15 CROM TFs tree and Gazebo simulation

4.2.2 Gazebo

Sources: crom_sim/crom_gazebo, crom_sim/crom_control

Packages developed to simulate the robot in Gazebo. Configuring the motor controllers types is a critical part to make actuators moving and behaving correctly in Gazebo. The arms have to be able to move correctly to the desired position, be able to pick and lift objects and be as similar as the real one.

For CROM, the great number of different actuators are properly configured to run all together as arms, head and torso (Fig. 15).

```
$ roslaunch crom_gazebo crom.launch
```

4.2.3 Moveit

Sources: crom_sim/crom_moveit

Moveit config package of the robot. All the components are distributed in different groups (arms, torso, head, arms+torso, etc.) to be used through MoveIt libraries and tools (Fig. 16).

```
$ roslaunch crom_sim_bringup gazebo_moveit.launch
```

4.2.4 Robotwebtools

On one hand, the front-end of the site is based on Bootstrap [12], a useful and faster framework for web development including responsiveness support. On the other hand, the browser connection with all the ROS infrastructure is performed by the library roslibjs. This library uses WebSockets to connect with rosbridge [11]. Finally, with all these elements it is possible to declare message publishers, subscribers, service calls, actionlib clients and many others essential ROS functionalities (Figs. 17 and 18).

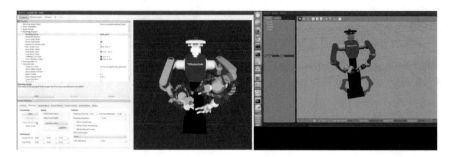

Fig. 16 CROM running RVIZ Moveit plugin

Fig. 17 CROM HMI

Fig. 18 Agvs robot

5 AGVS: Indoor Healthcare Logistics Transport Robot

Robots substitute humans in indoor transport of trolleys (roller containers) in modern hospitals. New hospitals are built taking into account the requirements of the transport robot fleet. The robot fleet autonomously transports storage, clothes, waste, food, etc., thus releasing the humans of the reduced added value task of transporting heavy trolleys for some decens of km per day.

This part describes the characteristics of this logistics transport robot, how ROS has been used in the robot software implementation (planning, control and navigation) and how the robot fleet is being simulated using the ROS framework and V-REP.

The main topics covered in this section are: gazebo [8], vrep [26], navigation.

The simulation packages for the Agvs robot can be found in:

https://github.com/RobotnikAutomation/agvs.git
ROS wiki: http://wiki.ros.org/agvs

5.1 Brief Description of the System

The Robot is composed by the following components:

- 1 traction motor
- 1 direction motor
- 2 SICK S3000 safety laser
- 2 safety bumpers
- 1 lifting unit
- 1 gyroscope

Main features:
The main goal of the robot is to carry autonomously heavy trolleys from one location to another one (mainly in hospitals). Localization is based on amcl and magnetic landmarks set through the path.

The robot is intended to work in crowded areas guaranteeing the safety of the people. Two safety lasers and bumpers provides this safety. Apart from avoiding robot collisions, lasers provide the robot with a 2D scan of the environment, necessary for the robot localization.

Several robots can work together sharing the same resources (paths, elevators, docking stations, etc.) without interfering with each other. This is controlled by the Fleet Control System.

Weight	250 kg
Payload	500 kg
Max. speed	1.25 m/s
Enclosure class	IP53
Autonomy	8 h
Communication	Wifi / ethernet

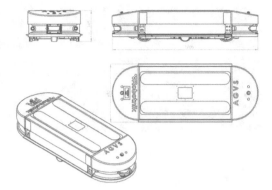

Fig. 19 Agvs specifications

5.2 Main Topics Covered

5.2.1 Gazebo

Sources: agvs/agvs_gazebo, agvs/agvs_control, agvs/agvs_robot_control

Packages developed to simulate the robot in Gazebo [8]. This package sets the motor controllers and configuration to run the platform in the simulator. It is an example of how the Ackermann kinematics can be set in Gazebo.

```
$ roslaunch agvs_gazebo agvs_office.launch
```

5.2.2 VREP

Sources: agvs/agvs_description/vrep, files to simulate the AGVS robot in V-REP

Additional packages needed: vrep_common, vrep_plugin, vrep_joy

This package contains the configuration files needed to run this robot in V-REP. It offers the possibility of running the same robot in two different simulators and compare the differences. For further information, follow the steps in http://wiki.ros.org/agvs/tutorials (Figs. 19 and 20).

5.3 Navigation

Sources: agvs/purepursuit_planner

Package that implements the Pure-Pursuit algorithm of a path. Since the robot is based on Ackerman kinematics, move_base package does not work for this kind of

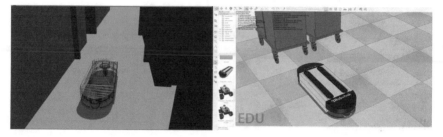

Fig. 20 Agvs simulated in Gazebo and VREP

Fig. 21 Pure Pursuit
algorithm concepts

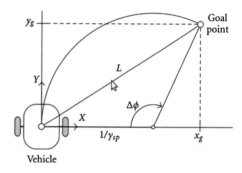

robots. The Pure-Pursuit algorithm finds a point that belongs to the path (sequence of waypoints) that is at a certain distance ahead from the robot (this distance is called lookahead distance). Once this Goal point is found, a curvature is computed to reach it as depicted in the Fig. 21.

In order to localize the robot in the environment, Agvs can use either amcl along with a map created previously or magnetic landmarks installed along the path. While amcl provides higher flexibility and easy and fast mapping, magnetic landmarks implies a tedious installation and the mapping of every mark. However, magnetic landmarks provide much higher accuracy and independence of environment or map modifications (Fig. 22).

5.4 Challenges

5.4.1 Fleet Control System

Coordination amongst robots sharing the same environment required a higher level component to control all the movements in order to manage the fleet planning and scheduling, and efficient resource sharing while avoiding system interlocks.

The planning problem is to some point similar to the planning and scheduling problem of an operating system. A set of n AGVS are commanded to perform a

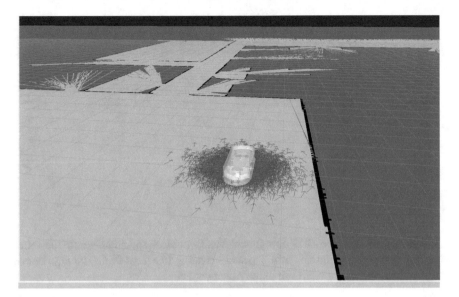

Fig. 22 RVIZ view of Agvs running amcl

number of transport missions inside a building. The set of missions have all the same structure: either a transport of trolley x from location Li to location Lj, or a displacement from current location to Li.

Each mission may have a priority level or a deadline (this is the case of food transport, that should be delivered on specific time frames). The missions are generated externally, in a way completely transparent to the AGV Fleet Control System (FCS), i.e. missions can be ordered by the users by setting directly a transport task in the queue, or programming a scheduled transport.

The AGV Fleet Control System (FCS) decides the missions to be performed according to the overall system state and the missions received. The overall system state includes the fleet state (the localization of the robots, their battery levels, etc.), but also the trolley locations (if a given dock is free or not, if there are no free positions to unload a trolley, etc.).

The FMS divides each mission into a set of tasks to be performed by each AGV. The fleet has therefore a centralized control.

A standard building has several floors and elevators to access them. The shared resources inside this type of building are: 1 Way Corridors, Elevators, Crossings, Doors and Locations.

The FCS communicates with components in the building like elevators or doors (via Ethernet or, most commonly, via decentralised periphery). Some of these systems may be temporarily out of order. Whenever possible, the system has to adapt to carry out the missions using different resources. This is the case e.g. if the building has several elevators and one of them is out of order, or if a corridor is blocked and there is an alternative path. If a resource is missing after the initial start of the mission, the system should be able to reschedule the tasks to fulfill the mission.

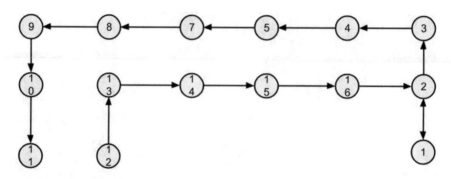

Fig. 23 Agvs path graph

Apart from that, the FCS component has to control and manage all the other hardware/software resources of the system (reading I/O from PLCs, taking control of elevators, switching ON/OFF battery chargers, etc.).

Manage Path

Routes in the environment are represented in a graph with nodes, and robots move from one node to another following the shortest route. Figure 23 shows an example of routes defined as a set of nodes (waypoints) and arcs (paths).

The FCS reserves a set of nodes for a robot and sends it to the last node in the list. Once the robot has passed a node, the FCS frees this node/group of nodes to be available for another robot that needs to pass through this path.

6 VINBOT: Robot for Precision Viticulture

Vinbot is an all-terrain autonomous mobile robot with a set of sensors capable of capturing and analyzing vineyard images and 3D data by means of cloud computing applications, to determine the yield of vineyards and to share information with the winegrowers. Vinbot responds to a need to boost the quality of European wines by implementing precision viticulture (PV) to estimate the yield (amount of fruit per square meter of the vine area). Wineries and vine growers will be able to make accurate yield predictions to organize the production and marketing their wines, coordinating the mixing of grapes of homogeneous quality to efficiently market a range of wines by quality and price. In addition, Vinbot estimates the amount of leaves, grapes and other data in the vine throughout the entire vineyard via computer vision and other sensors and generates online yield and vigour maps to help wine growers to optimize their management strategies.

This part describes how ROS is used in the implementation of the robotized solution, from simulation to the complete system development including localization and navigation as specific vineyard data acquisition solutions.

Vinbot is an EU project (Powerful precision viticulture tool to break traditional yield estimation in vineyards) funded under the FP7 SME program. Vinbot is formed by a consortium of developers and end-users (more information can be found in http://vinbot.eu/). The project is currently under development and the information presented in this chapter represents only the robotics developments that are not under protection and carried on during the first year of the project.

Main topics covered:

- navigation
- gazebo
- v-rep
- robot_localization
- matlab
- ros-io-package

The simulation packages for the Summit XL robot can be found in:

https://github.com/RobotnikAutomation summit_xl_sim, summit_xl_common

The specific Vinbot open source packages can be found in:

https://github.com/VinbotEU/vinbot_common
https://github.com/VinbotEU/vinbot_sim

6.1 Brief Description of the System

The VINBOT platform consists of:

- A robotic platform: durable, mobile, with1h ROS Hydro/Indigo and Ubuntu 12.04/14.04.
- Color cameras to take high-precision images of the vine
- 3D range finders to navigate the field and to obtain the shape of the canopies
- Normalized Difference Vegetation Index sensors to compute the vigour of the plants
- A small computer for basic computational functions and connected to a communication module
- A cloud-based web application to process images or create 3D maps

The Vinbot robot platform is based on a Summit XL HL mobile base able to carry up to 65 kg. Figure 24 shows the platform base with a 3D mapping sensor set and the robot with the specific precision viticulture sensing unit (RGB and near infrared cameras, outdoor Hokuyo lidar, SBAS GPS and inertial measurement unit).

Fig. 24 Vinbot Summit XL HL mobile robot platform testing in vineyards (*left*). Vinbot Summit XL HL platform with mounted precision viticulture sensing unit (*right*)

6.2 Challenges

6.2.1 Dynamic Stability Analysis

Vinbot has the objective to reach autonomous navigation at 1 m/s on vineyards. Besides the robot localization and navigation problem, that will be addressed next, the first problem to be considered is related with the platform stability. The center of gravity of the base platform is close to the ground, but adding cameras and sensors that must be mounted at heights of (>1 m) elevates the center of gravity. The overturning and tilting limits can be computed according to the center of gravity height, wheelbase, acceleration and decelerations, but the characteristics of the field (soil unevenness, gradient, etc.) cannot be neglected to have an accurate estimation of the dynamic vehicle response. For this purpose, the robot dynamics has been modeled in Gazebo and V-REP and the robot with the specific payload has been simulated. There is no special reason for testing the robot in both simulators, the results can probably be reproduced in both. However, to some extent V-REP is a bit more user friendly for the environment and robot modelling, so at the time the dynamic analysis was carried on the robot dynamic model was a bit more detailed in V-REP, this is why the robot dynamics were tested in both environments. The behavior of the simulated model matches the real one, thus allowing fine vehicle parameter tuning (wheelbase, wheel diameter, overall width, acceleration/deceleration limits, components mass distribution, etc.) (Fig. 25).

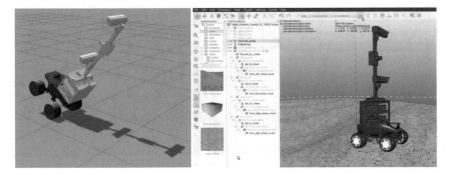

Fig. 25 Dynamic testing of the platform in Gazebo (*left*) and V-REP (*right*)

6.2.2 Sensor Suite Selection

Vinbot has as objective the development of a precision viticulture system that can be provided to farmers and winegrowers at an affordable price. Many of the project objectives are easier to achieve with expensive sensors, i.e. RTK-DGPS systems provide RT2 location estimations with ± 2 cm accuracy that could be used to navigate inside vineyards, and a similar principle can be applied to many of the devices (cameras, laser range-finders, etc.). So an important part of the research and development effort is related with the selection of the solutions able to do the task without exceeding the budget, in summary to find a compromise between price, reliability and simplicity (easy deployment) of the solution. For this specific task ROS and Gazebo are providing an important support. We have modeled vineyards in 5 different growing stages reducing a 90 % the number of polygons of commercial 3d vineyards. The same model is used for both visualization and collision in each vineyard since the simulated laser "sees" in the collision layer. Thus we can simulate a realistic environment and get a first estimation of the data that could be provided by the sensor suite.

In addition, we can test the specific sensors on the field and store the corresponding datasets. The range of sensors tested includes range-finders, cameras, gps devices, low cost dgps devices and associated antennas, inertial measurement units, 3D lasers, stereo cameras, optical flow devices, etc. The measured data is later characterized, analyzed and compared with other sensors and solutions (Fig. 26).

The obtained datasets in simulation are being used only for the development of the navigation and localization solution, but part of the information obtained with these sensors can also be used for the specific precision viticulture data analysis: canopy assessments (height, wideness, leaf layer number, exposed leaf area, etc.) and analysis of clusters and fruits.

Fig. 26 Robot navigating in a simulated vineyard in gazebo (*left*), robot navigating in a real field (*right*)

6.2.3 Navigation and Localization

Simultaneously with the sensor suite selection and performance evaluation under the real world field conditions, navigation and localization algorithms are being tested. The evaluation in this case affects not only the capability of the sensors to operate in the given conditions, but also the state of the art of algorithms that can use the provided information. Some of the selected sensors and algorithms can provide a better odometry estimation, others can also provide localization estimation and or mapping. The development process for the navigation solution follows a simplified version of the cycle described in [16], and is depicted in Fig. 27.

Most of the tests have been carried on in simulation or even in the lab. In some cases the algorithms have been tested using rosbags from data acquisition on the field. Some algorithms were considered at a point where the possibility of running the field data acquisition during the night was still under consideration. This is the case of all rgbd based algorithms, that have ultimately been discarded due to the need to operate in direct sunlight. Other algorithms in the table are still under evaluation and consideration (Table 2).

In parallel with the testing of the mentioned localization algorithms, we have been researching the idea of combining these with other low cost sensors and get benefit of the specific characteristics of the vineyard arrangement. Vinbot addresses vineyards raised in espalier, an agricultural practice of controlling the plant growth for the

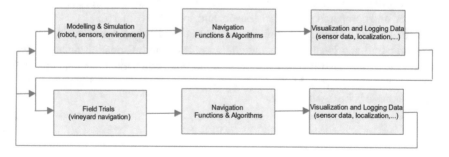

Fig. 27 Navigation development cycle

Table 2 Algorithms tested for localization and mapping or odometry estimation

Type	Sensor	Algorithms tested
slam3d	laser2d	gmapping, amcl, hector mapping
slam 6d	rgbd	rgbd-slam, ccny-rgbd, demo-rgbd
slam 6d	motorized laser2d	Loam (laser odometry and mapping) back and forth
monocular slam 6d	camera	Ptam (parallel tracking and mapping)
visual odometry	camera	Fovis (fast odometry from vision), svo (fast semi-direct monocular visual odometry)
Laser odometry	laser2d	laser_scan_matcher
slam 6d	stereo vision (or rgbd or 3d laser)	Rtabmap (real-time appearance-based mapping)
landmark	rgbd	ar_track_alvar (qr code tracking), visp_auto_tracker

production of fruit by pruning and tying branches to a fixed frame. This creates a kind of structured environment with organized rows and also some poles in the whole field. The following images show an example of the addressed environment.

The possibility of using previously created 2d/3d maps has to be considered carefully, as the data acquisition has to be done autonomously and continuously during the year, and the plant canopy evolution will completely change the maps due to the branches and leafs growth. A 2D/3D map close to the ground may provide a less changing environment, but at the same time with less information.

With the objective of avoiding the use of expensive RTK-DGPS devices, the following strategies have been tested:

- **gmapping and amcl based localization**: the performance is good even measuring close to the ground and seeing mostly the plant trunks and the field poles as in the following picture. However, the estimation does not provide the required reliability due to the environment changes and due to the characteristic particle filter uncertainty in long corridors. The particles (red arrows under the robot) can be seen in Fig. 28.
- **beacon based localization**: this localization solution makes use of a standard 2d range finder. By measuring not only the distance, but the reflected intensity, it is possible to identify reflective beacons previously installed in the field. This kind of beacon is used for localization by standard industrial navigation solutions as those provided by Sick [17], it has also been used as landmark (detection of the row end) for navigation in vineyards [18]. We have tracked them by using a standard hokuyo laser and processing the intensity values. For the development of the algorithms the Matlab ROS I/O Package has been tested. ROS I/O Package is a MATLAB® extension that provides the capability to interact with Robot Operating System (ROS) from MATLAB [19]. This software, allows an easy interface with ROS and is useful for fast testing of algorithms combined with existing software running on ROS prior to their implementation in C++ or Python. ROS I/O Package

Fig. 28 Robot following programmed waypoints (*green arrows*) using Amcl localization on a vineyard raised in espalier (map created with gmapping)

permits MATLAB programs to subscribe to ROS published data and to advertise and publish MATLAB data in ROS (Fig. 29).

- **reactive navigation**: reactive navigation has proven to work well and has been also tested by other research groups [20, 21]. In summary, the geometric map is replaced by a topological one (or geometric-topological). The navigation relies on the simple reactive behavior of row following. The rows are identified by means of the laser range finder and the corresponding lines extracted by RANSAC. The algorithm is able to follow the left or right row or both. The mission according to this scheme can be defined as a sequence of waypoints and behaviours (follow row, follow left, turn left on row end, etc.), implemented in a state machine (Fig. 30).

- **sbas localization and low-cost rtk-dgps devices**: different devices and antennas have been tested and evaluated. The GPS-SBAS device (GNSS uBlox LEA-6P) provides a location estimation with the standard SBAS accuracy of ± 0.5 m rms error. In order to converge to a differential solution with geostationary satellites, SBAS usually requires full sky view, which is the standard condition in vineyards. In order to measure the accuracy of different antennas, a 24 h tests is carried on keeping the gps antenna fixed on a given location. We have realized that the accuracy is highly dependent on the antenna quality, and that the original accuracy value of ± 0.5 m rms can be improved. In addition, it has to be noticed that the accuracy value is referred to a 24 h cycle, but SBAS can provide a higher precision for a specific time frame. Even with a reduced accuracy, GPS information can in this context be useful, first to determine the end of a robot behaviour (e.g. end of row) but also to geolocalize precision viticulture data. In the last years other low-cost, high-performance GPS receivers with Real Time Kinematics (RTK) have arrived on the market. These are able to provide centimeter level relative positioning accuracy at an affordable price. The first device of this class under evaluation is the Piksi (swift-nav) [22]. We have recently received one of the pre-serie devices and have been updating the firmware. To date the results are promising but the firmware still includes some errors that will sure be debugged in the coming firmware versions.

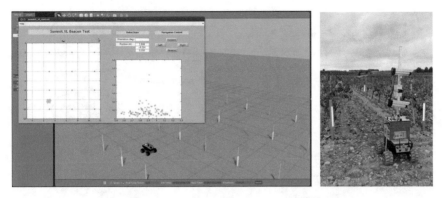

Fig. 29 Beacon based localization testing (*left*) and beacon based localization in real field (*right*)

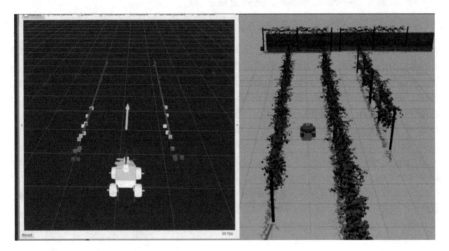

Fig. 30 Ransac row detection

- **hybrid approaches**: the final solution points to a combination of reliable localization inputs with reactive navigation. More field testing is needed but in summary there will be a localization solution and a reactive navigation solution coexisting. The following picture describes the proposed hybrid approach (Fig. 31).

An HMI implemented with interactive markers over the ROS visualization tool permits the definition of a set of waypoints and actions to configure and define the robot mission.

There is a map and an absolute robot localization to command waypoints in field areas where the robot cannot use reactive navigation (e.g. to move from one field to the next one, to move from one row to the next one if these are not contiguous, etc.). In this state, the robot is commanded by a high level planner that sends a sequence of waypoints to move_base (navigation stack). Alternatively the mission defines also

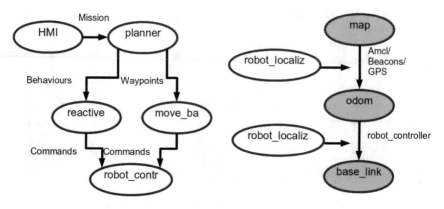

Fig. 31 VinBot navigation architecture (hybrid reactive)

points were reactive behaviours start and end (e.g. row following). For the localization estimation, the robot_localization [23] package is used. The outputs of amcl, beacon based localization and GPS (or other algorithm inputs) are fused by means of the robot_localization nonlinear state estimator, to provide the transform from map to odom (fusion of global absolute position data). At the same time, the same node is used to fuse the odometry input sources (imu, encoders, other odometry sources) to provide the transform from odom to base_link (fusion of continuous position data).

The proposed scheme permits the use of different solutions even for the same task (e.g. navigation in a row could be achieved by reactive navigation or by absolute localized waypoints). The final weight of reactive navigation versus absolute localization will ultimately depend on the reliability of the localization inputs and on the installation and deployment costs.

7   Summary and Conclusions

In the previous sections, a number of service robotics applications using ROS have been described. In most cases several robot units have been manufactured and delivered to specific customers and the robots are currently in operation.

The advantages of using ROS in these kind of applications have exceeded our initial expectations when we installed the *boxturtle* and decided to follow this path. At that time Robotnik was using custom software components for service robotics and PlayerStage [24] and OpenJAUS [25] for R&D robots. The high quality and reliability of the ROS architecture, the availability of powerful development tools, the possibility of using simulators and the wide range of devices and algorithms supported have increased the development speed considerably. All this has allowed us to provide hardware and service solutions that are modular, scalable and open source as the ones presented in this chapter.

The importance of the simulation tools deserves a special mention. The facilities to model a robot, its actuators and sensors and the environment allows getting realistic data streams and test control concepts in early stages of the projects. Despite the high specialization required, Gazebo (or V-REP) have demonstrated to be extremely useful in reducing the costs as they permit the parallel development of hardware and software, but most importantly because this development cycle allows to detect and correct many concept errors in early stages.

The number of advantages and the good acceptance of the ROS architecture from the customers makes difficult even to consider alternatives.

All these advantages have contributed to improve our efficiency and allowed us to provide a specialized service with a measurable added value. Nevertheless, a number of technical issues have to be also mentioned; at some point the fast version releases and incompatibilities have been a major cause of troubles in the continuous effort to adapt our software to the version most used by the community. Other issues related with the lack of documentation, limitations of the middleware (most will be solved in ROS2.0) and with the limitations of the physical engines of the simulators are also topics that forced us to make large efforts and that should be improved in the near future.

Despite the minor drawbacks, ROS has enabled a wide number of applications, increased the capabilities and fulfilled the promise of code reutilization.

References

1. M. Quigley, K. Conley, B.P. Gerkey, J. Faust, T. Foote, J. Leibs, R. Wheeler, A.Y. Ng, ROS: an open-source Robot Operating System, in *ICRA Workshop on Open Source Software* (2009)
2. ROS Conference. http://www.roscon.org (2015)
3. R. Guzman, R. Navarro, J. Ferre, M. Moreno, RESCUER: development of a modular chemical, biological, radiological, and nuclear robot for intervention, sampling, and situation awareness. J. Field Robot. (2015). doi:10.1002/rob.21588
4. rqt. http://wiki.ros.org/rqt (2015)
5. Google Earth. https://www.google.com/earth/ (2013)
6. MoveIt! http://moveit.ros.org/ (2015)
7. Kinematics and Dynamics Library KDL. http://www.orocos.org/kdl (2013)
8. Gazebo. http://gazebosim.org/ (2015)
9. PointCloud Library. http://pointclouds.org/ (2015)
10. multimaster_fkie, retrieved 2015, from http://wiki.ros.org/multimaster_fkie
11. rosbridge_suite, retrieved 2015, from http://wiki.ros.org/rosbridge_suite
12. BootStrap. http://getbootstrap.com/ (2015)
13. Django. https://www.djangoproject.com/ (2015)
14. URDF Unified Robot Description Format. http://wiki.ros.org/urdf (2015)
15. Robot Web Tools. http://robotwebtools.org/ (2015)
16. A. Linz, A. Ruckelshausen, E. Wunder, Autonomous service robots for orchards and vineyards: 3D simulation environment of multi sensor-based navigation and applications, in *ICPA* (2014)
17. A.G. Sick, NAV200 Operating Instructions. http://www.sick.com. Accessed 12 Feb 2007
18. M. Bergerman, S.M. Maeta, J. Zhang, G.M. Freitas, B. Hamner, S. Singh, G. Kanto, Robot farmers: autonomous orchard vehicles help tree fruit production. IEEE Robot. Autom. Mag. 54–63

19. R.S. Nah, Y. Zhang, R. Pillat, ROS Support from MATLAB, Presentation in ROSCon Chicago, Sep 2014
20. S. Marden, M. Whitty, GPS-free localisation and navigation of an unmanned ground vehicle for yield forecasting in a vineyard, in *Recent Advances in Agricultural Robotics, International workshop collocated with the 13th International Conference on Intelligent Autonomous Systems (IAS-13)*
21. H. Mousazadeh, A technical review on navigation systems of agricultural autonomous off-road vehicles. J. Terramech. **50**(3), 211–232 (2013). doi:10.1016/j.jterra.2013.03.004. ISSN: 0022-4898
22. Swift Navigation. http://www.swift-nav.com (2015)
23. ROS robot_localization
24. The Player Project. http://playerstage.sourceforge.net/ (2015)
25. OpenJAUS. http://openjaus.com/ (2015)
26. Coppelia Robotics V-REP. http://www.coppeliarobotics.com/ (2015)

Authors' Biography

Mr. Roberto Guzman (rguzman@robotnik.es) owns the degrees of Computer Science Engineer (Physical Systems Branch) and MSc in CAD/CAM, and has been Lecturer and Researcher in the Robotics area of the Department of Systems Engineering and Automation of the Polytechnic University of Valencia and the Department of Process Control and Regulation of the FernUniversität Hagen (Germany). During the years 2000 and 2001 he has been R&D Director in "Althea Productos Industriales". He runs Robotnik since 2002.

Mr. Roman Navarro (rnavarro@robotnik.es) owns the degree of Computer Science Engineer (Industrial branch) at the Polytechnic University of Valencia. He works in Robotnik since 2006 as software engineer and in the R&D department.

Mr. Marc Beneto (mbeneto@robotnik.es) owns the degree of Computer Engineer (Industrial branch) at the Polytechnic University of Valencia. He works in Robotnik since 2011 as software engineer and in the R&D department.

Mr. Daniel Carbonell (dcarbonell@robotnik.es) received the degree in electrical engineering (Industrial branch) from the Polytechnic University of Valencia in 2014. He also was a scholarship student in the ETH Zürich in 2012 where he started to feel interested in mobile robotics.

Standardization of a Heterogeneous Robots Society Based on ROS

Igor Rodriguez, Ekaitz Jauregi, Aitzol Astigarraga, Txelo Ruiz and Elena Lazkano

Abstract In this use case chapter the use of ROS is presented to achieve the standardization of a heterogeneous robots society. So on, several specific packages have been developed. Some case studies have been analized using ROS to control particular robots different in nature and morphology in some applications of interest in robotics such as navigation and teleoperation, and results are presented. All the developed work runs for Indigo version of ROS and the open source code is available at RSAIT's github (github.com/rsait). Some videos can be seen at our youtube: channel https://www.youtube.com/channel/UCT1s6oS21d8fxFeugxCrjnQ.

Keywords Heterogeneous robots · Old robot renewal · Standardization · Teleoperation · Human-robot interaction · Navigation · Speech recognition

1 Introduction

The Robotics and Autonomous Systems Lab (RSAIT) is a small research group that focuses its research on applying new machine learning techniques to robots.

The group was founded around year 2000 and inherited a B21 robot (RWI). Since then, the group has grown up and, in its development, has acquired different robots.

I. Rodriguez (✉) · E. Jauregi · A. Astigarraga · T. Ruiz · E. Lazkano
Faculty of Informatics, Robotics and Autonomous Systems Lab (RSAIT),
UPV/EHU, Manuel Lardizabal 1, 20018 Donostia, Spain
e-mail: igor.rodriguez@ehu.eus
URL: http://www.sc.ehu.es/ccwrobot

E. Jauregi
e-mail: ekaitz.jauregi@ehu.eus

A. Astigarraga
e-mail: aitzolete@gmail.com

T. Ruiz
e-mail: txelo.ruiz@ehu.eus

E. Lazkano
e-mail: e.lazkano@ehu.eus

© Springer International Publishing Switzerland 2016
A. Koubaa (ed.), *Robot Operating System (ROS)*, Studies in Computational
Intelligence 625, DOI 10.1007/978-3-319-26054-9_11

In our experience, robot maintenance is laborious, very time consuming, and does not provide immediate research results. Moreover, robots from different suppliers have their own control software and programming framework that require senior and incoming members to be trained once and again. This makes robot maintenance harder. That's why robotics labs often become scrap yards in the sense that old robots are often retired instead of upgraded and broken robots are discarded instead of repaired. But small research groups often do not have enough budget to invest in new robots, so that upgrading and repairing robots become mandatory.

We now own a heterogeneous set of robots, consisting of an old B21 model from RWI named *MariSorgin*; a *Kbot-I* from Neobotix; *Galtxagorri* a Pioneer 3DX and the PeopleBot *Tartalo*, both from MobileRobots; a humanoid NAO from Aldebaran; five *Robotino*-s from Festo (these ones used for educational purposes in the Faculty of Informatics). Each one came with its own API and software, most running on Linux. Thanks to ROS we now have a standard tool to uniformly use this society of heterogeneous robots.

Contributions of the book chapter: several case studies are presented in which some new ROS drivers and packages have been developed for navigation and gesture and speech based teleoperation that can be used for robots different in nature. Those applications are fully operative in real environments.

2 Robot Description

A brief description of the robots and the modifications and upgrades suffered during their operational life follows up, together with a reference to the software used to control them.

2.1 MariSorgin

Our heirloom robot is a synchro-drive robot that dates from 1996. It is a B21 model from Real World Interface provided with a ring of ultrasound, infrared and tactile sensors for obstacle avoidance. Opposite to its successor, the well known B21r model, it was not supplied with a laser sensor. In 2002 its motor controllers were damaged and sent to RWI for replacement, but they never came back to us. Ten years later, those boards were replaced with Mercury motor controllers from Ingenia Motion Control Solutions [8]. The two internal i386 PCs were replaced with a single newer motherboard. So, after 10 years *MariSorgin* became again fully operational.

In the beginning the original API, named *BeeSoft* [19] was replaced by a home made library that better suited to our control architecture development philosophy (*libB21*[1]). We combined it with *Sorgin* [2], a framework designed for developing

---

[1]Developed by I. Rañó at Miramon Technology Park, 2001 [17].

```
void main()
{
  /* Data declarations */
  io_data_t laser_readings;
  io_data_t motor_output;

  /* Behavior declaration */
  behavior_t avoid_obstacles;

  /* Data initialization */
  io_data_alloc(&laser_readings, 181, NULL, 0, 0);
  io_data_alloc(&motor_output, 2, NULL, 0 0);

  /* Behavior initialization */
  behavior_define(2, i, avoid_obstacles_start,
                        avoid_obstacles_stop,
                        avoid_obstacles_calculate);

  /* input/output connections */
  behavior_set_input(&avoid_obstacles, 0, running);
  behavior_set_input(&avoid_obstacles, 1, laser_readings);
  behavior_set_output(&avoid_obstacles, 0, motor_output);

  behavior_start(&avoid_obstacles);
  behavior_run(&avoid_obstacles);

  sleep(100);
  behavior_stop(&avoid_obstacles);
}
```

Fig. 1 *Sorgin*: example program

behavior-based control architectures. *Sorgin* allowed us to define and communicate behaviors in a way similar to ROS topics and nodes. Topics equivalents were arrays of floats defined as io_data structures, and nodes were behavior_t structures, with different associated functions (initialization, main loop and stop) launched in separated threads. Thus, *Sorgin*'s modular structure resembled ROS procedural organization and communication but in a more modest implementation. Figure 1 shows what a *Sorgin* program looks like.

After the "resurrection", we had no doubt: we must adapt it to ROS. So, we developed the necessary ROS drivers for the motor controllers and mounted a Hokuyo URG-30 laser on top of the enclosure, a Kinect camera and a Heimann thermal sensor (see Fig. 2).

(a)   **(b)**

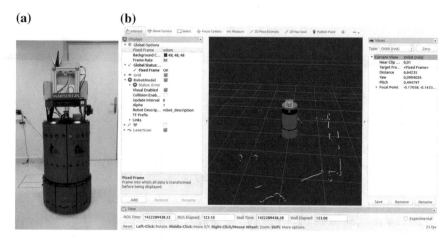

Fig. 2 *MariSorgin* and its URDF model. **a** Renewed B21, **b** URDF model visualized in Rviz

2.2 Tartalo and Galtxagorri

Two differential drive robots from MobileRobots. *Galtxagorri*, a Pioneer-3DX robot suffered some modifications from its initial configuration. On the one hand, a Leuze-RS4 laser sensor was mounted on top of its body (fed externally to extend the duration of the internal batteries and so, the robot's autonomy). Speakers have been added together with an amplifier (Fig. 3). Besides, *Tartalo* is a PeopleBot robot that facilitates human-robot interaction. Both platforms have a MAMBA VL-EBX-37A board with a 2.26 GHz Intel(R) Core(TM2) Duo CPU.

These two robots came with *Aria*, a framework that again did not fulfill our control architecture development schemata. Before ROS came up, our trend was to use Player/Stage (see [27]). Player/Stage offered us a wide set of drivers and a proper tool for developing our own algorithms without imposing restrictions in the type of control architecture being developed.

Player/stage shares with ROS the definition of what a robot is, i.e. a set of devices (sensors and actuators), each one with its own driver that gives access to the device data. Player offered an abstract layer of interfaces that allowed to access different devices similar in nature using the same code. Combining Player with *Sorgin* turned out to be straight forward (Fig. 4). This coupling allowed us to work with MobileRobots platforms in a flexible and suitable way for several years. But Player focused more on developing drivers than algorithms and stopped evolving when ROS appeared.

ROS provides the *P2OS* package that allows to control *Tartalo* and *Galtxagorri*'s base. Moreover, it also offers the appropriate driver for the Leuze RS4 laser scanner. Thus, only the URDF models were needed to set up for the two robots.

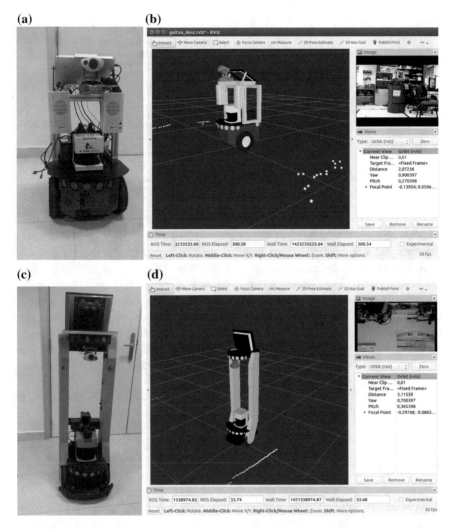

Fig. 3 P2OS robots and their URDF models. **a** Galtxagorri. **b** Galtxagorri in Rviz. **c** Tartalo. **d** Tartalo in Rviz

2.3 Robotino-s

Those are omnidirectional circular platforms from Festo Didactic that we mainly use for education. They are provided with several sharp GP2D12 infrared sensors, a bumper ring, a webcam and a Hokuyo URG-04LX sensor in order to be able to experiment with mapping and planning techniques. The control unit, placed on top of the wheeled platform, contains a 500 MHz PC104 processor that runs RTLinux.

Fig. 4 *Sorgin*+Player: organization

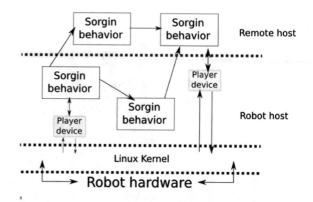

All the software is installed in a 4 GB Compact Flash card. The unit also offers an Ethernet port and a WLAN access point, 2 USB ports and a VGA connector.

RobotinoView is the interactive, graphic programming environment for Robotino. Besides, Robotino ships with an application programming interface (*RobotinoAPI2*) allowing the user to create programs using different programming languages like C, C++, Java and more. Communication between the control program and Robotino is handled via TCP and UDP and is therefore fully network transparent. The new API2 is based on a RPC like infrastructure. The REC-RPC library is an interprocess communication middleware similar to ROS. It is completely based on Qt and does not have any other dependencies.

Migration to ROS has been straight forward, since packages for *Robotino* can be found at wiki.ros.org/robotino.

2.4 NAO

NAO is an autonomous programmable humanoid robot developed by Aldebaran Robotics. NAO's human like shaped body is about 58 cm tall and weights about 4,8 kg. It is built in polycarbonate and ABS (a common thermoplastic) materials that allow better resistance against falls and it has a lithium battery with which it can get an autonomy of 90 min approximately. Its heart is composed by a 1.6 GHz Intel Atom processor running Linux. 25 servos enable to control the 25° of freedom of the robot. Regarding to robot motion, NAO can move in any direction (omnidirectional walking), it uses a simple dynamic model (linear inverse pendulum) and quadratic programming. It is stabilized using feedback from joint sensors. It can walk on a variety of floor surfaces, such as tiled and wooden floors, and he can transition between surfaces while walking.

NAO sees using two 920p cameras, which can capture up to 30 images per second. Also, it uses four microphones to track sounds and two loudspeakers to talk or play sounds.

Choregraphe is the original programming software of NAO. It is a multi platform desktop application that allows to create animations and behaviors, test them on a simulated robot, or directly on a real one and monitor and control the robot. *Choregraphe* allows to create very complex behaviors (e.g. interaction with people, dance, send e-mails, etc.) without writing a single line of code. In addition, it allows the user to add her own Python code to a *Choregraphe* behavior.

The behaviors created with *Choregraphe* are written in its specific graphical language that is linked to the *NAOqi* Framework, the main software that runs on the robot. NAO interprets them through this framework and executes them. *Choregraphe* also interacts with *NAOqi* to provide useful tools such as the Video monitor panel, the Behavior manager panel, the Toolbar, the Robot view or the Timeline Editor.

Again, transition to ROS was easy. ROS drivers for NAO can be found at http:// wiki.ros.org/nao.

2.5 Kbot-I

A differential drive robot built by Neobotix in 2004 for acting as a tour guide at the Eureka Museum of Science in San Sebastian. Supplied with a Sick S3000 laser scanner, the robot held a touch screen for receiving orders. An application specifically developed for the robot to act as a guide in the museum was running on the onboard Windows 2000 Professional machine. In 2006 the robot was damaged and stored in a garage until 2014. The robot was transferred to the University of the Basque Country (UPV/EHU). In spite of the lack of any detailed manual, our group managed to locate and repair broken connections. The onboard PC was replaced by a Zotax MiniPC with a NVIDIA graphics card, and the webcam on its head was removed and instead, a Kinect sensor has been mounted. The rigid arms that supported a huge touch monitor were also removed and replaced by plates built with a 3D printer (another invaluable tool for robot maintenance!).A smaller Getich monitor has been placed on the back side of the body. Fortunately, the source code of the drivers was available and only small modifications were needed to compile that code and make the necessary libraries under Linux. Albeit the time spent in code surfing, it was rather straightforward to implement the necessary ROS drivers to get it back running[2] (Fig. 5).

3 Working Areas of RSAIT Research Group

Navigation is a fundamental skill that mobile robots need in order to be autonomous. The navigation task has been approached in different ways by the main paradigms of control architectures. RSAIT has focused its navigation methodology within the

---

[2]Thanks to Marco Beesk from Neobotix for agreeing to make public our *Kbot-I*ROS nodes.

(a)   (b)

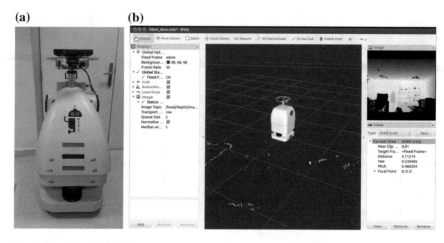

Fig. 5 *Kbot-I* and its URDF model. **a** Renewed Kbot-I, **b** Kbot-I visualized in Rviz

behavior-based philosophy (see for instance [9]) that focuses on biology to inspire its navigation strategies ([12]). But probabilistic approaches seem to increase adepts and some techniques are being distributed within the ROS community for mapping, localization and planning. No definitive solution exists nowadays but clearly, ROS navigation stack makes possible to compare different approaches. Therefore, it is worth to setup this stack for our robots.

Besides, in our research group we are working on different applications for natural human-robot interaction. On the one hand, we have developed two different ROS packages to enrich the teleoperation of robots: speech-based teleoperation in Basque Language (*Euskara*) and gesture-based teleoperation using the Kinect [18].

On the other hand, we have developed a system, called **Bertsobot**, which is able to construct improvised verses in Basque (named *bertsoak*) according to given constraints on rhyme and meter, and to perform them in public (see [1]). NAO is the robot that gives shape to the **Bertsobot** system. It is capable of understanding some "orders", composing and playing traditional Basque impromptu verses, also replicating the movements made by the impromptu verses singers. This project allowed us to combine diverse research areas such as body gesture expressiveness, oral communication and human-robot interaction in a single project.

Table 1 summarizes the developed ROS packages. The experiments described in the following sections will explain how these skills have been integrated in the different robots, according to their sensorial capabilities.

Table 1 Summary of basic ROS modules

	Used ROS modules	Adapted ROS modules	New ROS modules
General use			- Speech_eus: Basque TTS and ASR modules - heiman: thermopile driver
Galtxagorri	- p2os_driver - rotoscan_node - gscam	- galtxa_description: URDF model	- galtxa_teleop_speech_eus
Tartalo	- p2os_driver - sicklms - gscam	- tartalo_description: URDF model - tartalo_navigation: planner and costmap params	- tartalo_teleop_speech_eus
Robotino	- robotino_node - openni_node - openni_launch	- All packages catkinized - skeleton_tracker: openni_tracker modified	- robotino_teleop_gestures
Kbot-I	- sicks300 - openni_node - openni_launch		- kbot_description: URDF model - kbot_platform: drivers for driving motors, head tilt motor and integrated sensors (US) - kbot_teleop_joy: platform and head tilt motor control - kbot_guide_qt: interactive user interface for navigation
MariSorgin	- hokuyo_node - imu_um6 - openni_node - openni_launch		- mari_description: URDF model - canm: mercury motor controller driver - mari_teleop_joy: platform control - mari_teleop_speech_eus - mariqt: user interface for speech based teleoperation - heiman: thermopile driver

(continued)

Table 1 (continued)

	Used ROS modules	Adapted ROS modules	New ROS modules
NAO	- `naoqi_bridge` - `nao_robot` - `nao_meshes` - `nao_interaction` - `nao_extras`	- `skeleton_tracker`: `openni_tracker` modified	- `nao_teleop_gestures`: motion, hands and head controller - `nao_teleop_speech_eus` - `nao_bertsobot`

4   Case Study 1: Setup of the Navigation Stack

Navigation refers to the way a robot finds its way in the environment [13]. Facing this is essential for its survival. Without such a basic ability the robot would not be able to avoid dangerous obstacles, reach energy sources or return home after exploring its environment. Navigation is therefore a basic competence that all mobile robots must be equipped with. Hybrid architectures tackle the problem of navigation in three steps: mapping, localisation and planning. These are old problems from the perspective of manipulation robotics and are nowadays treated in a probabilistic manner. Hence the name of the field *probabilistic robotics* [26], that makes explicit the uncertainty in sensor measurements and robot motion by using probabilistic methods. ROS offers several stacks that use probabilistic navigation techniques and allow to empirically use, test and evaluate the adaptability of those techniques to different robot/environment systems. So for, and taking as starting point the navigation stack available for the P2OS robots, it has been setup in *Kbot-I*.

Since *Kbot-I* is now ready again for human-robot interaction, an interactive user interface has been developed using **rqt** (kbot_guide_qt) to retake the original task *Kbot-I* was designed for: be a guide within our faculty. The most frequently demanded sites of our faculty are located at the first floor. Hence, in this attempt a map of the first floor has been created with ROS mapping utilities and that map is being used as the floor plan of the developed GUI. This floor plan has been populated with several interaction buttons corresponding to the important locations people might be interested in, such as the administration, the dean's office, the lift, several labs and so on. Information about actual and destination locations is also displayed on the interface. Figure 6 shows what the GUI looks like.

The robot morphology makes door crossing insecure and thus, for the time being the GUI limits the robot guiding task to the front of the door that gives access to the desired location.

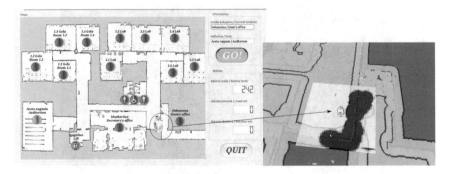

Fig. 6   Kbot navigation interaction window and costmap

System evaluation: Regarding the setup of the navigation stack, it is not a friendly process. Although the documentation has been improved, many parameters have to be set empirically, without any explicit methodology. The ROS navigation stack is based on probabilistic navigation techniques and it is known that they don't adapt well to dynamic environments. The system fails when there are severe mismatches between the current sensor readings and the stored map. Therefore, crowds should be avoided in front of the robot during the tours and all people must be advised to stay on the back of the robot so that the map remains reliable and the planner could find a way to the goal. Moreover, the application needs to know the robot's initial position in order to be able to plan routes to the goal. Hence, it is not able to face the global localization problem. It will be a great improvement to enhance the navigation stack with global localization capabilities to overcome this problem and to make it more robust and general to use.

But more important is to mention that, after ROSifying the robot, we got a navigation application running, working and prepared to be used in public in just a couple of weeks, but without the need of reimplementing the whole system. The application has been used for the first time in an open door event at our faculty on March 12 (2015). About 100 candidate students came to visit the faculty and they were divided on 6 small groups of 15–20 students. They were supposed to visit different labs and sites on different floors of the building. The robot was located on the first floor and guided the teams over the different places they should arrive to. Basque TV (EiTB) came to record the event and broadcasted it at the news (Fig. 7).

Fig. 7 Kbot making guided tours in the faculty

Still we can improve the system integrating door crossing abilities. Also, the system should be complemented with the maps of the second and third floors. Our plan is to set *Tartalo* in the second floor and *MariSorgin* in the third one, so that connection among floors will be done via the lift. Robots will not entry the lift but will communicate to be aware that they need to welcome "tourists" sent from other locations.

5 Case Study 2: Kinect Based Teleoperation

The term teleoperation is used in research and technical communities for referring to operation at a distance. Teleoperated robots are used in many sectors of society. Although those robots are not autonomous they are very useful e.g. in medicine for surgery [4, 5], for space exploration [3] or for inspection in nuclear power plants [16].

Different devices can be used for teleoperating a robot (joystick, smart phone, wii-mote) but gesture based teleoperation is increasing adepts ([6, 15, 25]) specially due to availability of cheap 3D cameras such as Microsoft's Kinect sensor. Real-time teleoperation of humanoid robots by detecting and tracking human motion is an active research area. This type of teleoperation can be considered as a particular way of interaction between a person and a robot, because it is a natural way to interact with robots. It is an interesting research topic and related work is abundant. For instance, Setapen et al. [21] use motion capture to teleoperate a NAO humanoid robot, using inverse kinematic calculations for finding the mapping between motion capture data and robot actuator commands. Matsui et al. [14] use motion capture to measure the motion of both, a humanoid robot and a human, and then adjust the robot motion to minimise the differences, with the aim of creating more naturalistic movements on the robot. Song et al. [22] use a custom-built wearable motion capture system, consisting of flex sensors and photo detectors. To convert motion capture data to joint angles, an approximation model is developed by curve fitting of 3rd order polynomials. Koenemann and Bennewitz [10] present a system that enables a humanoid robot to imitate complex whole-body motions of humans in real time, ensuring static stability when the motions are executed and capturing the human data with an Xsens MVN motion capture system consisting of inertial sensors attached to the body.

The above mentioned methods are limited in the sense that the human needs to wear different types of sensors in order to interact with the robot. This can be avoided with the Kinect sensor, moreover, the cost of the equipment is declined. That is why researchers have become more interested in Kinect. Song et al. [23] propose a teleoperation humanoid robot control system using a Kinect sensor to capture human motion and control the actions of remote robot in real-time. Suay and Chernova [24] present a new humanoid robot control and interaction interface that uses depth images and skeletal tracking software to control the navigation, gaze and arm gestures of a humanoid robot.

(a)   **(b)**

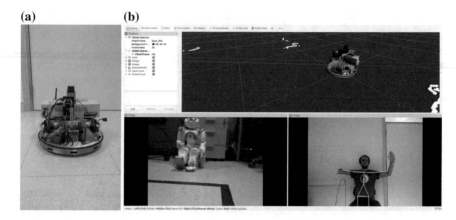

Fig. 8 *Robotino* and its teleoperation interface. **a** Robotino, **b** The teleoperation interface

ROS offers drivers for the Kinect together with a package that extracts and tracks the human skeleton from sensor data. Thus, taking as base tool these two packages (*openni_launch* and *openni_tracker*), a gesture-based teleoperation system has been developed for a holonomic wheeled robot and, afterwards, enriched to teleoperate a humanoid robot.

5.1 The `robotino_teleop_gesture` Package

The development of a gesture based teleoperation system requires first to identify the degrees of freedom that are going to be controlled and define the set of gestures that will control the robot. *Robotino*-s are holonomic wheeled robots and thus, can be moved along the plane in any direction without changing the robot heading. Also, a rotational velocity can be assigned. The defined gesture set is based on arm movements although internally is implemented through hand positioning (see Fig. 8). The gesture set consists of:

- Moving right arm tilt controls forward/backward movements (lineal velocity in x)
- Right arm pan movement controls side movements (linear velocity in y)
- Left arm yaw movement controls left/right rotation (rotational velocity)
- Lowering both arms at the same time stops the robot.

The developed teleoperation system has two sides: the user detection process and the robot motion control process. On the one hand, the user detection step is based on the `openni_tracker` package, but several changes have been introduced to produce an `skeleton_tracker`:

1. The node that tracks the skeleton now publishes the joint position information of the skeleton in the `skeleton` topic.
2. A new node makes available the Kinect image that includes the graphical representation of the skeleton links on it.

On the other hand, the `robotino_teleop_gesture` node contains a subscriber that receives messages published by `skeleton_tracker` in the `skeleton` topic. When a message is received, the operator's position is analyzed. And according to that position the robot executes the corresponding motion. Although each arm movement controls a velocity value, different gestures can be combined, i.e. move forward while turning.

System evaluation: The skeleton tracker performs properly when the only moving element of the scene is the teleoperator and so, the background needs to be static. Moreover, the system setup is designed for a single person sat on a chair in front of the kinect. But when the application is used with children, they must stand up so that size does not affect the calibration process. The application is fully operational for real indoor environments and is being used as a game/demo in several yearly events like the week of sciences (2013–2014), meetings with undergraduate students (2012–2015), robotics day (2013). Since its early development, it has been adapted for several ROS distros and *OpenRobotinoAPI* versions. Up to now, it is catkinized for Indigo and *OpenRobotinoAPI* version 0.9.13.

5.2 The `nao_teleop_gesture` Package

The gesture-based teleoperation system developed for the *Robotino*-s has been adapted and extended to be used with NAO. The skeleton tracking system is exactly the same, the only difference is that more degrees of freedom are to be controlled and, thus, the gesture set needs to be redefined and extended.

NAO's human like morphology allows not only the motion of the robot in the plane but also the movement of the arms. Thus, it is not adequate to use the operator arms to control the velocities of the robot. In this case, the selected gesture set is the following:

- If the operator steps forward/backward the robot walks forward/backwards.
- The lateral steps of the operator cause the side movements of the robot.
- Raising the left shoulder and lowering the right one causes clockwise rotation.
- Raising the right shoulder and lowering the left one causes ccw rotation.
- Left/right arm movements are used to control robot's left/right arm.
- Head pan and tilt movements are used to control NAO's head.

The new package, named `nao_teleop_gesture` contains three nodes:
1.- nao_motion_control: basically, this node has the same functionality as the node developed for the *Robotino*-s. It has to perform the following two main tasks:

- Receive messages published by the `skeleton_tracker` package.
- Publish NAO's walking velocities.

The `nao_motion_control` node has a publisher that publishes the omnidirectional velocity (x, y, and theta) in the `cmd_vel` topic for the walking engine. The velocity with which the robot moves has been set to a constant value. If the walking velocity is too high the robot starts to swing. Thus, the linear velocity and the angular velocities have been assigned low values.

2. - *nao_arm_control*: This node is in charge of sending to the robot the necessary motion commands to replicate the operator's arms motion. The node performs tasks by:

- Receiving the messages published by the `skeleton_tracker` package.
- Publishing NAO's joint angles with speed.

Therefore, *nao_arm_control* is subscribed to the `skeleton` topic in order to receive the operator's skeleton messages published by `skeleton_tracker`.

On the other hand, the `nao_arm_control` node has a publisher that publishes the joint angles with speed in the `joint_angles` topic, which allows the communication with the `nao_controller` node. The NAO's joints motion speed is set to a constant value appropriate for the robot to mimic the operator arms motion in "real" time.

3. - *nao_head_control*: This node is responsible of moving the robot's head. Similar to the way that `nao_arm_control` gets the arm joint angles, this node calculates the head joint pitch and yaw angles, and publishes them into the `joint_angles` topic.

The robot imitates human actions in real-time with a slight delay of less than 30 ms. This delay is approximately the time the system needs to capture the operator's arms/head motion, calculate the angles that make up the operator's arms/head joints (see [18]), and send motion commands to the robot via WiFi.

Only walking action movements (forward, backward, left, right) with rotational motions can be combined. No arm movement is allowed while walking so that the stability of the robot is not affected. Moreover, when the robot holds something on its arms the center of gravity (COG) of the walking robot needs to be lowered and backwarded, so that the COG is maintained within the support polygon (see Fig. 9). Thus the walking behavior has been modified for those cases in order to increase the stability.

A GUI has been created (this time with existing *rqt* plugins) in order to help the operator to know the system state. The GUI is divided into two main parts (Fig. 10). The top side is composed by the *Topic Monitor* and the *Rviz* interface. The *Topic Monitor* shows all the topics and messages sent by the nodes that are in execution. *Rviz* shows the NAO 3D model moving in real-time. The bottom side shows visual information from the cameras; *Image View* shows the image received from NAO's top camera and the right window shows the image captured by the Kinect together with the skeleton of the tracked body.

(a) (b)

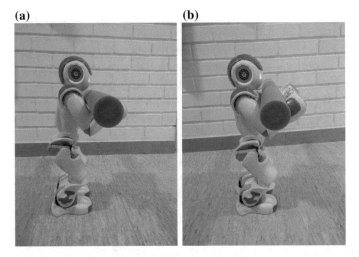

Fig. 9 Modified walking position. **a** Original. **b** Modified

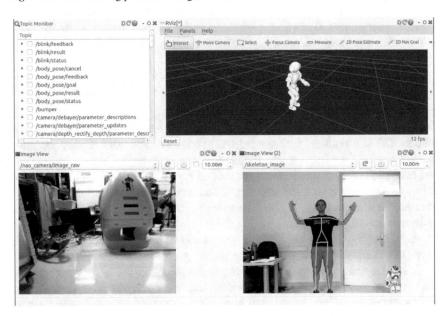

Fig. 10 Teleoperation display

The system starts with NAO in crouching position and when the operator enters the Kinect's view, the calibration process starts. NAO tells the operator that the calibration ended successfully saying "Kinect control enabled" and then, the operator can control the robot with his/her body.

System evaluation: Imitation is an important way of skill transfer in biological agents. Many animals imitate their parents in order to learn how to survive. It is also a way of social interaction. A sociable robot must have the capability to imitate the agents around it. In a human society, people generally teach new skills to other people by demonstration. We do not learn to dance by programming, instead we see other dancers and try to imitate them. Hence, our artificial partners should be able to learn from us by watching what we do. That idea pushed us to evaluate our application based on the imitation ability of the robot.

Two experiments were defined to evaluate the system. Those experiments involved several people that should give qualitative measures of the system performance by means of a questionnaire that participants completed after carrying out each experiment. Experiments were performed until each participant achieved the aim of the experiment at least once (see [18]). The experiments revealed three aspects that might be improved:

- The lack of side view makes more difficult the guidance of the robot. This problem is now alleviated with the addition of the head motion control.
- Although the selection of gestures is correct (natural) and the movements are quite precise, a short period of training is needed by the operator to get used to distances.
- The robot can loose balance when walking with the arms raised.

6 Case Study 3: Speech Based Teleoperation in Basque

Human-robot interaction (HRI) is the study of interactions between humans and robots. HRI is a multidisciplinary field with contributions from human-computer interaction, Artificial Intelligence, robotics, natural language understanding, design, and social sciences. A requirement for natural HRI is to endow the robot with the ability to capture, process and understand human requests accurately and robustly. Therefore it is important to analyse the natural ways by which a human can interact and communicate with a robot.

Verbal communication should be a natural way of human-robot interaction. It is a type of communication that allows the exchange of information with the robot.

To serve a human being, it is necessary to develop an active auditory perception system for the robot that can execute various tasks in everyday environments obeying spoken orders given by a human and answering accordingly. Several systems have been recently developed that permit natural-language human-robot interaction. Foster et al. [7] propose a human-robot dialogue system for the robot JAST, where the user and the robot work together to assemble wooden construction toys on a common workspace, coordinating their actions through speech, gestures, and facial displays.

A speech based teleoperation interface should provide the user the possibility to teleoperate the robot giving predefined orders [28]. The system also should give feedback to the operator when an instruction is not understood and this feedback should also be verbal. Three elements are identified in an architecture for speech-based teleoperation:

1. The automatic speech recognition system (ASR)
2. The text to speech (TTS) system
3. The robot control system

The first two elements are robot independent and, thus, have been integrated in a single package named **speech_eus**. This package contains two nodes, one responsible of the speech recognition step and the second one, responsible of the text to speech translation. Let's describe the nodes that compose the speech_eus package.

1.*gspeech_eus node*: Our robots are supposed to interact in **Euskara** (Basque, a minority language spoken in the Basque Country) and thus, a tool adapted to this requirement was needed. ROS *gspeech* package gives ASR capabilities to the robot and can be configured for many languages, including Basque. But this package needs some modifications in order to be useful in a real-time teleoperation scenario. These are the introduced changes:

- When the native gspeech runs the Google Speech Service, it is executed only once, i.e. when the user starts speaking, the audio is captured and sent to Google. There it is analysed and the text "corresponding" to the received audio is returned with a confidence level; then, the program ends. It could be tedious for the user to run the speech recognition node each time she/he wants to order something to the robot, or each time she/he receives an error message. Hence, the new node now runs iteratively avoiding the problem of having to launch the node each time the user wants to talk.
- When the Google Speech Service does not recognize the spoken words, it returns an error message and then the node is forced to quit. Now, error messages received from Google Speech Service are specially treated. If an error message is received, gspeech_eus publishes a Repeat message in the google_speech topic to advertise the user that his/her spoken words are not being recognized.
- The original gspeech node only prints the response received, it does not publish any messages or services, so it can not communicate with other nodes. After the modifications, the confidence level of the hypothesis received from *Google Speech* is processed and, if it is lower than a predefined threshold (0.15 for the performed experiments), the response is declined and treated as an error message.

2.*tts_eus node*: This is the node in charge of converting the text into speech using the *AhoTTS* tool [11]. That system, developed by the Aholab group in the University of the Basque Country, is a modular text to speech synthesis system with multithread and multilingual architecture. It has been developed for both, Euskara and Spanish languages. The TTS is structured into two main blocks: the linguistic processing module and the synthesis engine. The first one generates a list of sounds, according to the Basque SAMPA code [20], which consists of the phonetic transcription of the expanded text, together with prosodic information for each sound. The synthesis engine gets this information to produce the appropriate sounds, by selecting units and then concatenating them and post-processing the result to reduce the distortion that

Fig. 11 Setup for the experiments

appears due to the concatenation process. This tool is required for communicating in Basque Language, but it would not be required for English interlocution.

`tts_eus` has a subscriber that receives messages from the `text_speech` topic. When a text message is received, this node converts it into speech (an audio file) and plays the audio over robot's speakers.

6.1 Speech-Based Teleoperation in MariSorgin

Again, this teleoperation system has two sides: the instruction interpretation process and the motion controller. Regarding to the instruction interpretation part, and as mentioned before, our robots are supposed to interact in **Euskara**. The oral commands are captured by a microphone and sent to the Google Speech Service by the `gspeech_eus` node. Once the answer is received, the text is matched with our dictionary.

On the other hand, the robot control system must be defined, i.e. the meaning of the voice orders must be translated to actions. *MariSorgin* is a synchro-drive robot and as such, two degrees of freedom can be controlled: linear velocity and angular velocity. Thus, the orders that can be given are limited to moving forward/backward, rotating left/right, stopping and accelerating/decelerating. Figure 11 shows how the system is distributed and communicated over the net.

Although in a first attempt linear and angular velocities could be set independently, that is, setting the linear velocity wouldn't affect the current angular velocity (and vice versa), we found that controlling the robot in that manner was rather complicated and that a high level of expertise was needed. Thus, in the final propotype linear and angular velocities are not independently assigned. Modifications of the angular velocity imply that linear velocity is set to zero, and vice versa.

A Qt interface has been developed using `rqt` that shows the state of the speech recognition process and the velocity values at each time step. The interface includes minimum distances to obstacles at front, left and right sides, obtained from the laser readings, and the image captured by the robot so that the operator can see what the robot is facing to. Figure 12 shows what this simple interface looks like.

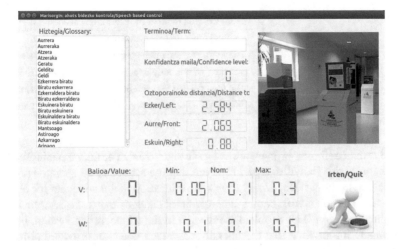

Fig. 12 *MariSorgin* teleoperation window

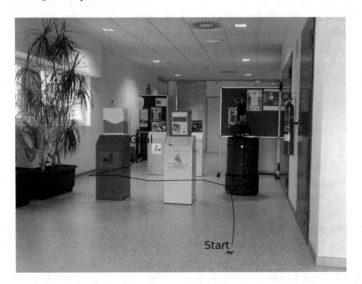

Fig. 13 Experimental setup

System evaluation: In order to measure the suitability of the system a experiment has been designed and performed in which 5 persons (3 males and 2 females), all but one not directly involved in the development of the system, were told to give the robot the oral instructions necessary to make the robot reach a predetermined goal from a starting position (see Fig. 13), and results can be seen in Table 2. The theoretical minimum number of instructions refers to the number of steps required by the designed trajectory (forward, left, forward, right, forward and stop). Besides, the empirical min number of instructions refers to the real minimum number of steps done by one of the volunteers.

Table 2 Results

Theoretical min number of instructions needed	6
Empirical min number of instructions	6
Mean num. of instructions per trip	9.2
Percentage of correctly understood instructions	79 %
Mean time needed to reach the goal	2 min 30 s
Minimum time required to reach the goal	1 min 38 s

The results of the experiment are not quite significant. The only meaningful thing that can be said is that after a period of training the robot can be operated properly in a real environment. But *MariSorgin*'s laser location is not adequate for obstacle avoidance. The robot will require hard structural changes to get the laser located in an optimal position. This problem is reflected in the teleoperation system, because the obstacle information that the operator can reach does not provide information about table and chair legs, for instance. This could be overcome setting the laser on the old pan-tilt unit and using the *laser_assembler* package to reconstruct the obstacles laying on the floor and offering the teleoperator the resulting pointcloud. But the main drawback is the delay between the speech identification and the robot action (about 2 s) that makes the system a bit dangerous specially when the robot is speeded up too much, or when the operator does not anticipate enough the order.

Note that it is straightforward to use this package in any of the wheeled robots.

6.2 The `nao_teleop_speech_eus` Package

MariSorgin is rather limited in its body expressiveness. It is not very appropriate for body language communication. NAO's morphology is much more suitable for HRI and has a huge potential for body language communication and, thus, for exploiting dialogues with humans.

Within the available ROS packages for NAO, the `nao_speech` node[3] provides the necessary tools for making NAO understand and speak in English. But this node is of no use when another language is required, as it is the case.

Again, the robot control system must be defined, i.e. the meaning of the voice orders must be translated to actions.

A new package named `nao_teleop_speech_eus`) has been developed. Within this package, the `nao_teleop_speech` node allows the user to control NAO's movements using several voice commands. The operator, situated in the tele-operation cab (the place where the remote PC is located), gives orders to the robot using a microphone. The robot is able to perform these movements: *Stand up*, *Sit down*, *Move forward*, *Move backward*, *Move left*, *Move right*, *Turn left*, *Turn right* and *Stop*.

---

[3]Developed by M. Sarabia, at the Imperial College London, 2012–2013.

As we previously said, commands are given in Euskara. With the intention to communicate with the robot in a more natural way, the user has more than one choice for each possible command. That is, if the operator wants the robot to stand up, he can say: "Altxatu", "Tente jarri", "Zutik jarri", etc. Therefore a dictionary of some predefined words has been created, including several synonymous for each command. When the user gives a voice command (it can be a long sentence), the voice is converted to text, and processed afterwards. The system tries to find matches between the dictionary and the text received from Google Speech Service. If a match is found, the robot performs the movement corresponding to the received command, otherwise the robot says that it could not understand the order and asks the user to repeat it.

Thus, `nao_teleop_speech` is in charge of receiving messages from `gspeech_eus`, finding any matches in the predefined commands dictionary and deciding which is the action that NAO must perform. It has a subscriber to receive messages that `gspeech_eus` publishes on the `google_speech` topic, and two publishers; one to set NAO's walking velocity according to the given command, and another one to publish the text messages that NAO has to say.

In order to test the speech capabilities in a HRI context, we integrated the ASR and TTS modules in our Bertsobot project [1]. The Bertsobot project was showed to the general public in a live performance entitled "Minstrel robot: science or fiction"[4] in wich NAO robot showed his verse-improvisation and speech-based communication capabilities. *ZientziaClub* or Club of Sciences is an initiative that aims to disclose science and technologies to the society. RSAIT showed some advances in human-robot interaction by presenting a monologue with NAO: https://www.youtube.com/watch?v=NEiDw\discretionary-JBER9M.

We are now working on improving NAO's body language while speaking in order to show a more human-like behavior and to be more emphatical. For the same reason, NAO's verbal communication capabilities should be improved so that it could give the same semantical answer using sentences of different gramatical structures, and of course, to perceive feedback from the public or the interlocutor and express accordingly.

7 Conclusions

In this paper, some ROS packages have been described and some of the applications given to those nodes were more deeply explained as case studies in concrete robot platforms. Of course, all the developed applications are setup for the rest of the robots. Some videos of life shows can be seen in RSAIT's youtube video channel (https://www.youtube.com/channel/UCT1s6oS21d8fxFeugxCrjnQ) and in our website.

---

[4]http://www.badubada.com/badubadatzen/es/robot-bertsolaria-zientzia-ala-fikzioa/.

ROS has provided us a tool to standardize this society of robots, different in nature and with different hardware, and has given us the opportunity to set up the same programming and control environment for all the robots. The decision to setup all our robots with ROS allows us to more easily understand, use and maintain them.

It has been hard to reach the actual state. It took time to setup all the robots, to develop the missing drivers and to establish a uniform configuration for all of them. ROS versioning has been a drawback. But it has been worth. Now, it is rather easy to adapt a behavior/application to a different robot. New lab members/students adapt rather quick to ROS basics and can work with any of the platforms. No need to learn several APIs and software environments, neither to know hardware differences among the robots further than movement restrictions and sensor nature.

Thus, rather than a programming tool, ROS has became a methodology for research in robotics. We are willing for ROS 2.0 to have a network of robots communicating to each other and performing operational work inside the faculty.

Acknowledgments This work was supported by the Basque Government Research Team Grant (IT313-10), SAIOTEK Project SA- 2013/00334 and the University of the Basque Country UPV/EHU (Grant UFI11/45 (BAILab).

References

1. A. Astigarraga, M. Agirrezabal, E. Lazkano, E. Jauregi, B. Sierra, Bertsobot: the first minstrel robot, in *Human System Interaction*, (2013), pp. 129–136
2. A. Astigarraga, E. Lazkano, B. Sierra, I. Rañó, I. Zarauz, Sorgin: A Software Framework for Behavior Control Implementation, in *14th International Conference on Control Systems and Computer Science (CSCS14)*, (Editura Politehnica Press, 2003)
3. J. Badger, M. Diftler, S. Hart, C. Joyce, Advancing Robotic Control for Space Exploration using Robonaut 2, in *International Space Station Research and Development*, (2012)
4. G. Ceccarelli, A. Patriti, A. Bartoli, A. Spaziani, L. Casciola, Technology in the operating room: the robot, *Minimally Invasive Surgery of the Liver* (Springer, Milan, 2013), pp. 43–48
5. C. Doarn, K. Hufford, T. Low, J. Rosen, B. Hannaford, Telesurgery and robotics. Telemed. e-Health **13**(4), 369–380 (2007)
6. G. Du, P. Zhang, J. Mai, Z. Li, Markerless kinect-based hand tracking for robot teleoperation. Int. J. Adv. Robot. Syst. (2012)
7. M.E. Foster, M. Giuliani, A. Isard, C. Matheson, J. Oberlander, A. Knoll, Evaluating Description and Reference Strategies in a Cooperative Human-robot Dialogue System, in *IJCAI*, (2009), pp. 1818–1823
8. Ingenia motion control solutions (2008). http://www.ingeniamc.com/En
9. E. Jauregi, I. Irigoien, B. Sierra, E. Lazkano, C. Arenas, Loop-closing: a typicality approach. Robot. Auton. Syst. **59**(3–4), 218–227 (2011)
10. J. Koenemann, M. Bennewitz, Whole-body Imitation of Human Motions with a NAO Humanoid, in *2012 7th ACM/IEEE International Conference on Human-Robot Interaction (HRI)*, (IEEE, 2012), pp. 425–425
11. I. Leturia, A.D. Pozo, K. Arrieta, U. Iturraspe, K. Sarasola, A. Ilarraza, E. Navas, I. Odriozola, Development and Evaluation of Anhitz, a Prototype of a Basque-Speaking Virtual 3D Expert on Science and Technology, in *Computer Science and Information Technology, 2009. IMCSIT'09*, (2009), pp. 235–242
12. H. Mallot, M.A. Franz, Biomimetic robot navigation. Robot. Auton. Syst. **30**, 133–153 (2000)

13. M.J. Matarić, *The Robotics Primer* (MIT Press, 2009)
14. D. Matsui, T. Minato, K. MacDorman, H. Ishiguro, Generating Natural Motion in an Android by Mapping Human Motion, in *2005 IEEE/RSJ International Conference on Intelligent Robots and Systems, 2005. (IROS 2005)*, (IEEE, 2005), pp. 3301–3308
15. F. Mohammad, K. Sudini, V. Puligilla, P. Kapula, Tele-operation of robot using gestures, in *7th Modelling Symposium (AMS)*, (2013)
16. K. Nagatani, S. Kiribayashi, Y. Okada, K. Otake, K. Yoshida, S. Tadokoro, T. Nishimura, T. Yoshida, E. Koyanagi, M. Fukushima, S. Kawatsuma, Emergency response to the nuclear accident at the Fukushima Daiichi Nuclear Power Plants using mobile rescue robots. J. Field Robot. **30**(1), 44–63 (2013)
17. I. Rañó, Investigación de una arquitectura basada en el comportamiento para robots autonomos en entornos semiestructurados, Ph.D. thesis, University of Basque Country, UPV/EHU, (2003)
18. I. Rodriguez, A. Astigarraga, E. Jauregi, T. Ruiz, E. Lazkano, Humanizing NAO robot teleoperation using ROS, in *Humanoids*, (2014), pp. 179–186
19. RWI (1995). Beesoft user's guide and reference. http://mobilerobotics.cs.washington.edu/docs/BeeSoft-manual-1.2-2/beemanx.htm
20. SAMPA, Speech assessment methods phonetic alphabet. EEC ESPRIT Information technology research and development program, (1986)
21. A. Setapen, M. Quinlan, P. Stone, Beyond Teleoperation: Exploiting Human Motor Skills with Marionet, in *AAMAS 2010 Workshop on Agents Learning Interactively from Human Teachers (ALIHT)*, (2010)
22. H. Song, D. Kim, M. Park, J. Park, Teleoperation Between Human and Robot Arm using Wearable Electronic Device, in *Proceedings of the 17th IFAC World Congress* (Seoul, Korea, 2008), pp. 2430–2435
23. W. Song, X. Guo, F. Jiang, S. Yang, G. Jiang, Y. Shi, Teleoperation Humanoid Robot Control System Based on Kinect Sensor, in *2012 4th International Conference on Intelligent Human-Machine Systems and Cybernetics (IHMSC)*, vol. 2 (IEEE, 2012), pp. 264–267
24. H.B. Suay, S. Chernova, Humanoid Robot Control using Depth Camera, in *2011 6th ACM/IEEE International Conference on Human-Robot Interaction (HRI)*, (IEEE, 2011), pp. 401–401
25. R.Y. Tara, P.I. Santosa, T.B. Adji, Sign language recognition in robot teleoperation using centroid distance Fourier descriptors. Int. J. Comput. Appl. **48**(2), 8–12 (2012)
26. S. Thrun, W. Burgard, D. Fox, *Probabilistic Robotics*, (MIT Press, 2005)
27. R.T. Vaughan, Massively multi-robot simulations in stage. Swarm Intell. **2**(2–4), 189–208 (2008)
28. B. Wang, Z. Li, N. Ding, Speech Control of a Teleoperated Mobile Humanoid Robot, in *IEEE International Conference on Automation and Logistics*, (IEEE, 2011) pp. 339–344

Part IV
Real-World Applications Deployment

ROS-Based Cognitive Surgical Robotics

Andreas Bihlmaier, Tim Beyl, Philip Nicolai, Mirko Kunze,
Julien Mintenbeck, Luzie Schreiter, Thorsten Brennecke,
Jessica Hutzl, Jörg Raczkowsky and Heinz Wörn

Abstract The case study at hand describes our ROS-based setup for robot-assisted (minimally-invasive) surgery. The system includes different perception components (Kinects, Time-of-Flight Cameras, Endoscopic Cameras, Marker-based Trackers, Ultrasound), input devices (Force Dimension Haptic Input Devices), robots (KUKA LWRs, Universal Robots UR5, ViKY Endoscope Holder), surgical instruments and augmented reality displays. Apart from bringing together the individual components in a modular and flexible setup, many subsystems have been developed based on combinations of the single components. These subsystems include a bimanual tele-manipulator, multiple Kinect people tracking, knowledge-based endoscope guidance and ultrasound tomography. The platform is not a research project in itself, but a basic infrastructure used for various research projects. We want to show how to build a large robotics platform, in fact a complete lab setup, based on ROS. It is flexible and modular enough to do research on different robotics related questions concurrently. The whole setup is running on ROS Indigo and Ubuntu Trusty (14.04). A repository of already open sourced components is available at https://github.com/KITmedical.

Keywords Cognitive robotics · Medical robotics · Minimally-invasive surgery · Modular research platform

1 Introduction

Research into the robot-assisted operating room (OR) of the future necessitates the integration of diverse sensor and actuator systems. Due to the rapidly progressing

A. Bihlmaier (✉) · T. Beyl · P. Nicolai · M. Kunze · J. Mintenbeck · L. Schreiter ·
T. Brennecke · J. Hutzl · J. Raczkowsky · H. Wörn
Institute for Anthropomatics and Robotics (IAR), Intelligent Process Control
and Robotics Lab (IPR), Karlsruhe Institute of Technology (KIT),
76131 Karlsruhe, Germany
e-mail: andreas.bihlmaier@kit.edu

H. Wörn
e-mail: woern@kit.edu

© Springer International Publishing Switzerland 2016
A. Koubaa (ed.), *Robot Operating System (ROS)*, Studies in Computational
Intelligence 625, DOI 10.1007/978-3-319-26054-9_12

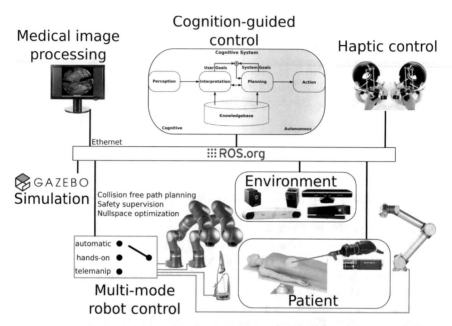

Fig. 1 Overview of our modular ROS-based research platform for robot-assisted minimally-invasive surgery (cf. [4]). The system contains several components to perceive the environment, the patient and the user. Physically actions can be executed by different robots. Interchangeability of components is essential, i.e. higher level algorithms should not have to know which robot, tracking system or camera is used. The whole setup also exists as a virtual model for the robotics simulator Gazebo. Thus algorithms can be evaluated in simulation on a single host without the necessity of accessing the real lab setup. This benefits both researchers and students by reducing the problem of scheduling lab access

state of the art, the volatility of project funding and the diversity of research questions, a modular and flexible platform is essential. Instead of each researcher developing a separate setup for his project, synergies can be taken advantage of if the core platform is used and extended by multiple researchers.

The ROS-based OP:Sense platform [9] tries to accomplish this for robot-assisted minimally-invasive surgery (MIS). Figure 1 shows the currently integrated components. Some components are directly used in the research applications. For these ROS provides a standardized and network transparent interface to use them—concurrently—from any computer within the lab network. This way many lab resources can be shared and do not have to be under exclusive control, e.g. sensors. Other resources require arbitration, e.g. control of robot manipulators. Yet, it is still advantageous that they are not bound to a specific control host. However, the more interesting platform subsystems are those relying on multiple resources and present these resources as more capable components to research applications. For example, it is either possible to control the redundant lightweight robots directly from the application or to interface them through a subsystem. This subsystem uses the environment sensors to optimize their redundant degree of freedom automatically. Examples of research applications include an automated endoscopic camera guidance robot [2], intuitive human robot interaction in the OR based on multiple fused Kinect cameras [1] and probabilistic OR situation recognition [7].

The remainder of the chapter consists of the following topics:

- First, a brief introduction to robot-assisted surgery and current research topics in this field are provided.
- Second, the general lab setup in terms of computing and networking hardware is described.
- Third, an overview of the diverse components that are integrated into the platform. For some components we rely on packages provided by the ROS community. These will be referenced and briefly discussed. The components that were either heavily adapted or created in our lab and released to the community will be treated in greater detail. Our components comprise different perception components (Kinects, Time-of-Flight Cameras, Endoscopic Cameras, Marker-based Trackers, Ultrasound), input devices (Force Dimension Haptic Input Devices), robots (KUKA LWRs, Universal Robots UR5, ViKY Endoscope Holder), surgical instruments and augmented reality displays.
- Fourth, following the single components, this part focuses on subsystem, which provide higher level functionality through the combination of multiple components. The subsystems include a bimanual telemanipulator, multiple Kinect people tracking, knowledge-based endoscope guidance and ultrasound tomography.
- Fifth, we will focus on organizational and software engineering aspects of the overall setup. In our case the robot actually is the whole lab. Therefore, change and configuration management are particular problems to be addressed.

2 Background

Research in the field of surgical robotics has seen a similar shift to research in industrial robotics. Instead of aiming for fully automated systems with a minimum of human involvement, the goal is to provide the surgeon and more general the operating room (OR) staff with cooperating assistance systems. These systems require information about the current situation within the OR, in particular about the tasks and activities of the OR staff. The robots in the OR must also be sensitive to their immediate surroundings instead of simply executing a pre-programmed task. Research questions pertain to improve the robot's capabilities of perception, planning or action. In order to facilitate research that relies on progress in multiple of these dimensions, we designed a ROS-based modular platform for cognitive surgical robotics.

Fig. 2 An example slave
side of the telemanipulation
scenario featuring three
different types of robots:
Two manipulating robots
(LWR, UR5) with attached
articulated tools in the
minimally-invasive
configuration and the ViKY
system holding the
endoscopic camera

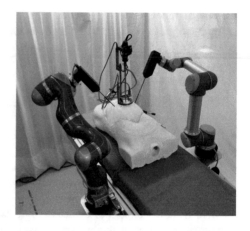

3   ROS Environment Configuration

Although industrial robot manipulators are part of our lab setup, the actual robotic
system is the whole operating room. For that reason, each component and subsystem
of the platform will be described on its own in the following chapters. Beforehand,
only a few basic points about the complete computing and networking hardware are
presented here. The first part of the computational infrastructure are the core machines
that provide a ROS abstraction to the components and subsystems. The second part are
the individual researchers' client machines, which run the high-level applications.
In the core platform a large Gigabit Ethernet switch serves as central networking
hub. Some large components contain an additional internal network hierarchy that is
connected to the central switch (e.g. Sects. 4.4 and 4.5). Core computing is distributed
across about 15 computers. For the most part these consist of commodity desktop
machines, some with high-end graphics cards (see Sect. 4.6), some small form-factor
PCs and a few single-board computers. All core machines run Ubuntu Linux.

4   Components

4.1   Robots

KUKA LWR IV Each of the KUKA LWR arms (cf. left robot in Fig. 2) has seven
degrees of freedom with position and torque sensors in each joint. Both robots are
controlled from a single PC. The communication protocol between the robots and
the PC—the Fast Research Interface (FRI)—is provided by KUKA and is based on
UDP. We control the robots with 1 kHz update rate, so the computer has to respond
within 1 ms. If the computer does not meet the cycle deadline, i.e. does not respond
in time to the measured joint data with new target parameters, the robot control per-

forms an emergency stop of the robot. Thus, for each robot there are two threads. One thread is running at the highest Linux kernel priority handling the FRI communication. FRI joint positions and torques are stored and converted into Cartesian position and forces/torques. Then one interpolation step towards the current target is performed, either only in joint space or first in Cartesian space and then validated and truncated within joint space. The other thread with normal priority manages the ROS communication. Its update rate is decoupled from the robot and flexible, depending on the task.

We decided to use the publisher/subscriber mechanism since the `actionlib` is not suited for our scenario where we have constantly changing target positions. These are due to the fact that we operate in a highly dynamic and unpredictable environment. The robot is controlled online by a human instead of moving on a predefined trajectory. Messages to the robot will not be forwarded directly, instead they update the target for the interpolation, so velocity and acceleration constraints are handled by the controller and the pose update rate is flexible. For robot control in joint space we use `sensor_msgs/JointState`, both for reading the current robot position and commanding a new target. In the second case, the semantics of the velocity field are adapted to command the maximum joint velocity and the effort field is used to command the maximum joint acceleration. For Cartesian control we use `geometry_msgs/Pose` and specify fixed velocity and acceleration limits in an initialization file. The robot controller PC handles some further low level control strategies like hands-on mode and an optimization for the elbow position (the robot redundancy). In hands-on mode, each joint accelerates into the direction of external torque (the compensation of gravity and dynamic torques is done by the KUKA controller), while heavily decelerated by viscous friction, so the robot comes to halt quickly after being released. To optimize the elbow position, a cost function is created which considers the following aspects: distance of each joint to its limits, distance of each hinge joint to its stretched position (to avoid singularities), distance of the elbow to an externally specifiable target. The last point is used for interacting with surgical personnel, which is further described in Sect. 5.3.

Universal Robots UR5 The UR5 robot is supported out of the box through the universal_robot stack provided by the ROS Industrial community. However, in addition to the actionlib interface, we added the same topic interface as described for the LWR IV above. Furthermore, a Gazebo plugin was developed that exposes the manufacturer proprietary network format (see Sect. 5.6).

Trumpf ViKY A ROS interface for the motorized endoscope holder ViKY was developed that allows servo position control over the network. The interface is the same as for the two other types of robots. But since the ViKY only has three degrees of freedom (DoF) a Cartesian topic is not provided, instead trocar coordinates are exposed. The latter consist of spherical coordinates centered in the remote center of motion, which is defined by the small trocar incision in the patient's abdominal wall.

Path Planning and Collision Avoidance For both lightweight manipulators, to be more precise for any combination of manipulators and tools attached to the OR table, a URDF robot description has been created.[1] Based on the robot's URDF we use MoveIt! for path planning and collision avoidance. Path planning is an optional component in some configurations of the system (cf. Sect. 5.1). However, collision avoidance is always active with respect to all static parts of the environment.

4.2 Endoscope Cameras

Endoscopic cameras are a prerequisite for minimally-invasive surgery, thus the platform provides not only one but two kinds of endoscope cameras. The first camera is a medical device, R. Wolf Endocam Logic HD, which is interfaced through a HD-SDI connection with a PC-internal video capture card, Blackmagic Design DeckLink. A ROS driver, decklink_capture, was developed which provides the images using image_transport. The second camera is an industrial GigE Vision camera, Allied Vision Manta G-201, which is compatible with the community provided prosilica_driver package. Proper use of ROS name remapping and of the information provided in `sensor_msgs/Image` ensures that higher level applications transparently work with both cameras.

4.3 OR Perception System

One major focus of OP:Sense is the perception of the robotic system and its environment, especially people acting around and interacting with the robots. As the environment in the OR is often very crowded, we developed a dense sensor system to avoid occlusions. The sensing system consists of the following components that each publishes its data under a separate topic namespace:

- an optical tracking system (ART): `/art/body1..n`
- a time-of-flight (ToF) 3D camera system (PMD): `/pmd/S3/camera1..n`
- a structured light 3D camera system (Kinect): `/kinect/camera1..n`

Figure 3 gives an overview of the realized network topology for the whole perception system. Figure 4 shows a picture of the real setup as realized in the laboratory. In the following, the components and their implementation as parts of our ROS network are explained in more detail.

---

[1]To cope with the many possible combinations, we defined the models in a hierarchical manner using the Gazebo SDF format, which we convert to URDF using the sdf2urdf converter provided by our pysdf package.

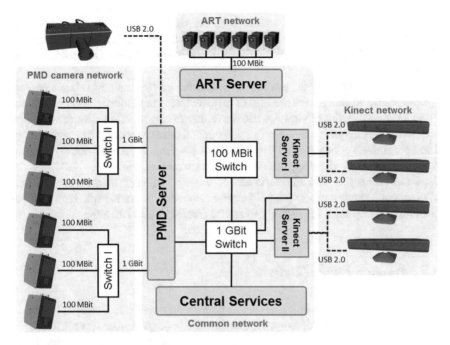

Fig. 3 Network topology of perception subsystems

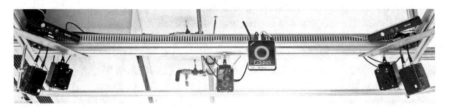

Fig. 4 One side of our sensor rig with different cameras: Kinect (*top left* and *top right*), ToF cameras (PMD S3, outermost), optical tracking system (ART, with two visible *red* LEDs), Kinect server (*center*, with visible *green circle*)

4.4 Marker-Based Optical Tracking

For high-accuracy 6D pose tracking e.g. of medical instruments, the ARTTrack2 system by ART is used. We use a six-camera configuration in order to reliably track rigid bodies in a volume over the OR table in presence of occlusions, e.g. by robots or humans. The tracking system provides a network stream of the tracked data in a simple proprietary network protocol. For usage in the ROS-based OP:Sense environment, a ROS node was implemented that provides the pose of tracked objects both as a 'raw' data stream with all values provided in original ART format (`Float64MultiArray`) and as a `PoseStamped` message with correct `frame_id` to be directly accessible in ROS nodes and via the ROS command line tools.

The scientific value of the tracking system is the precise tracking of rigid bodies, e.g. to record and analyse the motions of surgical instruments during an intervention. From a technical point of view, we use it as a day-to-day tool for conveniency tasks, such as quickly acquiring an object's location, and routine tasks such as registration. The coordinate system defined by the optical tracking system is used as a world coordinate system, where applicable. Registration methods have been developed for the different sensors and actors present in our system, such as cameras (3D and 2D, e.g. endoscopic), robots (end effectors equipped with marker spheres), medical phantoms and systems for augmented reality (projector and LED array).

4.5 Time-of-Flight Cameras

For low latency, low resolution 3D scene perception a multi-ToF-camera system is used. It consists of six pmd[vision] S3 cameras (64 × 48 px) and one pmd[CamCube] 2.0 (204 × 204 px). This camera system is for applications where speed matters more than high resolution, e.g. safety critical applications such as human-robot-interaction in a shared workspace and collision avoidance.

As ToF cameras are prone to interference when used in the same space, we implemented a time- and frequency-based multiplexing synchronization. The cameras are controlled via their proprietary API from a dedicated PC over Ethernet segments separate from the ROS network (and USB in case of the CamCube). The data acquisition and processing/publishing have been split into two different threads, thus a slower processing/publishing (e.g. publishing to many subscribers via TCP) does not affect the rate with which the cameras are triggered.

Acquired data is lightly preprocessed (filtering typical ToF artifacts such as jumping pixels) and published onto the ROS network with the according data types, e.g. `sensor_msgs/PointCloud2` for point clouds and using image_transport for the amplitude, depth and (in case of CamCube) greyscale images. As the low-level details of the camera control such as time-multiplexed triggering are hidden from the ROS network, there is also no direct low-level per camera control exposed via ROS (such as offered in the pmd_camcube_3_ros_pkg package). Instead, control is implemented as a service that allows to reconfigure the camera system to different preconfigured modes, e.g. a mode with high frame rate and reduced integration time (resulting in higher noise) or with high integration times and subsequent triggering that offers the best data quality.

4.6 RGB-D Cameras

In order to allow a high resolution supervision of the scene, which is of great importance for human action detection within the operating room, RGB-D cameras are used in OP:Sense. More specifically, we utilize the Microsoft Kinect 360 cameras of which four are attached to a ceiling-mounted camera rig (see Fig. 4). The Kinect camera depth sensing is based on a structured light pattern which is projected onto the scene. This pattern in the infrared spectrum is observed using an infrared camera. The principle is based on stereo vision using an active component. Through the use of an astigmatic lens system, the light dots of which the structured light is composed are observed as ellipses with orientation. Their geometric relation and size is dependent on the position of the surface the light is projected, relative to the camera sensor and can therefore be used to reconstruct the 3D scene information. The output data of the Kinect cameras is a 11 bit depth map with 640 × 480 pixels resolution. In our system, this depth map is directly used for people tracking (Primesense NITE framework) in addition to the calculation of point clouds. The four Kinect cameras are mounted in a rectangular configuration of approximately 1.80 × 2.10 m over the OR table. The center of the field of view of each camera is in the center of the rectangle to observe the operating table and it's surrounding with the highest possible resolution that can be achieved with Kinect cameras.

The Microsoft Kinect 360 cameras are integrated into the ROS environment using the OpenNI package which allows for using the Kinect cameras from within a ROS system. To keep the Kinects' footprint in the OR as small as possible, we split the access and processing of Kinect data between different computers. Each two Kinects are connected to a small form-factor PC (Zotac ZBOX nano AD10, featuring two USB host controllers and a 1 GBit Ethernet port). This Kinect server runs the driver component of the OpenNI package and publishes the depth map and unprocessed RGB image on the network. The processing of all connected Kinect cameras is done on a central server located further away from the OR table, where sterility and space usage are not an issue. This approach allows for a flexible number of Kinect cameras which can easily be adapted to local requirements. The current small Kinect servers, which have been integrated in 2011, still use internal fans for cooling, which is problematic in the sterile OR environment. By today, the same system could be easily realized with passive-cooled Kinect servers.

In addition, a new Kinect-based camera system using the Time-of-Flight based Kinect One is currently being integrated. This system uses a four camera configuration like the system described above. However, for the upcoming Kinect One system we use a small computer for each camera running Windows 8.1 that already performs initial data processing such as point cloud calculation and user tracking based on the Microsoft Kinect SDK 2.0.

The processed data is published to ROS using a custom bridge based on win_ros. At time of implementation, win_ros was available only for Visual Studio 2010 whereas Kinect SDK 2.0 requires Visual Studio 2012, so we split the ROS bridge into two components that exchange their data using shared memory. Data is published using the following message types:

- `sensor_msgs/PointCloud2`: Organized, 512 × 424 px RGB point cloud (extended with a "uid" field that encodes the user id if the according point corresponds to a tracked user).
- `sensor_msgs/Image`: 1920 × 1080 px RGB image.
- `geometry_msgs/PoseArray`: Joint poses per tracked human.
- `UInt8MultiArray`: Tracking state for each joint per tracked human (as defined by Kinect SDK 2.0: not tracked, inferred, tracked).

To deal with the expected data volume, a new 10 GBit Ethernet segment is currently added to our ROS network through which the published data will be transferred to a Ubuntu-based workstation for further processing.

4.7   Input Devices

One of the main research topics in the scope of OP:Sense is human robot cooperation. Therefore interfaces that allow for a natural interaction with the system are required. In the telemanipulation scenario the surgeon directly controls the robotic arms using haptic input devices accessible from a master console. The input system is composed of a left hand device, a right hand device and a foot pedal. Monitors are included in the master console providing an endoscopic view together with additional information about the environment as well as the system and intervention state. The haptic input devices are 7 axis devices from Force Dimension. In OP:Sense we use the Sigma.7 device (right hand) and the Omega.7 device (left hand). The devices are composed of a delta kinematic for translational movement in Cartesian space coupled to a serial kinematic for rotational input. Additionally, a gripper is attached at the handle of the input device to actuate medical grippers or coagulators attached to the medical robot. The Omega.7 device (Fig. 5) is capable of displaying translational forces and gripper forces to the user. The Sigma.7 device additionally allows to render torques on the rotational axis.

Fig. 5 The Omega.7 haptic input device mounted at the master console

For the integration of these devices, a ROS wrapper based on the available Linux driver of the haptic input devices has been written. This wrapper is a ROS node which is run for each device. It allows to access the pose of the device using a topic of type `geometry_msgs/PoseStamped`. The gripper position can be accessed via a topic of type `std_msgs/Float32`. In order to render forces and torques on the devices, a topic of type `std_msgs/Float32MultiArray` with 3 translational entries, 3 rotational entries and 1 value for the gripper opening is used. The foot pedal component includes two pedals and is used in the telemanipulation scenario as a clutching system (deadman switch) and to change the scaling factor. The pedal component is a medical pedal that has been connected to a USB I/O adapter that can be accessed using an open source driver. Our ROS wrapper publishes all I/O channels as `std_msgs/Float32MultiArray`.

4.8 *OpenIGTLink-ROS-Bridge*

OpenIGTLink (Open Network Interface for Image Guided Therapy) is a standardized network protocol, used for communication among computers and devices in the operating room. The protocol provides a simple set of messaging formats, e.g. pose data, trajectory data, image data or status messages. Reference implementations are available in C/C++, Matlab and Java. OpenIGTLink is supported, among others, by 3D Slicer, IGSTK, MeVisLab, MITK and Brainlab. In order to connect components using OpenIGTLink with OP:Sense a bridge between OpenIGTLink and ROS was implemented. This bridge provides user-defined duplex communication channels between ROS nodes and OpenIGTLink connections. OpenIGTLink clients and servers are supported. The channels convert the messages into the required target format. Each channel is able to

- subscribe messages from ROS topics, convert them and send the converted messages to OpenIGTLink connections;
- receive messages from OpenIGTLink connections and publish them on ROS topics.

Table 1 shows the currently supported OpenIGTLink message types and the mapping to ROS message types. Due to the complexity of the OpenIGTLink `IMAGE` message, it had to be split into 3 ROS messages.

Table 1 The mapping between OpenIGTLink and Ros messages

OpenIGTLink message type	ROS message type
POSITION	geometry_msgs/PoseStamped
TRANSFORM	geometry_msgs/TransformStamped
IMAGE	sensor_msgs/Image tf/tfMessage geometry_msgs/Vector3Stamped

4.9 Ultrasound Imaging

The OP:Sense platform provides an ultrasound system for intraoperative imaging and navigation. The functionality of the system allows tracked live ultrasound imaging, acquisition of 3D volumes, ultrasound-guided interventions, intraoperative navigation and preoperative imaging fusion [5]. As hardware components the Fraunhofer DiPhAS ultrasound research platform, a 2D transducer with tracker marker mount, a tracking system and a workstation computer for data processing are used. The transducer can be mounted to a robot. As software components, the Plus toolkit for navigated image-guided interventions and 3D Slicer for visualization and planning are used. The components are connected via OpenIGTLink. Any OpenIGTLink compatible device can be used for tracking system or imaging. Alternatively, any device supported by Plus can be used.

The integration with OP:Sense is realized by the OpenIGTLink-ROS-Bridge component. Plus as the central component receives pose data from the bridge node which acts as an OpenIGTLink server. Pose data can be received from any ROS topic that provides either messages of type `geometry_msgs/PoseStamped` or `geometry_msgs/TransformStamped`. The tracked image stream is provided by Plus via an OpenIGTLink server. The OpenIGTLink-ROS-Bridge connects to the Plus server and makes the stream available in the ROS network.

4.10 Surgical Instruments

OP:Sense can be used with different types of surgical instruments

- Rigid instruments: Instruments that can be used for open surgery and are rigidly attached to the end effector of the robots, such as scalpels.
- Articulated minimally-invasive instruments: Standard minimally-invasive instruments with a motor for opening/closing of the instrument and one for the rotation along the instrument shaft.
- Flexible Surgical Instruments: Research instruments with flexible shafts that allow for better manoeuvrability and dexterity.

Articulated Minimally-Invasive Instruments The instruments are standard laparoscopic instruments with a modified gripping mechanism and a motorized rotation along the instrument shaft. These instruments are used in the minimally-invasive telemanipulation configuration of the OP:Sense system. Faulhaber brushless DC-servomotors with an integrated motion controller and a CAN interface provide motorization. An instrument control node serves as a bidirectional wrapper between ROS messages and appropriate CAN messages. After calibration, the node accepts a gripper opening angle on a `std_msgs/Float64` topic (in percentage of maximum angle) and publishes its current opening angle on another topic of the same type. The same also holds for the rotation angle of the instrument shaft rotation.

Fig. 6 Instrument holder
with flexible instruments

Flexible Instruments   Instrument holder

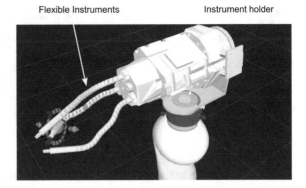

Flexible Surgical Instruments For interventions in the abdominal area, a higher dexterity of the instruments than that achievable by traditional laparoscopic instruments can be benefitial. For this case, an instrument holder with three flexible instruments (see Fig. 6) was designed, that can be attached to the end-effector of a robot [8]. Using this holder, all three instruments can rotate around a common axis and move individually on a translational axis. The flexible part of the instrument, 180 mm in length and with a diameter of 10 mm, is composed of two segments. Each segment with a length of 90 mm consists of a stack of alternating rigid and soft elements, which are equipped with two Degree of Freedom (DoF) that are actuated individually by cables. The rigid elements are stereolithographically printed, whereas the soft elements are made out of vacuum casted silicone. By pulling the cables, the silicone elements are deformed and the segment bends into the according direction. Additionally, in the center axis of the flexible structure a channel is realized where an optical sensor is integrated, capturing the current shape of the instrument. A gripper with one DoF is fixed at the tip of two instruments and a chip-on-the-tip camera module is attached to the third instrument. From the kinematical point of view, each flexible segment features 54 DoF. These are reduced to two main axis of motion, represented as revolute DoFs around the x- and y-axis per silicon element. The control of one instrument via ROS is realized by a `sensor_msgs/JointState` topic for the instrument holder and each instrument. Due to the negligible velocity of the joints, only the position field contains the x- and y-axis values for both segments. In the same way, there are topics for the grippers.

4.11 Augmented Reality

With respect to the field of human-machine-interaction, OP:Sense integrates modalities for spatial augmented reality, where information is directly provided in the scene without assistant devices such as AR glasses. For projecting information onto the situs, a full HD short-distance projector (Benq TH 682ST) is mounted over the OR

Fig. 7 Exemplary live scenario for spatial augmented reality for minimally-invasive interventions with projected instrument shaft, tip point and view cone of the endoscopic camera

table and is connected to one of the Kinect servers (see Sect. 4.6). Connection via HDMI is possible by the configuration of the vendor-specific graphics card driver in order to have a non-modified 1920×1080 pixel output signal.[2]

Rendering of content is performed using openframeworks, which we extended with a ROS interface. For displaying basic geometric information like projecting the trocar points and instrument poses onto a patient, we use the scalable vector graphics (SVG) format. We implemented a SVG preprocessor as part of our augmented reality node that parses the SVG string for a custom tf extension and replaces the tf frame ids with the correct coordinates from the point of view of the projector. If for example an application needs to highlight the robot's end effector (which pose is available on tf), it can simply send an SVG graphics that contains `<circle *tf [cx|cy] RobotToolTip /tf* r=100 fill=blue (...) />`. The augmented reality node will receive this message, continuously evaluate it based on the tf transformations available on the ROS network and project the according image. Figure 7 shows a demonstration scenario with live projection based on current system data available on tf. To keep the network load low, we realized a continuous reactive projection, e.g. onto a target that provides position updates with 100 Hz, by sending only one initial message to the projection node.

5 Subsystems

5.1 Telemanipulation

Telemanipulation is a concept to use robots in a master-slave mode. This means that input devices (masters) are used to control the robots (slaves) of a system. The concept can be used in hazardous environments such as chemistry labs. In medicine, telemanipulation is usually applied to minimally-invasive surgery (MIS) to allow the

---

[2]In our case, for an AMD graphics card on Ubuntu, the required command is: `aticonfig-set-dispattrib=DFP2,sizeX:1920` and `,sizeY:1080` as well as `,positionX:0` and `,positionY:0`.

Fig. 8 Overview of a large part of the components (see Sect. 4) in our lab setup in a configuration for telemanipulation (see Sect. 5). *Front* Master console with haptic input devices, foot pedals and screens. *Back* Different robots mounted to the OR table. *Top* Perception and AR system

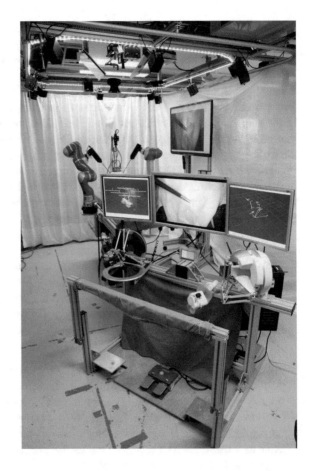

surgeon to work more similar to open surgery than is the case in traditional MIS. Additional goals are improvement of ergonomics and accuracy, e.g. by providing a scaling factor between human input and instrument motion.

Our system is composed of a master console with haptic input devices and slave robots, mounted to the operating bed, to which the surgical tools are attached. The system is depicted in Figs. 8 and 2. A third robot such as the UR5 or the ViKY is utilized to hold the endoscope providing vision to the situs. The instruments of the robots are introduced to the patient's body through trocars which are the remote center of motion for the telemanipulation task. This remote center of motion creates the fulcrum effect which causes a mirrored motion of the tip of the instruments compared to the surgeon's hand motion. In non-robotic surgery, this effect can only be compensated by the surgeon through practical experience. With a telemanipulated robot, this effect can be compensated in software. Additionally, the remote center of motion restricts the movement of the tooltip to four degrees of freedom (rotation around the tool shaft, three translations of the tooltip), which means that two rotations (hand wrist movement) are lost as long as no additional joints at the tooltip are used.

Finally, a problem in MIS is that the image from the endoscopic image is displayed on a monitor which is in most cases not registered to the surgeon's hand, thereby producing an additional cognitive load. In OP:Sense the complete system is registered with respect to the endoscopic camera whose image is displayed at the master console. Thus, the surgeon's vision of the instruments is in the same coordinate frame than his/her eyes, i.e. the instrument's position and orientation corresponds to the surgeon's hand position and orientation. The configuration of the telemanipulation system, e.g. setting which input device controls which robot, is completely controlled by launch files and can therefore be adapted to various combinations of input device and robot. The telemanipulator directly uses the robot controlling mechanisms of OP:Sense described above, i.e. the input and output of the telemanipulator is a `geometry_msgs/Pose`. This means that for a bimanual setup the telemanipulation system is launched twice with different robot-input device configurations.

The telemanipulation system is registered optically using the optical tracking system and the endoscopic image, working in the base frame of the tracking system. In addition, tracking is used to acquire the current position of all components during the operation. This allows to move the robots and the camera during the procedure while providing a stable coordinate system to the surgeon. The telemanipulation system is composed of the nodes listed below, which are implemented as nodelets to minimize the latency and the (de)serialization overhead:

- Input device node: The haptic device node described in Sect. 4.
- Telemanipulation node: This node receives the input device position, the robot position and in case of articulated instruments also the instrument position. Additionally, the node reads the calibration and periodically published tracking poses (`geometry_msgs/Pose`). The system is triggered through the update rate of the haptic input devices, that thereby serves as clock generator of 1 kHz. Each cycle new positions for the instrument tooltip are computed, which are published as `geometry_msgs/Pose` messages.
- Trocar node: A trocar point can be defined by means of an optically tracked pointing device. The device is pointed at the desired position at which moment a `std_srvs/Empty` service is called. The trocar node reads the current position of the pointing device and uses its position as trocar point. When the service is called for the first time, the trocar algorithm is activated. As soon as the trocar algorithm is running, it continuously calculates the necessary pose of the robot end effector while abiding to the trocar constraint (for each instrument pose, the trocar point has to be along the shaft of the instrument). This results in a robot pose that holds the instrument tip at the desired position, but neglects the desired orientation of the instrument. In case of the use of an articulated instrument, the trocar node also computes the required pose for the instrument. The final output of the trocar node is a topic of type `geometry_msgs/Pose` which is passed to the robot controller and a topic of type `std_msgs/Float64` representing the desired rotation of the rotation axis of the instrument in radian.

  The telemanipulator can also be used in an open surgery mode where no trocar is used. In this case the trocar algorithm is bypassed and the output of the telemanipulation mode is directly used to compute the new robot end effector pose based on the calculated tooltip pose. Articulated instruments are not used in this case as in open mode no trocar point is present and all 3 rotations and translations of the robot can be used.

- Robot controller: The robot controller directly drives the robot to the desired pose computed by the trocar node without using path planning. This is possible as the increments of the path in telemanipulation mode are small enough given an appropriate update rate of the telemanipulator.

As described above, the pedal serves as a dead man switch (which disconnects the robot when not pressed) and for setting the motion scaling. The pedal's current configuration is passed to a pedal node which passes the state of the clutching pedal as a `std_msgs/Bool` to the telemanipulation node. Whenever the clutch is closed, the telemanipulation node performs a registration routine after which any movement of the input devices and the robot is computed relative to the position of the robot and the haptic input device at the time the clutch closed. The second pedal switches between a set of scaling factors. Every time the scaling is changed, the current scaling factor is passed to the telemanipulation node as a topic of type `stdmsgs/Float32` where it is used to scale the motion of the haptic input device with respect to the robots movement.

5.2 Multi-RGBD People Tracking

Another important part in the scope of OP:Sense is the semantic perception of the environment. To allow for sophisticated human robot interaction such as situation detection or the optimization of the robot pose with respect to the human, information about the human pose is needed. For this purpose we utilize the Kinect system where each Kinect camera is registered to a reference one. The approach is based on OpenCV and PCL. For registration, we first detect a checkerboard in both RGB cameras (reference camera and camera to be registered), then we use the depth camera to find the 3D positions of the checkerboard corners in both Kinect frames. The resulting correspondences are then used to estimate the transform between the cameras. This is repeated until all cameras are registered to each other. In a second registration step we use the reference camera and the RANSAC plane detection algorithm to detect the floor plane which is later on used for the computation of people positions.

The aim of the people detection system is to integrate information about the position of the humans, the points representing the humans and the skeletal configuration of the humans. As written in Sect. 4.6, one workstation is dedicated to perform all computations on both the RGB and the depth images from the Kinect. A filter pipeline implemented as nodelets is used to perform a fusion on the heterogeneous data from the Kinect. The components of this pipeline are:

- People detection node: The Primesense NITE algorithm is used to detect and track people in the camera frames. The algorithm runs for every camera and extracts a depth image that only includes the pixels representing a human (other pixels are black). NITE is capable of detecting up to 16 humans at a time so we compute 16 depth images. We then perform an erosion operation on the depth images to remove some of the noise introduced by the use of multiple Kinect cameras. For each camera we then publish 16 topics whereas each topic is used for one of the users depth images.[3] Additionally we add the skeletal configuration of each human that is computed by the NITE algorithm with respect to the camera frame where the human is detected to the tf tree. Finally, we publish topics of type `std_msgs/Float32` to provide the keys of users detected and the keys of users of which tracking information are available for following computations.
- User image processor node: This node computes a 3D pointcloud for each human detected in a Kinect camera using the known transfer function of the Kinect. In the human detection system, this node runs once for every Kinect camera and subscribes to the 16 depth image topics representing the users. After the computation of a point cloud for every user, we perform an additional noise removal step using the statistical outlier removal algorithm from PCL. This produces a low noise point cloud for every human detected in the scene. For every possible detected human a `sensor_msgs/PointCloud2` topic is created which is used to publish the point clouds representing the users. When four Kinect cameras are used, this creates 64 topics that are either empty or contain information associated to a detected human.
- Fusion: This node is not triggered by the camera driver as the Kinect cameras are not running synchronously. Instead, it uses its own computation loop for clock generation. It subscribes to all point clouds representing users and to the array of detected users published by the NITE node. The purpose of this node is to combine the clouds of corresponding users that are detected by multiple cameras. As a measure to determine these correspondences, the centroid of each point-cloud representing a human is computed using the according PCL algorithm and projected to the floor. To allow for a more robust correspondence estimation, we project the centroids of the point clouds to the floor. The Euclidean distance between the centroids is used to concatenate the point clouds of humans which were detected in multiple cameras. The final point clouds are published as `sensor_msgs/PointCloud2`. Additionally we use a custom message holding a 2D array that provides the information about corresponding humans in multiple camera views.
- Distance computation: This node computes features that can be used to infer information about the current state in the OR. Features include distances between humans and different parts of the robots or between human and human. One particular feature is the distance between the human closest to the two robots. It is

---

[3]Unfortunately, the NITE nodes cannot be run as nodelets.

used as a simple low-dimensional input to the robot cost function in Sect. 4.1. The node subscribes to the combined point clouds of the Fusion node and uses CUDA to perform brute force Euclidean distance computation between the point clouds and CAD models of scene items. An example for such a CAD model is the robot in its current joint configuration.

5.3 Human-Robot-Interaction

In the design of a large modular surgical robotic platform, ease of use and intuitive interaction has to be considered, but also safety for the medical staff, the patient as well as the technical devices. Therefore, we combine probabilistic models and rule based functions in order to interpret the context information in an ongoing surgery. Context information can be modelled by using workflows. These workflows are composed of individual workflow steps presented as Hidden Markov Models (HMM). For each workflow step, a different HMM was trained using previously acquired training data. In Fig. 9 a representation of the target workflow (autonomous switching to the hands-on mode) is given. During the training phase we recorded various features to fit each HMM individually.

In the following example we employ four workflow steps and thus four HMMs ('s' = start, 'a' = touch EE, 'b' = move robot, 'l' = release EE). Additionally, we annotate each step by capturing the corresponding keyboard input, including the ROS time to synchronize the recorded features and the keyboard input afterwards. The quality of the online classification strongly depends on the selected feature vector. One representative feature for the workflow is the minimal distance between the human and the robot end effector as calculated by the Multi-RGB-D People Tracking subsystem (cf. Sect. 5.2) and provided as a `std_msgs/Float`. Another feature is the robot's velocity which is included in the `sensor_msgs/JointState` messages provided by its controller (see Sect. 4.1). The last feature "approach" indicates if the human is moving towards or away from the robot derived from minimal distance feature. The implementation was done in Python, for which a Gaussian HMM

| (a) | (b) | (c) | (d) |

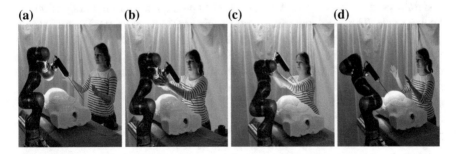

Fig. 9 Intuitive switching of a robot from position mode into Hands-On mode and back, based on the perception subsystem using Hidden Markov Models. **a** Start position 's'. **b** Touch EE 'a'. **c** Move robot 'b'. **d** Release EE 'l'

model is provided within the package scikit. After training of the HMMs, they can be used for realtime online classification in OP:Sense. The feature vector is passed to each HMM and evaluated by using the log likelihood probability implemented in scikit. The HMM with the highest likelihood will be set as the most likely state. The result state is published as a `std_msgs/String` for other subcomponents.

5.4 Endoscope Guidance

The teamwork between the surgeon and the human camera assistant in minimally-invasive surgery poses many challenges and opportunity for improvement. Therefore, we research having a robot instead of a human assistant guiding the endoscope. Motorized endoscope holders, such as ViKY (see Sect. 4.1), are already commercially available. However, they require that the surgeon manually controls every repositioning, thus further increasing his cognitive load. A cognitive endoscope guidance system is currently being researched that provides autonomous endoscope repositioning by means of a knowledge base. Since this assistance system is not developed standalone, but based on our ROS platform, many synergies can be exploited and additional functionality is provided to the surgeon. For further information, we refer to the detailed description in [2].

5.5 Ultrasound Tomography

The ultrasound imaging component can be used to create 3D volumes using the freehand-3D-ultrasound ability. To acquire more detailed volumes with fewer distortions a robotic ultrasound tomography was implemented. Using a robot mounted transducer it is possible to record equidistant and parallel slices which allows a direct volume reconstruction. In addition, the area of interest can be scanned from several directions, which can decrease speckle noise. Due to the force control scheme of the LWR robot, it is possible to acquire scans by following uneven surfaces with the probe such as the human body. As a tracking system, the pose data from the optical tracking system is subscribed. Alternatively it is possible to use the robot position as tracking data. The robot and ultrasound imaging system, using the OpenIGTLink-ROS-Bridge, is controlled from a Python ROS node. As an HMI, a GWT GUI using rosjs was implemented, so any device with a browser can be used for inputs. The results are displayed in 3D Slicer. The reconstructed volume can be published to the ROS network using the OpenIGTLink message type `IMAGE`.

5.6 Simulation

Due to the cost and complexity of the complete platform, only a single instance of it exists in our institute. It is shared by multiple researchers and their students. Naturally this leads to a bottleneck in scheduling time for complex experiments. The modularity and network transparency provided by ROS mitigate the problem to some extent, because sensor components and subsystems can be used in parallel. However, in the common case that multiple researchers require exclusive access to some components, e.g. the OR table mounted robots, another solution is required. Providing an accurate simulation of large parts of the setup, of which multiple instances, running on office desktop computers, can be used independently from each other, helped us a lot. More details about our use of the Gazebo simulator for this purpose and how the simulation can help to perform advanced unit and regression testing (Robot Unit Testing) is published in [3].

5.7 Software Frameworks

Table 2 provides an overview of important libraries and frameworks, besides ROS, that are used in our setup.

6 Organization and Software Engineering

6.1 Registration and Calibration

In OP:Sense, calibration and registration are organized in a separate ROS package. The usual reference for calibration is the optical tracking system. For calibration

Table 2 Major software frameworks used in our surgical robotics platform

Name	Website	Version(s)
3D Slicer	http://www.slicer.org/	4.4
Chai3D	http://www.chai3d.org/	
Gazebo	http://gazebosim.org/	4.0; 5.0
Matlab	http://de.mathworks.com/products/matlab/	
MITK	http://mitk.org/	
MoveIt!	http://moveit.ros.org/	
OpenCV	http://www.opencv.org	2.4; 3.0-beta
Openframeworks	http://openframeworks.cc/	
OpenIGTLink	http://openigtlink.org/	2
Plus	https://www.assembla.com/spaces/plus/wiki	2.2
Point Cloud Library (PCL)	http://www.pointclouds.org	
Scikit-learn	http://scikit-learn.org/stable/	0.15.2

purposes we use a tracking pointer with a rigid tip location which is found using pivotisation. Using the transformation acquired by pivotisation, we can now compute the position of the pointer's tip with reference to the optical tracking system. The pointer is then used to register other devices such as cameras to the optical tracking system using landmarks in the scene that can be touched with the pointer and can at the same time be detected within the camera's image. Using OpenCV we compute the transformation from the camera coordinate frame to the optical tracking system. If the camera body itself is equipped with markers, e.g. in case of the endoscopic camera, we can compute the transformation between the optical frame and the frame of the markers, allowing to precisely calibrate the optics to the tracking system. The calibrated pointer is also used for online calibration tasks such as setting new trocar points for the telemanipulation system. If a new transformation is computed through registration, the translation and the quaternion representing the pose are saved to file. These registration files can be loaded through a pose publishing node that is controlled via a launch file to publish the registration information on the tf tree and/or on a dedicated `geometry_msgs/Pose` topic.

6.2   TF and Pose Topics

OP:Sense uses both the tf tree and a `geometry_msgs/Pose` based mechanisms to send transformations. tf is usually selected if a transformation has to be visualized. Additionally, tf is beneficial if the publishing frequency of a transformation is low or average. For complex transformation chains this can easily deliver the transformation from a frame to any other frame in the tree. Unfortunately, tf is not as performant as publishing transformations on Pose topics and degrades further when transformations are published with high frequency. For fast transformation publishing such as in the telemanipulation use case (1000 Hz) we solely rely on sending transformations on dedicated `geometry_msgs/Pose` topics and perform the frame multiplications in the node. During the development phase, we usually keep the publishing frequencies low and use tf. At the end of a development cycle when the frequency is increased we switch to topics which enables most of the nodes to use both mechanisms. Additional benefits of this method are that transformations can at any time be visualized using RViz and the load on the tf is reduced to a minimum level.

6.3   Windows/Matlab Integration

The research and development process for the realization of math intensive components such as flexible instruments (see Sect. 4.10) or inverse kinematics for robots requires the use of a rapid prototyping environment. In our case Matlab/Simulink is the tool for development of the robotic system kinematics, workspace analyses and control design. However, the problem in our setup is, that Matlab runs on a

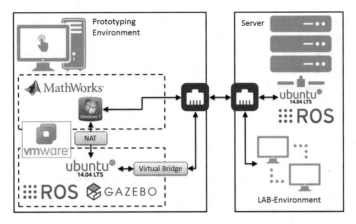

Fig. 10 Infrastructure to connect Gazebo via ROS with Matlab on Windows through a virtual Linux machine

"Windows-Desktop-PC" and the roscore of the lab on a Linux system. In order to control and interact with both systems, the lab setup, where the final hardware runs, as well as the prototyping system and simulation in Matlab on Windows, we decided to use virtualization (see Fig. 10). Inside the Linux guest system, Ubuntu 14.04 is installed including ROS (Indigo) and Gazebo (4.0). The communication from Windows to the lab environment is established by a physical network adapter with a dedicated static IP. Additionally, there is a virtual bridge from the virtual machine also using the physical network adapter of the host system. Now, it is possible to send ROS messages from Windows to the lab's ROS network in the same way as to the virtual one. The configuration inside the virtual machine is limited to starting a local roscore and setting up the environment variables accordingly. Regarding the Matlab part in this setup, the ROS I/O package for Matlab 2013b is used to communicate with ROS. Therefore, a Matlab class-object is implemented to create nodes, topics and the associated publishers and subscribers. For future work, the latest Matlab release 2015a with full ROS support via the Robotic System Toolbox will be used, thereby achieving the integration of ROS, Gazebo, Simulink and real-time hardware.

6.4 Software Repositories and Configuration Management

The balance between independence of each researcher on the one hand and keeping all components and subsystems working together on the other is difficult to get right. After trying out many different ways to organize the software and configuration of our platform, we suggest the following best practices for such a large setup: First, maintain one source code repository for each part that is usable on its own. Second, changes to core components that change an interface visible to the ROS network must be documented and agreed upon beforehand. Third, have one repository with a hierarchical set of launch files to bring up different parts of the core components and subsystems. Fourth, always mind the software engineering principles of "Single Point of Truth" (SPOT) or "Don't Repeat Yourself" (DRY) [6] with regard to code and configuration information. Fifth, maintain a (virtual) blackboard listing all available system functionality and links to documentation on how to use it.

References

1. T. Beyl, P. Nicolai, J. Raczkowsky, H. Worn, M.D. Comparetti, E. De Momi, Multi Kinect People Detection for Intuitive and Safe Human Robot Cooperation in the Operating Room. In: *2013 16th International Conference on Advanced Robotics (ICAR)* (2013), pp. 1–6
2. A. Bihlmaier, H. Wörn, Automated Endoscopic Camera Guidance: A Knowledge-Based System towards Robot Assisted Surgery. In *Proceedings for the Joint Conference of ISR 2014 (45th International Symposium on Robotics) and ROBOTIK 2014 (8th German Conference on Robotics)* (2014), pp. 617–622
3. A. Bihlmaier, H. Wörn, Robot Unit Testing. In *Proceedings of the International Conference on Simulation, Modelling, and Programming for Autonomous Robots (SIMPAR 2014)*, (2014), pp. 255–266
4. A. Bihlmaier, H. Wörn, ROS-based Cognitive Surgical Robotics. In *Workshop Proceedings of 13th International Conference on Intelligent Autonomous Systems (IAS-13)* (2014), pp. 253–255
5. T. Brennecke, N. Jansen, J. Raczkowsky, J. Schipper, H. Woern, An ultrasound-based navigation system for minimally invasive neck surgery. Stud. Health Technol. Inf. **196**, 36–42 (2014)
6. A. Hunt, D. Thomas, *The Pragmatic Programmer: From Journeyman to Master* (Addison-Wesley, Boston, 1999)
7. L. Schreiter, L. Senger, T. Beyl, E. Berghöfer, J. Raczkowsky, H. Wörn, Probabilistische Echtzeit-Situationserkennung im Operationssaal am Beispiel von OP:Sense. In: *13. Jahrestagung der Deutschen Gesellschaft für Computer- und Roboterassistierte Chirurgie e.V*, (2014), pp. 177–180
8. J. Mintenbeck, C. Ledermann, R. Estana, H. Wörn, EndoSnake–A Single-Arm-Multi-Port MIRS System with Flexible Instruments. In: *13th International Conference on Intelligent Autonomous Systems (IAS-13)* (2014)
9. P. Nicolai, T. Brennecke, M. Kunze, L. Schreiter, T. Beyl, Y. Zhang, J. Mintenbeck, J. Raczkowsky, H. Wörn, The OP: sense surgical robotics platform: first feasibility studies and current research. Int. J. Comput. Assist. Radiol. Surg. **1**(8), 136–137 (2013)

Authors' Biography

Andreas Bihlmaier Dipl.-Inform., obtained his Diploma in computer science from the Karlsruhe Institute of Technology (KIT). He is a Ph.D. candidate working in the Transregional Collaborative Research Centre (TCRC) "Cognition-Guided Surgery" and is leader of the Cognitive Medical Technologies group in the Institute for Anthropomatics and Robotics-Intelligent Process Control and Robotics Lab (IAR-IPR) at the KIT. His research focuses on cognitive surgical robotics for minimally-invasive surgery, such as a knowledge-based endoscope guidance robot.

Tim Beyl Dipl.-Inform. Med., received his Diploma in Medical Informatics at the University of Heidelberg and University of Heilbronn. He is working as a Ph.D. candidate in the medical group at the IAR-IPR, KIT. His research focuses on human robotic interaction for surgical robotics, telemanipulation, knowledge based workflow detection and scene interpretation using 3D cameras.

Philip Nicolai Dipl.-Inform., received his a Diploma in computer science at Universität Karlsruhe (TH). He is working as a Ph.D. candiate in the medical group at the IAR-IPR, KIT. His research focuses on safe applications of robots to medical scenarios, safety in human-robot interaction based on 3D scene supervision and application of augmented reality for intuitive user interaction.

Mirko Kunze M.Sc., received his master's degree in mechatronics at Ilmenau University of Technology. He is working as a Ph.D. candidate in the medical group at the IAR-IPR, KIT. His research focuses on workspace analysis for redundant robots.

Julien Mintenbeck M.Sc., received his master degree in mechatronics at the university of applied science of Karlsruhe. He is working as a Ph.D. candidate in the medical group at the IAR-IPR, KIT. His research focuses on active controllable flexible instruments for minimally invasive surgery and autonoumous microrobots for biotechnological applications.

Luzie Schreiter Dipl.-Inform. Med., received her degree at the University of Heidelberg and University of Heilbronn. She is working as a Ph.D. candidate in the medical group at the IAR-IPR, KIT. Her research focuses on safe and intuitive human-robot interaction as well as workflow identification in a cooperative systems, such as surgical robotics systems for minimally-invasive surgery.

Thorsten Brennecke Dipl.-Inform., received his Diploma in computer science at TU Braunschweig. He is working as a Ph.D. candidate in the medical group at the IAR-IPR, KIT. His research focuses on sonography aided computer assisted surgery.

Jessica Hutzl Dipl.-Ing., received her Diploma at the Karlsruhe Institute of Technology (KIT). She is a Ph.D. candidate working in the TCRC "Cognition-Guided Surgery" and is a member of the Cognitive Medical Technologies group in the IAR-IPR, KIT. Her research focuses on port planning for robot-assisted minimally invasive surgery with redundant robots.

Jörg Raczkowsky Dr.rer.nat, is an Academic Director and Lecturer at the Karlsruhe Institute of Technology. He received his Diploma in Electrical Engineering, and Doctoral Degree in Informatics from the Technical University of Karlsruhe in 1981 an 1989 respectively. From 1981 to 1989 he was a Research Assistant at the Institute for Process Control and Robotics working in the field of autonomous robotics and multi sensor systems. Since 1995 he is heading the Medical Robotics

Group at the Intelligent Process Automation and Robotics Lab at the Institute of Anthropomatics and Robotics. His current research interests are surgical robotics, augmented reality and workflow management in the OR.

Heinz Wörn Prof. Dr.-Ing., is an expert on robotics and automation with 18 years of industrial experience. In 1997 he became professor at the University of Karlsruhe, now the KIT, for "Complex Systems in Automation and Robotics" and also head of the Institute for Process Control and Robotics (IPR). In 2015, he was awarded with the honorary doctorate of Ufa State Aviation Technical University, Russia. Prof. Wörn performs research in the fields of industrial, swarm, service and medical robotics. He was leader of the Collaborative Research Centre 414 "Computer- and Sensor-assisted Surgery" and is currently in the board of directors and principal investigator for robotics in the Transregional Collaborative Research Centre (TCRC) "Cognition-Guided Surgery".

ROS in Space: A Case Study on Robonaut 2

Julia Badger, Dustin Gooding, Kody Ensley, Kimberly Hambuchen
and Allison Thackston

Abstract Robonaut 2 (R2), an upper-body dexterous humanoid robot, was developed in a partnership between NASA and General Motors. R2 has been undergoing
experimental trials on board the International Space Station (ISS) for more than
two years, and has recently been integrated with a mobility platform. Once post-
integration checkouts are complete, it will be able to maneuver around the ISS in
order to complete tasks and continue to demonstrate new technical competencies
for future extravehicular activities. The increase in capabilities requires a new software architecture, control and safety system. These have all been implemented in the
ROS framework. This case study chapter will discuss R2's new software capabilities,
user interfaces, and remote deployment and operation, and will include the safety
certification path taken to be able to use ROS in space.

Keywords Space robotics · Safety architecture · Human-robot interaction

1 Introduction

Robonaut 2 (R2), an upper-body dexterous humanoid robot, has been undergoing
experimental trials on board the International Space Station (ISS) for more than two
years [1]. Thus far, R2 has been restricted to working from a stanchion on orbit [2],

J. Badger (✉) · D. Gooding · K. Ensley · K. Hambuchen
NASA- Johnson Space Center, Houston, TX 77058, USA
e-mail: julia.m.badger@nasa.gov

D. Gooding
e-mail: dustin.r.gooding@nasa.gov

K. Ensley
e-mail: kody.g.ensley@nasa.gov

K. Hambuchen
e-mail: kimberly.a.hambuchen@nasa.gov

A. Thackston
Oceaneering Space Systems, Houston, TX 77058, USA
e-mail: allison.thackston@nasa.gov

© Springer International Publishing Switzerland 2016   343
A. Koubaa (ed.), *Robot Operating System (ROS)*, Studies in Computational
Intelligence 625, DOI 10.1007/978-3-319-26054-9_13

Fig. 1 Robonaut 2 with
mobility upgrade

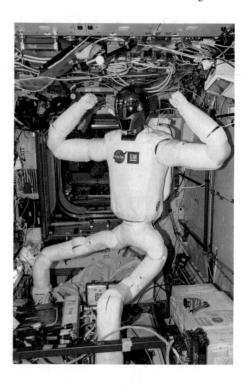

but it has recently been integrated with a mobility platform, shown in Fig. 1. Once
post-integration checkouts are complete, it will be able to maneuver around the ISS.
Its objectives are twofold. First, R2 will strive to reduce time spent by astronauts on
routine duties by completing maintenance and cleaning tasks inside the ISS. Some
example tasks are inventory management, handrail cleaning, vacuuming filters, and
data collection, such as air flow measurements. Second, the operational experience
R2 gains inside the ISS will be essential in guiding mechanical, operational, and
control system designs for a Robonaut unit to operate outside for extravehicular
activities (EVA) in the future.

There were many important and unique challenges associated with creating a
software control system for Robonaut 2. First, with its 58° of freedom and sophisticated set of sensors, any controller must be able to handle the complexity while still
creating a stable and useful platform for the challenging applications planned for
R2 [3]. Second, stringent safety requirements are in place for any system on board
the International Space Station, and R2 is no exception. Because R2 will be moving
around inside of the ISS in close proximity to crew members, a two fault-tolerant
software safety architecture has been certified to provide three levels of safing actions
when required [4]. Finally, software deployment and robotic operations to a remote
system using a latent, bandwidth-limited and uniquely configured network require
specialized solutions.

All of this work was done using native ROS tools and capabilities. In particular, Orocos [5] is used within the ROS framework to coordinate the many degrees of freedom while maintaining precision and dynamic controller performance that is required for the application. The safety system involves four processors plus all embedded controllers, hundreds of sensors, and over 100 different checks and monitors, nearly all of which are driven by the ROS messaging environment. Software deployment and operational commanding use more open-source tools that have been shown to work well with the ROS environment, such as OpenVPN (https://openvpn. net/).

The case study chapter will cover the design and implementation of the software architecture, user interfaces, and software deployment scheme of Robonaut 2, with an emphasis on the centrality of ROS to the entire system.

- The overall architecture will be presented in Sect. 2.
- The novel control software will be discussed in Sect. 3.
- The safety system will be discussed in Sect. 4. Section 4 will also discuss safety certification of R2's software, including the certification path for ROS and the other open source tools used.
- Higher level control elements, including machine vision capabilities, will be presented in Sect. 5.
- The R2 simulation and other custom user interfaces developed for or integrated with the system will be described in Sect. 6.
- Deployment and operations challenges and solutions will be presented in Sects. 7 and 8, with a discussion on the benefits and drawbacks of using ROS with respect to the maintainability of the overall software load.
- Finally, results of experiments with R2 will be presented in Sect. 9, along with a discussion of the merits and concerns for implementing an EVA robotic control system in ROS.

Throughout the chapter, best practices and lessons learned will be emphasized.

2 Architecture Overview

2.1 System Architecture

The Robonaut 2 upper body has two seven degree of freedom arms with complex, tendon driven hands that feature another twelve degrees of freedom each. R2's neck has three joints for expanded viewing capabilities, particularly for teleoperators, and the upper body has a joint at its waist. R2 features series elastic actuators that are torque-commanded in the waist and the five major joints of each arm. Force and torque can also be measured in the six-axis load cells in each shoulder and forearm. The head features several vision sensors, including analog "teleoperation" cameras, machine vision cameras, and a time of flight sensor. The fingers have several strain gauges to simulate a sense of touch.

The mobility upgrades to R2 brought many changes to the original system. R2 was outfitted with two seven degree of freedom legs that were designed to be long enough to climb between nodes inside the ISS. Each leg has a gripping end effector capable of grasping handrails internal to the ISS. Each of the seven leg joints are series elastic actuators that are capable of being torque-controlled. The end effector is an over-center locking mechanism with a manual release in case of emergency. The end effector has a sensor package that includes a machine vision GigE camera, a 3D time-of-flight sensor, and a six-axis load cell. In both the upper body and the legs, the joints all communicate with the central processors (or "Brainstem") via a communication protocol loosely modeled after the M-LVDS standard.

R2 now has three Core i7 processors running Ubuntu as part of the mobility upgrades. The processors each have different functions in the control and safety monitoring of the robot. There is one processor whose primary job is control of the robot, the "pilot;" a second processor, the "captain," monitors joint telemetry and runs the same calculations as the primary control processor as a check. The third processor, the "engineer," connects to the other two via ethernet and compares these calculations as another monitor. The third processor is also used for vision processing and other high-level supervisory control functions. All robot code on these Brainstem processors runs in the ROS framework.

R2 will be upgraded at a future time with a battery backpack to allow for wireless climbing through the ISS. R2 will connect to the ISS wireless network via an embedded processor called the Cortex, which speaks ROS in a Digi Linux operating system. This Cortex guards the ISS network from the traffic on the robot's internal network, limiting what data can be passed in or out to the ground control station. The Cortex also adds some safety monitoring capabilities to R2.

2.2 Software Architecture

The overall software architecture is depicted in Fig. 2. The control architecture is unique in many ways. The architecture consists of model-based, joint-level (embedded) impedance control components designed to accurately track command inputs from the centralized coordinated dynamic trajectory controller, RoboDyn. The Robonet I/O system directs the high-speed communication protocol between the coordinated control Brainstem processor and the joints. The JointApi section acts as an abstraction layer between the different data and control needs of different types of joints (i.e., series elastic versus rigid, or simple versus complex) and the coordinated controller. Above the control layer sits the safety system, which consists of a set of controls and monitors that protect for two major hazards, excessive force and the inadvertent release of both gripping end effectors. Also in this layer is an arbiter that manages which processes have control of the robot. The supervisory control layer consists of controllers and estimators (such as machine vision processing) that are resident to the robot and provide autonomy to R2. These processes are outside of the safety-controlled functionalities and must use a specific API to access the

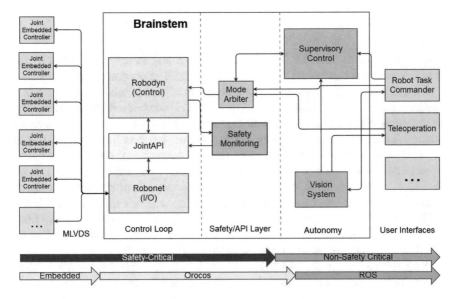

Fig. 2 Software architecture diagram

robot's control interfaces. This is similar to the set of user interfaces that exist off-board the robot. These user interfaces could be scripts, supervisory control elements, teleoperation, or other interfaces as needed.

A key feature of this architecture is that it allows a separation between safety-critical and non-safety critical code, even when they are resident on the same processor. This separation is largely constructed as the difference between Orocos components and ROS capabilities. For example, any command that can be received via ROS topic or service is not safety-critical, but instead part of the API to the robot. Commands that could affect the safety system of the robot are limited to Orocos or rttlua, its real-time script engine environment, interfaces only (i.e., ports, properties, component configuration). This is useful since elaborate safety certification testing is required for most changes to the safety-critical code base, which would severely hinder any technology development process using R2. Instead, technology development can happen in the non-safety critical supervisory and user interface layers without requiring any special testing or procedures.

3   Control Software

This section outlines the basic control system for R2. This software is considered safety-critical, and generates the general control API for the robot. The RoboNet I/O system, the JointApi abstraction layer, and the novel Robodyn controller are all described in detail.

3.1   RoboNet

Robonaut 2 utilizes an M-LVDS hardware bus and custom protocol, RoboNet, for communication between the Brainstem and embedded joint controllers.

RoboNet is a packet-based request-reply protocol and is used to atomically set and get values on the joint controllers. The M-LVDS bus master device (specifically, an Acromag XMC-SLX150 in the Brainstem) is programmed with a round-robin communications loop, continuously reading and writing data from each M-LVDS bus node device (specifically, FPGAs on the joint controllers). The bus master can be put into one of two modes: data synchronization or streaming. In "data mode" (used for nominal runtime), the communications loop synchronizes 512B blocks of memory between the bus master and each bus node, and can support up to 12 M-LVDS channels with 15 bus nodes on each channel. In "stream mode" (used for joint controller configuration), the master communicates with a single node, transferring an arbitrary number of named values to and from the node.

Brainstem access to the M-LVDS memory blocks is handled using a two-layer driver. At the "access driver" layer, a Linux kernel module provides basic read/write and ioctl capabilities to the bus master block device. Then at the "policy driver" layer, user-land C++ classes provide constrained calls to the access driver, giving a well-described, always-safe interface to the bus master. The bus master is provided with a description of what nodes exist and which memory offsets to synchronize. Figure 3 describes the RoboNet component architecture that is implemented in Orocos [5].

Component Details The DataWarehouse is a name-value abstraction for accessing data from RoboNet nodes while the RoboNet master is in data synchronization mode. It provides a way for high-level processes to easily access exactly the information

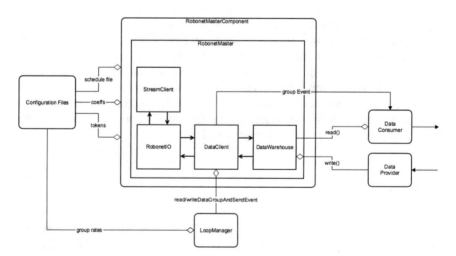

Fig. 3 RoboNet component-level architecture diagram

they need without having to use the arcane process of directly offsetting into a block of memory. Essentially a C++ std::map, the DataWarehouse is accessible across all processes running on R2, giving a uniform, easy-to-use view into low level data.

The DataWarehouse is populated via the DataClient. The DataClient is periodically triggered to synchronize groups of data between the RoboNet master and the DataWarehouse. Data elements are grouped by direction (e.g., read, write) and function (control, monitoring, etc.). In this way, "frames" of data are created, allowing for consistent status and control of joint controllers. The DataClient itself is provided with a description of each data element and what memory offset it lives at.

The StreamClient, upon request from a high-level process or the robot operator, reads a set of named values from a configuration file and sends them to a specific joint controller. The StreamClient temporarily places the RoboNet master into streaming mode during its execution, and returns the RoboNet master to data synchronization mode when it is finished.

A manager named RobonetMaster instantiates all of the above (the driver layers, the DataWarehouse and DataClient, and the StreamClient) and handles their interconnections and moding requests. It is a standard C++ class that can be utilized by any high-level process that needs to interact with RoboNet nodes. R2 makes use of the Orocos Real-Time Toolkit (RTT) and has an Orocos component wrapper around RobonetMaster. This component makes available many of the RobonetMaster methods via OperationCallers as well as a number of Orocos DataPorts for triggering RobonetMaster actions and streaming status information.

Another Orocos component, the LoopManager, used during nominal runtime of R2, has a description of which data groups exist and at which rates they should be synchronized. The LoopManager sends triggering messages to RobonetMaster, which causes DataClient to synchronize data between the RoboNet master and the DataWarehouse. Simultaneously, the LoopManager notifies the RTT environment (via DataPorts) that new data is available in the DataWarehouse. Ultimately, via the LoopManager, the control loops in R2's Brainstem are triggered by RoboNet data synchronization events, providing end-to-end execution synchronization for the robot.

Benchmarks Table 1 lists benchmarks for RobonetMaster that were conducted on a quad-core Intel i7-2715QE running at 2.10 Ghz with 8 Gb RAM. Each measurement was taken as an average of 3 runs of 1000 iterations.

There is a significant amount of overhead associated with the DataClient synchronizing the RoboNet master and DataWarehouse. The computational price has a tangible benefit though, namely that access to RoboNet data is human understandable which reduces debugging time and enables rapid development of new software.

Table 1 RoboNet benchmark results

Benchmark	Primary call	Iteration frequency (Hz)	Data transfer rate (MB/s)
Single read	readMemory()	393442.62	1.5
Single write	writeMemory()	932835.82	3.56
Channel read	readChannelMemory()	88.09	1.29
Channel write	writeChannelMemory()	174.75	2.56
R/W "command" group	transferDataGroup()	176.9	0.89
R/W all groups	transferDataGroup()	106.32	0.85

3.2 JointApi

R2 has various types of joints that make up its 58° of freedom (DOFs). Some of R2's joints are simple rigid joints, some are series elastic, and some are complex mechanisms. Some of the embedded joint controllers are fully capable of achieving desired positions and torques on their own, while some are limited to pulse-width modulation (PWM) control of an actuator. Some of the embedded controllers were designed during the original torso design, and newer ones, with different capabilities, were designed during the mobility upgrade. The need for a standard, consistent interface to all these DOFs is essential, and thus R2 employs the notion of two "Joint APIs" that makes common both the control/moding and the commanding of each degree of freedom.

Controlling the mode of a joint in R2 is achieved by setting bits in 16-bit registers that are sent to the joint from the Brainstem. Unfortunately, R2's upper body was designed separately from its lower body, and the embedded joint controllers that manage the two halves have different features and require unique control register values to achieve similar modes. Additionally, keeping track of which bits and masks are what across the different controller types is tedious.

As a solution, a utility was created that maps user-friendly names for bits to their bit offsets and sizes. Then, by convention, developers gave matching names to bits that performed the same tasks across joint controllers (e.g., engaging the brake), thus abstracting away how the joint in question performed the control task. Building on this naming convention, a set of modes were identified that all of R2's joints supported: OFF, PARK, DRIVE, NEUTRAL, etc. A set of layered state machines then manages the safe transition from one mode to another, ensuring safety-related prerequisites are met and users are notified when transitions fail.

Finally, an Orocos component responsible for managing the state of every joint listens for cues from users and other processes to safely change joint modes across the robot. This Joint Control API component can listen to ROS messages or data on Orocos ports, giving flexibility to how R2 is moded.

In a similar fashion to managing modes of different joint controllers in a common way, dissimilar joints in R2 are commanded in a common way. R2 has four different styles of joints (simple rigid, series elastic, various complex mechanisms where actuator space and joint space are not easily mapped, and mechanisms), and some joints require low-level control loop closure at the Brainstem. Forcing the overall control system to uniquely command each type of joint is unrealistic, so a Joint Command API was designed to provide a common commanding interface, regardless of joint-type specifics.

Actual joint commands are sent to the joints from the Brainstem in the form of integers and floats (e.g., desired position, desired torque). A utility was created to give user-friendly names for these values and value memory locations to easily map between "control space" and "I/O space" (see Sect. 3.1). On top of this utility, a set of "API" C++ classes exist, one for each joint type, that accepts standard sensor_msgs/JointState commands and translates them into joint type-specific I/O commands. For the complex finger and wrist mechanisms, the API class also instantiates a Brainstem-side version of the embedded control loop found on R2's rigid and series elastic joints, so these joints can achieve positions (as their joint controllers only accept actuator commands).

Finally, a pair of Orocos components (one for joints with embedded control loops and one for joints that need Brainstem-side control loops) are instantiated that manage the API classes. The "embedded" Joint Command API Orocos component is triggered at 30 Hz and the "Brainstem" component is triggered at 210 Hz. These two rates provide a consistent command flow that maps from generic high-level control loop output to hardware-specific embedded joint controller inputs.

The ros_control packages (http://wiki.ros.org/ros_control) provide very similar functionality to R2's Joint Command API. The functionality of ros_control's "Controller" and "hardware_interface" layers are approximated in the individual API C++ classes previously discussed and Joint Command API's Orocos component layer is similar to ros_control's "Controller Manager" layer.

The choice between using ros_control or writing the Joint API components was decided in two ways. First, there was a concern, at the time of development, with how real-time safe the ros_control code would be, so writing Orocos components would mitigate that risk. Second, ros_control was very new at the time, and ROS itself was in a phase of large changes to now-stable interfaces. Since software changes for space systems are difficult, the decision was made to write a custom layer for mapping the high-level control system to the I/O layer. R2's Joint APIs have worked well for multiple configurations of joints and joint capabilities and has proven itself safe.

3.3   Robodyn

The Multi-Loop embedded controller forms the foundation upon which the overall control system is built by ensuring that each joint tracks its commanded trajectory while conforming to desired dynamic performance characteristics and safety thresholds. It achieves this by employing four cascaded high rate (5 kHz) control loops,

each consisting of sensor-fed, model-based feed forward control terms and traditional PID controllers to achieve tight tracking performance. The feed forward terms
have been designed to largely compensate for the non-linear physical characteristics
of the R2 joints, thus minimizing and linearizing the contributions needed from the
PID control loops.

The Robodyn controller handles the coordinated control calculations necessary
for smooth, integrated Cartesian control of R2. It sends synchronous commands to
the joint embedded Multi-Loop controllers such that, if the embedded controllers
achieve those commands, then the desired Cartesian trajectory is accomplished. A
block diagram of the controller is shown in Fig. 4. Robodyn can be broken up into
kinematic and dynamic components. These components are tightly coupled via inputs
and outputs as well as in execution order for smooth coordinated motion.

The kinematic part takes joint and Cartesian commands from the user or from
a supervisory control component and creates joint trajectories to achieve the commands. It generates position, velocity, and acceleration references per joint. These
velocity and acceleration references are fed to the dynamic part, which creates feed-
forward torques to compensate for inertia and gravity. The dynamic part also outputs
a stiffness and damping for each joint based on each joint's calculated effective inertia and desired natural frequency and damping ratio. Finally, it outputs joint torque
limits. The Synchronator component is used to synchronize all the commands to be
sent to the joint each time step.

All components are written in the Orocos framework and component triggering
is designed such that all calculations are completed in a tight sequence each time
step. Several ROS topics are available to interface with the Robodyn controller,

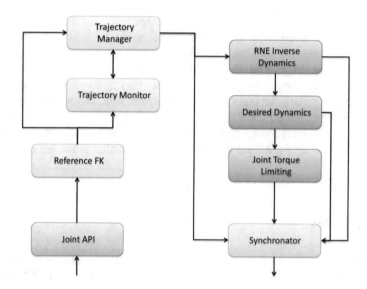

Fig. 4 Robodyn: Brainstem control architecture for Robonaut 2

in both sending commands and receiving data. Trajectory monitoring accepts joint and Cartesian reference commands and tracks the progress of these requests. The Monitor can also be configured to customize the tracking performance required by the specific task. If trajectory tracking does not occur within the specified parameters for any reason, the commanded trajectory is smoothly canceled.

The maximum velocity and acceleration used by the trapezoidal trajectory generator for both joint and Cartesian moves can be set. External forces can be added to the inverse dynamics calculation as necessary. This functionality is very useful for testing in gravity offload test facilities. Finally, joint desired dynamics data (natural frequency, damping ratio, and integrator windup) as well as joint torque limits can be commanded.

4   Safety System and Certification

The safety monitoring system involves nearly all of the available torque and position sensing on R2, as well as all of the processors. For the ISS safety requirements involving excessive loads, there are three different types of force monitoring that are present in this system. Each of those (static, dynamic, and crushing) will be detailed in this section. Two other types of safety monitoring are also described. System health monitoring is required by the ISS safety community to ensure that the robot has redundancy equivalent to the criticality of the perceived risk of R2's excessive force capabilities. An overall component-level illustration of the safety system required to achieve two fault tolerance in excessive force is shown in Fig. 5. Trajectory monitoring is not a required part of the safety monitoring system, but is used to reduce the occurrences of potential robot safing actions due to other force monitors reaching a limit.

Three Brainstem processors (pilot, captain, and engineer) are used for creating two fault tolerance to excessive force. All three processors are connected via ethernet and communicate using the same roscore, resident on the engineer, as shown in Fig. 5. All safety monitors and controls are implemented as Orocos components.

The other major control requirement imposed by the ISS safety community restricts R2 to always have at least one attachment to structure (handrail or seat track). In order to avoid inadvertent release, a sophisticated control system is required to avoid commanding an end effector open if there is no other verified attachment point (the manual release mechanism will still function, however, in case of emergencies). The controller essentially filters all commands to enable or release a gripper through several conditions; if all the conditions are not met, the gripper remains parked and locked. This control will be outlined in this section as well.

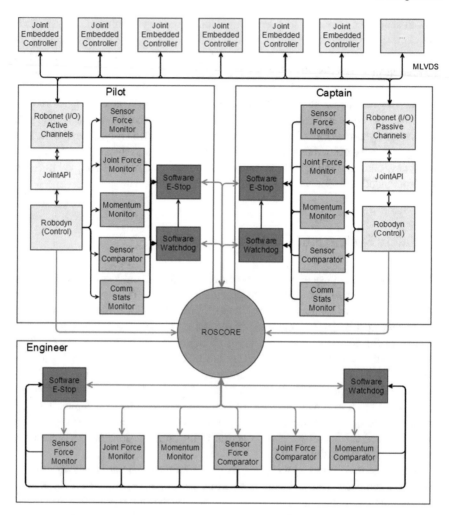

Fig. 5 Safety architecture for the excessive force catastrophic hazard

4.1 Static Force

The ISS safety community has imposed a requirement that R2 may not create a static load of more than 45.45 kg (100 lbf) on any structure, equipment, or crew on board the ISS. R2 monitors static force by calculating the force that each limb is exerting at any time, as well as calculating the system or body force. This force is calculated using two different sets of sensors. First, the joint sensor package in the series elastic actuators is used to calculate the torque on each joint and a forward dynamics calculation finds the effective force and moment at each end effector (Joint Force Monitor). These forces are translated and combined in the body frame to ensure that the body is not imposing a force greater than the limit in the case that more than one end effector is contacting structure. Because of the many sensors that can be used to determine the torque and position in each joint, it is possible to consider this calculation redundant with health monitoring, which will be described later.

A similar calculation is performed using the six-axis load cells in each appendage. These direct force and moment measurements are transposed to the end effector frame for both arms and both legs (Sensor Force Monitor). Also, the measured forces are transposed and summed in a body frame to get the system force. As these measurements are not affected by singularities in the kinematic chain as the previous calculations are, this calculation is the sole source of force data on an appendage if that appendage becomes singular.

If any of the calculated force or moment values exceed the required limit, a safing event occurs. Once a safing event has been triggered, the motor power is removed from the system in three redundant ways.

4.2 Dynamic Force

Instances of dynamic force must also be accounted for and controlled in a safe manner. To limit the amount of dynamic force that can be applied by R2, the amount of momentum that is generated is monitored. An accurate mass model is used to calculate the inertia of each link (which comes from the Robodyn dynamic components), which is then multiplied by the velocity of each link, calculated from the forward kinematics of the robot, to produce momentum. To create a meaningful momentum, a base reference frame for the velocity calculation must be chosen. For R2 there are two scenarios that would produce high momentum, the end effectors separately and the body as a whole. Therefore the momentum is calculated on each end effector chain, as well as from a base world frame. For the end effector chains, the base is taken at the center body frame. For the body momentum, the base must be set at a frame connected to world. For R2, this frame coincides with the gripper that is attached to station, and therefore changes as R2 traverses throughout ISS (Momentum Monitor). In order to determine an appropriate limit for momentum, the robot was commanded to impact a force plate at various speeds. The impulse produced from the impact was then analyzed. The requirements for the ISS limit R2's peak load to 1000 lbf for short durations. This translates into approximately 34 N-s for metal on metal contact. By controlling the amount of momentum an individual end effector as well as the whole body can produce, the maximum dynamic force that can be produced by the system is limited to within safe parameters. If any calculated momentum value exceeds the required limit, a safing event occurs.

4.3 Crushing Force

The end effectors on the legs, shown in Fig. 6, must clamp tightly around handrails in order to support moving the robot's large inertia. The hands are not used for support of R2 and so they are not a part of this discussion. Because the clamping mechanism requires some amount of motor torque to drive it over center, caution must be taken with ensuring that the end effectors are only using the minimal force required during closing and clamping in order to not crush pipes, conduits, and wires that may be mistakenly grabbed while reaching for a handrail. There are two major ways that crushing force is limited. The force at the end effector is limited to a constant value while the jaws are not closed. Once the jaws reach a position that has been defined as closed, this force limit is increased to allow the motors to drive the mechanism over center. In order to achieve this constant force limit, current to the motors are limited based on jaw position. The second protection is a pulse width modulation (PWM) duty cycle limit that is constant throughout all jaw positions.

Fig. 6 Robonaut 2 end effector assembly

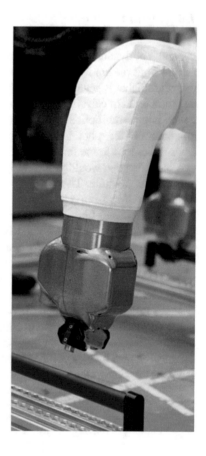

4.4   Health Monitoring

In order to ensure that the sensors required for all the essential force, moment, position, and momentum calculations are working correctly, system health monitoring has been implemented as part of the safety monitoring system. The health monitoring is very complex, with many comparisons between more than 100 sensors happening on various processors (Sensor Comparator) and on various calculated values from the measurements (Comm Stats Monitor, Sensor Force Comparator, Joint Force Comparator, Momentum Comparator). The essential theme of the system is, however, that each sensor is compared against others to ensure that values are consistent. If any comparisons fail, a safing event occurs, stopping the robot so that more investigation can occur. The position sensors (two absolute position sensors, an encoder, and a hall effect sensor per series elastic joint) are compared in several places as calculated position and velocity. The load cells are compared against the calculated force and moment at each end effector while the corresponding limb is not singular. As control of the limb is difficult near singularities, most operation occurs in regions where the force comparison is valid. Critical Orocos components on all three processors are checked for health using the Software Watchdog components.

The health monitoring aspect of the safety system also serves as protection for random bit flips due to radiation hits, as sensor values and important calculations are checked and computed on two processors, published on ROS topics, and then compared on the third processor. If any calculation or sensor value is affected on one processor, that will be caught by another processor and this discrepancy will cause the robot to safe.

4.5   Trajectory Monitoring

Because the monitors discussed thus far are required by the ISS safety community to satisfy the risk assessment for R2 and the force requirement on board the ISS, when one of those monitors are triggered, the robot must safe itself by removing motor power. Once a safing event occurs, there is a process that must be followed in order to clear the safing event. This process often involves contacting several people in the operational chain, including the flight director. In order to reduce the number of safing events to increase the up-time of the robot, a soft stop has been implemented at warn limits that are slightly inside the limits that trigger the safing events for the static force, dynamic force, and health monitors. This soft stop does not remove motor power, but instead only smoothly interrupts and arrests any commanded trajectories. This is very useful to guard against commanding the robot into a safing event while not affecting the safety monitors that will still guard against any failure situation that may trigger run-away conditions.

Fig. 7 Trajectory
monitoring allows R2's arm
to be stopped with little effort

Another "unofficial" protection, similar to the warn limits, is trajectory monitoring, that is described in Sect. 3.3 above. Because the joint torque limits can be set very low due to the accuracy of R2's mass model, this protection enables R2 to be stopped using very little effort. Figure 7 shows trajectory monitoring in action.

4.6 Inadvertent Release

Figure 2 shows the overall control architecture. There is an arbitration component between the user and the supervisory level inputs and the coordinated control system, Robodyn. Since enabling the gripper is considered to be unsafe when it is the sole attachment point to structure, the commands to enable or park the gripper motors are handled only through a gripper supervisor in the coordinated controller. Any command to change the state of the gripper motors are rejected by the Mode Arbiter.

The user or supervisory controllers can command the grippers to open, close, and lock, however. The command to lock is always allowed, as long as the gripper is not currently in the locked state. Commands to open or close the grippers, however, are only allowed when certain conditions have been met. The non-commanded gripper must be locked, there must be no faults in its joint-level controller, the motor state must be parked, and the gripper must have been verified as a base frame. Commands to grippers that are sent together in one package are handled separately and sequentially, so that commands to open two grippers that both satisfy the requirements will result in one gripper opening and the command to open the other gripper being rejected.

Grippers can be verified as a base frame through a special maneuver designed to determine that the gripper is indeed locked to structure. This command, called a "push-pull" command, is actually a supervisory controller that runs through a state machine to verify the attachment points of the robot. The push-pull command is only valid when both grippers are locked, healthy, and parked. The supervisor, once enabled, locks out and changes settings such as the desired dynamics and joint torque limits of the leg joints, and then sends up to two trajectories to the legs. The first trajectory is a roll maneuver about the gripper coordinate frame. If a gripper is not truly over-center and locked, this would cause the jaws to open. If the trajectory succeeds or if the jaws open, the base frame verification fails and the push-pull supervisor releases control. If the trajectory fails and the grippers remain locked, a second trajectory is sent. This trajectory includes both a roll in the opposite direction about the gripper coordinate frame as well as a short "pull" away from the attachment point. If this trajectory succeeds or if the gripper jaws open, the base frame verification fails, and the push-pull supervisor releases control. If the trajectory fails and the jaws remain locked, the verification succeeds. Both grippers are marked as verified base frames and the push-pull supervisor releases control. Trajectory success and failure are judged using the trajectory monitoring described in the last section. This supervisor is the only supervisor implemented as an Orocos component due to its criticality. All other supervisors are implemented as ROS nodes.

Finally, in order to have a redundant system throughout, one more protection on the gripper joint enable is needed. When a successful gripper command is issued, the gripper supervisor on two separate processors must identify the command as a valid one and both processors must send a gripper enable request to the third processor. That processor will issue a positive reply when two such requests happen within a short time frame, which then allows the gripper supervisor on the controlling processor to issue a gripper enable command to the joint-level controller. The third processor, the engineer, also monitors the state of the gripper motors via data on ROS topics and issues a safing event if it sees a gripper enabled without recently approving a request. This avoids the joint-level controllers from enabling the motors without a command.

4.7 Certification

The safety requirements on R2 were dictated by the ISS Safety Review Panel, which also provides the oversight on all software safety testing and changes that happen. The safety solution was created by assuming that all outside commands and code (including the operating system and the ROS framework) provides hazardous commands and conditions to the robot. The robot in turn must be able to reject all hazardous commands, even with one to two separate failures, and still protect against inadvertent release and excessive force respectively. This is achieved using a fail-safe philosophy. The monitors all trigger safing events that remove motor power for the excessive force events; the over-center locking mechanism requires motor power to release the grippers. So in either case, the removal of motor power ensures that the robot remains safe. The software safety certification testing involves more than 100 separate tests to show that all safety monitors and controls will correctly and safely react or act. An extensive fault matrix analysis proves that after any two classes of failures, at least one hazard control remains for excessive force, and any one failure for inadvertent release. Hardware and processor errors (such as runaway threads or inexplicable failure to calculate correct numbers) are all modeled in the fault matrices.

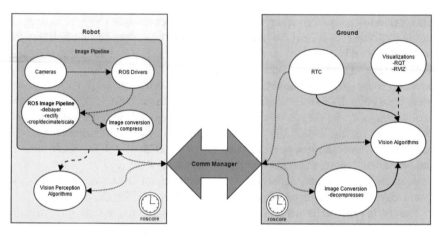

Fig. 8 Overall vision architecture

5   Vision and Supervisory Elements

An important part of the software architecture is the separation between safety critical and non-safety critical software. The supervisory controllers and machine vision elements, which are expected to change frequently as tests are run on board the ISS, are part of the non-safety critical software. This allows the development team to rapidly iterate on these components and uplink them to the robot as required. Since this framework is based on ROS, it also provides a robust interface to the robot controller which aids in integrating research from outside groups quickly and efficiently. By utilizing the ROS framework, Robonaut 2 is able to easily use any visual processing algorithm, supervisor, or other peripheral package developed in the ROS community.

In general, the vision architecture is based on nodelets and deployed as shown in Fig. 8. Nodelets were chosen as the base framework because they provide a zero copy transport between nodelets within the same manager, and because they are dynamically loadable at runtime.

Cameras that reside on the robot (locations shown in Fig. 9) are triggered to start by loading the corresponding driver nodelet. This reduces bandwidth usage and required processing power by only loading the component when it is needed. Then the imagery can go through the ROS-provided image pipeline[1] to perform common tasks such as rectification. This saves time and development resources since these common tasks are often done for any perception algorithm. From there, specific perception algorithms can be developed and deployed. Since the perception algorithms are developed to use the ROS framework, they can be deployed anywhere, on the robot or on the ground machine, as shown in Fig. 8. Although the algorithms lose the advantage of zero copy transport if loaded outside of the main vision manager, this

---

[1]http://wiki.ros.org/image_pipeline.

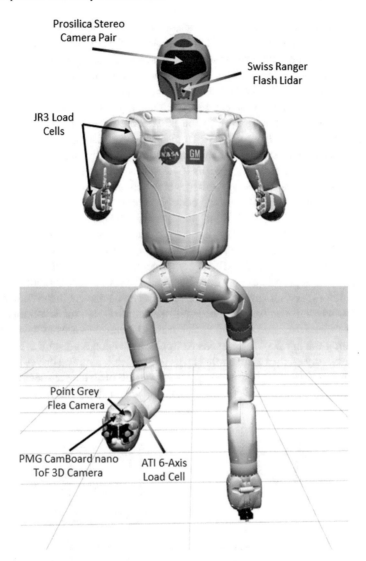

Fig. 9 Sensor locations on Robonaut 2

flexibility is helpful in debugging and testing algorithms. The loading/unloading of these algorithms is controlled via ROS service calls from the ground work station using a custom Robot Task Commander (RTC) software [6] or through supervisors through the standard nodelet interface.

One example of a vision algorithm is a handrail detection algorithm which identifies and locates obstacle free handrail sections. This algorithm uses a Point Grey Research Flea camera and a Camboard Nano depth sensor, both of which are located on R2's climbing leg end effectors. These sensors are used in combination with functions in the open source Point Cloud Library (PCL) and the OpenCV library to determine the position and orientation of a handrail, along with the sections of the handrail that are obstacle free. This information is published on a ROS topic and is fed to related user interfaces and supervisory controllers.

The supervisory control architecture, depicted in Fig. 10, allows for behavior management, autonomy, and complex task and skill development for Robonaut 2. Supervisors can be triggered from users or from other supervisors. Basic supervisorial components can be thought of as behavioral units. These behavioral units can then be chained together in other supervisors to form more complex actions.

A key component is the Mode Arbiter, which activates and deactivates the supervisors in a controlled way. The ground control computer starts controlling the robot by setting itself as the "master." At that point, the ground computer is able to transfer control of the robot to supervisors, but also to interrupt supervisory control when necessary. Supervisors are called via ROS messages from the current controlling node (or master) to the Mode Arbiter component. The Mode Arbiter component then triggers the appropriate supervisor by name. Control can be temporarily transferred to other supervisors using the same topic but by also using in the "incumbent" field. Then, when the called supervisor releases control, control is returned to the incumbent supervisor. If no incumbent is specified, control returns to the master.

Continuing the handrail rendezvous example, the vision algorithm is associated with a supervisory control element that takes in the vision data and drives the appropriate point of reference on the robot to interface with obstacle-free sections of the handrail. The supervisor, when called, waits for vision data, and drives the robot

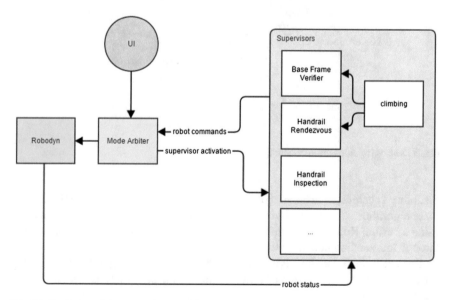

Fig. 10 Depiction of the supervisory architecture

to either reposition for reviewing (in the case of no obstacle-free handrail regions detected) or drives into the handrail. The supervisor relinquishes control when either it has driven to the handrail or has been unsuccessful in its second attempt to find a clear section of handrail. This simple supervisory action takes in vision and robot state data and sends Cartesian position commands to the robot. A more complex supervisor that completes a full automatic handrail docking procedure, with gripper locking and verification tasks, could call this handrail rendezvous supervisor as part of the greater state machine, with its name in the incumbent field so that control returns appropriately.

6 Simulation and User Interfaces

6.1 Simulation

A simulation of R2 (both upper and lower body) that works with Gazebo [7] has been provided at https://bitbucket.org/nasa_ros_pkg. The simulation is compatible with ROS Fuerte and Hydro only, but provides the API to the robot control system. Different world and object models have also been provided, such as the US Lab module of the ISS, handrails, and the taskboard that R2 has interacted with on board the ISS. The R2 simulation has been an invaluable tool in collaborations with outside groups as a framework in which to integrate their technology before attempting integration on the actual robot. Documentation and a description of the ROS topic API can be found at the wiki (https://bitbucket.org/nasa_ros_pkg/nasa_r2_simulator/wiki/Home).

6.2 Affordance Templates

The operations concept for Robonaut 2 on ISS involves control from ground operators over a latent and bandwidth-restricted communication network that experiences loss of signal (LOS) periods. Operations of a semi-autonomous robot, such as R2, under these circumstances could be enhanced by supervisory control interfaces that allow a human operator to monitor the robot and command it at a high level [8]. A predictive, interactive visualization approach was selected for supervisory control of R2, similar to previous work with NASA-Johnson Space Center mobile robots [9]. The approach involves the Affordance Template framework, developed for use with the Valkyrie robot during the DARPA Robotics Challenge [10, 11]. Affordance Templates are based on RViz and tools provided for that visualization, plus other ROS-based packages to enrich the operator's experience.

An affordance defines the relation between an object/environment and an agent that affords the opportunity for an action to be performed by that agent [12]. Affordance Templates provide a robot operator with an interactive interface to define an affordance for a robot for particular tasks. For example, a valve affords turning behaviors. The affordance template for a valve should define for the robot how to accomplish a task for turning said valve. This can be accomplished in a multitude of ways, from defining joint angles for the robot's manipulators to turn the valve, to defining waypoints for the robot's end effector to move to turn the valve, to defining the internal force required to keeping the end effectors on the valve to complete the task. Currently for this framework, affordances are defined as end effector waypoints, whether they are hands, grippers or feet.

Interactive markers provided for use with RViz are the basis of affordance templates. Interactive markers are used in combination with visual meshes for an object template. Each end effector waypoint is also an interactive marker, complete with no-to 6-DOF controls. Parameters for the task afforded by the template are associated with it and can be altered by the user in real-time. The interactive marker controls allow the operator to move templates about the 3D visualization environment, while the interactive marker menus provide both standard template options and customized options based on the parameters for that template. To handle multi-component tasks, multiple affordance templates can be combined together. For example, an item such as an electrical or communication cable can exist as a single template, while its connection location can be defined separately. Combining these templates provides an overall affordance to the robot of how to connect a cable to its outlet. Along with defining affordances, the templates can also provide predictive planning for the operator. MoveIt! is currently integrated into the framework to visualize plans from a robot end effector's current position to a desired waypoint. MoveIt! sends this information as a MarkerArray to RViz for visualization of the plan. This allows an operator to see a robot's motion before it is executed, and decide whether or not a specific plan should be sent on to the robot. If the predicted plan is acceptable, the affordance template can then be used to command the robot. Currently, two modes of control are available: 6DOF Cartesian waypoint commands or joint trajectory commands for single end-effectors.

An affordance template can be created easily using a pre-defined JSON format. This format allows for stringing together multiple templates to create a single complex template. It also allows for multiple end effector waypoints to be defined and whether or not those waypoints have 6DOF controls. The framework exists so that end effector definitions can be generalized, allowing for a single affordance template to work with many varying robots. RViz Panels are also provided with the templates to allow for stepping between waypoints (both forward and backward), for commanding an entire series of waypoints and for monitoring where the robot is with respect to the affordance template.

Figure 11 shows an example of an affordance template for an ISS IVA handrail. Figure 11a shows a single handrail with R2 leg end effector waypoints, while Fig. 11b shows the same handrail with R2 arm end effector (hand) waypoints. The waypoints define locations to which the end effectors should move to ensure efficient grasping. The waypoints themselves can be relocated in real-time using their controls. In Fig. 11c, the handrail is shown in atop visualized data from R2's depth sensor, which

(a)   **(b)**

(c)   **(d)**

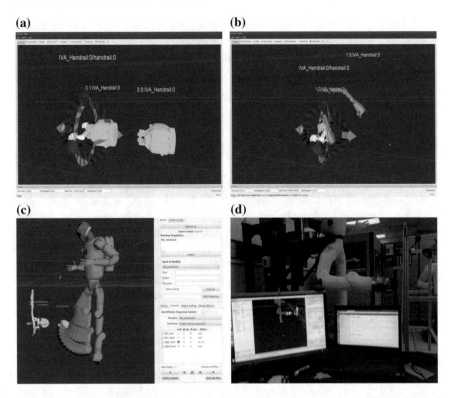

Fig. 11 **a** ISS IVA Handrail affordance template with R2 leg gripper waypoints. **b** ISS IVA Handrail affordance template with R2 hand waypoints. **c** Handrail affordance template matched to R2's sensory data showing predicted motion to grasp waypoints (using MoveIt!). **d** R2 grasping handrail as commanded using handrail affordance template

is represented as PointCloud data. This figure demonstrates how using a human-in-the-loop can handle perception inabilities that the robot may not be able to manage autonomously. The human can locate the handrail based on noisy data, and during real-time operations, move the template based on feedback from other sensors (i.e. cameras and image_pipeline data). This also allows for operations with imperfectly calibrated sensors. Also shown in this figure is the predicted path to the initial waypoint, as provided by the MoveIt! planner. The predicted locations are shown as a MarkerArray message type. Finally, Fig. 11d shows R2 using the handrail affordance template to grasp the actual handrail.

Currently, the affordance templates are being tested on the ground-based R2 for use with single manipulators. Waypoints for end effectors are used as the affordance definitions for template-related tasks. Tunable parameters currently available are scaling of the templates and the state of the end effector. Several features are planned in the near future. These include providing perceptual registration of a template to sensory data, so that initial placement of the template is decided by the robot itself; providing a plug-in component for planning of robot motion in relation to a template, specifically so that whole-body motions can be considered (MoveIt! does not currently support this); and adding force- and contact-based parameters to templates for managing compliance and stiffness of the robot during real-time task execution.

6.3   Vanguard

The ISS community imposes very strict rules regarding the operational safety of the systems on board. To abide by these rules, R2 was designed to accept "potentially hazardous commands" from the ground and to continue to be safe with help from a tool named Vanguard. In particular, all potentially hazardous commands would be sent from Vanguard clients on the ground and the Vanguard service running on R2 would screen all commands received and reject unsafe ones. An example of a potentially hazardous command is changing a safety-critical text file on the robot; Vanguard is meant to prohibit such operations while allowing the operator the flexibility to do similar operations that do not affect safety-critical components or files.

Vanguard was designed from the ground up to be service-based. A service-based architecture lends itself to writing small, simple services that can be composed to achieve complex tasks. Additionally, these small services can individually manage safety concerns they uniquely have authority over. Finally, Vanguard's services use a request-reply communications paradigm, so R2 operators can easily know whether commands were successful, failed, or were rejected (and why). The underlying technology behind Vanguard is ZeroMQ's Majordomo Protocol (MDP). Vanguard's MDP clients and workers exchange strongly-typed Google Protobuf-based requests and replies, and were written in Python and C++. In addition to the ability to control different quality of service parameters not covered by the current ROS transport, the choice was made to implement Vanguard outside of ROS in order to separate the robot commands from the other potentially hazardous commands, since ROS communication has no notion of authentication or authorization to prohibit or restrain certain messages or message senders.

Vanguard's service-based architecture and strongly-typed messages allowed it to support multiple UI choices for operators. Simple command-line clients as well as simple graphical clients are obvious choices. But, just as easily, Vanguard requests can be embedded into otherwise complex UIs to provide a more seamless user experience, without the need to run a potentially hazardous command from a special UI and normal commands from another UI.

Vanguard helps manage R2's overall state, both when R2 is simply a computer (with no robot control software running, "OS mode") and when R2 is actively controlling its joints ("robot mode"). To be both safe and flexible for users, Vanguard operates differently in those two states. In OS mode, Vanguard uses a blacklist to prevent known-hazardous commands from executing (generally, changing safety critical software and parameters). But in robot mode, Vanguard uses a whitelist to only allow known-safe commands (a very small set). Vanguard also manages the transition between those two states, and only allows robot mode to start if approved safety-critical software is present on R2's computers.

7   Software Deployment

R2 operates in a very unique environment, so it was very important to carefully design its network configuration and software deployment process to meet the ISS requirements. Network communication on the Robonaut 2 platform is split between two devices, the Brainstem and the Cortex. The primary processing center for the robot is called the Brainstem. Housed within the Brainstem are three Concurrent Technologies second generation Intel Core i7-2715QE Single Board Computers and two Kontron CP932 5+1 Port Gigabit Ethernet Switches. Each of these single board systems, or cards, are based on the Eurocard PCB spec and interface with a Compact Peripheral Component Interconnect (CPCI) backplane. The devices in the Brainstem primarily communicate through the gigabit ethernet interface on each card.

The Cortex is a Digi ConnectCore Wi-i.MX53 embedded system-on-module attached to a custom breakout board. The Cortex has two 10/100 Base-T Ethernet ports, one designated "internal to the robot," the other designated "external to the robot." The Cortex additionally offers a 802.11g/n (2.4/5 GHz) interface which has been bonded to the "external to the robot" Ethernet interface. This allows the Cortex to automatically switch between wired and wireless depending on the presence of a physical Ethernet connection.

The internal Cortex Ethernet interface connects to the Brainstem. The external Cortex interfaces connect to the outside world. This abstraction of internal versus external is enforced by more than convention. In order to alleviate concerns that rogue commands on the ISS network could adversely affect the integrity of the robot, a virtual private network (VPN) is used to isolate network traffic. The Cortex runs a VPN server that allows users on the external network to securely access the network devices on the internal Cortex network through it. Each VPN client computer is issued a key from the Cortex VPN Server guaranteeing another controlled layer of access to the robot on the ISS.

R2 employs Ubuntu 12.04 Server as the primary operating system on the Brainstem. Ubuntu was selected primarily because ROS supports Ubuntu, though it also offered the Robonaut team a fully featured OS to develop against. Ubuntu is built from Debian and utilizes Debian packages to distribute and install libraries and executables. In an effort to ensure traceability and standardization across robots, the R2 code is compiled into Debian packages that are hosted on a private Debian repository. Debian packages allow quick validation of what version of the packages are installed on a given machine. Additionally, the md5sums file in the header of each Debian package allows verification of the state of every file contained within the Debian package.

The software development process for R2 is doggedly organized, documented, and controlled. The Robonaut team employs the git distributed revision control system. Git is enhanced with Atlassian Stash, a software that enforces a high granularity permission set to each branch of code. The standard git branching model is followed. In this model, there are two primary branches of code, the master branch and the develop branch. The master branch contains the last stable tested release of code, whereas the develop branch contains the next stable version of code. Additionally, this branching model allows software developers to create any number of feature branches. Feature branches are clones of another branch of code in which developers may make unrestricted changes.

The granular control offered by Atlassian Stash ensures software developers may only clone code from a safety critical branch, and developers cannot push their changes back into the develop (or master) branch of code. In order for a software developer to push their code back into a safety critical branch of code, it must first go through a peer review, then the developer must document changes to the code and what systems and configurations he or she used to test their code. Once this process is complete, designated repository heads merge the feature branch back into the safety critical code branches.

After internal testing and verification are completed on the release candidate, all release candidate branches are committed to the master branch of code. The final master branch code is then compiled into Debian packages and an official, comprehensive safety test is conducted with Quality Assurance/Quality Engineering supervising. The guidelines for the official safety test are outlined in the Safety Data Package, though the implementation is revised for each official test.

Following the successful completion of the official safety test, the ISS Safety Review Panel (SRP) must review the results of the official safety test, and ultimately approve the deployment of the latest version of code to the R2 on the ISS. To ensure the Debian packages under review by the SRP are the same ones delivered to the R2 on the ISS, a copy of the md5sums file in each safety critical Debian is delivered to the SRP.

Once the SRP signs off, the Debian packages are sent to the Robonaut 2 onboard the ISS. However, before these packages can be installed, each safety critical Debian package is verified against the copy of the md5sums given to the SRP. This verification is performed by Vanguard, introduced in Sect. 6.3. Vanguard additionally performs this md5sums verification every time the R2 software is enabled.

8   Remote Operations

Remote operation of complex research hardware such as R2 poses significant challenges in areas of communication and configuration management. Communication methods between ground operators and R2 must handle latency, variability, and loss of signal in graceful ways. Round-trip communication can be as bad as 3 s, and loss of signal can last for 40 min. Keeping R2 operating in a safe manner with an unreliable communications path requires a number of key technologies. Likewise, keeping both R2 and operator consoles in a working configuration at all times requires diligence and planning.

To gracefully handle loss of signal, R2 utilizes a UDP-based OpenVPN[2] service. Even though stateful connections are established through the VPN, OpenVPN keeps them alive through long network disconnections. This ensures that ROS topic streams, like joint states and camera views, continue through disconnections without requiring a restart of topic subscribers. Over the VPN, the ISS network safety for R2 is ensured in that all R2 communication happens over 2 ports: one for direct SSH connections and one for VPN traffic.

To account for delayed communication and also handle loss of signal, R2 uses a multi-master ROS network, MultiMaster FKIE.[3] One roscore exists on the ground and one exists on R2 itself. This distributed layout ensures that communication between on-robot processes do not suffer from space-to-ground latencies during configuration and can query their rosmaster at any time, regardless of network conditions. Similarly, ground-based user interfaces do not need to manage ground-to-space network latencies or disconnections, as their rosmaster is local. The multimaster_fkie packages offer operators insight into what nodes, topics and services are alive on each roscore and give the ability to inspect their contents and settings. Deploying ROS nodes locally and remotely is also supported, which helps operators manage task-specific software on R2 as needed.

Because of the overall complexity of both R2 and the network configuration in which it operates, having a well-tested, stable configuration for both the robot and the operator console is essential for R2's success. NASA's Johnson Space Center (JSC) has facilities in its Mission Control Center (MCC) for commanding ISS and some payloads, but the MCC could not meet the needs of the R2 project. The Robonaut Control Center (RCC) was created to house R2's specific software configurations and network requirements. The RCC's network can be configured to connect to R2 on ISS, or to the R2 certification unit on the ground. When in "certification configuration," the RCC's consoles can be tested with a flight-like robot prior to actual space operations, to ensure robust operation and that operator preferences are met. Similarly, the R2 certification unit is loaded with the same software that will eventually be installed on R2 on ISS, and operators are able to verify that safety requirements are met and operational milestones are achievable. When in "flight configuration," RCC consoles are only allowed to communicate through the VPN and SSH services to R2 on ISS – all other network traffic is blocked. This ensures that no outside systems can interfere with operations, and operators have a distraction-free environment to work in. The RCC has hosted multiple successful operations for R2 on ISS and more are planned for the future.

---

[2]https://openvpn.net/.
[3]http://wiki.ros.org/multimaster_fkie.

9 Discussion

The control and safety system for R2 combine to create a responsive, high-performance, safe system. Two experiments to illustrate this performance were conducted at different maximum Cartesian velocities. In one experiment, the peak force when the end effector of the robot runs full-speed into a force plate was recorded. In the second, deviation from commanded positions over a 1-m long rectangular trajectory was measured. Results from both are shown in Fig. 12.

One of the major reasons that the control system has this performance while still maintaining low torque limits is the quality of the model that informs the inverse dynamics algorithm that calculates the feed-forward torque value per joint. Figure 13 shows a plot of the modeled and actual joint torque of two arm joints over a scripted motion. The difference between the modeled and actual joint torque values dictate the joint torque limit needed to accomplish this move at the given speed. For this move, joint torque limits as small as 10 Nm could be used to command this motion. For the trajectory monitoring system using nominal configuration parameters, that would then require only 15 Nm of torque to overcome motion on any joint.

On-orbit deployment of the software has been successfully accomplished. Network connection to the robot over the VPN works well to protect data and commands to the robot over latent and low bandwidth connections. Commanding to the robot has occurred, already proving the need for things like latched ROS topics and reliable communication constructs. An important lesson learned from trying to command a robot over a lossy, latent connection is the ability to connect separate roscores, as the distance between the robot's roscore and the user interfaces on the ground causes

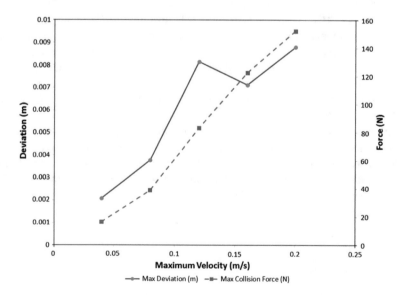

Fig. 12 Force and deviation from commanded position versus maximum Cartesian velocity

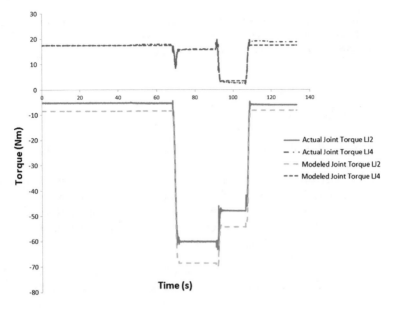

Fig. 13 Comparison of actual versus modeled joint torque for two R2 arm joints

major communication issues. The MultiMaster FKIE package is a promising solution to this problem.

Another lesson is the need for supervisory interfaces to command the robot, such as the affordance templates introduced in Sect. 6.2. It is difficult to transfer machine vision algorithms, such as localizing features on objects, to cameras that have been vibrated by a rocket, have pixel loss due to radiation, and experience unique lighting conditions. Therefore, it is important to give supervisory control to human users to adjust for these inaccuracies in the vision system.

Lessons learned from the previous control and safety system on the R2 torso that was demonstrated through several tasks on the ISS show that it is important to maintain as much up-time on the robot as possible. This meant spending development time trying to reduce the false positives that trigger the safety system and creating a more robust I/O system to ensure reliable data transfers in both directions. All of these improvements have dramatically increased up-time on the ground-based certification robot. Testing on the on-orbit robot is forthcoming, depending on crew schedule constraints.

The fail-safe nature of the current safety system (if a monitor triggers, the robot safes) is important in certifying ROS, Orocos, and Ubuntu for safety-critical space software. Essentially, the open source software is treated like the other non-safety critical robot software. The safety certification proves that any error in these (i.e., partial shutdown, run away processes, segmentation faults) will only trigger a safing action from the robot, never an unsafe action. Since an extravehicular version of Robonaut would likely have to have some ability to continue to operate through one or more failures, it would be much more difficult to complete that certification. However, the experience gained from the R2 experiment will be invaluable for the future of ROS in safety-critical robotic operations.

10 Conclusions

The Robonaut 2 IVA Mobility system is a unique case study for ROS usage. R2 has stringent requirements placed on its control, safety, deployment, and use that are, as a whole, different from any other system on Earth. However, the solutions and lessons learned on this system are certainly applicable to many other robotic systems. The use of ROS in space gives it a whole new dimension and set of solutions that has promise for robotic systems that could be used to aid in disaster relief, military operations, or other applications.

Acknowledgments The authors would like to acknowledge Joshua Mehling, Vienny Nguyen, Philip Strawser, and the many former members of the software team for their contributions to this work.

References

1. M. Diftler, J. Mehling, M. Abdallah, N. Radford, L. Bridgwater, A.M. Sanders, R.S. Askew, D. Linn, J. Yamokoski, F. Permenter, B. Hargrave, R. Piatt, R. Savely, R. Ambrose, Robonaut 2—the first humanoid robot in space, in *Proceedings 2011 IEEE International Conference on Robotics and Automation* (2011), pp. 2178–2183
2. M. Diftler, T. Ahlstrom, R. Ambrose, N. Radford, C. Joyce, N. De La Pena, A. Parsons, A. Noblitt, Robonaut 2 initial activities on-board the ISS, in *Aerospace Conference, 2012 IEEE.* (IEEE, 2012), pp. 1–12
3. J.M. Badger, A.M. Hulse, R.C. Taylor, A.W. Curtis, D.R. Gooding, A. Thackston, Model-based robotic dynamic motion control for the Robonaut 2 humanoid robot, in *Proceedings of IEEE-RAS International Conference on Humanoid Robots* (IEEE, 2013)
4. J.M. Badger, A.M. Hulse, A.Thackston, Advancing safe human-robot interactions with Robonaut 2, in *Proceedings of the 12th International Symposium on Artificial Intelligence, Robotics and Automation in Space* (2014)
5. The Orocos Project (2013), http://www.orocos.org/
6. S. Hart, P. Dinh, J.D. Yamokoski, B. Wightman, N. Radford, Robot task commander: a framework and IDE for robot application development, in *IEEE/RSJ International Conference on Intelligent Robots and Systems (IROS 2014)*, (IEEE, 2014) pp. 1547–1554
7. O.S.R. Foundation, Gazebo (2014). http://gazebosim.org/
8. T.B. Sheridan, *Telerobotics, Automation, and Human Supervisory Control* (The MIT Press, Cambridge, 1992)
9. R.R. Burridge, K.A. Hambuchen, Using prediction to enhance remote robot supervision across time delay, in *IEEE/RSJ International Conference on Intelligent Robots and Systems, IROS 2009* (IEEE 2009), pp. 5628–5634
10. N.A. Radford, P. Strawser, K. Hambuchen, J.S. Mehling, W.K. Verdeyen, A.S. Donnan, J. Holley, J. Sanchez, V. Nguyen, L. Bridgwater et al., Valkyrie: NASA's first bipedal humanoid robot. J. Field Robot. **32**(3), 397–419 (2015)

11. S. Hart, P. Dinh, K. Hambuchen, The affordance template ROS package for robot task programming, in *2015 IEEE International Conference on Robotics and Automation* (IEEE, 2015)
12. J.J. Gibson, *The theory of affordances* (Hilldale, USA, 1977)

Authors' Biography

Julia Badger is the Software and Applications Lead and Deputy Project Manager for the Robonaut Project at NASA-Johnson Space Center in Houston, TX. She develops applications and tests for Robonaut 2 on board the International Space Station and also leads the software effort for the Robonaut Mobility System flight payload. She has previously worked at developing autonomous control and planning algorithms for the various robotics projects in her group, including the Space Exploration Vehicle and Robonaut 2. Julia has a BS in Mechanical Engineering from Purdue University, a MS in Mechanical Engineering from the California Institute of Technology, and earned her PhD in Mechanical Engineering from Caltech in 2009.

Dustin Gooding has a bachelors degree in Applied Mathematics from Texas A&M University. He has worked at NASA-Johnson Space Center for over ten years in areas of robotics software development, IT security response, and project management. His professional interests include solving complex problems with elegant software solutions and improving utility and reliability in software development processes. Dustin is currently the Software Architect for the Robonaut 2 project, advising project leadership on software development best practices and steering software design decisions to help ensure holistic success.

Kody Ensley graduated from Salish Kootenai College in 2012 with a BS in Computer Engineering. Kody completed over two years of internships with the Robonaut 2 group before joining the team full-time in 2012. He is currently the Flight Software Lead for Robonaut 2. This role includes advising and coordinating safety tests on new releases of code as well as deployment and installation of code to R2. His personal interests are education outreach, making complex systems easier to interact with, and configuration management.

Kimberly Hambuchen is the Deputy Manager for the Human Robotic Systems project, funded through NASA's Space Technology Mission Directorate. Since 2004, she has been a robotics engineer in the Software, Robotics and Simulation division at NASA-Johnson Space Center. She received a B.E. in Biomedical Engineering and Electrical and Computer Engineering (1997), and M.S. (1999) and Ph.D. (2004) degrees in Electrical Engineering, all from Vanderbilt University. She is a former NASA Graduate Student Research Program fellow and previously held a post-doctoral position at NASA through the National Research Council. Dr. Hambuchen is an expert in developing novel methods for remote supervision of space robots over intermediate time delays. She was the User Interface Lead for JSC's entry into the DARPA Robotics Challenge, using her expertise in remote supervision of robots to guide operator interface development for the DRC robot, Valkyrie.

Allison Thackston is the Lead Engineer for Robotic Perception on Robonaut 2, the first humanoid robot on the International Space Station. She has a degree in Electrical Engineering from Georgia Tech and a graduate degree in Mechanical Engineering from the University of Hawaii at Manoa. Her thesis focused on collision avoidance in supervised robotic manipulation. She currently works at the Johnson Space Center on the Robonaut project where she is responsible for software development and applied vision research to facilitate cooperation between robots and humans.

ROS in the MOnarCH Project: A Case Study in Networked Robot Systems

João Messias, Rodrigo Ventura, Pedro Lima and João Sequeira

Abstract Networked Robot Systems (NRS) have a wide range of potential real-world applications. However, these systems have functional requirements that lie outside of those considered in the typical use cases of ROS. This chapter describes the use of ROS in the context of the ongoing MOnarCH FP7 project on social robotics. We describe the software architecture used in the MOnarCH NRS, focusing on the decentralized information sharing framework we developed called Situational Awareness Module (SAM), and present some of the current results of our project that showcase the applicability of our ROS packages in real-world environments.

Keywords Networked Robot Systems · Social robots · Multi-master · Sensor fusion

1 Introduction

The MOnarCH[1] project is about having (social) robots in real social environments, behaving in a socially acceptable way, and studying the establishment of relationships between humans and robots [1, 7].

The project main test environment is the Pediatric ward of an Oncological Hospital (see Fig. 1) in which the premises include the main corridor, a playroom, and a class-

---

[1] Multi-Robot Cognitive Systems Operating in Hospitals. Project reference: FP7-ICT-2011-9-601033. Web: http://monarch-fp7.eu.

J. Messias · R. Ventura (✉) · P. Lima · J. Sequeira
Institute for Systems and Robotics, Instituto Superior Técnico, Universidade de Lisboa,
Av. Rovisco Pais 1, 1049–001 Lisbon, Portugal
e-mail: rodrigo.ventura@isr.ist.utl.pt

J. Messias
e-mail: jmessias@isr.ist.utl.pt

P. Lima
e-mail: pal@isr.ist.utl.pt

J. Sequeira
e-mail: jseq@isr.ist.utl.pt

© Springer International Publishing Switzerland 2016
A. Koubaa (ed.), *Robot Operating System (ROS)*, Studies in Computational
Intelligence 625, DOI 10.1007/978-3-319-26054-9_14

375

Fig. 1 Camera images from the pediatric ward of the IPOL hospital, showing the deployment of a MOnarCH robot during tests

room. The robots will engage in activities such as interactive playing, and performing basic teaching assistant activities in the classroom. The ultimate objective is the improvement of the quality of life of inpatient children. A key scientific thesis underlying the MOnarCH research is that current technologies make possible the acceptance of robots by humans as peers.

Supporting these objectives are the extensive work on (i) autonomous and Networked Robot Systems (NRS) [6], enabling sophisticated perception and autonomous decision-making and navigation, and (ii) interfaces for human-robot interaction, allowing robots to be expressive and natural. MOnarCH addresses the link between these two areas, having robots playing specific social roles, interacting with humans under tight constraints and coping with the uncertainty common in social environments. The project builds on the success of pioneer projects involving NRS, namely FP6 URUS [5] and Orebro Universitys PEIS Ecology [4], as well as the Japanese NRS project [9].

The hospital environment imposes a number of constraints on the usual engineering development cycle, namely (i) at no time the experiments can interfere with the regular operation of the ward, (ii) admissible sensing is constrained by regulations, and (iii) the specific social rules enforced at the hospital are to be verified at all times.

The MOnarCH hospital setup comprises up to two robots, static cameras covering the common areas the robots can use, and a central server machine running multiple algorithms, namely those related to the image acquisition and processing. The robots include two computers, RFID reader, two RGB-D cameras, two laser range finders, among other devices, namely those used for human-robot interaction purposes. The devices onboard the mobile robots and static on the environment are thus nodes of a network endowed with its own specific dynamics.

The social robots in MOnarCH (shown in Fig. 2) are required to interact in a natural way with the children, parents, staff, and visitors. Even though humans are often resilient to changes in their social environments, preliminary results clearly suggest that small changes in the dynamics of the robots can have major effects in what is

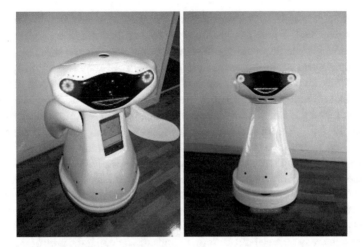

Fig. 2 The two versions of the MOnarCH robots: the Socially Oriented (SO) one on the *left* and the Perception Oriented (PO) one on the *right*. The core platform is essentially the same, differing only on the set of actuation devices. The SOs are targeted for human-robot interaction, while the POs are targeted to perceive the environment from a mobile platform

perceived by an observer. Small changes such as having the robot turning the head as it moves are immediately noticed by the observers. For example, in the preliminary experiments this small change was often referred by observers as "the robot begin more interactive". This sensitivity has thus to be taken into account when designing a system such as MOnarCH, namely by insuring small latencies in the communications such that the dynamics of the behavior is made dependent, exclusively, on the behavior. In the aforementioned example a delay in moving the neck would have a behavioral meaning, e.g., the robot is not worried with the interaction. In a sense, this suggests that adequate control of each component in the network is a key issue for the success of the system.

The aforementioned resiliency of humans to change is expected to filter out some of the less adequate behaviors the robots may eventually generate. However, in scenarios such as the MOnarCH project, strict social norms are enforced, meaning that it is expected that both humans and robots strictly comply with them, and that it cannot simply be expected that humans adapt to the potential inadequacies of robots. Therefore, adapting the behaviors of the robots over time, as the characteristics of the inpatient population changes, is also of paramount importance for the success of the whole project.

Designing robots for social purposes has been a trendy topic in the last decades. The literature in this area is vast and yields valuable lessons (see for instance [11]). However, experiments where robots and people coexisted for long periods of time, outside lab environments, i.e., for periods longer than the transient in the dynamics of human expectations, have seldom been reported [2]. Robustness and dependability issues are critical in such a hospital application and further justify the use of a framework able to account for rapid prototyping, high flexibility in adjustments.

Even though system development is ongoing, it is already possible to make a positive assessment. Figure 1 is just one example of the interest one of the systems mobile robots is generating among the children and adults at the hospital.

Developing and running such a system dependably requires carefully-thought software integration procedures, since a multitude of subsystems for perception, decision-making and control that operate at different levels of abstraction are involved and interact among them. These subsystems must be able to communicate information reliably, both at a local level—for subsystems running on the same physical machine; and at the network level—for subsystems running in different network locations. As the complexity of the system (and its respective subsystems) increases, it becomes progressively harder to define and implement each of the communication mechanisms that must exist between different software modules. For a large-scale system, an unmanaged, decentralized implementation of the communication between different processes is not feasible, since this not only potentiates serious design errors and simultaneously makes them difficult to track down (e.g. if the type of information sent by process A is unknowingly misinterpreted by process B), but it also makes the networked system brittle to communication failures, if they can happen at any point in the software architecture.

In terms of implementation, the choice of ROS as middleware framework is a natural one. We followed an implementation strategy using COTS[2] components whenever possible, in order to minimize development time. In particular, we use AMCL[3] for single robot localization, SMACH[4] for behavior implementation, TF[5] for geometric transformations management, and *actionlib*[6] for behavior/functionality activation and supervision, and *rosbag*[7] for data logging.

Even though data streaming in ROS is peer-to-peer, node/topic/service registration still requires a centralized node, the ROS Master. In large NRSs, a centralized Master can be a liability, since it is a single point of failure for the entire system (*i.e.* if the Master fails then the functionality of the whole system is compromised). This is an especially relevant concern for NRSs that are intended to operate over long periods of time (days, weeks) as it is the case in MOnarCH. To attain a both efficient and decentralized architecture, we have chosen to use separated ROS Master instances in each NRS agent (including robots and servers). To support information sharing among agents, a novel module was developed, called **Situational Awareness Module (SAM)**. Since this module was developed from scratch to address the needs of the MOnarCH NRS, while being generic enough to be used outside this project, we will devote the bulk of this chapter to the technical description of SAM. We claim that SAM provides a powerful framework on top of which robust and fully decentralized NRS can be implemented.

---

[2]Components Off-The Shelf.

[3]http://wiki.ros.org/amcl.

[4]http://wiki.ros.org/smach.

[5]http://wiki.ros.org/tf.

[6]http://wiki.ros.org/actionlib.

[7]http://wiki.ros.org/rosbag.

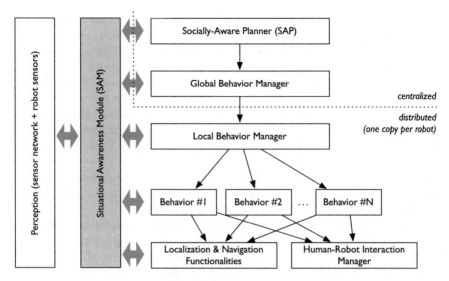

Fig. 3 System architecture of MOnarCH, showing the data flow between SAM and all the other modules as *grey arrows*, together with the decomposition of a plan, coming out from the planner (SAP) down to the robot functionalities (flow shown as *dark arrows*)

This chapter is structured as follows: the architecture of the MOnarCH NRS is described in Sect. 2, followed by a thorough description in Sect. 3 of SAM. Then, we will report the current results of the project, focusing on the usage of SAM in Sect. 4. Section 5 discusses related work and Sect. 6 wraps up the chapter with some concluding remarks.

2 System Architecture

The NRS in use in MOnarCH comprises a centralized part, consisting in a central server, and a distributed part, consisting of robots and distributed sensors, e.g., video cameras. The overall system architecture is shown in Fig. 3.

Each computational entity (robot onboard computers and central server) is based on the ROS framework running a single, individual ROS Master. The data flow among these entities is supported by the Situational Awareness Module (SAM), which will be thoroughly described below. This module functions on top of a multi-master ROS solution,[8] thus allowing seamless communication among local ROS environments.

All perception modules, comprising not only the processing of the distributed sensors on the central server, but also sensor data processing onboard the robots relevant

---

[8]In the MOnarCH NRS, we have chosen the *multimaster_fkie* package for this purpose (http://wiki. ros.org/multimaster_fkie). However, any other package that allows for the sychronization of topics between different ROS Masters could be used.

to the NRS, feed the SAM. In addition, SAM contains internal data processing for, e.g., sensor fusion and symbolic grounding. In both of these cases, data is read and written to SAM (more concrete examples will be given further ahead).

The Socially-Aware Planner (SAP) is responsible to generate a consistent plan, incorporating social aspects that the system as a whole is required to comply, e.g., social norms [10]. This plan is then broken down into behaviors that are dispatched to individual robots. These behaviors can either be purely individual, e.g., a robot going from a point A to a point B, or cooperative, e.g., a team of robots looking for a specific child in the environment. The dispatch of individual behaviors is carried out jointly by the Global Behavior Manager (GBM), which receives them from SAP, and the Local Behavior Manager (LBM) that activates them in individual robots. The GBM runs on the central server, while each robot runs one instance of the LBM. All data flow between GBM and LBM is supported by SAM.

At the robot level, the LBM instance is responsible for activating behaviors to be executed internally. If a behavior requires a coordinating among several robots, e.g., formation control of a fleet of robots, each one runs one instance of that behavior, coordinating with other instances in other robots through SAM.

At the lowest level of the architecture we have the core functionalities of robots, namely in what concerns autonomous navigation [3] and human-robot interaction. Autonomous navigation requires accurate localization of the robot, being tightly coupled with the navigation algorithms via local ROS communication. Localization accuracy may be improved using data fusion methods, combining onboard sensor readings, e.g., from the laser-range finders, with external sensors, e.g., the distributed video camera network. The sensor fusion can be performed by SAM, receiving data from different network nodes, and distributed among the NRS via SAM.

3 Situational Awareness Module (SAM)

The goal of the Situational Awareness Module is to provide the MOnarCH software architecture with a managed, (conceptually) centralized information sharing framework. It can provide sensor-fusion, decision-making and control algorithms in the NRS with a consistent global picture of the whole system. Each client process and its corresponding software module can access the SAM to retrieve information from other modules, which possibly reside elsewhere in the network, and can then process the data in a way that it is suitable for its use. When (and if) that data is written back into the SAM, future clients will already know its form, purpose, and producer, and will only be able to access the data in some conditions that are known to be error-free.

SAM can manage intranetwork communication, so that end-users do not need to worry about which network node is providing them with their data; and handles communication failures so that client software modules can have (as much as possible) a consistent *history* of the data that is flowing in the NRS.

As an example use-case of SAM, the Observer in the Constraint-Satisfaction Planner (CSP) [10], could implement ROS nodes that compute the values of logical predicates based on information provided by SAM slots containing perceptual data, use that information to describe its own view of the world state, and possibly write the logical values back in the SAM and share it across the network so that other modules can also access these values for their own use (e.g. for HRI purposes).

In short, the most relevant differences between the use of the SAM as an information/communication manager, and the use of the native publish/subscribe mechanisms of ROS, are the following:

- The ROS publish/subscribe paradigm is **unmanaged and anonymous.** Any node can communicate with any node without restrictions, and it is difficult to guarantee system-wide properties with respect to communication (for example, that there are no information loops, or that all messages should contain timesteps). The communication paradigm implemented by SAM is **managed and not anonymous.** All of the producers and consumers of data are known to the SAM, and they can only communicate with each other after being authorized to do so. This allows the architecture designer to place restrictions on the rate and contents of the communication channels between different nodes.
- **ROS publish/subscribe does not work natively for multi-master systems**, and even if specialized packages are used for that purpose, these do not manage communication on their own. SAM operates over the multi-master communication layer that is implemented by those modules, and abstracts it, providing a transparent way of sharing data between nodes running under different ROS Masters with minimal configuration. It also executes consistency checks to ensure that communication failures are handled in such a way as to minimize the inconsistency of the data being accessed by client software modules in different network locations.

3.1 Definitions and Overview of the Concepts

We assume a network of sensors and mobile robots (with sensors and actuators), as well as other possible actuators. The Situation Awareness Module (SAM) is a distributed information management system that is composed of:

- A set of Situation Awareness Repositories (SARs), each of them uniquely associated to a ROS Master (each ROS Master is assumed to contain at most one SAR);
- A set of *slots*, within each SAR, each associated a semantic description of a feature of the NRS (e.g., "Robot N Localization Estimate", "Person B Localization Estimate", "Group A Recognized Activity"). Slots can be *read from* and *written to*. A slot in a SAR can be *shared* to other repositories, in which case a "read-only" copy of that Slot is created in all other SARs, making its data accessible to other network locations;
- A set of SAR *Managers*, one for each SAR. Managers mediate the flow of read/write operations to SAR slots, by imposing a set of restrictions on those operations that are designed to ensure that there is a valid data flow between all

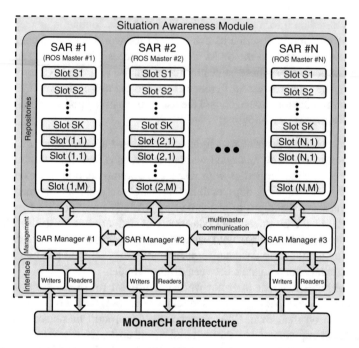

Fig. 4 An overview of the proposed organization for the SAM. SA Slots S1 through SK are shared, in the sense that they are replicated in multiple (potentially all) SARs. Slots $(i, 1)$ through (i, M) contain local information that is only present in SAR #i

client software modules. Managers also handle the communication of shared slot data between multiple ROS Masters.

A graphical overview of these concepts, and their organization, is shown in Fig. 4. To operate the SAM, we assume the following:

- A multi-master communication module that enables the communication between different ROS Masters, and that supports both fixed-rate and approximately asynchronous communication modes;
- That all network nodes are synchronized with respect to an external clock.

SAR Slots. SAR Slots are abstract "containers" of the data that is intended to flow across different MOnarCH modules, and that provides, as a whole, a coherent view of the "state" of the MOnarCH networked robot system and its environment. As such, SAR slots may represent any type of information that is deemed to be relevant to the operation of the MOnarCH software modules. In the typical case, SAR Slots will represent:

- Processed sensor data, or features representing the data at any point of the data processing pipeline, that are deemed to be relevant to the operation of one or more software modules besides their producer (e.g. the position of a robot or a human as detected by one of the overhead cameras);

- Perceptual information resulting from the fusion of data from two or more sources that may reside in different network locations (e.g., position of person_B as the result of the visual observation by static cam 3, mobile robot 2 cam, and information provided by person_B RFID);
- High-level symbolic information, such as logical predicates, that can be obtained by abstracting the lower-level perceptual data; or by appropriately defined transformations over other existing data at that level (e.g. predicates defined through propositional formulas containing other predicates, such as whether there is a robot in the vicinity of a detected human).

Note that, even though Slots are viewed by data producers and consumers as readable/writable data containers, they do not necessarily store data within themselves (except eventually for synchronization purposes). Functionally, Slots most typically act as mediators of the communication of the associated data between its producers and consumers.

Shared Slots. Each slot is assumed to be a unique object within the SAM. By default, slots that are created and maintained in a given network location are visible and accessible only by the processes that are being executed in that location (i.e. it represents locally writable/readable data). However, slots can be *shared* by one SA Manager to any others. In this case, the data represented by that slot can be read by clients that reside elsewhere in the network.

When the same data slot is present in more than one SAR, the SAM will enforce that the data contained in each copy is *consistent*, in the sense that:

- With hypothetical (ideal) instantaneous communication between the various ROS Masters, and negligible processing delays, the information contained in each copy would always be exactly the same;
- In the realistic case that there is potentially unreliable and sporadic communication, and there are processing delays when updating the information in each SAR slot, the error between the different copies of the same slot is minimized whenever communication is available.

Readers & Writers. SAM Readers and Writers can be instantiated by each producer or consumer of data in software modules that are external to SAM. They provide a structured way to access the SAM data, and through their identification, they make it possible to analyze the flow of information across different software modules, even if they are located in different ROS Masters. This abstraction also warns users about any information loops in the system, which may have been inadvertently created (e.g. fusing Slots A and B into Slot C, and then using Slot C to update Slots A or B).

Slots can have multiple Readers, but they can only have one Writer. Shared Slots can also only have one Writer, regardless of its location in the network. SAM clients that intend to fuse multiple sources of information can register as Readers of multiple Slots, and should write the result of that fusion process back into the SAM under a different Slot.

After being registered in the SAM, Readers and Writers interact with each other in much the same way as a standard ROS subscriber and publisher (respectively).

That is, the communication rate is driven by the Writer, and the Reader is asynchronously woken up through a callback whenever new data arrives. If the writer is also asynchronous (for example when writing predicate values), then the entire execution loop will not have a fixed rate.

Managers. SAR Managers provide the services that client software modules need to use in order to access the SAM, and they supervise the access of Readers and Writers to Slots. They also abstract the communication mechanisms involved in the access of a network node to the data contained in the SAM. For example, Robot *A* can instantiate a SAR Reader to access the position of Child *X*, *without needing to know which ROS Master is actually keeping that information* (i.e. which other robot or server is observing Child X).[9] SAR Managers also carry out the consistency checks of shared Slots, and can initiate synchronization procedures that update the contained information if it is detected to be inconsistent.

What this Means to the User. From the perspective of the software modules that will make use of the SAM, SARs can be thought of as repositories of abstract information that can be written to/read from in a controlled manner. However, at its core, the most important function of the SAM is not to *contain* data within itself, but rather to manage the flow of information from producers to consumers in a way that is not anonymous (contrarily to the default ROS publish/subscribe system); and that is agnostic with respect to the network locations of the clients and their respective communication protocols (meaning that a client does not need to know where the data is hosted nor how it is communicated, which is not natively possible in a multi-master layout of ROS).

Using the above concepts, the software modules that are external to the SAM and that implement sensor fusion algorithms can receive information from sensors located in the same device or from sensors distributed across several devices. To do this, they must simply instantiate multiple SAR Readers for each of the relevant Slots, irrespectively of where they are being written. Alternatively, local instances of these sensor fusion algorithms can be run in each ROS Master where the fused information is to be available, in which case the respective algorithms should be exactly the same, or at least they must they must produce the same output given the same sequence of inputs; or they can be run in a central network location (for example a server), from where their result can be distributed to an arbitrary number of clients.

Although one of the main purposes of SAM is to facilitate sensor fusion in the NRS, SAM itself does not place any restrictions on the sensor fusion algorithms that can be implemented for that purpose. These algorithms will run on self-contained ROS Nodes that act as SAM Readers/Writers, and that may consume and produce any type of data that is required for their operation. For example, cooperative perception algorithms can take into account data recency (e.g., when a time-tag is much older than current time, the information is discarded) and information agreement (e.g., two

---

[9]The information that the Robot receives back will allow the agent to know who produced that data, since that can be important for decision-making purposes. However, the agent does not need to know that information *a priori* in order to request it.

estimates of the same object position at a Mahalanobis distance larger than a given threshold are not fused).

SAM requires a considerable number of design options to be taken at the NRS design-time, following from the specifications for the desired NRS performance (e.g., desired accuracy for person location):

- devices that support SAM distributed components;
- slots to be located in each device's local SAR;
- cooperative perception (sensor fusion for NRS) algorithms and their implementation in software modules for all the cases where information fusion is required.

3.2   An Example of the Use of SAM

The following scenario exemplifies the use of the Situational Awareness Module. Consider the situation shown in Fig. 5 in which two robots, **A** and **B**, observe the same object, a ball, which lies on a flat surface over which two topological areas are defined, rooms **L** and **R**. Both robots are equipped with the same hardware and software. Each robot can locally estimate the following variables that are relevant for the purposes of this example:

- The pose of the robot in the world frame (returned by its self-localization algorithm), with an associated description of its uncertainty;
- The position of the object, both in the robot frame and in the world frame (returned by its object detection/tracking algorithm), with an associated description of its uncertainty;
- The time elapsed in the system (returned by the local clock of each robot);
- The value of a logical *predicate* that can be used for decision-making purposes, *IsBallInRoomL*, which should be true when the ball is in Room **L**;

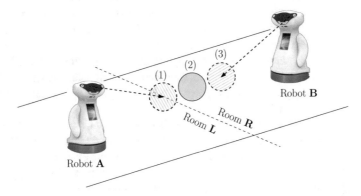

Fig. 5 A situation in which robots have contradictory information. (*1*) The position of a ball as seen by Robot A; (*2*) The actual position of the ball (in Room R); (*3*) The position of the ball as seen by Robot B

The purpose of SAM, in this scenario, is to mediate the process of fusing the local estimates of that each robot has regarding these variables. This is intended to mitigate the effect of local estimation errors and disambiguate possible contradictory information from local sources. To see why this is relevant, suppose that the object detection subsystem of each robot has an unmodeled bias with respect to the object distance, that is, it returns a position for the object that is closer to (or farther from) the robot than the actual ground truth measurement of that position. Then, the local estimates of the values of the predicate can be contradictory. That is, for Robot **A**, *IsBallInRoomL* is true, while for Robot **B** it is false. A centralized logic-based planner cannot rely on this contradictory information.

To resolve this problem, we need to produce a *joint* or *fused* estimate of the position of the ball in the world frame—that is, the estimated position of the ball when taking into account the local estimates of each robot and their associated uncertainties. The value of the logical predicate should then be computed over this joint ball position estimate, instead of being evaluated over the local estimates of either robot.

As described in the previous section, a SAR is expected to run locally in each Robot, as each of them runs its own ROS Master. In accordance with the expected use cases of the MOnarCH NRS, a third Master is expected to run on another network location, hereafter referred to as the "SAP Server", where the Social Aware Planner (SAP) will be executed. In this scenario, the SAR that is running in the SAP Server has the following slots, writers and readers:

The Readers and Writers identified above, and shown in Table 1, are actual ROS Nodes that implement an associated data transformation/fusion process. The network locations where the nodes are being executed are also identified in the table (for example, both robots have their own instance of the localization and ball detection algorithms).

The crux of this particular sensor fusion problem is addressed by a specialized algorithm that fuses the locally estimated ball positions of both robots by taking into account the poses of each robot and the respective uncertainties (here named "Ball Fusion Algorithm Z"). The most important concept that this example attempts to demonstrate is that, although the data regarding robot poses and ball positions is written locally by each robot into their own SARs, by *sharing* those Slots when they are created, they become visible in the SAP Server's own SAR, from where they can be read. These Slots cannot be written to by the Server, since they already have one writer in the network (either of the robots). Recall that a Slot can only have a single writer. If that Slot is shared, this property also holds regardless of where that writer is located in the network.

The sensor fusion algorithm is implemented in as a self-contained ROS Node running on the SAP Server, and its result is written back into the SAM ("Joint Ball Position"). Using that information, another ROS Node evaluates the Predicate "*IsBallInRoomL*", which the Social-Aware Planner can then use directly.

This organization of SA Slots, Readers and Writers is shown graphically in Fig. 6.

Note that this is but one example scheme to solve this particular sensor fusion problem. Other equally valid alternatives would be:

Table 1 Slots and respective readers/writers of the SAR of the SAP Server in the example above

Name	Type	Shared	Written by	Read by
Robot A Pose	Pose + Uncertainty	Yes	Localization Algorithm X (Robot A)	Ball Fusion Algorithm (SAP Server)
Robot A Ball Position	3D Position + Uncertainty	Yes	Ball Detection Algorithm Y (Robot A)	Ball Fusion Algorith (SAP Server)
Robot B Pose	Pose + Uncertainty	Yes	Localization Algorithm X (Robot B)	Ball Fusion Algorithm (SAP Server)
Robot B Ball Position	3D Position + Uncertainty	Yes	Ball Detection Algorithm Y (Robot B)	Ball Fusion Algorith (SAP Server)
Joint Ball Position	3D Position + Uncertainty	No	Ball Fusion Algorithm Z (SAP Server)	Predicate Update (SAP Server)
IsBallInRoomL	Predicate (Boolean)	No	Predicate Updater (SAP Server)	SAP (SAP Server)

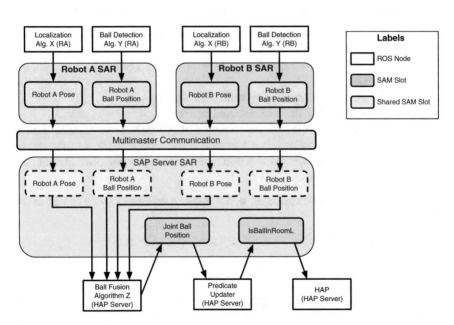

Fig. 6 The organization of SAR Slots and sensor fusion/processing ROS nodes in the ball fusion example

- To share poses and ball positions between robots, and calculate locally in each robot an estimate of the fused target position, according to that robot. That estimate could be sent back to the SAP Server or used locally for other purposes (e.g. navigation).

However, keep in mind that in this case the SAP would receive N fused estimates for N robots (in N different Slots), which it would need to post-process into a single quantity;

- To evaluate the predicates locally based on each robot's estimated *joint* ball position (using the info of all other robots), and communicate the predicates to the SAP Server, where a consensus function could then be used (e.g. majority ruling) to determine the final value of the predicate.

3.3   Technical Description

In this section, we will describe the operation of SAM and its implementation as a ROS module for the MOnarCH project.

The simplified UML class diagram represented in Fig. 7 shows how the main concepts of SAM (Managers, Repositories and Slots) are related to each other from an implementation standpoint. In each of the following subsections we will discuss each of these components.

Slots are objects that can be described by the following main attributes:

- a name, that should be descriptive of the information that is associated to the Slot (e.g. Robot 1 Pose);
- a readable, short explanation of the slot's purpose (e.g. The local estimate of the pose of the robot);
- an associated data type (can be any valid ROS message);
- an indication of whether the Slot is shared or not;
- an indication of whether the Slot is *latched*, that is, whether new Readers should be provided with the last data that was sent to the Slot, regardless of the age of that data (otherwise, new Readers will only be woken up at the next time that the Slot receives new data from its Writer).

In addition to the above attributes, a Slot instance maintains (not exposed to the user):

- an *alias*, which is a ROS-compatible representation of its name;

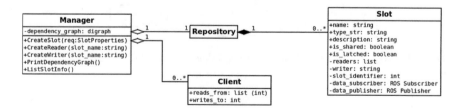

Fig. 7   A simplified UML class diagram of SAM. Private methods are omitted

- a list of Readers (the names of the nodes that are authorized to access the Slot), and the identity of a Writer;
- an input topic, to which the Writer should publish the Slot's data;
- an output topic, to which the Slot's Readers should subscribe;

Functionally, a Slot acts as a controllable latch in ROS' publish/subscribe system. That is, instead of having the publisher sending data directly to its subscribers, this data is sent *through a SAM Slot*. This allows us to manage the access of the SAM clients to that Slot and control the flow of information therein, possibly restricting it, storing it for communication reliability purposes, or modifying it as needed. Note that, as in the case of the standard ROS publish/subscribe system, the data flow through a SAM Slot is asynchronous. Whenever data is written into a Slot via its input topic, it will immediately appear in in its output topic, and its respective Readers will be consequently "woken up".[10]

A Slot can only be successfully created if no other Slot with the same name *or the same alias* exists. There are two possible ways of creating a Slot: either by calling a Service that is provided by the associated Manager; or by specifying the above attributes in a configuration file.

Shared Slots. From a technical standpoint, the only functional difference between a shared Slot and a regular (private) Slot is that data written into shared Slots is distributed to all ROS Masters running SAM, through a multi-master communication module that is not exposed to the user. That is, an auxiliary ROS node, which mediates inter-master communication, subscribes or publishes to the output or input topics (respectively) of that Shared Slot. From the standpoint of the user, the process of creating/writing to/reading from shared Slots is exactly the same as it is for regular Slots (this is one of the major requirements of the SAM implementation).

Slot Groups. In multiagent systems, it is often necessary to represent and share information that characterizes the state of each agent (e.g. the pose of each robot), or information that possesses a global meaning but that is subjective, depending on the perception of each agent (e.g. the pose of a certain child according to each robot). In order to use SAM for these cases, it is necessary to instantiate multiple Slots that are functionally equivalent, in that they possess the same characteristics, differing only in their identification, and possibly on their sets of Readers and their Writers. In these cases, the user can define Slot *Groups*, to facilitate the creation of, and access to, multiple Slots that are semantically equivalent but subjective to multiple agents. A Slot group is simply a set of slots that all share the same public attributes except for their name. A group is defined through a *representative* Slot that specifies the characteristics of all of the Slots in that group, as well as a base name for the group (e.g. " Pose"); and a set of *hosts* which represent those agents for which the representative Slot should be instantiated (e.g. Robots 1 and 2). Afterwards, the slot can be accessed by using both its base name and its host (e.g. "Pose" of Robot 1). Note that this is functionally the same thing as explicitly creating individual Slots

---

[10]Evidently, if a Writer writes data synchronously into a Slot (i.e. at a fixed rate), its Readers will also be woken up at a fixed rate, so the whole process can be made synchronous by the Writer.

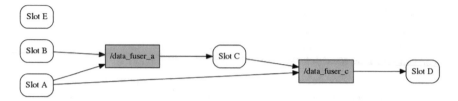

Fig. 8 An example of an SAR dependency graph. Nodes in *gray* represent client ROS Nodes

for each agent. Furthermore, a Slot group is a different concept than that of a shared Slot. Slot groups do not provide any additional functionality to SAM, they simply represent sets of similar slots. It is possible to have a private (non-shared) group of Slots entirely within a single SAR.

Repositories. A repository is fundamentally a container of Slots (see Fig. 7). We isolate its implementation from that of the Manager in order to precisely control the processes of creation, deletion, and access to Slots. In particular, we ensure that each Slot that is added to the repository is associated with an unique integer identifier, and that no two Slots can have the same name or alias.

Managers. As it was previously discussed, each Manager is associated with a particular Repository. Managers provide the services that SAM users can use in order to interact with their respective repositories, namely, services to:

- create new Slots;
- register new Readers to a Slot;
- register a Writer to a Slot;
- list out the contents of the Repository in a format that can be parsed and that is also human-readable (YAML);

Internally, Managers maintain at all times a graph of dependencies between Slots and SAM clients. The Manager ensures that this graph is always acyclic. Before a new Reader or Writer is registered, the Manager determines if that would form an information loop in the system, which cannot be allowed due to error propagation issues. This *dependency graph* can also be printed out for visualization (see Fig. 8).

Managers also distribute the information flowing through shared slots through any active masters running SAM.

3.4 Using SAM in Your Own NRS

SAM is publicly available as a *Catkin* ROS package at: https://github.com/larsys/monarch_situational_awareness. A *wiki* is maintained alongside the code repository. There, it is also possible to find documentation and examples on how to use SAM from your own code.

The interface between SAM and its client nodes can be accomplished through a set of native ROS services; or by using the contained *sam_helpers* library, which provides C++ and Python APIs to facilitate the creation of Readers and Writers.

4 Current Results

In order to provide empirical support of the validity of the MOnarCH software architecture as an adequate solution for the control of a NRS, the operation of the full system was evaluated while in use in the MOnarCH project demos. As the second yearly review meeting of the project, the following demos were ran at the testbed of IST (see Fig. 9):

Human-aware patrolling—A *mbot* (short for MOnarCH robot) is visiting in sequence a list of waypoints; when a person is visible by the networked cameras, the robot approaches and interacts with him/her in different ways depending on whether that person is holding a RFID tag. This interaction include speech synthesis, e.g., a verbal salutation appropriate to the person. The people localization tracker is running on the central server, being fed by the networked cameras, and updating a SAM slot with person detection and location data. This location is then used by the mbot to navigate towards the detected person.

Interactive game—A mbot plays the popular Flow Free game with a human on a 3×4 or 4×4 board projected on the floor. There are two modes: *tutorial mode,* where the next square the human player should go is blinking, and *reactive mode,* where the human should decide where to go on his/her turn, while the system checks for the rules compliance. SAM is used in this demo to close the loop between the NRS and the human player: it stores an updated location of the human player, used by the game engine to determine the mbot next move, which is then conveyed to the mbot to take the appropirate behavior.

Fig. 9 Photos of three of the demos described in the text, from *left* to *right* human-aware patrolling, interactive game, and formation control

SAP integration—A mbot is patrolling the environment, while a low battery situation is simulated; when this happens, the robot interrupts patrolling and goes to the docking for charging at once. SAM is used to store the mbot battery status; when the SAP detects a low battery situation, it triggers, through SAM, the robot to interrupt the current behavior and go to the charging station.

Formation control—Two mbots navigate along the environment in a leader-follower formation, that is, a leader mbot is performing patrolling while the follower aims at staying on the back of the leader at a specified distance, while avoiding obstacles. SAM is here used to convey updated location of each robot, that is, each robot runs a formation control algorithm with each robot location as input, taken from SAM, and the robot velocity as output.

All of these demos fully exploited SAM to share information among the robots involved and the central server. The modules involved were the following, while denoting with [C] of [L] whether the module runs Centrally or Locally on each robot: the global [C] and local [L] behavior managers, the patrolling behavior [L], the people localization tracker [C], the game engine [C], the robot game interface [L], the SAP [C], and the graph-based formation control [L].

In Fig. 10, we show a plot of the rate of messages being passed through SAM on the formation control demo, and respective throughput. The relatively low maximum throughput (ca. 8 kb/s) is indicative of the type of information that is being passed between robots—that is, high-level data such as robot posture and control signals, but not raw sensor data.

A second set of experiments was carried out with the purpose of evaluating the effect of the use of SAM on the latency of the communication between ROS nodes, when compared with standard ROS topics. A set of 1000 round-trip message pairs (also known as *ping*) containing timestamped headers were exchanged at a rate of 1 Hz between two ROS nodes in the following conditions and the round-trip time was collected. A pair of SAM slots (or standard ROS topics) were used, one for the "ping"

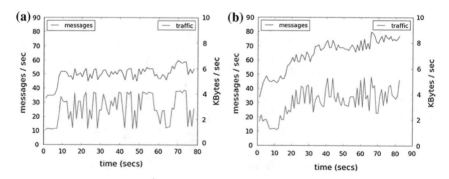

Fig. 10 Rate and throughput of messages passed through SAM in the formation control demo for **a** the leader robot and **b** the follower robot

message and the other for the reply. All experiments were performed simultaneously in order to factor out wireless throughput variability.

- Using a single ROS Master, and running both nodes on the same machine;
- Using a single ROS Master, but running each node at a different network location, with a single ROS Master running on one of the machines;
- Using a single SAM instance, and running both nodes in the same machine;
- Using a single SAM instance, but running each node at a different network location, with the SAM instance running on one of the machines;
- Using two SAM instances, running each node at a different network location, with the multimaster layer mediating both SAM instances.

These results are shown in Table 2. As can be expected, direct connection between nodes via standard ROS outperforms SAM in terms of latency, since the latter must validate each message, and it involves the overhead of multimaster synchronization. However, these results show that the latency values—order of magnitude of 100 ms—are compatible with the time-scale of data flowing through SAM, e.g., people detection events, behavior activation, robot and people tracking data.

Finally, the throughput in these network setups was also tested, by running a publisher at a high volume of data: a 1MB random string was sent between nodes at 100Hz. The respective results are shown in Table 3, and show that SAM had no impact on the throughput, even in multi-Master mode.

Current experiments focus on the basic human-robot interaction, namely the baseline assessment of the acceptance of the robot by the children. These experiments last for several hours, with the robot navigating autonomously in the Pediatrics ward of the hospital and with interactions with the children being triggered (some of them autonomously other using a Wizard-of-Oz technique) at sparse instants (see [8] for a summary of the experiments). The assessment includes (i) metrics related to dependability and (ii) to acceptance of the robot. The notion of dependability used in MOnarCH requires that the robot do not disturb the normal operation of the ward,

Table 2 Network latency, in miliseconds, in five different node-to-node network setups

Scenario	Min	Median	Mean	Max	Stddev
Local ROS Master	0.177	0.352	0.347	0.794	0.035
Remote ROS Master	2.163	6.483	18.807	241.867	34.822
Local SAM	39.230	59.706	59.645	79.829	9.824
Remote SAM	12.770	49.774	61.845	490.569	40.368
Shared SAM	70.980	119.538	126.969	432.701	38.191

Table 3 Throughput, in MB/s, in four different node-to-node network setups

	Local ROS	Local SAM	Remote ROS	Remote SAM (Multimaster)
Throughput (MB/s)	98.10	74.25	0.674	0.690

in addition to the common requirement of zero software/hardware failures. In both operation modes, i.e., fully autonomously and semi-autonomously (Wizard-of-Oz), running over the networked environment, no software failures were registered over a total running time in excess of 70 h (as of June 2015), with uninterrupted runs lasting between 2 and 4 h).

5 Related Work and Alternatives

Recent efforts towards the design of ROS 2[11] aim at addressing some limitations of the current version of ROS . Although the design of ROS 2 is not yet finished, current proposals aim at the use of the industry standard DDS communication infrastructure, a fully distributed communication mechanism. Another feature proposed for ROS 2, of interest for NRS, is the auto-discovery of network nodes. Both of these features are already addressed by the multimaster module being used in SAM. However, their integration into a new ROS infrastructure would be preferable.

Another related framework, currently under development, is Robotics in Concert (Rocon).[12] This framework provides an abstraction layer over a multirobot and devices ecology. It is based on a pool of services, provided by anonymous network nodes (robot of other devices). However, even though it follows a zeroconf philosophy, it depends on a central hub, unlike SAM.

6 Conclusion

In this chapter, we have described our case-study on using ROS in the MOnarCH NRS. The most characteristic requirements of this system are its reliance on multiple ROS Masters as a way of increasing system-wide robustness to failures; and the large number of software modules that must be designed in parallel throughout the lifetime of the project, and interact across different network locations in a reliable way.

We have introduced the SAM, a communication manager for NRSs that was developed in order to satisfy these requirements, and that has been made publicly available for use in other NRSs.

Finally, we have presented some of our current results in the MOnarCH project, which have shown that our complete NRS can operate reliably; and that the use of SAM has minimal impact on the communication latency and throughput between nodes.

The ultimate goal of the MOnarCH project is to develop a socially intelligent NRS that can operate real-world domains for the benefit of the public. Future work in MOnarCH will see the deployment of our NRS in its target environment, a pediatric

---

[11] http://design.ros2.org.

[12] http://www.robotconcert.org.

hospital, over extended periods of time. This will not only provide further validation of our software architecture in practice, but it is expected to produce valuable results in the domain of social robotics.

References

1. MOnarCH project website. http://monarch-fp7.eu/
2. I. Leite, Long-term Interactions with Empathic Social Robots. Ph.D. thesis, Instituto Superior Técnico, Universidade de Lisboa, 2013
3. J. Messias, R. Ventura, P. Lima, J. Sequeira, P. Alvito, C. Marques, P. Carrico, A Robotic Platform for Edutainment Activities in a Pediatric Hospital, in *Proceedings of the IEEE International Conference on Autonomous Robot Systems and Competitions*, (2014)
4. A. Saffiotti, M. Broxvall, PEIS ecologies: Ambient Intelligence Meets Autonomous Robotics, in *Proceedings of the 2005 Joint Conference on Smart objects and Ambient Intelligence: Innovative Context-aware Services: Usages and Technologies*, (ACM, (2005)), pp. 277–281
5. A. Sanfeliu, J. Andrade-Cetto, Ubiquitous Networking Robotics in Urban Settings, in *Workshop on Network Robot Systems. Toward Intelligent Robotic Systems Integrated with Environments. Proceedings of 2006 IEEE/RSJ International Conference on Intelligence Robots and Systems (IROS2006)*, (Beijing, China, October (2006))
6. Alberto Sanfeliu, Norihiro Hagita, Alessandro Saffiotti, Special issue: network robot systems. Robot. Auton. Syst. **65**(10), 791–791 (2008)
7. J. Sequeira, P. Lima, A. Saffiotti, V. Gonzalez-Pacheco, M.A. Salichs, Monarch: Multi-robot Cognitive Systems Operating in Hospitals, in *ICRA 2013 Workshop on Many Robot Systems*, (2013)
8. J. Sequeira, I. Ferreira, Deliverable: D8.8.4—long-run monarch experiments at ipol, Technical Report, MOnarCH (FP7-ICT-2011-9-601033), (2015)
9. M. Shiomi, T. Kanda, H. Ishiguro, N. Hagita, Interactive humanoid robots for a science museum. IEEE Intell. Syst. **22**(2), 25–32 (2007). March
10. S. Tomic, F. Pecora, A. Saffiotti, Too Cool for School—Adding Social Constraints in Human Aware Planning, in *Proceedings of the 9th International Workshop on Cognitive Robotics, CogRob 2014 (ECAI-2014 Workshop)*, (2014)
11. Kazuyoshi Wada, Takanori Shibata, Toshimitsu Musha, Shin Kimura, Robot therapy for elders affected by dementia. Eng. Med. Biol. Mag. IEEE **27**(4), 53–60 (2008)

Case Study: Hyper-Spectral Mapping and Thermal Analysis

William Morris

Abstract A study of the development of a car-mounted system for mobile hyper-spectral mapping that integrates thermal cameras, near-infrared cameras, and 3D LiDAR data. The data produced by this system is used for city scale thermal energy analysis, which allows property owners to determine the most cost effective energy efficiency improvements for their buildings. The data collection system uses ROS to record several terabytes of data during each night of operation. This case study will consider our internal best practices and lessons learned during development of a robust system running ROS for thousands of miles.

Keywords Thermal · Mapping · Perception · Best practices

1 Introduction

ROS was originally designed as a framework providing tools for the development and control of robots and autonomous systems, however in our experience it has been shown that ROS is also a capable platform for high volume multi-sensor data acquisition. This case study will consider the commercial use of ROS for hyper-spectral mapping and thermal analysis; one that operates in both the longwave and near-infrared sections of the electromagnetic spectrum. Furthermore, it will provide information on our internal best practices and lessons learned deploying a system which has an uptime requirement measured in thousands of miles. This system is the commercial implmentation of Ph.D. thesis [1, 2] research from the Field Intelligence Lab at Massachusetts Institute of Technology.

W. Morris (✉)
Essess, Inc., 51 Melcher Street, 7th Floor, Boston, MA 02210, USA
e-mail: bill.morris@essess.com
URL: http://www.essess.com/

© Springer International Publishing Switzerland 2016
A. Koubaa (ed.), *Robot Operating System (ROS)*, Studies in Computational Intelligence 625, DOI 10.1007/978-3-319-26054-9_15

1.1 Company Information

Essess (http://www.essess.com) is a Boston, Massachusetts based startup, providing a *Software as a Service* platform that enables city-scale data analytics for the energy industry. Reports generated by this platform provide energy companies and their customers a novel insight into the energy efficiency of their building envelopes. Using this information, homeowners can prioritize home improvements that will maximize return on investment.

1.2 Vehicle Overview

The imaging vehicle (shown in Fig. 1) is used to acquire city-scale datasets which are post-processed to produce energy efficiency reports. The data acquisition system was built on a Toyota RAV4 sport utility vehicle and while the vehicle itself is operated by a human driver and navigator, the system is otherwise fully automated and utilizes many of the same technologies found on self-driving cars.

1.3 Operating Environment

Scanning is performed from sunset to sunrise during the winter months to optimize the performance of the thermal sensors. Generally the best performance is achieved on clear nights with low air temperatures. While some work has been done to enable operation under less than ideal conditions, the system is not currently operated if there

Fig. 1 Energy diagnostics imaging vehicle (EDIV)

is any precipitation. This constraint is mostly to simplify operation and removes the need for humans to clean the sensitive optics. The car is driven at a constant 10 miles/h to reduce motion blur and provide sufficient exposure time for the near-infrared cameras.

2 Hardware

Much of the hardware platform was developed prior to the decision to use ROS, and one of the critical factors in deciding to use it was the comprehensive out-of-the-box hardware support for many of the sensors already in use. Other benefits include a comprehensive calibration infrastructure for sensors and the wide range of simulation and visualization tools available, these are discussed further in Sect. 3.

2.1 Imaging

The Energy Diagnostics Imaging Vehicle (EDIV) is equipped with several gigabit Ethernet machine vision cameras for continuous imaging of building facades. An example of this output is shown in Fig. 2. The radiometric thermal camera was

Fig. 2 Example thermal output

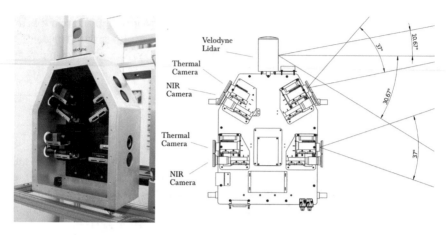

Fig. 3 Sensor locations and coverage

selected to provide a wide enough field of view to image 80 % of the target structures within a single image while still maximizing image quality.

For the system to get sufficient coverage of buildings that are multiple stories, additional pairs of cameras are aimed upwards 30° on both sides of the car. Figure 3 shows the sensor position and orientation. It also shows the coverage areas for the Velodyne [3] LiDAR along with two of the thermal cameras on the passenger side of the car. Hardware and software synchronization of all of these cameras was one of the key challenges for stitching and registering the multiple thermal and near-infrared images into a single output. The left half of Fig. 4 shows two of the thermal cameras from one side of the car stitched together, while the right half of the image shows the registration of the thermal image (red) and the near-infrared image (blue).

The system is equipped with four radiometric FLIR [4] long-wave infrared thermal cameras running a custom driver. The optics provide a field of view of

Fig. 4 Sitched(l) and registered(r) images

45° (H) × 37° (V). Environmental protection is provided by an anti-reflective coated germanium window. The thermal cameras were only capable of providing accurate radiometric measurements when operating at 30 Hz.

In addition to the thermal cameras, there are four Allied Vision [5] Manta G-223B Near-Infrared Cameras. Given the i/o constraints and the sparse support for 16 bit images in OpenCV, the cameras were configured to output 8 bit images.

The near-infrared cameras are used by the system to provide texture information for computer vision algorithms to detect features such as brick walls. Camera selection was optimized for maximum sensitivity while avoiding interference from the Velodyne LiDAR operating at 905 nm.

The near-infrared cameras utilize the `prosilica_driver` [6] which provided all necessary functionality out of the box.

The near-infrared cameras use Kowa JC10M series fixed focus optics that provide a 54° (H) × 41.9° (V) field of view. It was found that a circular polarization filter helped reduce high-intensity reflections from roadsigns being illuminated.

The imaging system uses a hardware sync signal to capture near-infrared images at 10 Hz, phase aligned with the thermal cameras running at 30 Hz.

2.2 Laser Rangefinder

3D reconstruction is made possible by utilizing multiple LiDARs. On the front bumper, a Sick LMS111 2D LiDAR is mounted in a *jazz hands* configuration for *pushbroom scanning*. This sensor has proven extremely tough, having survived a collision with the curb during testing. The Sick LiDAR is complimented by a Velodyne HDL-32E 3D LiDAR which uses 32 lasers built into a rotating assembly. Point-clouds generated by the Velodyne are dense horizontally, but fairly sparse in the vertical direction normal to rotation. An example of data generated by the LiDAR can be seen in Fig. 5 where it is projected on to a thermal image. LiDAR data has been useful for semantic segmentation of images allowing for removal of obstacles such as trees and cars from thermal analysis.

The Sick LMS111 uses the `ros-indigo-lms1xx` driver and the Velodyne HDL-32 uses the `ros-indigo-velodyne` driver for pointcloud generation.

2.3 Position and Orientation Sensors

The car has been equipped with an iMAR iMWS-V2 magnetic wheel encoder attached to a brake caliper which avoids exposing the mechanism to the elements. This is a significant consideration for operating outdoors in the winter where roads are often covered with snow and ice.

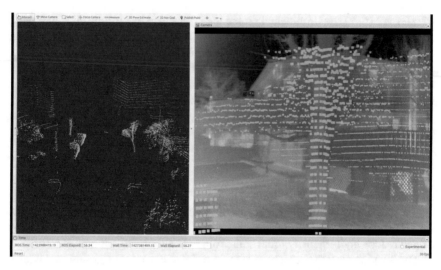

Fig. 5 Velodyne LiDAR output with ground plane removal

An in-house driver provides access to the Velodyne IMU and GPS information. A SparkFun 9 Degrees of Freedom—Razor IMU provides additional information via the `ros-indigo-razor-imu-9dof` package.

2.4 GPS

After testing several asset tracking and survey grade GPS units, we found that the NovAtel [7] ProPak6 equipped with an additional L-Band antenna met our needs. The `ros-indigo-novatel-span-driver` worked well after some configuration and modifications [8] to disable SPAN and enable RTK. This provided us with not only a high precision position estimate, but also a continuously updated estimate of the accuracy of the sensor. This allowed us to confirm when RTK was active providing centimeter accuracy and enabled diagnostic notification if the GPS dropped in accuracy due to canyoning or other sources of interference.

2.5 Environmental Sensors

The system is also equipped with a variety of basic environmental sensors to monitor the condition of the system and to allow for baseline calibration. Several DS18S20 and MCP9808 temperature sensors are attached to an Arduino via `ros-indigo-rosserial`, along with a SHT15 humidity sensor and a AM2315 humidity and temperature sensor to determine the effects of humidity on the system.

2.6   Processing and Storage

The primary on-board computer utilizes an Intel Core i7-5960X processor which provides 16 Cores with Hyperthreading enabled. The CPU is cooled with a Corsair H75 Liquid Cooler, this was chosen to reduce the amount of weight attached to the CPU to avoid stress each time the car hits a pothole. The CPU is attached to an EVGA X99 Micro Motherboard with 16 GB DDR4 3000 Memory and SATA3 support.

2.7   Network

In the current configuration the average continuous network utilization is well over 1 Gbps. However as the cameras are synchronized to transmit frames at the same time, every 1/30th of a second the instantaneous bandwidth requirements come very close to fully saturating multiple gigabit Ethernet connections.

We have tried multiple methods but have not been able to directly control the timing of the image data transmission. This appears to be a limitation of the GigE Vision SDK [9] being used.

The roof mounted sensor box is equipped with an industrial Ethernet switch and multiple waterproof M12 x-coded gigabit Ethernet connections that provide the required bandwidth to the computer stored in the trunk.

A network management interface is provided by a Cradlepoint [10] 4G WiFi router.

2.8   Data Storage

The operating system and applications are stored on a 256 GB Samsung 850 Pro Series solid-state disk, while all of the sensor and diagnostic data generated by the system is written as ROS bag files to a 12 terabyte striped RAID 0 disk array using two 6 terabyte enterprise hard drives.

One of the most useful features of ROS is the ability to record and playback sensor data while maintaining an internally consistent sense of simulated time [11] referenced to the system clock when the data was being recorded. This allows bags to be played back in slow motion, while API calls will return the time the sensor data was recorded instead of the current time. The bag file format is well documented [12] and for some of our applications we found it helpful to unpack data directly into a Mongo database for processing.

During a nominal night of scanning our system records over 6 terabytes of data during 12 h of continuous operation, at this point we have cumulatively recorded several hundred terabytes of sensor data in the ROS bag format. This data is recorded as a series of 2 GB files along with a separate bag file for GPS data that is used as an index during the data import process. We have also tested recording at higher data rates and have encountered a few interesting *features*.

If the data is being written to bag files with unlimited buffering, and it exceeds the disk i/o bandwidth or if compression is enabled and the CPU is unable to keep up, the system can fail silently. As the system queues old messages it will end up running out of memory, and once that happens the OOM (out of memory) Killer [13] will kill the rosbag process. So if your robot stops logging data mysteriously this might be something to consider checking. For this system to maintain stability over 12 h, the bandwidth utilization was reduced after initial testing by switching the near-infrared cameras to 8-bit mode and using on-camera hardware binning to produce smaller images.

Despite CPU upgrades and several attempts at various compression levels, the system was unable to compress data fast enough for the volume of data being produced. The increased storage requirements of uncompressed data are balanced by the performance improvement in post-processing by removing the need for decompression.

2.9 Hardware Configuration and Introspection

Experience attempting to utilize data from previous experiments has shown that it is important to store the hardware configuration in the data stream to allow for later introspection. Publishing camera serial numbers allows us to determine when hardware changes were made simply by reading the bag files.

Thermal camera calibrations are saved to files named by serial number and are version controlled in a separate package. The near-infrared cameras store calibration information in nvram onboard the cameras.

The relative position of sensors is stored in URDF [14], which enables visualization in RViz [15] and other tools such as tf [16].

3 Software

The ROS robotics framework provides an extensive set of tools for rapid development of robotics software, data visualization tools including RViz and the rqt tools [17] for plotting and data analysis. In addition to the tools provided by ROS, we also found the *glances* [18] system monitoring software to be valuable for analyzing system performance.

3.1 Simulation and Development

Robotics development is often constrained by limited resources and the conflicting needs of the hardware and software development teams. Given the cost of some sensors, there is often not be enough equipment to provide software developers access to physical hardware until the hardware team is finished. ROS was utilized to record raw sensor output via rosbag [12], this allowed the software team to use hardware simulations to work on algorithms without waiting for access to physical hardware. Productivity was also improved by building a test cart (as shown in Fig. 6) to allow for indoor testing of hardware under development. The left half of Fig. 6 shows the cart with the sensor box attached to the cart, whereas the right half of the image shows how the sensor box can be removed from the cart for attachment to the car.

3.2 Diagnostics

We extensively use diagnostic aggregators [19] to provide information to the human operators. These diagnostic tools monitor topic frequency and provide basic data validation to check that sensors are working properly. Figure 7 shows system status indicators, red boxes show subsystems that are operating outside of specification, green boxes show subsystems that are operating nominally, warnings are shown in yellow. In this example, the *Car* system is in an error state due to the *Environmental*

Fig. 6 *Descartes* development cart: sensor box attached(l) and removed(r)

Fig. 7 Diagnostic interface running on a Nexus 9 tablet

subsystem showing an alarm with the humidity sensor. In this case the diagnostic message is used importing the data into the analytics engine allowing data of marginal quality to be flagged for human review due to high ambient humidity.

3.3 Operator Interface

The interface (Fig. 8) for the system is provided by a *Nexus 9* tablet using the Robot-WebTools [20] JavaScript framework. A scaled and throttled video stream from the cameras is sent over WiFi using the `ros-indigo-web-video-server` package. Additional user interface components allow the human operator to start/stop logging and reboot the system at the click of a button. The buttons along the top of Fig. 8 allows the operator to switch between longwave and near-infrared camera views shown below.

4 Best Practices and Lessons Learned

Our experiences working with ROS have led us to develop a set of internal best practices that help us maintain robust performance and minimize miscommunication.

Fig. 8 Operator interface showing thermal camera output

4.1 Data Storage and Indexing

The rosbag files are automatically named by passing the −o option to rosbag, (e.g. ediv2_2015-02-14-02-17-31_0.bag) which allows us to differentiate between vehicles. Test data that is sent for full processing is flagged with the prefix 'testrun' to prevent confusion.

Each data drive contains as set of 2 GB rosbags which contain the output of the sensors and diagnostics. An info file is produced from the output of rosbag info for each bag file. Field engineers can also save text files to the drives to provide notes to the software engineers for specific data to be reviewed during the data import process.

After each night of field operation the drives are shipped for processing. After the drives are received, the rosbag info file is used to provide validation that all of the data survived shipment. Once imported the diagnostic messages recorded in the bags are used for initial data validation.

Given the amount of data stored, a separate rosbag containing only timestamped GPS data is used as an index. This allows the system to generate a variety of reports for customers on a per zipcode basis without having to extract and process all of the data collected.

4.2 Configuration Management

We have used Chef [21] with ROS on our production systems for configuration management with mixed results. Some colleagues are using Ansible with ROS and they seem to think it may be a better choice. Some of the problems we encountered may be fixable with more experience using Chef.

On a positive note, our software team has found that Chef + ROS works very well for cloud based processing. It makes it very easy to spin up an unlimited number of known working instances for parallel off-line processing.

Unfortunately we experienced several problems with using Chef with ROS. The biggest problem is that system configuration is often opaque to everyone except the operations team, whereas operations is often unfamiliar with hardware related issues.

For example, at one point we had sensor calibration files installed via Chef, this caused significant confusion as to which calibration should be installed on a given set of hardware. It also required additional overhead when the calibrations were updated, as roboticists were making pull requests to an unfamiliar repository. Eventually, we found it easier to store the configurations in a ROS package and load them dynamically based on the serial number.

Network configuration management also proved to be problematic. The hardware platform had multiple Ethernet interfaces that needed IP address configuration based on MAC addresses. When the hardware team needed to replace a two-port network card with a four-port card, it also temporarily broke the Chef configuration until someone on the operations side could update Chef.

To provide a separate environment for production and testing we had Chef install `production.rosinstall` and `development.rosinstall` files and used rosws [22] to install to `~/development_ws` and `~/production_ws`. This helped segment ROS development from Chef and made everyone happier and made tagging git releases easier.

Once everything was working it was fairly easy to generate an ISO installer that contains the Chef tools needed to bootstrap a system. Debian packages were copied to the apt cache on the USB drive to cut down on network traffic for the install.

While Chef maintenance and testing was tedious, static production deployments to multiple systems is where it starts to become invaluable.

4.3 Startup

System startup is managed via upstart scripts based on the ROS System Daemon [23] package. The ROS launch files are stored in a seperate package along with the configurations of individual drivers. This allowed for version controlled management of drivers installed via the system package manager.

Critical sensors utilize watchdog timers to restart drivers for system recovery in the event of a transient error condition. The thermal camera for example is launched with the `respawn = "true"` option [24] and the watchdog timer is checked against timestamp of the last image transmitted. If the driver has not transmitted an image within the period specified by the timer, the driver exits and is relaunched by the ROS master. Otherwise it resets the timer.

5 Conclusion

Utilizing ROS has reduced the product development time and improved the quality and reliability of the data produced by the system. Beyond providing basic functionality, ROS has provided a solid technical foundation for continued development and proven stable across thousands of miles and hundreds of terabytes.

References

1. L.N. Phan, *Automated rapid thermal imaging systems technology*. Ph.D. thesis, Massachusetts Institute of Technology, 2012
2. E.C. Shao, *Detecting sources of heat loss in residential buildings from infrared imaging*. Ph.D. thesis, Massachusetts Institute of Technology, 2011
3. Velodyne Acoustics Inc, Velodyne lidar (2015), http://www.velodynelidar.com. Accessed 3 Aug 2015
4. FLIR Systems Inc, Flir systems (2015), http://www.flir.com. Accessed 3 Aug 2015
5. Allied Vision Technologies GmbH, Allied vision (2015), http://www.alliedvision.com. Accessed 3 Aug 2015
6. Github, Ros driver and the sdk for avt/prosilica cameras (2015), https://github.com/ros-drivers/prosilica_driver. Accessed 3 Aug 2015
7. NovAtel Inc, Novatel (2015), http://www.novatel.com. Accessed 3 Aug 2015
8. Github, Fork of novatel ros driver (2015), https://github.com/essessinc/novatel_span_driver. Accessed 3 Aug 2015
9. Pleora Technologies Inc, Pleora ebus sdk (2015), http://www.pleora.com/our-products/ebus-sdk. Accessed 3 Aug 2015
10. CradlePoint Inc, Cradlepoint (2005), http://cradlepoint.com. Accessed 3 Aug 2015
11. Ros wiki, Clock (2015), http://wiki.ros.org/Clock. Accessed 3 Aug 2015
12. Ros wiki, Bags/format (2010), http://wiki.ros.org/Bags/Format. Aaccessed 3 Aug 2015
13. Malcolm Parsons, Oom killer (2009), http://linux-mm.org/OOM_Killer. Accessed 3 Aug 2015
14. Ros wiki, urdf (2014), http://wiki.ros.org/urdf. Accessed 3 Aug 2015
15. Ros wiki, rviz (2014), http://wiki.ros.org/rviz. Accessed 3 Aug 2015
16. Ros wiki, tf (2015), http://wiki.ros.org/tf. Accessed 3 Aug 2015
17. Ros wiki, rqt (2013), http://wiki.ros.org/rqt_common_plugins?distro=indigo. Accessed 3 Aug 2015
18. N. Hennion, Glances (2015), https://nicolargo.github.io/glances/. Accessed 3 Aug 2015
19. Ros wiki, Diagnostics (2015), http://wiki.ros.org/diagnostics. Accessed 3 Aug 2015
20. Robot web tools (2015), http://robotwebtools.org/, Accessed 3 Aug 2015
21. Chef (2015), http://docs.chef.io/. Accessed 3 Aug 2015
22. Ros installation tools (2011), http://docs.ros.org/independent/api/rosinstall/html/, Accessed 3 Aug 2015
23. Ros system daemon (2013), https://github.com/TurtleBot-Mfg/ros-system-daemon-hydro. Accessed 3 Aug 2015
24. Ros wiki, Roslaunch/xml/node (2015), http://wiki.ros.org/roslaunch/XML/node. Accessed 3 Aug 2015

Part V
Perception and Sensing

A Distributed Calibration Algorithm for Color and Range Camera Networks

Filippo Basso, Riccardo Levorato, Matteo Munaro
and Emanuele Menegatti

Abstract In this tutorial chapter we present a package to calibrate multi-device vision systems such as camera networks or robots. The proposed approach is able to estimate—in a unique and consistent reference frame—the rigid displacements of all the sensors in a network of standard cameras, Kinect-like depth sensors and Time-of-Flight range sensors. The sensor poses can be estimated in a few minutes with a user-friendly procedure: the user is only asked to move a checkerboard around while the ROS nodes acquire the data and perform the calibration. To make the system scalable, the data analysis is distributed in the network. This results in a low bandwidth usage as well as a really fast calibration procedure. The ROS package is available on GitHub within the repository `iaslab-unipd/calibration_toolkit` (https://github.com/iaslab-unipd/calibration_toolkit). The package has been developed for ROS Indigo in C++11 and Python, and tested on PCs equipped with Ubuntu 14.04 64 bit.

Keywords ROS · Calibration · Camera · Depth · Camera network · Distributed system · Kinect · RGB-D

F. Basso (✉) · R. Levorato · M. Munaro · E. Menegatti
IAS-Lab, Department of Information Engineering (DEI), University of Padova,
Via Ognissanti 72, 35131 Padova, Italy
e-mail: bassofil@dei.unipd.it
URL: http://robotics.dei.unipd.it

R. Levorato
e-mail: riccardo.levorato@dei.unipd.it

M. Munaro
e-mail: matteo.munaro@dei.unipd.it

E. Menegatti
e-mail: emg@dei.unipd.it

© Springer International Publishing Switzerland 2016
A. Koubaa (ed.), *Robot Operating System (ROS)*, Studies in Computational
Intelligence 625, DOI 10.1007/978-3-319-26054-9_16

413

1 Introduction

Robotic systems and camera networks consist of many heterogeneous vision sensors. The estimation of the poses of all such sensors with respect to a unique, consistent world frame, is a challenging and well-known problem. As a matter of fact, a good calibration of these sensors can be a useful starting point for several applications in the computer vision field (e.g. 3D mapping, people recognition and tracking [4, 14], microphone calibration for audio localization [13]) as well as in many robotics applications (e.g. simultaneous localization and mapping (SLAM) applications, grasping and manipulation).

However, even most of the time a good calibration is mandatory for the success of the application, there are still no tools that permit to easily calibrate multiple vision sensors together in a uniform way. In fact, most of the existing tools are for specific applications or specific sensors (e.g. stereo cameras); there are only few methods developed to simultaneously calibrate an heterogeneous sensor network. As stated by Le et al. [12], the most followed approach is to divide the sensors into pairs and calibrate each pair independently, even using different algorithms for each one.

Our idea, i.e., the one behind the development of this package, is to go beyond this calibration technique, and develop an easy-to-use and easy-to-extend calibration package for ROS, such that users can add their own sensor types, their own error functions and perform the calibration. Moreover, since calibration is a time-consuming task, a fast procedure would be a very useful tool, especially when the involved sensors need often to be moved—and therefore re-calibrated.

We are still far from having a sensor-independent calibration toolbox like the aforementioned one, however, as we have already demonstrated [3, 13, 15], the very same calibration procedure can be used to calibrate standard cameras, Kinect-like and Time-of-Flight depth sensors as well as omnidirectional cameras and actuated laser scanners. So, we took what we had learned during the development of our previous works and packed it in a completely new package, heavily based on Eigen [11] and Ceres Solver [1] libraries.

The package addresses the problem of calibrating networks composed by cameras and depth sensors like the one in Fig. 1. The approach we followed is an extension of the classical single camera calibration procedure: users are asked to move a checkerboard pattern in front of each camera and depth sensor and, as soon as any of the sensors see the checkerboard, the calibration starts. Then, whenever the pattern is visible by at least two sensors simultaneously, a constraint is added to the calibration problem. This process goes on until all the sensors are connected to the others. At the end, all the data are processed inside an optimization framework that improves the quality of the initial estimation. This procedure has already been tested on camera-only networks as part of the OpenPTrack[1] project [15]. Here we explain how to configure and use the package as well as the theoretical background behind the developed algorithm. The remainder of the chapter is organized as follows:

---

[1] http://openptrack.org/.

Fig. 1 Example of sensors in a PC network that the proposed package aims to calibrate

- First, we review some of the existing works on camera and sensor network calibration.
- Second, we give a short overview of the calibration problem for camera networks.
- Third, we explain in detail how the calibration is performed from a theoretical point of view.
- Fourth, we introduce the readers to the package with a real-world example.
- Fifth, we describe how to configure and use the nodes provided by the package to calibrate a user-defined sensor network.
- Finally, we draw some conclusions.

2   Related Work

In the robot vision field, RGB cameras have been a key technology in the development of visual perception. In the last few years, the introduction of RGB-D sensors contributed deeply to the advancement of data fusion in practical applications. Auvinet et al. [2] proposed a new method for calibrating multiple depth cameras for body reconstruction using only depth information. Their algorithm is based on plane intersections and the NTP protocol for data synchronization. The calibration achieves good results: even if the depth error of the sensor is 10 mm, the reconstruction error with 3 depth cameras is, in the best case, less than 6 mm. However, they have to manually select the plane corners and, above all, they only deal with depth sensors.

Another approach to solve the calibration problem for a network of cameras and depth sensors, is the one proposed by Le and Ng [12]: they proposed to jointly calibrate groups of sensors. More specifically, the groups were composed by a set of sensors able to provide a 3D representation of the world (e.g. a stereo camera, an RGB camera and a depth camera, etc.). First of all they calibrated the intrinsics of each sensor, secondly they calibrated the extrinsic parameters of each group, then they calibrated the extrinsic parameters of each group with respect to all the others. Finally the calibration parameters were refined in one optimization step. Their experiments show that this method not only reduces the calibration error, but also requires a little human intervention. An advantage of having groups that output 3D data is that the same calibration object can be used to calibrate a group with respect to all the others, regardless of the sensor type. Also, a joint calibration does not accumulate errors like a calibration based on sensor pairs do. However, they state that they should combine this two steps and jointly calibrate all parameters at once, as we proposed in our works [3] and propose here. In fact, the main drawback of this approach is that they always need to group the sensors beforehand in order to have 3D data outputs.

Finally, Furgale et al. [10], recently developed a similar ROS package, *Kalibr*, that tackles the spatio-temporal calibration of multi-sensor systems composed of cameras and an IMU. To the best of our knowledge, this is the work most similar to ours, even though it does not deal with depth sensors.

3   Background

3.1   The Calibration Problem

Definition 1 Let $\mathfrak{S} = \{S_1, S_2 \ldots S_K\}$ be a set of sensors. For each sensor $S_i \in \mathfrak{S}, i = 1 \ldots K$, the goal is to find its pose $^W S_i$ with respect to a common reference frame $\mathcal{W}$, namely the *world*.

To solve such problem, it is important to know the concept of reference frame and how affine transformation (in a 3D space) works. Moreover, since we are dealing mostly with cameras, it is mandatory to know how 3D points are converted to pixels and vice-versa.

3.2   Affine Transformations

Let $\mathbf{x} = (x, y, z)^\mathsf{T}$ be a point in 3D space. For any non-zero real number w, $(xw, yw, zw, w)^\mathsf{T}$ is the set of homogeneous coordinates associated to the point $\mathbf{x}$. In particular, when $w = 1$, the resulting 4D vector $\tilde{\mathbf{x}} = (x, y, z, 1)^\mathsf{T}$ is called the *normalized homogeneous form*.

A rigid transformation in 3D space is represented by a linear transformation $\mathbf{T} \in \mathbb{SE}(3)$ on normalized homogeneous vectors

$$\mathbf{T} = \begin{bmatrix} \mathbf{R} & \mathbf{t} \\ \mathbf{0}^\mathsf{T} & 1 \end{bmatrix}$$

where $\mathbf{R} \in \mathbb{SO}(3)$ is the *rotation matrix* and $\mathbf{t} \in \mathbb{R}^3$ the *translation* vector.

To transform a 3D point $\mathbf{x}$ it is sufficient to left-multiply its normalized homogeneous form $\widetilde{\mathbf{x}}$ by the transformation matrix $\mathbf{T}$

$$\widetilde{\mathbf{y}} = \mathbf{T} \cdot \widetilde{\mathbf{x}} .$$

The resulting vector $\widetilde{\mathbf{y}}$ is the normalized homogeneous form of the desired 3D point $\mathbf{y}$. This operation is equivalent to perform an affine transformation in $\mathbb{R}^3$

$$\mathbf{y} = \mathbf{R} \cdot \mathbf{x} + \mathbf{t} .$$

3.3 Reference Frames

A *reference frame* $\mathcal{F}$ is a coordinate system used to represent and measure position and orientation of objects. A transformation matrix ${}^{\mathcal{T}}_{\mathcal{S}}\mathbf{T}$ from a source reference frame $\mathcal{S}$ to a target frame $\mathcal{T}$ is a transformation matrix that allows to convert the coordinates of a point from $\mathcal{S}$ to $\mathcal{T}$.

Let, for example, $\mathbf{x}$ be a point and let ${}^{\mathcal{S}}\mathbf{x}$ be its coordinates in $\mathcal{S}$, $\mathbf{x}$'s coordinates in $\mathcal{T}$, namely ${}^{\mathcal{T}}\mathbf{x}$, can be computed as

$$^{\mathcal{T}}\widetilde{\mathbf{x}} = {}^{\mathcal{T}}_{\mathcal{S}}\mathbf{T} \cdot {}^{\mathcal{S}}\widetilde{\mathbf{x}} .$$

3.4 Pinhole Camera Model

Let $\mathbf{C}$ be a camera and $\mathcal{C}$ its reference frame. The pinhole model describes the mathematical relationship between the coordinates of a 3D point and its projection onto the image plane. The model assumes that the camera has 4 parameters, namely *intrinsic parameters*:

- $(c_x, c_y)^{\mathsf{T}}$ is the *principal point* that is usually at the image center;
- f_x, f_y are the *focal lengths* expressed in pixel units;

usually arranged in a 3×4 matrix $\mathbf{K}_{\mathbf{C}}$

$$\mathbf{K}_{\mathbf{C}} = \begin{bmatrix} f_x & 0 & c_x & 0 \\ 0 & f_y & c_y & 0 \\ 0 & 0 & 1 & 0 \end{bmatrix} .$$

The relationship between the coordinates of a 3D point ${}^{\mathcal{C}}\mathbf{x}$ and its projection onto the image plane $\mathbf{x}'$ (in pixels) is

$$s \cdot \widetilde{\mathbf{x}}' = \mathbf{K}_{\mathbf{C}} \cdot {}^{\mathcal{C}}\widetilde{\mathbf{x}}, \tag{1}$$

at least theoretically. Unfortunately real lenses usually have some distortion. So, the pinhole model is extended with a vector of *distortion coefficients* $\mathbf{d}_C$ and a distortion function $d_C(\cdot)$. Equation (1) thus becomes

$$s \cdot \widetilde{\mathbf{x}}' = \mathbf{K}_C \cdot d_C({}^C\widetilde{\mathbf{x}}) \ . \tag{2}$$

For more details on the distortion function please refer to [6].

In the following, given a camera C, the *reprojection* of a 3D point onto C's image plane, i.e. (2), will be denoted by the function $r_C(\cdot)$, that is

$$\mathbf{x}' = r_C\left({}^C\mathbf{x}\right) \ . \tag{3}$$

3.5 Notations

We use non-bold characters x to represent scalars, bold lower case letters $\mathbf{x}$ to represent vectors with no distinction between cartesian coordinates and homogeneous coordinates. Bold upper case letters $\mathbf{M}$ represent matrices. Note that matrices can be seen as ordered lists of vectors, one for each column.

The reference frame of an object B is in calligraphic style $\mathcal{B}$. The coordinates of an entity e with respect to the reference frame $\mathcal{F}$ are denoted by ${}^{\mathcal{F}}e$. According to this notation, the pose of a body A in B's coordinate system $\mathcal{B}$ is denoted as ${}^{\mathcal{B}}A$ and the relative homogeneous transformation matrix is ${}^{\mathcal{B}}_{\mathcal{A}}\mathbf{T}$.

4 Camera-Only Network Calibration

4.1 Pose Estimation

To solve the calibration problem (Definition 1) for a camera-only network, we use a *checkerboard pattern*. So, let B be an $R \times C$ checkerboard and let $\mathcal{B}$ be its reference frame. Let also C be a camera with reference frame $\mathcal{C}$, intrinsic parameters $\mathbf{K}_C$ and distortion coefficients $\mathbf{d}_C$. We can estimate the checkerboard pose ${}^{\mathcal{C}}_{\mathcal{B}}\mathbf{T}$ in the camera reference frame by finding its corners in the image and solving the correspondent *Perspective-n-Point* (PnP) problem [9].

That is, let $\mathbb{I} = \{1 \dots R\} \times \{1 \dots C\}$ be the corner indices, let also

$${}^{\mathcal{B}}\mathbf{B} = \left\{{}^{\mathcal{B}}\mathbf{b}_{r,c}\right\}_{(r,c)\in\mathbb{I}}$$

be the checkerboard corners and

$$\mathbf{B}' = \left\{\mathbf{b}'_{r,c}\right\}_{(r,c)\in\mathbb{I}}$$

the correspondent locations in the image. The checkerboard pose $_B^C\mathbf{T}$ can be estimated by means of a single function called *solvePnP*, i.e.

$$_B^C\mathbf{T} = \text{solvePnP}\left(\mathbf{K}_C, \mathbf{d}_C, {}^B\mathbf{B}, \mathbf{B}'\right) , \tag{4}$$

and is the one that minimizes the *reprojection error*

$$\mathbf{e}_{rc}\left(_B^C\mathbf{T}, {}^B\mathbf{B}, \mathbf{B}'\right) = \sum_{(r,c)\in\mathbb{I}} \left\| \mathbf{b}'_{r,c} - \mathbf{r}_C\left(_B^C\mathbf{T} \cdot {}^B\mathbf{b}_{r,c}\right) \right\|^2 . \tag{5}$$

Now, let $\mathbf{C}_1$ and $\mathbf{C}_2$ be two different cameras, with reference frame $\mathcal{C}_1$ and $\mathcal{C}_2$ respectively and suppose they both can see the checkerboard at the same time. To be more precise, let's define $k \in \mathbb{N}$ as an *acquisition step*, that is, a progressive number that is incremented each time the checkerboard is moved to a different location and an acquisition for every camera is triggered at the same instant. Let also $_B^W\mathbf{T}^{(k)}$ be the checkerboard pose at step k. Using (4) we can estimate both $_B^{C_1}\mathbf{T}^{(k)}$ and $_B^{C_2}\mathbf{T}^{(k)}$. Starting from these two poses (and the fact the the checkerboard is in the same location), we can estimate the pose of one sensor with respect to the other $_{C_2}^{C_1}\mathbf{T}$ with a closed formula

$$_{C_2}^{C_1}\mathbf{T} = {}_{C_2}^{C_1}\mathbf{T}^{(k)} = {}_B^{C_1}\mathbf{T}^{(k)} \cdot {}_B^{C_2}\mathbf{T}^{(k)-1} . \tag{6}$$

Then, recalling that affine transforms can be chained

$$_A^B\mathbf{T} = {}_X^B\mathbf{T} \cdot {}_A^X\mathbf{T} ,$$

and that

$$_A^B\mathbf{T} = {}_B^A\mathbf{T}^{-1} ,$$

we can find a solution to our calibration problem for a network composed by N cameras: we just need to estimate (or set) the pose $_{C_i}^W\mathbf{T}$ of one camera $\mathbf{C}_i$ with respect to the world reference frame $\mathcal{W}$ and move the checkerboard around until every camera pose is computed, using any of the aforementioned equations.

To estimate the pose of a camera $\mathbf{C}_i$ with respect to the world reference frame $\mathcal{W}$, first of all we must know whether we need to define a world reference frame $\mathcal{W}$ in the environment or not. Actually, if it really does not matter where such reference frame is, any camera reference frame $\mathcal{C}_i$ can be set as the world, that is $\mathcal{W} = \mathcal{C}_i$ (or equally $_{C_i}^W\mathbf{T} = \mathbf{I}$) for one $i \in \{1 \dots N\}$. Otherwise the pose can be set manually: $_{C_i}^W\mathbf{T} = \mathbf{W}$ for some transformation matrix $\mathbf{W}$ and $i \in \{1 \dots N\}$; or estimated by moving the checkerboard to the desired position and setting $_B^W\mathbf{T}^{(k)} = \mathbf{I}$, for some $k \in \mathbb{N}$.

At this stage we have good estimations of the sensor poses, however, due to errors in the measurements, usually

$$_{C_2}^{C_1}\mathbf{T}^{(k)} \neq {}_{C_2}^{C_1}\mathbf{T}^{(l)}$$

for two different steps k and l. Therefore we must perform an optimization step to refine the estimated camera poses, such that the error on the estimated poses is reduced as much as possible.

4.2 Optimization

Taking a step back to the acquisition part, we can organize the calibration data in a matrix, like the one in Fig. 2. Following the *bundle adjustment* approach [19], we refine both the camera poses ${}^{\mathcal{W}}_{C_i}\mathbf{T}$, $i = 1 \ldots N$ and the checkerboard poses ${}^{\mathcal{W}}_{\mathcal{B}}\mathbf{T}^{(k)}$, $k = 1 \ldots K$. Indeed, even if we perfectly know the pose of every camera with respect to the world, the pose of a checkerboard B estimated at step k using two different cameras, say C_i and C_j, is likely to be different:

$$ {}^{\mathcal{W}}_{C_i}\mathbf{T}^{(k)} \cdot {}^{C_i}_{\mathcal{B}}\mathbf{T}^{(k)} \neq {}^{\mathcal{W}}_{C_j}\mathbf{T}^{(k)} \cdot {}^{C_j}_{\mathcal{B}}\mathbf{T}^{(k)} . $$

To achieve a satisfying solution, a good candidate to be minimized is the reprojection error defined in (5). The complete error function E_{C} that we minimize is therefore

$$ \begin{aligned} E_{\mathsf{C}} &= \sum_{k=1}^{K} \sum_{i=1}^{N} u_{ik} \cdot \frac{1}{\sigma^2_{\mathsf{C}_i}} \cdot \mathrm{e}_{\mathrm{r}_{\mathsf{C}_i}} \left({}^{\mathcal{W}}_{C_i}\mathbf{T}^{-1} \cdot {}^{\mathcal{W}}_{\mathcal{B}}\mathbf{T}^{(k)}, {}^{\mathcal{B}}\mathbf{B}, \mathbf{B}_i'^{(k)} \right) \qquad (7) \\ &= \sum_{k=1}^{K} \sum_{i=1}^{N} u_{ik} \cdot \frac{1}{\sigma^2_{\mathsf{C}_i}} \cdot \sum_{(r,c)\in\mathbb{I}} \left\| \mathbf{b}_{i,r,c}'^{(k)} - \mathrm{r}_{\mathsf{C}_i} \left({}^{\mathcal{W}}_{C_i}\mathbf{T}^{-1} \cdot {}^{\mathcal{W}}_{\mathcal{B}}\mathbf{T}^{(k)} \cdot {}^{\mathcal{B}}\mathbf{b}_{r,c} \right) \right\|^2 , \end{aligned} $$

Steps

Cameras		${}^{\mathcal{W}}_{\mathcal{B}}\mathbf{T}^{(1)}$	${}^{\mathcal{W}}_{\mathcal{B}}\mathbf{T}^{(2)}$	${}^{\mathcal{W}}_{\mathcal{B}}\mathbf{T}^{(3)}$	${}^{\mathcal{W}}_{\mathcal{B}}\mathbf{T}^{(4)}$	${}^{\mathcal{W}}_{\mathcal{B}}\mathbf{T}^{(5)}$	${}^{\mathcal{W}}_{\mathcal{B}}\mathbf{T}^{(6)}$	${}^{\mathcal{W}}_{\mathcal{B}}\mathbf{T}^{(7)}$	${}^{\mathcal{W}}_{\mathcal{B}}\mathbf{T}^{(8)}$
$\mathbf{K}_{\mathsf{C}_1}, \mathbf{d}_{\mathsf{C}_1}$	${}^{\mathcal{W}}_{C_1}\mathbf{T}$				$\mathbf{B}_1'^{(4)}$				$\mathbf{B}_1'^{(8)}$
$\mathbf{K}_{\mathsf{C}_2}, \mathbf{d}_{\mathsf{C}_2}$	${}^{\mathcal{W}}_{C_2}\mathbf{T}$	$\mathbf{B}_2'^{(1)}$	$\mathbf{B}_2'^{(2)}$			$\mathbf{B}_2'^{(5)}$		$\mathbf{B}_2'^{(7)}$	
$\mathbf{K}_{\mathsf{C}_3}, \mathbf{d}_{\mathsf{C}_3}$	${}^{\mathcal{W}}_{C_3}\mathbf{T}$	$\mathbf{B}_3'^{(1)}$					$\mathbf{B}_3'^{(6)}$		$\mathbf{B}_3'^{(8)}$
$\mathbf{K}_{\mathsf{C}_4}, \mathbf{d}_{\mathsf{C}_4}$	${}^{\mathcal{W}}_{C_4}\mathbf{T}$				$\mathbf{B}_4'^{(4)}$	$\mathbf{B}_4'^{(5)}$	$\mathbf{B}_4'^{(6)}$		
$\mathbf{K}_{\mathsf{C}_5}, \mathbf{d}_{\mathsf{C}_5}$	${}^{\mathcal{W}}_{C_5}\mathbf{T}$		$\mathbf{B}_5'^{(2)}$	$\mathbf{B}_5'^{(3)}$				$\mathbf{B}_5'^{(7)}$	

Fig. 2 Matrix view of the calibration data. Each row is associated to a camera C_i and contains both the camera parameters $\mathbf{K}_{\mathsf{C}_i}$ and $\mathbf{d}_{\mathsf{C}_i}$, and the camera estimated pose ${}^{\mathcal{W}}_{C_i}\mathbf{T}$. The columns are instead associated to the steps and contain the poses of the checkerboard at every step k, namely ${}^{\mathcal{W}}_{\mathcal{B}}\mathbf{T}^{(k)}$. A cell (i, k) contains the corners locations (in pixels) $\mathbf{B}_i'^{(k)}$ of the checkerboard at step k in the image provided by camera C_i, if the checkerboard is visible

where u_{ik} is an indication function equal to 1 if camera C_i sees the checkerboard at step k (otherwise it is 0) and the fraction $\frac{1}{\sigma_{C_i}^2}$ is instead a normalization factor. The term σ_{C_i} is usually set to 1 or 0.5 and indicates the error on the corner's estimated position (in pixels) in the images provided by camera C_i.

4.3 Additional Constraints

In one of our first applications of this calibration algorithm, OpenPTrack [15], we needed to calibrate a camera network in a big room against the floor, i.e. extract the floor equation and set the world frame somewhere on it. We decided to estimate the floor coefficients during the calibration procedure, exploiting the fact that positioning a checkerboard on the floor would have allowed us to define the plane equation as well as the world reference frame. However, since the room was quite big and the checkerboard far from every camera, the results were not satisfactory: the plane had often a non-negligible rotation with respect to the real one. To overcome this issue, printing a bigger checkerboard was not a viable solution. Instead, we imposed that two or more checkerboards were lying on the same plane and added the geometrical constraints to the error model. In fact, if we fix the plane π on which a checkerboard can move, the checkerboard pose can be defined by a 2D transform $_B^P\mathbf{T}_2$ with respect to the reference frame of plane π, namely $\mathcal{P}$.

So, let define a plane by means of its reference frame $_\mathcal{P}^\mathcal{W}\mathbf{T}$, such that the x- and y-axes are on the plane and the z-axis is its normal, as depicted in Fig. 3. The pose of a checkerboard lying on π at step k is

$$_B^\mathcal{W}\mathbf{T}^{(k)} = {}_\mathcal{P}^\mathcal{W}\mathbf{T} \cdot {}_B^\mathcal{P}\mathbf{T}_2^{(k)}, \tag{8}$$

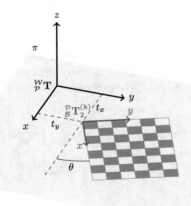

Fig. 3 Reference frames of the plane π and a checkerboard B lying on it

where the 2D transform ${}^{\mathcal{P}}_{\mathcal{B}}\mathbf{T}_2^{(k)}$ is wrapped into a 3D one to perform the matrix multiplication

$$
{}^{\mathcal{P}}_{\mathcal{B}}\mathbf{T}_2^{(k)} = \begin{bmatrix} \cos(\theta) & -\sin(\theta) & 0 & t_x \\ \sin(\theta) & \cos(\theta) & 0 & t_y \\ 0 & 0 & 1 & 0 \\ 0 & 0 & 0 & 1 \end{bmatrix}.
$$

We can now substitute (8) into (7) to refine both the plane and the checkerboard pose.

5 Extension to a Depth Sensor-Camera Network

5.1 Pose Estimation

In Sect. 4 we have presented a calibration procedure for camera-only networks. Such procedure works well with cameras, but how can we calibrate a network composed by both cameras and depth sensors?

A depth sensor D provides an $R \times C$ point cloud view of the scene, i.e. an indexed set of 3D points

$$
{}^{\mathcal{D}}\mathbf{P} = \left\{ {}^{\mathcal{D}}\mathbf{p}_{r,c} \right\}_{(r,c)\in\mathbb{I}},
$$

where $\mathbb{I} = \{1\dots R\} \times \{1\dots C\}$, reflecting the shape of the scene in the sensor's field of view. Obviously, we cannot directly estimate the checkerboard pose using the corners, as we do for the camera-only network calibration, since they are not visible. However, we can exploit the 3D data to extract the pattern plane and perform a *plane-to-plane calibration* [20]. That is, supposing we have already estimated the checkerboard pose ${}^{\mathcal{W}}_{\mathcal{B}}\mathbf{T}^{(k)}$ at step k, we can define the checkerboard plane ${}^{\mathcal{W}}\pi_{\mathsf{B}}^{(k)}$ using three non-collinear corners. We can also estimate the pattern plane ${}^{\mathcal{D}}\pi_{\mathsf{B}}^{(k)}$ from the point cloud, using a for example a RANSAC-based [9] plane fitting algorithm.

So, let $\{k_1\dots k_n\}$, with $n \geq 3$, be the intersection between the steps in which depth sensor D sees the checkerboard and those in which the checkerboard pose has been estimated, and let the plane equations be of the form $\mathbf{n}^{\mathsf{T}} \cdot \mathbf{x} - d = 0$. Following [20], we define

$$
{}^{\mathcal{W}}\mathbf{N} = \begin{bmatrix} {}^{\mathcal{W}}\mathbf{n}_{\mathsf{B}}^{(k_1)} \dots {}^{\mathcal{W}}\mathbf{n}_{\mathsf{B}}^{(k_n)} \end{bmatrix}^{\mathsf{T}} \qquad {}^{\mathcal{W}}\mathbf{d} = \begin{bmatrix} {}^{\mathcal{W}}d_{\mathsf{B}}^{(k_1)} \dots {}^{\mathcal{W}}d_{\mathsf{B}}^{(k_n)} \end{bmatrix}^{\mathsf{T}}
$$

$$
{}^{\mathcal{D}}\mathbf{N} = \begin{bmatrix} {}^{\mathcal{D}}\mathbf{n}_{\mathsf{B}}^{(k_1)} \dots {}^{\mathcal{D}}\mathbf{n}_{\mathsf{B}}^{(k_n)} \end{bmatrix}^{\mathsf{T}} \qquad {}^{\mathcal{D}}\mathbf{d} = \begin{bmatrix} {}^{\mathcal{D}}d_{\mathsf{B}}^{(k_1)} \dots {}^{\mathcal{D}}d_{\mathsf{B}}^{(k_n)} \end{bmatrix}^{\mathsf{T}}
$$

and compute

$$\begin{matrix}^W\\_\mathcal{D}\end{matrix}\mathbf{T} = \begin{pmatrix} ^W_\mathcal{D}\mathbf{R} & ^W_\mathcal{D}\mathbf{t} \\ \mathbf{0}^\mathsf{T} & 1 \end{pmatrix}$$

as

$$\begin{matrix}^W\\_\mathcal{D}\end{matrix}\mathbf{t} = \left(^W\mathbf{N}^\mathsf{T} \cdot {^W}\mathbf{N}\right)^{-1} \cdot {^W}\mathbf{N}^\mathsf{T} \cdot \left(^W\mathbf{d} - {^\mathcal{D}}\mathbf{d}\right) \ ,$$

$$\begin{matrix}^W\\_\mathcal{D}\end{matrix}\mathbf{R} = \mathbf{V} \cdot \mathbf{U}^\mathsf{T} \ ,$$

where $\mathbf{U} \cdot \mathbf{S} \cdot \mathbf{V}^\mathsf{T}$ is the SVD decomposition of $^\mathcal{D}\mathbf{N}^\mathsf{T} \cdot {^W}\mathbf{N}$.

5.2 Optimization

For what concerns the error function, we calculate it as a sort of distance between the plane defined by the checkerboard $^W\pi_\mathsf{B}^{(k)}$ at step k, and the plane fitted to the checkerboard depth data of sensors D_j, $j = 1 \ldots M$, namely $^{\mathcal{D}_j}\pi_\mathsf{B}^{(k)}$. So, let $\mathsf{p}_\pi(\mathbf{x})$ be the *line-of-sight* projection of a point $\mathbf{x}$ onto plane π as described in [17], we can define the error function E_D as

$$E_\mathsf{D} = \sum_{k=1}^{K}\sum_{j=1}^{M} u_{jk} \cdot \frac{1}{\sigma_{\mathsf{D}_j}^2} \cdot e_{\mathsf{p}_{\mathcal{D}_j\pi_\mathsf{B}^{(k)}}}\left(^W_{\mathcal{D}_j}\mathbf{T}^{-1} \cdot {^W_\mathcal{B}}\mathbf{T}^{(k)}, {^\mathcal{B}}\mathbf{B}\right)$$

$$= \sum_{k=1}^{K}\sum_{j=1}^{M} u_{jk} \cdot \frac{1}{\sigma_{\mathsf{D}_j}^2} \cdot \sum_{(r,c)\in\mathbb{I}} \left\| {^{\mathcal{D}_j}}\mathbf{b}_{r,c}^{(k)} - \mathsf{p}_{\mathcal{D}_j\pi_\mathsf{B}^{(k)}}\left(^{\mathcal{D}_j}\mathbf{b}_{r,c}^{(k)}\right) \right\|^2 \ ,$$

where

$$^{\mathcal{D}_j}\mathbf{b}_{r,c}^{(k)} = {^W_{\mathcal{D}_j}}\mathbf{T}^{-1} \cdot {^W_\mathcal{B}}\mathbf{T}^{(k)} \cdot {^\mathcal{B}}\mathbf{b}_{r,c} \ ,$$

and u_{jk} is an indication function equal to 1 if sensor D_j sees the checkerboard at step k (otherwise it is 0) and the fraction $\frac{1}{\sigma_{\mathsf{D}_j}^2}$, as for cameras, is a normalization factor. But, while for cameras, σ_{D_j} is usually set to corner estimation error in pixels, for depth sensor it describes the error on the depth estimation. This means that each depth sensor may have a different normalization factor. Moreover, there are sensors, like Kinects v1, for which this error depends on the depth value, that is

$$\sigma_{\mathsf{D}_j}(z) = a + b \cdot z + c \cdot z^2 \ , \tag{9}$$

for some coefficients (a, b, c). Typical values[2] for a Kinect v1 are: $a = 0, b = 0, c = 0.0035$. Taking into account this fact, the error function E_D becomes

---

[2]http://wiki.ros.org/openni_kinect/kinect_accuracy.

$$E_\mathsf{D} = \sum_{k=1}^{K} \sum_{j=1}^{M} u_{jk} \cdot \sum_{(r,c) \in \mathbb{I}} \frac{1}{\sigma_{\mathsf{D}_j}^2 \left(^{\mathcal{D}_j} z_{r,c}^{(k)} \right)} \cdot \left\| ^{\mathcal{D}_j} \mathbf{b}_{r,c}^{(k)} - \mathbf{p}_{^{\mathcal{D}_j} \pi_\mathsf{B}^{(k)}} \left(^{\mathcal{D}_j} \mathbf{b}_{r,c}^{(k)} \right) \right\|^2 \qquad (10)$$

where $^{\mathcal{D}_j} z_{r,c}^{(k)}$ is the z component of $^{\mathcal{D}_j} \mathbf{b}_{r,c}^{(k)}$.

6 ROS Environment Configuration

6.1 Dependencies

The package depends on some external libraries, well integrated in ROS, such as:

- Boost
- OpenCV
- Eigen 3.2
- PCL 1.7

They are all available in the Ubuntu 14.04 repository and easily installable. For what concerns the optimization algorithm, instead, the package relies upon Ceres Solver [1] a library to solve non-linear least squares problems. In Ubuntu 14.04, it is possible to install version 1.8 of such library by selecting the apt package `libceres-dev`, however, due to some bugs on that library version, our package does not compile. To overcome this issue, we have prepared a script to download the latest tested version and install it. Just type

```
roscd calibration_toolkit/../scripts
./install_ceres.sh
```

on a terminal, and Ceres Solver as well as its dependencies will be installed on your system.

The algorithm assumes that all the sensors' intrinsic parameters are already estimated and expects that each camera publishes its own calibration parameters as a `sensor_msgs/CameraInfo` message. This is a common assumption for cameras, for which a lot of specialized tools (e.g. the `camera_calibration` ROS package) exist, while it is a bad assumption when dealing with depth sensors. In fact, for depth sensors, even if it has been demonstrated [5, 7, 8, 16, 18, 22] that depth measurements are not reliable, this operation is not really common. Unfortunately, up to this time, there is not an established way of calibrating such sensors, not even a ROS way to deal with their distortion. Users must therefore rely on external packages, like the ones presented in [5, 8] or [18].

Finally, it's worth noticing that the package is developed in C++11 and needs a compatible compiler to work. In particular we are currently compiling with gcc 4.8.

6.2 Basic Configuration

In order to allow communication between nodes in different computers, the environment variable ROS_MASTER_URI on every client PC must be set to the IP address of the PC where the master node is launched, namely the *master PC*. Additionally, the ROS_IP and ROS_PC_NAME environment variables must be set. This can be done temporarily by typing

```
export ROS_MASTER_URI=http://<MASTER_IP>:11311/
export ROS_IP=<MACHINE_IP>
export ROS_PC_NAME=<MACHINE_NAME>
```

on a terminal. Note that the PC names assigned can be whatever the user wants, not necessarily related to the real names of the PCs. To set them definitively, they can be added directly to the end of the .bashrc file in the home folder. As an example:

```
echo "export ROS_MASTER_URI=http://192.168.1.1:11311/
export ROS_IP=192.168.1.5
export ROS_PC_NAME=Phoenix" >> ~/.bashrc
```

7 Real-World Example

Suppose all the PCs are configured as explained in Sect. 6, and that we want to calibrate a network of two cameras connected to two different PCs that are called, respectively, Phoenix and Gemini. The multi-sensor calibration is performed by running a *master node* in a so called master PC, in our case Lyra, and a device driver in every PC attached to a device (one driver node for each device). First of all, we have to run the ROS drivers for all our devices. As an example, for a PointGrey camera, type on a terminal:

```
roslaunch pointgrey_camera_driver camera.launch
```

Then, we have to wrap these drivers in our calibration environment so that they can communicate with the master node in Lyra. To this aim, we run on both Phoenix and Gemini:

```
roslaunch multisensor_calibration camera_node.launch \
camera_name:=camera image_topic:=/camera/image \
camera_info_topic:=/camera/camera_info
```

where image_topic and camera_info_topic are the real topics on which the camera drivers publish their data. If everything is fine, we will see in our terminal:

```
[/Gemini/camera_node] All messages received.
[/Gemini/camera_node/get_device_info] Service started.
[/Gemini/camera_node/extract_checkerboard] Action
    server started.
```

At this point, Phoenix and Gemini are working as expected, we can therefore move to Lyra. Here we must let the master node know the network. We create a file called `network.yaml` in the `multisensor_calibration/conf` directory, open our favourite text editor, and fill `network.yaml` with the two PCs and the two cameras connected to them:

```
# Network configuration
network:
  - pc: "Phoenix"
    devices: ["camera"]
  - pc: "Gemini"
    devices: ["camera"]
```

Note that the strings in the `devices` arrays exactly match the value of the argument `camera_name` we set when launching `camera_node.launch`. We must then define the calibration pattern, i.e. the checkeboard, that we will use during the calibration procedure. So, let's create a file named `checkerboard.yaml` in the `multisensor_calibration/conf` directory and fill it with our checkerboard parameters:

```
# Checkerboard configuration
checkerboard:
  cols: 6
  rows: 5
  cell_width: 0.12
  cell_height: 0.12
```

We can now start the calibration procedure:

```
roslaunch multisensor_calibration master_node.launch
```

If all the nodes are launched correctly we'll have an output similar to the one below:

```
[/Lyra/master_node] Connected to [/Phoenix/camera_node/
    get_device_info] service.
[/Lyra/master_node] Connected to [/Gemini/camera_node/
    get_device_info] service.
[/Lyra/master_node] Connected to [/Phoenix/camera_node/
    extract_checkerboard] action server.
[/Lyra/master_node] Connected to [/Gemini/camera_node/
    extract_checkerboard] action server.
[/Lyra/master_node] Getting device infos...
[/Lyra/calibration] Sensor [/Phoenix/camera] added.
[/Lyra/calibration] Sensor [/Gemini/camera] added.
[/Lyra/master_node] Initialization complete.
```

We can then start the data acquisition phase. So, we take our checkerboard and move it around letting all the sensors see it. Every time we are in a good position we can publish an empty message on the topic `/Lyra/master_node/acquisition` to get one instance of the checkerboard (if visible) from each sensor:

```
rostopic pub /Lyra/master_node/acquisition std_msgs/
    Empty -1
```

Note that the, instead of publishing a message each time we need, we can let the publisher run at a fixed rate, substituting −1 with the desired rate −r <rate>. In this case, we must pay attention not to move the checkerboard too quickly, to avoid blur calibration errors due to the non perfect synchronization of the sensors.

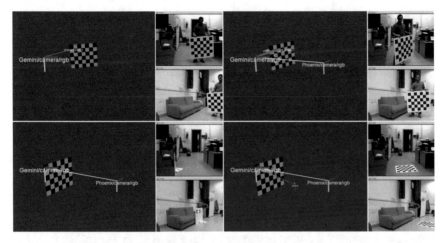

Fig. 4 Screenshots acquired during the calibration procedure. *Top-left* As soon as one sensor detects the checkerboard pattern, it becomes part of the tf tree and its pose is published. *Top-right* Then, every sensor that sees the checkerboard is added to the tree. *Bottom-left* As new detections arrive, the pose of the sensors are refined with the optimization algorithm, the checkerboard visualized is the last one. *Bottom-right* When the program is asked to estimate a plane, its pose is published and can be visualized using Rviz

We can monitor the whole calibration procedure via Rviz (Fig. 4).

It is sufficient to set the world frame as fixed frame, add the *tf* view and a *marker* view on topic /Lyra/master_node/markers and every time the checkerboard is detected or a sensor pose is estimated we will see it on the screen. In Fig. 4 some screenshots acquired during the calibration are shown. Before finishing the calibration, we lay the checkerboard on the floor and publish:

```
rostopic pub /Lyra/master_node/action std_msgs/String "
    begin plane" -1
```

From now on, the calibration algorithm assumes that all the checkerboards are lying on the same plane (see Sect. 4.3). We perform some acquisitions with the checkerboard lying on the floor and then publish an end plane instruction:

```
rostopic pub /Lyra/master_node/action std_msgs/String "
    end plane" -1
```

Finally, to get the results of the calibration, we use the service get_results offered by the master node. On a terminal we run:

```
rosservice call /Lyra/master_node/get_results
```

and get the estimated poses, similar to the ones below:

```
poses:
  - frame_id: '/world'
    child_frame_id: '/Gemini/camera'
    pose:
      position: {x: 0.0, y: 0.0, z: 0.0}
```

```
   orientation: {x:  0.0,  y:  0.0,  z:  0.0,  w:  1.0}
- frame_id: '/world'
   child_frame_id: '/Phoenix/camera'
   pose:
      position: {x:  1.12064,  y:  0.321081,  z:  0.565662}
      orientation: {x:  0.0471913,  y:  -0.377825,  z:
         -0.0340694,  w:  0.924046}
```

8 ROS Package

8.1 Architecture

The main purpose of the here-presented package is to allow the calibration of all the sensors (for now cameras and Kinect-like depth sensors) within a ROS network, no matter where they are. That is, suppose to have a set of sensors *distributed* in a PC network as in Fig. 1, the typical approach for the calibration procedure is to develop a *calibration node* that grabs the data generated by all the sensors, elaborate them and then estimate the sensors' rigid displacement in the scene (Fig. 5). This approach is clearly not scalable: more sensors means more bandwidth yet more computational power needed. To overcome such problem, we propose a different, distributed, architecture that allows us to both drastically reduce the bandwidth usage and distribute the computational cost over the network. In fact, we separate the data analysis from the calibration procedure. As depicted in Fig. 6, there are two sorts of calibration nodes: *device nodes* and the *master node*. A device node is responsible for elaborating the data provided by its device, it extracts the corners from every image, creates a message and sends it to the master node that executes the calibration procedure.

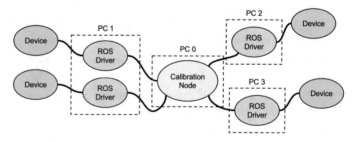

Fig. 5 Typical approach of network calibration algorithms. All the driver nodes are directly connected to a central node that performs the computation. The bandwidth usage is high

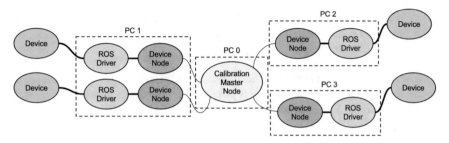

Fig. 6 Our approach to the network calibration procedure. The data provided by a device are elaborated by a node on the same PC and only the necessary calibration features are sent to the calibration node. Here, the thin lines that connect the master node with the device nodes mean that the quantity of data on the network is limited, with respect to the quantity of the typical approach (Fig. 5)

Remembering that ROS device drivers typically stream data at a defined frame-rate, with the proposed architecture we are able to work in a request-reply way: it is the master node that asks for new data when needed. In particular, since ROS service calls are blocking, the communication relies on the `actionlib` stack.

8.2 Device Node

8.2.1 Node Parameters.

A device node is responsible for getting the data from a device and, upon request, elaborate and send them to the master node. We first need to distinguish between the word *device* and *sensor*. We define a device as an item that can be connected to a PC, while a sensor, rather physical or virtual, is the item whose pose will be estimated in the calibration procedure. For example, a Kinect is a device composed by three different sensors: the RGB camera, the IR camera and a virtual depth sensor [7].

So, from a ROS perspective, a device node subscribes to the image and/or depth topics of the device sensors and keeps listening to an action topic until a request from the master node arrives. Then, the last images received are processed and the extracted calibration features are packed into a message and sent back to the master.

To explain how to configure and run a device node, we suppose we have to create a launch file for a Kinect v1. We first need to set the name of the device and its serial in case two or more Kinects are launched on the same PC:

```xml
<?xml version="1.0"?>
<launch>
  <arg name="device_name"   default="kinect1" />
  <arg name="device_serial" default="#1" />
```

Then, to avoid conflicts between nodes launched from different PCs, we group everything inside the namespace $ROS_PC_NAME:

```
<group ns="$(env ROS_PC_NAME)">
```

Now we include the launcher for the Kinect driver. The argument camera let us
define the namespace for all the topics published by the driver as well as the reference frames in the Kinect messages: here we add the _driver suffix just to avoid
confusion.

```
<include file="$(find openni_launch)/launch/openni.
    launch">
  <arg name="camera" value="$(arg device_name)
      _driver" />
  <arg name="device_id" value="$(arg device_serial)
      " />
  <arg name="publish_tf" value="false" />
</include>
```

The last node we need to add is our device node. The package multisensor_cal-
ibration already provides the node as a binary called device_node. So, we
rename it to match our current device, paying attention that, for communication
reasons, the node name needs the suffix _node.

```
<node pkg="multisensor_calibration" type="
    device_node"
        name="$(arg device_name)_node" output="screen
            ">
```

We then define the Kinect sensors that we want to calibrate. They have to be defined
within the ~device namespace, in particular:

name sets the device name, only for logging purposes;
sensors defines the list of sensors to calibrate, it is divided into:
>   intensity the sensors that will be treated as pinhole cameras;
>   depth the sensors that will be treated as depth sensors;
< sensor >] sets, for each sensor defined in sensors:
>   frame_id its unique frame id;
>   error the error polynomial defined in (9), where [depth sensors only:
>   min_degree/max_degree the minimum and maximum degree of the
>   polynomial (in the example below $\sigma(z) = c_0 \cdot z^0 + c_1 \cdot z^1 + c_2 \cdot z^2$);
>   coefficients the polynomial coefficients c_i;
transforms defines the known transforms between the sensors:
>   < sensor >] the child sensor;
>   parent the parent sensor, that is, the frame to which the transform is
>   defined;
>   translation the translation between the two sensors;
>   rotation the quaternion defining the rotation between the two sensors.

```
<rosparam param="device" subst_value="true">
  name: "$(arg device_name)"
  sensors:
    intensity: ["rgb"]
    depth: ["depth"]
  rgb:
    frame_id: "/$(env ROS_PC_NAME)/$(arg
        device_name)/rgb"
  depth:
    frame_id: "/$(env ROS_PC_NAME)/$(arg
        device_name)/depth"
    error:
      min_degree: 0
      max_degree: 2
      coefficients: [0.0, 0.0, 0.0035]
  transforms:
    depth:
      parent: "rgb"
      translation: {x: -0.025, y: 0.0, z: 0.0}
      rotation: {x: 0.0, y: 0.0, z: 0.0, w: 1.0}
</rosparam>
```

Finally, we have to connect the device node to the topics published by the driver. We use the *remapping* feature of ROS to set the sensor nodes listen to the right topics. Both for depth sensors and cameras, the default topics they listen to are of the form ~/device/<sensor>/<topic type>, where <topic type> is image for either images or depth images, camera_info for the camera calibration parameters and cloud for the point clouds:

```
<remap from="~device/rgb/image"
       to="$(arg device_name)_driver/rgb/
          image_color"/>
<remap from="~device/rgb/camera_info"
       to="$(arg device_name)_driver/rgb/
          camera_info"/>
<remap from="~device/depth/cloud"
       to="$(arg device_name)_driver/depth/points
          " />
<remap from="~device/depth/camera_info"
       to="$(arg device_name)_driver/depth/
          camera_info" />
    </node>
  </group>
</launch>
```

8.3 Master Node

8.3.1 Node Parameters.

Let's take a look to the launch file for the master node to see the parameters it needs to run. The launch file for the master node is simple:

```
<?xml version="1.0"?>
<launch>
  <arg name="network_file" default="$(find
     multisensor_calibration)/conf/network.yaml" />
  <arg name="checkerboard_file" default="$(find
     multisensor_calibration)/conf/checkerboard.yaml" /
     >
  <group ns="$(env ROS_PC_NAME)">
    <node pkg="multisensor_calibration" type="
       master_node"
         name="master_node" output="screen">
      <param name="network_file" value="$(arg
         network_file)" />
      <rosparam command="load" file="$(arg
         checkerboard_file)" />
    </node>
  </group>
</launch>
```

Firstly, we need to set the file containing the network description (`network_file` parameter) to let the master node know which are the sensors that are being calibrating. The network configuration is expected to be in a *yaml* file of the form:

```
# Network configuration
network:
  - pc: "<ROS_PC_NAME_1>"
    devices: ["<DEVICE_NAME_1>", "<DEVICE_NAME_2>"]
  - ...
  - pc: "<ROS_PC_NAME_N>"
    devices: ["<DEVICE_NAME_1>", ..., "<DEVICE_NAME_N>"
       ]
```

typically stored in the `multisensor_calibration/conf` directory of the master PC. Here, the `pc` parameters must match the names previously given to the PCs via the `export` command, while the strings in the `devices` array are the device names. They have to match the first part of a device node, that is, all but the suffix _node. Note that each device node is expected to be reachable in the network using the PC name as a namespace. As an example, the device node camera that runs in the PC named `Gemini`, is expected to be a node called /Gemini/camera_node.

The second parameter is `checkerboard_file`. It must be a *yaml* file and contain the checkerboard pattern specifications:

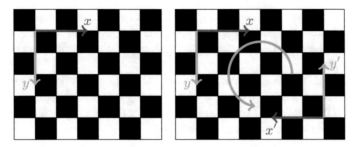

Fig. 7 The reference frame of a checkerboard is located internally with respect to one of the *black corner-cells* (if present), according to the checkerboard size, that is, with the *x*-axis along the columns and the *y*-axis along the rows. *Left* The reference frame of a 5 × 6 checkerboard has only one possible location. *Right* Here, due to a rotational symmetry, the reference frame of the 5 × 7 checkerboard can be positioned in two different locations, leading to pose estimation problems

```
# Checkerboard configuration
checkerboard:
  cols: <internal corners along the x dimension>
  rows: <internal corners along the y dimension>
  cell_width: <cell size along the x dimension in
     meters>
  cell_height: <cell size along the y dimension in
     meters>
```

Because of symmetry (see Fig. 7), it is mandatory that one of `cols` and `rows` is odd and the other even, no matter which. Otherwise two different sensors can assign to the same checkerboard two different reference frames, invalidating the calibration procedure. In fact, according to our experience, the OpenCV corner detector starts enumerating the corners from one of the black corner-cells (if present), in row-major order. We rely on this order to set the reference frame of the checkerboard: the *x*-axis along the columns and the *y*-axis along the rows.

8.3.2 Services and Messages.

As already shown in Sect. 7, it is possible to interact with the master node with messages and services. The most important topic the node is listening on is ~acquisition. Publishing a message on such topic would result in a data acquisition. Note that, the message queue on the node is set to 1, so even if the node receives multiple requests while still elaborating previous data, only the last one will be taken into account. Note also that `rostopic pub` permits to publish messages at a defined rate with the flag `-r <rate>`. Users must pay attention to use such feature since the sensors may be not perfectly synchronized. In fact, if the data are acquired while the pattern is still moving, the resulting calibration might be wrong.

The node is also listening to an action topic named ~action. It is used to send special calibration commands in form of strings. For now the only two commands accepted are `begin plane` and `end plane`. The former command tells the calibration algorithm that, from that instant on, the checkerboards are all lying on the same plane (see Sect. 4.3). The latter, instead, makes the algorithm go back to its standard behavior.

Finally, to get the results of the calibration, the master node offers a service called `~get_results`. It can be invoked with an empty request and returns a vector of messages of type `calibration_msgs/ObjectPose`. In the future we envision to improve this service with some options like:

- ask for the poses of checkerboards and/or planes;
- set a fixed transform between the world and a sensor in the request, the response will be filled with the poses transformed according to it.

9 Conclusions

We have presented a ROS package that lets users calibrate networks of sensors composed by cameras and Kinect-like depth sensors. A checkerboard pattern is used to estimate the relative poses between sensors and everything is optimized with a state-of-the-art non-linear least squares solver [1]. The choice of using ROS as the developing framework gives our implementation lots of advantages with respect to, for example, Matlab-based toolboxes [21]. The main advantage is the way the sensors and their synchronization are managed. On the other side, one of the main drawbacks of the presented approach is that it highly depends on the intrinsic parameters provided. If they are not well estimated, the calibration results will not be accurate. In the future we can think of adding the possibility to refine also the intrinsic parameters of cameras while estimating their pose.

From the implementation point of view, we think that in the future, it will be really useful to provide a tool to assess the quality of the obtained calibration, in order to let users verify if the obtained results are reliable or not. Another important improvement will be at code level. In our idea, the package will be part of a *calibration ecosystem* for ROS, where contributors can implement their own calibration algorithms or contribute to the existing ones by adding, for example, new sensors. Hence, it will be necessary to define some standard interfaces between nodes and classes to lower the entry barrier for new developers and let the ecosystem grow.

Acknowledgments The authors would like to thank Prof. Mohamed Chetouani, Salvatore Maria Anzalone and Stéphane Michelet from Université Pierre-et-Marie-Curie (UPMC) and the Institut des Systèmes Intelligents et de Robotique (ISIR) for their support and help.

The authors would also like to thank Jeff Burke, Alexander Horn and Randy Illum from University of California, Los Angeles (UCLA) for the extensive collaboration in designing and testing the calibration methods during the development of OpenPTrack [15]. OpenPTrack has been sponsored by UCLA REMAP and Open Perception. Key collaborators include the University of Padova and Electroland. Portions of the work have been supported by the National Science Foundation (IIS-1323767).

References

1. S. Agarwal, K. Mierle et al., Ceres solver, http://ceres-solver.org
2. E. Auvinet, J. Meunier, F. Multon, Multiple depth cameras calibration and body volume reconstruction for gait analysis, in *11th International Conference on Information Science, Signal Processing and their Applications (ISSPA)*, pp. 478–483, July 2012
3. F. Basso, R. Levorato, E. Menegatti, Online calibration for networks of cameras and depth sensors, in *Proceedings of the 12th Workshop on Non-classical Cameras, Camera Networks and Omnidirectional Vision (OMNIVIS)*, Hong Kong, China, June 2014
4. F. Basso, M. Munaro, S. Michieletto, E. Menegatti, Fast and robust multi-people tracking from RGB-D data for a mobile robot, in *Proceedings of the 12th Intelligent Autonomous Systems (IAS) Conference.* vol. 193, Jeju Island, Korea, pp. 265–276, June 2012
5. F. Basso, A. Pretto, E. Menegatti, Unsupervised intrinsic and extrinsic calibration of a camera-depth sensor couple, *IEEE International Conference on Robotics and Automation (ICRA)*, Hong Kong, China, pp. 6244–6249, June 2014
6. G. Bradski, The OpenCV library. Dr. Dobb's Journal of Software Tools (2000)
7. A. Canessa, M. Chessa, A. Gibaldi, S.P. Sabatini, F. Solari, Calibrated depth and color cameras for accurate 3D interaction in a stereoscopic augmented reality environment. J. Vis. Commun. Image Representation **25**(1), 227–237 (2014)
8. M. Di Cicco, L. Iocchi, Grisetti, G.: Non-parametric calibration for depth sensors, in *Proceedings of the 13th International Conference on Intelligent Autonomous Systems, (IAS-13)*, Padova, Italy (2014)
9. M.A. Fischler, R.C. Bolles, Random sample consensus: a paradigm for model fitting with applications to image analysis and automated cartography. Commun. ACM **24**(6), 381–395 (1981)
10. P. Furgale, J. Rehder, R. Siegwart, Unified temporal and spatial calibration for multi-sensor systems, in *IEEE/RSJ International Conference on Intelligent Robots and Systems (IROS)*, pp. 1280–1286, Nov 2013
11. G. Guennebaud, B. Jacob et al., Eigen (2010), http://eigen.tuxfamily.org
12. Q. Le, A. Ng, Joint calibration of multiple sensors. *IEEE/RSJ International Conference on Intelligent Robots and Systems (IROS)*, pp. 3651–3658, Oct 2009
13. R. Levorato, E. Pagello, Probabilistic 2D acoustic source localization using direction of arrivals in robot sensor networks. In: D. Brugali, J.F. Broenink, T. Kroeger, B.A. MacDonald (eds.) *Simulation, Modeling, and Programming for Autonomous Robots.* Lecture Notes in Computer Science, vol. 8810, (Springer International Publishing, 2014), pp. 474–485
14. M. Munaro, F. Basso, E. Menegatti, Tracking people within groups with RGB-D data, in *Proceedings of the International Conference on Intelligent Robots and Systems (IROS)*, Vilamoura, Portugal, pp. 2101–2107, Oct 2012
15. M. Munaro, A. Horn, R. Illum, J. Burke, R.B. Rusu, OpenPTrack: People tracking for heterogeneous networks of color-depth cameras, in *IAS-13 Workshop Proceedings: 1st Intl Workshop on 3D Robot Perception with Point Cloud Library*, Padova, Italy, pp. 235–247, July 2014
16. J. Smisek, M. Jancosek, T. Pajdla, 3D with kinect, in *IEEE International Conference on Computer Vision Workshops (ICCV Workshops), ICCVW 2011*, pp. 1154–1160 (2011)
17. E. So, F. Basso, E. Menegatti, Calibration of a rotating 2D laser range finder using point-plane constraints. J. Autom. Mob. Robot. Intell Syst. **7**(2), 30–38 (2013)
18. A. Teichman, S. Miller, S. Thrun, Unsupervised intrinsic calibration of depth sensors via SLAM, in *Proceedings of Robotics: Science and Systems*, Berlin, Germany, June 2013
19. B. Triggs, P.F. McLauchlan, R.I. Hartley, A.W. Fitzgibbon, Bundle adjustment-a modern synthesis, in *Vision Algorithms: Theory and Practice*, vol. 1883, Lecture Notes in Computer Science, ed. by B. Triggs, A. Zisserman, R. Szeliski (Springer, Berlin Heidelberg, 2000), pp. 298–372
20. R. Unnikrishnan, M. Hebert, *Fast extrinsic calibration of a laser rangefinder to a camera*, Technical report, Carnegie Mellon University (2005)

21. M. Warren, D. McKinnon, B. Upcroft, Online calibration of stereo rigs for long-term autonomy, in *IEEE International Conference on Robotics and Automation (ICRA)*, Karlsruhe, Germany (2013)
22. C. Zhang, Z. Zhang, Calibration between depth and color sensors for commodity depth cameras, in *IEEE International Conference on Multimedia and Expo (ICME)*, pp. 1–6 (2011)

Acoustic Source Localization
for Robotics Networks

Riccardo Levorato and Enrico Pagello

Abstract This chapter presents a technical research contribution in the audio for robotics field where ROS was used for validating the results. More specifically it deals with 2D Audio Localization using only the Directions of Arrival (DOAs) of a fixed acoustic source coming from an audio sensor network and proposes a method for estimating the position of the acoustic source using a Gaussian Probability over DOA approach (GP-DOA). This method was thought for robotics purposes and introduces a new perspective of the audio-video synergy using video sensor localization in the environment for extrinsic audio sensor calibration. Test results using Microsoft Kinects as DOA-sensors mounted on robots within the ROS framework, show that the algorithm is robust and modular and prove that the approach can be easily used for robotics applications. The second part is dedicated to the detailed description of the implemented ROS package.

Keywords Acoustic Source Localization (ASL) · Direction Of Arrival (DOA) · Audio-sensor network · Robot audition · Microsoft kinect · Robot Operating System (ROS)

1 Introduction

Audio for robotics is a field that is not well explored yet, as vision for robotics is. Probably it happens because acoustic skills are considered less important than visual ones. It is a matter of fact that the visible light frequency range is very important for living beings (e.g. for navigation purposes or object recognition) but it cannot cover some particular events which can be detected analysing only the acoustic signals

R. Levorato (✉) · E. Pagello
Department of Information Engineering (DEI), IAS-Lab, University of Padova,
via Ognissanti 72, 35131 Padova, Italy
e-mail: riccardo.levorato@dei.unipd.it
URL: http://robotics.dei.unipd.it

E. Pagello
e-mail: enrico.pagello@dei.unipd.it

© Springer International Publishing Switzerland 2016
A. Koubaa (ed.), *Robot Operating System (ROS)*, Studies in Computational
Intelligence 625, DOI 10.1007/978-3-319-26054-9_17

(e.g. horn machines, exploding bombs or barking dogs). Considering only audio skills, we can retrieve also other important information about the environment (e.g. classification of sounds, room shape construction using echoes and reflections, direction of arrival (DOA) of the sound source); but, on the other hand, this information is not always enough to localize properly the acoustic source position in the environment. For example, the estimation of the distance of an acoustic source requires further skills that take into consideration also sound reflections and echoes (i.e. bats echolocation) [4]. Hence, there is the need for fusing data coming from different sensors in order to have more information of the surroundings. In our robotics field we imagine that a robot should find itself in a map (SLAM), hear some sounds and fuse these data to find what and where the acoustic source is. Moreover, if we deal with connected robots, we can in addition share the data of each robot with the others in order to have a better understanding of the environment by the robots within the network. To avoid bottleneck problems sharing audio-video data among robots, we propose a method that shares only the pose of the robot and the DOAs of the acoustic sources heard. So, from the considerations above and in conjunction with the spreading use of ROS, we propose a new way of thinking the audio for robotics: to use video information for robot audio localization.

2   State of the Art

The current state of art offers lots of works that refer to Acoustic Source Localization (ASL) (i.e. for a detailed and complete overview look up the Ph.D. Thesis of Pertilä [10] and Salvati [12]). The majority of these works focuses on various techniques that use the audio signals coming from all the microphones and estimate the position of the sources fusing all these data. In a robotic environment, this approach cannot be applied because collecting all the audio signals in a unique master node for the calculations can cause a network block; furthermore, synchronizing all audio signals in time [14] could be very challenging.

In the acoustic field, a DOA sensor consists specifically of an array of at least two microphones. Common tested acoustic sources are human speakers [9], gun shots and human screams [13], clapping hands and so on. There are various techniques for calculating the DOA of an acoustic source such as Angle of Arrival (AoA), Time Difference of Arrival (TDOA), Frequency Difference of Arrival (FDOA), and other similar techniques [10, 12].

Recently there was an increasing use of sensors networks, that permit to share audio and video data in a cooperative way, achieving more precise knowledge of the environment. In an Audio Sensor Network (ASN), by knowing the position and pose of each DOA sensor, it is possible to better estimate the position of the acoustic source by sharing and synchronizing the DOA estimations [1, 3, 9]. Following this idea, Hawkes et al. [5] proposed an analytic Weighted Least Square method (WLS-DOA) that minimizes the distance of the estimated point to the estimated DOA line and achieves good results in 3D space. In his Ph.D. thesis, Pertilä [10] also focused on

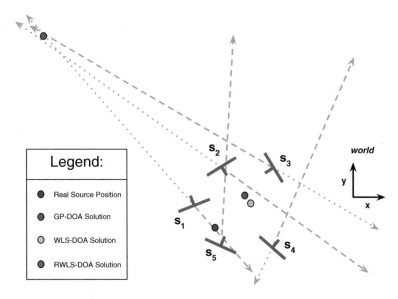

Fig. 1 Example of failure of the Robust DOA-Based Localization proposed in [10]

2D DOA-Based Localization and introduced a new approach that tried to eliminate the sensors with bad acoustic DOA estimations starting from the WLS-DOA method (Robust WLS-DOA (RWLS-DOA)) [5].

Although it was shown that this method performed better than WLS-DOA in outlier situations, we found that it is error prone, because in some situations it can discard the DOA sensors that are better than others. A proof of the failure of this method can be easily given if we consider the example in Fig. 1: we can see that the Robust WLS-DOA solution (in magenta) is very far from the real one (in blue). On the other hand the WLS-DOA solution (in green) and our GP-DOA approach (in red) solutions are closer to the real solution. This is due to the fact that the DOA estimations of sensors s_1, s_2 and s_3 intersect very close to each other (near the R-DOA solution), not considering that only sensors s_1 and s_5 are the only good estimations and the others (s_2, s_3 and s_4) are outliers. So, it is not advisable to discard any of the sensors because the real position of the source is unknown and it is not possible to detect which DOA sensors are outliers if they have the same probability error to be outliers. In addition, the simple WLS-DOA approach can also lead to errors because a WLS-DOA estimation is considered as a line without a precise direction. As an example, in Fig. 1, even if sensor s_2 DOA estimation is pointing towards south-east, the WLS-DOA approach considers also the north-west direction (RWLS-DOA fails also for this reason).

In our approach, proposed for the first time in [7], we use all the DOA sensors and propose a new way of thinking DOAs that is based on the probability error over the angle of each DOA estimation. The difference with all the WLS-DOA approaches is that our method minimizes the angle (instead of the distance) of the estimated point

Fig. 2 Acoustic Source
Localization Problem in an
indoor environment. In the
example there are a fixed
Microsoft Kinect, a Pioneer
3-AT equipped with a
Microsoft Kinect and a NAO
robot localizing the position
of an alarm clock

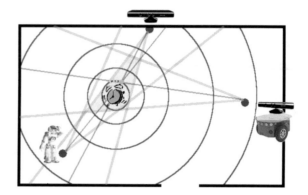

to the estimated DOA. This new approach is not affected by the angle ambiguity
problems of the WLS-DOA approaches. A further contribute is the new concept
of Audio-Video data fusion. More specifically, Audio Localization will be strictly
correlated to Video Calibration: working with mobile robots it would be very hard
to calibrate the relative position of all microphones of all moving robots with noninvasive techniques (i.e. using an acoustic high-frequency periodic signal as audio
reference) at each time instant. For this reason, using SLAM techniques and knowing
the relative position of the microphones with respect to the visual sensors, will help
robots to share also audio information.

In a Mobile Robot Sensor Network (Fig. 2), to avoid bottleneck problems sharing audio-video data among robots, we proposed in [7] a method that shares only
the pose of the robot and the DOAs of the acoustic sources heard introducing the
GP-DOA approach that uses a probabilistic model of the sensor for finding the
acoustic source position and outperforms the WLS-DOA method. Starting from the
GP-DOA approach, we analysed in [6] also a faster algorithm for calculating the maximum probability point using an adapted binary search, achieving the same result in
$\Theta(\log^2 n)$ time instead of $\Theta(n^2)$.

The chapter is structured as follows: Sect. 3 will summarize the DOA-Based Localization Problem and the Sect. 4 will explain the GP-DOA approach. Sect. 5 will show
the proposed fast algorithm both theoretically and practically. The two Sections of
Simulation and Results will follow, ending with a detailed explanation of the ROS
package we implemented and the final conclusions.

3 DOA-Based Localization Problem

This problem is well described in [5] and [10]. Let S be a set of DOA sensors, with
$|S| = N_s \geq 1$. In the audio field, A DOA sensor $s \in S$ consists in a microphone array
able to compute and estimate the DOA of a generic acoustic source r with respect
to the intrinsic reference system of s. Let $s_k^p = [s_k^x, s_k^y]^T \in \mathbb{R}^2$ and $s_k^o \in [-\pi, \pi]$ represent respectively the two-dimensional position and orientation of the kth sensor,

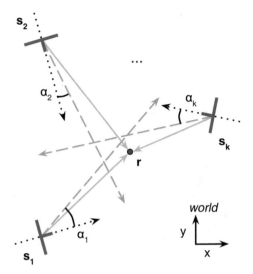
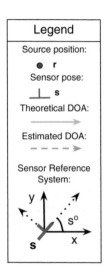

Fig. 3 DOA—Based Localization Problem

$k \in [1, N_s]$, with respect to the Cartesian plane reference system *world*. A DOA estimation of s_k is the angle $\alpha_k \in [-\pi, \pi]$. The problem consists in finding the closest estimation of the acoustic source position $\mathbf{r} = [r^x, r^y]^T \in \mathbb{R}^2$ assuming that all s_k^p and s_k^o are known and all α_k are estimated using a DOA estimation algorithm (Fig. 3).

The problem is very complex and has a lot of variables. From now on, in order deal with a simplified version of the problem, the following assumptions are considered:

1. all sensors are connected together within a network and can share their pose and their DOA estimations that can both be error-prone;
2. all DOA estimations are synchronized;
3. there is only one fixed (not moving) acoustic source at time;
4. the precision of all α_k estimations depends only on the accuracy of each DOA sensor and on the DOA estimation methods used;
5. echoes and sound reflection effects are considered to be already managed by the DOA estimation algorithm.

Assumptions 1 and 2 are provided by the ROS framework by default. Assumption 3 limits the problem to the single source localization problem and points 4 and 5 leave the precision problems to the DOA estimation methods of the sensors.

The problem solution given by Hawkes et al. [5] is calculated with the following one-shot formula:

$$\hat{\mathbf{r}} = \left[\left(\sum_{k=1}^{N_s} w_k \right) \mathbf{I} - \mathbf{A}\mathbf{W}\mathbf{A}^T \right]^{-1} \mathbf{B}\mathbf{w}, \qquad (1)$$

where $\mathbf{w} = [w_1, w_2, \ldots, w_{N_s}]^T$ are the weights, $\mathbf{W} = diag(\mathbf{w})$, $\mathbf{I}$ is the identity matrix, $\mathbf{A} = [\alpha_1, \alpha_2, \ldots, \alpha_{N_s}]^T$ and
$$\mathbf{B} = \left[(\mathbf{I} - \alpha_1 \alpha_1^T) \mathbf{s}_1, (\mathbf{I} - \alpha_2 \alpha_2^T) \mathbf{s}_2, \ldots, (\mathbf{I} - \alpha_{N_s} \alpha_{N_s}^T) \mathbf{s}_{N_s} \right].$$

4   Gaussian Probability over DOA Approach

The Gaussian Probability over DOA approach was first introduced in [7]. A brief explanation of the main concept is in the following. Each real DOA estimation α_k has an intrinsic error that depends mainly on the accuracy of the kth sensor. This error can be modelled as a Gaussian probability error in the angle domain with zero mean and variance σ_k using only values in the range $\phi \in [-\pi, \pi]$. So the angular probability sensor model $\mathcal{M}_k$ is defined as follows:

$$\mathcal{M}_k \sim \mathcal{N}(0, \sigma_k)_{(-\pi, \pi]} = \frac{1}{\sigma_k \sqrt{2\pi}} e^{-\frac{\phi^2}{2\sigma_k^2}} \quad , \quad \phi \in [-\pi, \pi] \tag{2}$$

At this step it is needed a change of domain from the angular domain to the Cartesian coordinate system $G = (n \times n) \in \mathbb{Z}^2$. G is as a spatial 2D grid with a fixed spatial $range$ (m) and a fixed precision parameter $prec$ (m) such that $n = range/prec$. For each generic point $\mathbf{q} = [q^x, q^y]^T \in G$ it is calculated the angle $\beta_k^{\mathbf{q}}$ that considers $\mathbf{s}_k^p$ as vertex and it is included between a first line that passes through $\mathbf{s}_k^p$ and $\mathbf{q}$ and a second line given by the axis s_k^y (Fig. 4):

$$\beta_k^{\mathbf{q}} = atan2(q^y - s_k^y, q^x - s_k^x) - s_k^o \quad , \quad \mathbf{q} \in G \wedge k \in [1, N_s] \tag{3}$$

The probability $\mathcal{G}_k(\mathbf{q})$ for each point $\mathbf{q}$ in the grid G is given by evaluating each angle $\beta_k^{\mathbf{q}}$ in the angular probability sensor model $\mathcal{M}_k$ with respect to the DOA estimation α_k:

$$\mathcal{G}_k(\mathbf{q}) = \mathcal{M}_k(\beta_k^{\mathbf{q}} - \alpha_k) \quad , \quad \mathbf{q} \in G \wedge k \in [1, N_s] \tag{4}$$

Fig. 4 Representation of all considered angles in a DOA sensor

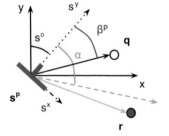

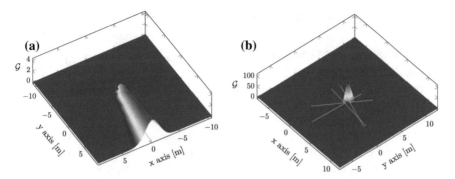

Fig. 5 Probability $\mathcal{G}$ over G with one and four DOA sensors **a** $\mathcal{G}_k$: Probability sensor model $\mathcal{M}_k$ of a DOA sensor in the 2D space. **b** Probability $\mathcal{G}$ over G with four DOA sensors

All results of operations among angles in Eqs. (3) and (4) take a value in the range $(-\pi, \pi)$. The graphical representation of a single DOA sensor estimation probability over the Cartesian plane can be seen in Fig. 5a where red and blue regions have higher and lower likelihoods respectively.

At this point all the $\mathcal{G}_k(\mathbf{q})$ are multiplied point-wise obtaining the multiplication of all the probabilities of all sensors $\mathcal{G}(\mathbf{q})$ in the 2D space domain (Fig. 5b):

$$\mathcal{G}(\mathbf{q}) = \prod_{k=1}^{Ns} \mathcal{G}_k(\mathbf{q}) \ , \quad \mathbf{q} \in G \tag{5}$$

Finally, the point $\hat{\mathbf{r}} \in G$ with the maximum value of $\mathcal{G}$ is considered the estimation of the solution with the proposed Gaussian Probability over DOA approach:

$$\hat{\mathbf{r}} = \underset{G}{\mathrm{argmax}} \ \mathcal{G} \tag{6}$$

In Eq. (5) it is used the product instead of the sum for the fact that if the likelihoods from different DOA sensors are independent, the intersection of sets equals their product, as stated in [10]. It is worth noting that even though the probability axiom $P(\Omega) = 1$ is no longer satisfied in Eq. (2), the omission of this axiom will not compromise the correctness of the procedure. Dealing with multiplication (and not with sums) of probabilities in Eq. (5), all unused values can be omitted because the multiplication of numbers $\in [0, 1]$ still takes a value $\in [0, 1]$. On the other hand it is important to set an *a priori* experimental validated variances σ_k in order to represent correctly each angular probability sensor model $\mathcal{M}_k$.

5 Algorithm Analysis and Optimization

The algorithm showed in Sect. 4 has clearly a computational complexity of $\Theta(n^2)$ because it has to calculate all the probabilities in $G = (n \times n) \in \mathbb{Z}^2$ in order to find the maximum value in a two-dimensional matrix. In the following we show a smarter and faster approach proposed in [6] for searching for the maximum value in G which takes into consideration the shape of $\mathcal{G}$.

Observing the shapes of the probability density function $\mathcal{G}$ (e.g. Fig. 5b) it seems that $\mathcal{G}$ is a Gaussian probability function. Unfortunately, each $\mathcal{G}_k$ is no longer a Gaussian like $\mathcal{M}_k$ although $\mathcal{G}_k(\mathbf{q}) = \mathcal{M}_k(\beta_k^{\mathbf{q}} - \alpha_k)$. The reason is due to the formula for calculating $\beta_k^{\mathbf{q}}$ that translates the values from the Cartesian domain to the angle domain. This change of domain causes the loss of the property for $\mathcal{G}_k$ of being a Gaussian probability function, preventing from using the property that states that multiplication of Gaussians is still a Gaussian [11]. On the other hand, in the majority of problems, it can be noted that $\mathcal{G}_k$ is unimodal, hence it has only one absolute maximum:

Definition 1 A function $f(x)$ is an **unimodal** function if for some value m, it is monotonically increasing for $x \leq m$ and monotonically decreasing for $x \geq m$. In that case, the maximum value of $f(x)$ is $f(m)$ and there are no other local maxima.

This property occurs when the DOA estimations are quite accurate. But if we are not under this specific condition it can happen that the maximum is no longer absolute. Fortunately, tests with increasing error over the DOA estimations will show that this unpleasant event happens very rarely and in most of that cases the solution provided by the new search algorithm is still good (see Sect. 6.1.2). Before explaining the details of this algorithm we suggest a funny riddle:

The Top of the Unimodal Mountain

Suppose you are a special climber that wants to reach the top of an unimodal mountain. You always know which is your geographical position and altitude but you don't know where is the top because there is too much fog. Fortunately you can move safely to any geographical position you want with your special intelligent and autonomous jet-pack. Which is the method that lets you reach the top of the unimodal mountain with the smallest number of use of your special jet-pack?

The solution of this riddle is not very immediate so let's first concentrate on a single dimension of a mountain n meters long. If we start from the foot of the mountain (e.g. from the left) and "walk" towards the climb (on the right), we could find the top in $\Theta(n)$ time. In a smarter way we can take advantage of the unimodality of the mountain and of our special jet-pack by applying a special Unimodal Binary Search (UBS) that takes into consideration also the slope of the mountain: for each point we explore using the binary search, we will explore also the adjacent point (i.e. the one on the left) and measure their respective altitudes in order to decide in which direction to jump, in a binary search way. This method will reasonably take $\Theta(\log_2 n)$ time.

Fig. 6 Example of a
solution of the *Top* of the
Unimodal Mountain riddle
using the Unimodal Binary
Search (UBS) approach

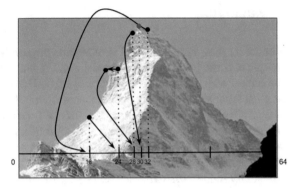

Before going on, let's see now a complete simple example in one dimension: the unimodal mountain in Fig. 6 is long $n = 64 = 2^{log_2 n}$ where $l = log_2 n = 6$ is the number of levels. Level $l = 6$: first we check the altitude of position 32 (2^{l-1}) and its adjacent on the left 31 and we see that the top should be on the left. Level $l = 5$: now we jump to the left to the position 16 $(32 - 2^{l-1})$ and check also position 15. We see that the altitude in 16 is higher than the one in 15 hence we jump right to position 24 $(16 + 8)$ (Level $l = 4$). When we check position 23 we see that it has the same altitude than position 24. We now start walking on the left till we find a different altitude. Since in position 20 we find a lower altitude, we jump right to position 28 $(24 + 4)$ (Level $l = 3$). Note that we can "walk" on the left only for 2^{l-1} position because the other left position are already been checked. At Level $l = 2$ we see that we have to jump to position 30 $(28 + 2)$ where we have to check positions 29 and 31 because we are at the end of the algorithm. Finally the above positions give minor altitudes so we can state that the top of the mountain is in position 30.

For a two-dimensional unimodal mountain we can use a two-level Unimodal Binary Search: the first level is a UBS over the rows; for each visited row we apply another UBS (second level) that finds the position of the maximum of that row in a UBS way. With this approach we reach the top of the mountain in $\Theta((\log_2 n)^2) = \Theta(\log_2^2 n)$ time instead of the first naive method that took $\Theta(n^2)$: a very big amount of time saved! Working with probabilities and floating-point numbers it can happen that two adjacent points have the same probability. In this case the above algorithm cannot decide which way to go. A solution of this problem is, remembering the mountain, to "walk" always in the same direction until we find a different altitude. So if we have plain areas that are all $m < n$ long the overall complexity of the algorithm is $\Theta((m + \log_2 n)^2) = \Theta(m^2 + \log_2^2 n + m \cdot \log_2 n)$. Since we found that in the experimental tests $m \ll n$, the computational time is still $\Theta(\log_2^2 n)$. Furthermore, if we want to deal with 3D sound localization, this approach could be even more advantageous because it would give a solution in $\Theta(\log_2^3 n)$ time instead of $\Theta(n^3)$.

6 Simulation

Simulation was an helpful tool to compare the localization performances of our approach to the traditional Weighted Least Square method (WLS-DOA) [5]. The simulation scenarios consisted in groups of N_s DOA sensors in a virtual room G. At each iteration, all sensors positions and orientations were positioned randomly in G. All tests had only one acoustic source at time positioned randomly in G. For all simulations it has been assumed that all sensors were similar, so they had the same probability error over the DOA estimation and hence same σ_{k_sim}. So each DOA estimation was modified from the real one, for simulation purposes, using the probability model $\mathcal{M}_k$ with the fixed σ_{k_sim} for all k. The simulation was repeated t_{sim} times for each group of sensors. The cardinality of the sensor group was varied in the range $[3, N_{s\_max}]$. The case with $Ns = 2$ was not considered for the reason that the solution is the trivial intersection of the only two existing DOAs. Simulation code was implemented in MATLAB. The simulations done were four and are listed here in chronological order, each encouraging the further inspection of the next one:

1. Simulation for testing the accuracy and precision of the basic proposed algorithm;
2. Simulation for testing the accuracy and precision of the faster proposed algorithm;
3. Simulation for testing the accuracy and precision of the faster proposed algorithm augmenting only the error over the DOA estimations;
4. Simulation for testing the accuracy and precision of the faster proposed algorithm augmenting only the error over the position of the DOA sensors.

The localization performance metric used is the Distance Error (DE) that is two-dimensional distance between the estimated source and the real source positions in meters. The validation metrics calculated were the mean and variance values of all DEs.

For each simulation type we used these fixed parameters: $range = 10$ (m), $\sigma_{k\_sim} = 0.1$, $N_{s\_max} = 20$, $t_{sim} = 10,000$. Conversely, Table 1 summarizes all the choices of the parameters for each simulation. $[x : y : z]$ means that the parameter is varied from x (included) to z (included) with a step of y.

Table 1 Simulation parameters

# Simulation	1	2	3	4
GP-DOA basic *prec* (m)	0.001	0.01		
GP-DOA Fast 1 *prec* (m)		0.01		
GP-DOA Fast 2 *prec* (m)		0.0001	0.0001	0.0001
e_{α^k} (degrees)	5	5	[5:5:50]	5
e_{s^k} (m)	0	0	0	[0:0.1:1]

6.1 Simulation Results

1. **GP-DOA versus WLS-DOA**

   In Fig. 9a we see that GP-DOA performs better in mean than WLS-DOA approach and the error diminishes as the number of sensors grows in the environment. On the other hand, the WLS-DOA variance performs better than GP-DOA variance (Fig. 9b) that means that GP-DOA has an higher accuracy but lower precision with respect to WLS-DOA.

2. **Slow versus Fast**

   In Fig. 9c, d we can see the results of the mean and the variance of the first simulation test. Looking first at the mean in Fig. 9c, we can see that both GP-DOA Fast approaches are better than the WLS-DOA method. Although having the same precision of the grid of the GP-DOA Basic, GP-DOA Fast 1 is less precise and less accurate, as we can see also looking at the variance of Fig. 9d. Augmenting the precision of the grid to 0.0001 m it allows to reach practically the same results obtained with the GP-DOA Basic method both in mean and in variance with much less time. The difference of the computational time taken by both algorithms can be computed as follows:

   $$n^2_{GP-DOA\_Basic} = (range/prec_{GP-DOA\_Basic})^2 = 10^6;$$

   $$n^2_{GP-DOA\_Fast\_2} = range/prec^2_{GP-DOA\_Fast\_2} \approx 276.$$

3. **DOA Angle Error**

   Comparing the results of the approaches with different maximum errors of the angle of the DOA sensors reveals that GP-DOA Fast 2 always outperforms the WLS-DOA[1] (Fig. 9e). Moreover the source estimation error seems to be linearly dependent on the angle error. On the other hand, the number of the DOA sensors maintains its inverse proportionality with respect of the source estimation error.

4. **DOA Sensor Position Error**

   The results varying the accuracy of the DOA sensors reported in Fig. 9f show that GP-DOA Basic and GP-DOA Fast 2 behave similarly both in mean and in variance[1]. Comparing these data with the WLS-DOA approach it can be easily noted that WLS-DOA starts to outperform the GP-DOA approaches when the estimation of the position of the sensors in the map has a maximum error of 0.4–0.5 (m). This threshold can be used in a real environment in order to select the approach to be used for ASL purposes. Since we are dealing with rooms with a *range* = 10 (m), a robot position error of 0.5 (m) can be reasonably considered as borderline for actual SLAM algorithms.

---

[1]The variance chart is omitted because resembles the one of the mean.

(a)   **(b)**

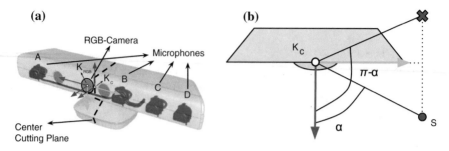

Fig. 7 Simulation results of GP-DOA versus WLS-DOA **a** Kinect sensors: Microphones (A-D) and RGB-camera positions in 3D. **b** Example of limitation of the Kinect DOA sensor. S is the right available source position, the red crossed position is the wrong available position

Table 2 Kinect sensors—Microphones (A–D) and RGB-camera—positions over x axis

Mic A (m)	RGB sensor (m)	Center (m)	Mic B (m)	Mic C (m)	Mic D (m)
−0.1150	−0.0140	0	0.0350	0.0750	0.1150

7 Real Tests

We tested the algorithms in a small real environment with two experiments. We always used three Microsoft Kinect[2] as DOA-sensors. Microsoft Kinect has one RGB-camera, a 3D depth sensor and four microphones positioned as shown in Fig. 7.[3] Each DOA estimation came from *HARK* software[4] implemented in a ROS package with an error of ± 0.0873 (rad) = 5° [8]. For the sensor calibration, we used the ROS *multisensor_calibration* software[5] [2] developed in our laboratory that helped us to easily calibrate and find the extrinsic parameters among Kinect RGB-cameras. Starting from this calibration, we translated each RGB-camera reference point to the reference point of the DOA estimation given by *HARK* with respect to the reference system *world*. The reference point of the DOA estimation given by *HARK* is exactly the center point over the x axis of the Kinect as shown in Table 2.

For the reason that the four microphones of the Kinect are all positioned on its K_{RGB}^x axis (see Fig. 7a and Table 2), a DOA estimator that analyses the four audio signals can mathematically give only the rotation angle (azimuth) of the plane with the normal perpendicular to the K_{RGB}^y axis (i.e. the normal can be thought as belonging to the plane created by K_{RGB}^x and K_{RGB}^z axes), taking the K_{RGB}^x axis as zero axis and K_C as zero point for the angle estimation. For example, if the sound source is in front of the device and its position belongs to the plane created with the point K_C and the

---

[2]http://www.xbox.com/kinect.

[3]Note: all figures has the following well-known common notation for 3D axes: x axes are in red, y axes are in green and z axes are in blue.

[4]http://winnie.kuis.kyoto-u.ac.jp/HARK.

[5]https://github.com/iaslab-unipd/multisensor_calibration.

Table 3 Sensors positions and orientations in real test

Sensor type	k	s_k^x (m)	s_k^y (m)	s_k^o (rad)
Kinect	1	−1.05296	0.897504	0
Kinect	2	0	0	0
Kinect	3	0.0314011	0.0343866	0

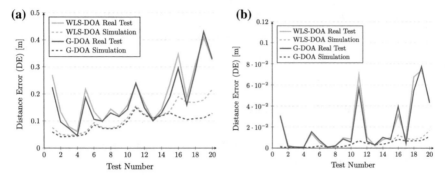

Fig. 8 Real test results of GP-DOA versus WLS-DOA **a** Mean **b** Variance

two vectors K_{RGB}^y and K_{RGB}^z, then the estimated azimuth should be $\alpha = 0$ degrees. Note that in this way it is impossible to understand if the sound comes from the front or behind the device (Fig. 7b), like in the well-known Cone of Confusion problem.

The first real test consisted in setting three Kinects in a line with the positions and orientations shown in Table 3 and clapping hands in different positions. A total of twenty different positions with $r^x \in [-1, 1]$, $r^y \in [0.5, 2]$ and a step of 0.5 (m) are visible in Fig. 10. Another simulation was done using the same parameters used in the real test in order to compare the results. For the GP-DOA simulation we used $prec = 0.01$ and $t_{sim} = 1000$ and the mean of the number of samples collected for each real test number is 35.

In the second test we used three Microsoft Kinect as DOA-sensors, with one of them mounted over a Pioneer 3-AT with a laser scanner. The real test consisted in having two fixed kinects and moving the Pioneer in the environment towards the clapping hands position (Fig. 11). The experiment was repeated twenty times with different starting positions of the robot.[6]

---

[6]Short video that shows one of the twenty tests made for the second test visible at url https://www.youtube.com/watch?v=TVAQ-sFSpF8.

7.1 Real Tests Results

Looking to real tests results and its related simulation results shown in Fig. 8a, b it is possible to see that GP-DOA performs always better than WLS-DOA both in mean and in variance and that real results does not differ so much to the ones of its relative simulation. It is evident that errors becomes bigger as the distance between the acoustic source and the sensors grows.

Results for the second real test (Table 4) showed that the algorithm performed well with a mean error of 0.37 (m). The error was mainly due to the SLAM algorithm that didn't estimate well the initial robot pose in the map.

8 ROS Package Documentation

This section will explain why we used ROS, how to correctly install the package and how to use it for real experiments.

8.1 Why ROS?

The choice of using the Robot Operating System for developing all the features explained in the chapter is due to the following factors:

- simplification of sensors signal connections in a network;
- easy data sharing;
- robust data synchronization;
- abstraction of the concept of device and no further driver implementation;
- modular approach using logical nodes;
- orientation towards the large community development and reuse.

Undoubtedly the ROS framework is a very useful tool because it avoids wasting time with other issues that does not properly concern the field of robotics.

8.2 Package Download

Our package has been implemented in C++ and can be downloaded at the following url: https://github.com/iaslab-unipd/DOA_acoustic_source_localization

It is written in order to be extended with other features like 3D Audio Localization and Multiple Sources Localization. The basic version allows to estimate only one acoustic source at time. A map of the environment can be linked if the user has a fixed reference.

Table 4 Distance error (cm) for the second real test

Test number	1	2	3	4	5	6	7	8	9	10	11	12	13	14	15	16	17	18	19	20
Distance error (cm)	52	58	25	15	24	31	68	57	42	39	31	11	57	24	36	63	42	12	46	19

Table 5 HARK Packages

Package name	Version	Package name	Version
harkfd*	2.0.0.6582	harktool4-cui	2.1.0
harkfd-dev*	2.0.0.6582	julius-4.2.3-hark	2.1.0
harktool4*	2.0.0.5427	julius-4.2.3-hark-plugin	2.1.0
flowdesigner-0.9.1-hark	2.1.0	libhark-netapi	2.1.0
hark-designer	2.1.0	libharkio-dev	2.1.0
hark-kinect	1.2.0.4440	libharkio1	2.1.0
hark-music	2.1.0	libharkio2	2.1.0
hark-ros-hydro	2.0.0.6430	libharkio2-dev	2.1.0

8.3   Installation and Required Packages

The current version of the *sound_localization* ROS package works only under **Ubuntu 12.04 (Precise)** with **ROS Hydro Medusa**. The reason is due to the HARK-Kinect ROS package we used for the implementation that is compatible only with the above configuration. After installing ROS environment the HARK-ROS package should be installed first following the instruction from the HARK site[7] and being careful to select the Ubuntu 12.04 (Precise) part. Then it should be installed the HARK-KINECT package from the instruction of the HARK site[8] and tested with a real Microsoft Kinect plugged in. The user must verify that the following packages are installed with the correct version. Other versions of the following packages are not tested. The packages signed with an * must be installed with a version that is different from the current one and can be downloaded from the repository of our package or from the HARK repository (Table 5).[9]

For installing the *DOA_acoustic_source_localization* package do the following:

1. copy *DOA_acoustic_source_localization* package in your */rosbuild* directory;
2. *cd ./your_rosbuild_path/rosbuild*
3. *rosws set DOA_acoustic_source_localization* and then answer "yes"
4. *~ /.bashrc*
5. *roscd DOA_acoustic_source_localization*
6. *make*

If there are no errors, the package is correctly installed.

---

[7]http://www.hark.jp/wiki.cgi?page=HARK+Installation+Instructions.

[8]http://www.hark.jp/wiki.cgi?page=HARK-KINECT+Installation+Instructions+%28as+a+USB+recording+device%29.

[9]http://www.hark.jp/harkrepos/dists/precise/non-free.

8.4 Input Parameters

In this part are detailed all available input parameters:

- **is_simulation**: type *boolean*. If set to "*true*" it disables the subscription to the topic containing the DOA estimations published by the HARK-Kinect ROS nodes. In this mode the DOA estimation data can only be set in a file.
- **data_simulation_file_path**: type *string*. Absolute path of the file containing the angles (in degrees) of the DOA estimations of the sensors. Each row contains an angle and the value in the first row is assigned to the first sensor, the second row is assigned to the second and so on. Example of file "*/home/riccardo/workspace/ros/angles_sim.txt*" containing three angles:

```
45
90
-45
```

- **n_acoustic_DOA_sensors**: type *int*. Number of DOA sensors in the network.
- **audio_signal_power_threshold**: type *double*. Threshold that helps selecting the acoustic sources with a minimum signal power. Common values goes from 25 to 40. A good choice is for example 35.
- **algorithm_type**: type *int*. There are three available algorithms:

  1. GP-DOA fast
  2. GP-DOA slow
  3. WLS-DOA

  WLS-DOA solution is always calculated (one shot formula) and it is used as starting point for the GP-DOA research and refinement. If there are only two DOAs in the 2D space—trivial problem—all the three algorithms give reasonably the same solution, so it is calculated only the WLS-DOA one because it has a one-shot-formula.
- **precision_grid**: type *double*. Precision in meters of the grid. Common values are 0.01 or 0.001 for GP-DOA Slow and 0.001 or 0.0001. Note that the precision value of the grid can be very small if it is used with the GP-DOA fast algorithm because of its high reduction of the computational time with respect to the slower one (see Sect. 5).
- **range**: type *double*. Exploration range in meters. Note that the center point of the grid G is the WLS-DOA solution.
- **tf_world_name**: type *string*. Name of the reference frame to which all the poses are referred. Example: *world*
- **tf_solution_name**: type *string*. Name of the reference frame of the published solution. Example: *sound_source_estimated_position*

- **rviz_DOA_line_lenght**: type *double*. Length of the lines that start from the DOA sensor and represent the DOA. A good value is 20 meters.
- **sensor_3D_pose_topic_** $< \mathbf{i} >$: type *string*. Name of the topic of the *i*th-DOA-sensor pose.
- **DOA_topic_** $< \mathbf{i} >$: type *string*. Name of the topic of DOA estimation of the *i*th-DOA-sensor.

8.5   Simulation Tests

For testing the package in simulation mode you have first to set the pose (position and orientation) of the DOA-sensors. You can easily do it by modifying the launch file that publishes the tf. Here is an example:

```
<node pkg="tf" type="static_transform_publisher" name="DOA_sensor_1"
args="1 2 0 0 0 1 /world /DOA_sensor_1 100" />
```

where the *DOA_sensor*_1 is in the position $s_1^{x,y,z} = [1, 2, 0]$ and the orientation, written in quaternions, is 180 degrees around z axis. Note that the program takes into consideration the DOA in the xy-plane and a DOA with 0 degrees is considered to be on the x axis (see Fig. 7). During the simulation test the user can vary the values of the data contained in the *data_simulation_file*; after saving the file the package will update the values and will show the new solution. For a quick test, launch the *hark_kinect_simulation.launch* file that will show an example with three DOA sensors.

8.6   Real Tests

Real tests need the calibration of the DOA-sensors. Doing it manually can be a very long process and usually is error prone. Finding the extrinsic calibration parameters between DOA-sensor is not so easy because the user have to measure the position of all the microphones and then calculate the position of the DOA starting point. As we deal with mobile robots, this procedure cannot be done considering only the microphones because the robot changes its pose and consequently the pose of its microphones. For this reason it can be helpful to use the vision sensors that normally do the Simultaneous Localization And Mapping (SLAM) that update the pose of the robot and provide a link between video and audio sensors. In order to achieve this aim we used another ROS package that allows to calibrate multiple vision sensors.[10] We used this package with the RGB images coming from the Microsoft Kinect.

---

[10]https://github.com/iaslab-unipd/multisensor_calibration.

8.6.1   Microsoft Kinect Configuration

Using HARK with Microsoft Kinect is very simple but there are some details to take into consideration. First of all, once the RGB calibration is done, we need to consider the following transform that refers to Fig. 7 and Table 2:

```
<node pkg="tf" type="static_transform_publisher" name="hark_to_rgb"
args="-0.0140 0 0 -0.5 0.5 -0.5 -0.5 /kinect_rgb /hark 100" />
```

Second we need to link each kinect RGB sensor to the right DOA HARK estimation. This procedure is very important and must be done with high attention. HARK ROS files contain all the information that HARK needs to localize the sounds. These data are collected in files with extension ".n". In our package we provided also some example files that can be modified for the user needs. The parameters that the user has to modify are substantially three. One is the *plughw* value that refers to the audio hardware to which the Kinect is associated. This value can be found in the computer by typing:

```
cat /proc/asound/cards
```

and looking to the number of the device assigned to the kinect. Here is an example of output:

```
0 [MID  ]: HDA-Intel - HDA Intel MID
   HDA Intel MID at 0xd3710000 irq 47
1 [PCH  ]: HDA-Intel - HDA Intel PCH
   HDA Intel PCH at 0xd3714000 irq 48
2 [Audio]: USB-Audio - Kinect USB Audio
   Microsoft Kinect USB Audio at usb-0000:00:14.0-2.1, high speed
```

so in the file ".n" you have to put *plughw* : 2. Note that if you plug more than one Kinect to the same computer, the *plughw* will increase and the unique way to associate the right HARK signal to the right Kinect is to plug the Kinects one at time, being careful to note the *plughw* value in order to set the right topic in the launch files. The second parameter that the user needs to modify is the *TOPIC_NAME_HARKSOURCE* inserting the name of the topic of the DOA. Finally the user have to modify the *A_MATRIX* parameter of the node *node_LocalizeMUSIC_1* and put the absolute path of the file *kinect_loc.dat* that contains the information for localizing sounds in the kinect (Figs. 9, 10 and 11).

Summarizing, the steps for a real test are the following:

1. plug the usb cable of the Kinects to the computer one at time and label them with a name or a number that links to the *plughw* number;
2. calibrate the Microsoft Kinects and find the extrinsic parameters;
3. launch *roscore*;

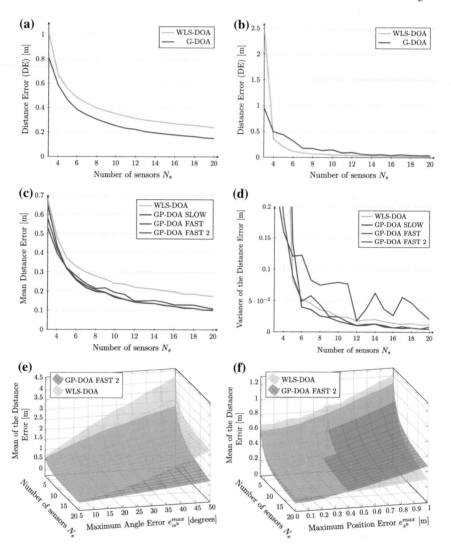

Fig. 9 Simulation results of WLS-DOA approach versus GP-DOA approach in all four simulation types. WLS-DOA is always in *green* and GP-DOA in *red*. In the first two simulations the x-axis is the number of the sensors N_s and the y-axis is the Distance Error (DE) in meters. In Simulations 3 and 4 the x-axis is the number of the sensors N_s, the z-axis is the Distance Error (DE) in meters and the y-axis is the Maximum Angle Error $e_{\alpha^k}^{max}$ in degrees and the Maximum Position Error $e_{s^k}^{max}$ in meters respectively **a** Simulation 1—Mean **b** Simulation 1—Variance **c** Simulation 2—Mean **d** Simulation 2—Variance **e** Simulation 3—Mean **f** Simulation 4—Mean

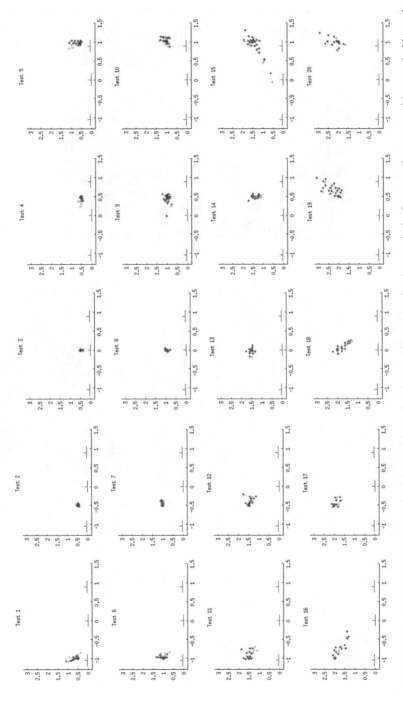

Fig. 10 Real Tests results. WLS-DOA and GP-DOA position estimations are in *green* and *red* respectively. Source and sensors positions are in *blue* and *red* respectively. The axis represents are consistent to the *world* and their scale is measured in meters

(a)

(b)

Fig. 11 Real experiment with three Microsoft Kinect as DOA-sensors, with one of them mounted over a Pioneer 3-AT with a laser scanner. Video available at url https://www.youtube.com/watch?v=TVAQ-sFSpF8 **a** Audio Localization with map in Rviz **b** Screenshots of the video

4. launch the ROS transform publishers and visualize them in rviz;
5. modify the .n files and the launch file of the ROS *DOA_acoustic_source_localization* package being careful to assign the correct topics to the correct values;
6. launch the .n files that start to output the DOAs from the Kinects;
7. launch the ROS *DOA_acoustic_source_localization* package.

9 Conclusions and Future Work

In this chapter it is described a new ROS package for solving the DOA-Based Localization Problem using the Gaussian Probability over DOA approach. The new approach has been validated with many tests, both in simulation and in a real environment. The chapter also offers a detailed description of the package it provides different solutions to the problem using three different algorithms (WLS-DOA, GP-DOA Slow and GP-DOA Fast) and has several input parameters, each one detailed in Sect. 8. Further research and implementation will focus on managing multiple sound sources scenarios also in the 3D space.

Acknowledgments We strongly thank the Université Pierre-et-Marie-Curie (UPMC) and the Institut des Systèmes Intelligents et de Robotique (ISIR) for hosting Ph.D. Student Riccardo Levorato during the first phases of the project. A special thank goes to Prof. Mohamed Chetouani, Post-Doc Salvatore Maria Anzalone and Ph.D. student Stéphane Michelet for their indispensable support and help.

References

1. P. Aarabi, The fusion of distributed microphone arrays for sound localization. EURASIP J. Adv. Sig. Process. **2003**(4), 860465 (2003)
2. F. Basso, R. Levorato, E. Menegatti, Online calibration for networks of cameras and depth sensors, in *OMNIVIS: The 12th Workshop on Non-classical Cameras, Camera Networks and Omnidirectional Vision—2014 IEEE International Conference on Robotics and Automation (ICRA 2014)* (2014)
3. J.H. Di Biase, H.F. Silverman, M. Brandstein, Robust Localization in Reverberant Rooms, in *Microphone Arrays: Signal Processing Techniques and Applications* (Springer, New York, 2001)
4. D.R. Griffin, *Listening in the Dark: The Acoustic Orientation of Bats and Men* (Yale University Press, New Haven, 1958)
5. M. Hawkes, Arye Nehorai, Wideband source localization using a distributed acoustic vector-sensor array. IEEE Trans. Sig. Process. **51**(6), 1479–1491 (2003). June
6. R. Levorato, E. Pagello, DOA acoustic source localization in mobile robot sensor networks, in *2015 IEEE International Conference on Autonomous Robot Systems and Competitions (ICARSC)*, pp. 71–76, Apr 2015
7. R. Levorato, E. Pagello, Probabilistic 2D Acoustic Source Localization using Direction of Arrivals in Robot Sensor Networks, in *Simulation, Modeling, and Programming for Autonomous Robots*, Lecture Notes in Computer Science, ed. by D. Brugali, J.F. Broenink, T. Kroeger, B.A. MacDonald (Springer International Publishing, Switzerland, 2014), pp. 474–485
8. K. Nakadai, T. Takahashi, H.G. Okuno, H. Nakajima, Y. Hasegawa, H. Tsujino, Design and implementation of robot audition system "hark"; open source software for listening to three simultaneous speakers. Adv. Robot. **24**(5–6), 739–761 (2010)
9. M. Omologo, R. De Mori, Acoustic Transduction, in *Spoken Dialogue with Computers* (Academic Press, New York, 1998)
10. P. Pertilä, Acoustic source localization in a room environment and at moderate distances. Ph.D. thesis, Tampere University of Technology, 2009
11. K.B. Petersen, M.S. Pedersen, The matrix cookbook, Nov 2012. Version 20121115

12. D. Salvati, Acoustic source localization using microphone arrays. Ph.D. thesis, Department of Mathematics and Computer Science, University of Udine, 2012
13. G. Valenzise, L. Gerosa, M. Tagliasacchi, E. Antonacci, A. Sarti, Scream and gunshot detection and localization for audio-surveillance systems, in *IEEE Conference on Advanced Video and Signal Based Surveillance, 2007. AVSS 2007*, Sept 2007, pp. 21–26
14. D.B. Ward, E.A. Lehmann, R.C. Williamson, Particle filtering algorithms for tracking an acoustic source in a reverberant environment. IEEE Trans. Speech Audio Process. **11**(6):826–836 (2003)

Part VI
Software Engineering with ROS

ROS Web Services: A Tutorial

Fatma Ellouze, Anis Koubâa and Habib Youssef

Abstract This tutorial presents how to integrate the Service-Oriented Architecture (SOA) paradigm into Robot Operating System (ROS). The main objective consists in exposing ROS ecosystem as a service that can be invoked by Web Services (WS) clients. This integration enables end-users and client applications to seamlessly interact with the ROS ecosystem via common WS interfaces while hiding all implementation details of the applications deployed in the ROS middleware. By the end of this tutorial, the reader will be able to develop web services that expose ROS topics and services to the end-users and client applications. This tutorial was developed under Ubuntu 12.4 and for ROS Hydro version.

Keywords ROS · RESTful/SOAP Web Services · Rosjava

F. Ellouze (✉) · A. Koubâa
Cooperative Networked Intelligent Systems
(COINS) Research Group, Riyadh, Saudi Arabia
e-mail: fatma.ellouze@coins-lab.org

A. Koubâa
College of Computer and Information Sciences,
Prince Sultan University, Riyadh, Saudi Arabia
e-mail: akoubaa@coins-lab.org

F. Ellouze
Research Laboratory on Development and Control of Distributed Applications
(ReDCAD Laboratory) ENIS, University of Sfax, Sfax, Tunisia

A. Koubâa
CISTER/INESC-TEC, ISEP, Polytechnic Institute of Porto, Porto, Portugal

H. Youssef
PRINCE Research Unit, University of Sousse, Sousse, Tunisia
e-mail: habib.youssef@fsm.rnu.tn

© Springer International Publishing Switzerland 2016
A. Koubaa (ed.), *Robot Operating System (ROS)*, Studies in Computational
Intelligence 625, DOI 10.1007/978-3-319-26054-9_18

463

1 Introduction

Robotics is a hot research area. It has brought a significant economic impacts to human lives and is expected to become increasingly integrated into our daily life by supporting human activities and performing tedious and dangerous tasks. However, robotic systems are not available to public at large scale because they are usually very complex to learn and to use, and require a deep knowledge on programming at low and high level to develop service robots applications, which make them heavily dependent on robotic expert developers. The objective of this chapter is to provide a viable solution to software developers and users to be able to develop applications for service robots at a larger scale, without the need to be an expert on robotics. To address this issue, we take benefit of web services technologies and propose a Service-Oriented Architecture approach for on demand access to robotic resources for enabling non technical users to remotely access, interact, manipulate and develop robots through the web interfaces. One motivation of this work is to develop a software architecture for service robots in the context of the myBot project [1] that allows easier interaction with the robot through abstract interfaces. A video demonstration of the myBot service robot is available in this link [2].

The concept of ROS Web services was recently proposed in [3] with the objective to leverage the use of Web services to provide new programming abstraction layers to ROS based on SOAP and REST Web services. The paper presented object-oriented software meta-models for the integration of SOAP and REST Web services into ROS. In this tutorial, we present a step-by-step tutorial on how to implement ROS Web services with ROSJAVA, and we present a different implementation methodology of the meta-model as compared to [3]. The main objective consists in exposing ROS ecosystem as a service that can be invoked by Web Services clients. This integration enables end-users and client applications to seamlessly interact with the ROS ecosystem via common Web Services interfaces while hiding all implementation details of the applications deployed in the ROS middleware. This avoids using complex message serialization between heterogenous systems such as the one proposed in [5]. There are two types of Web Services commonly available [4]: (1) SOAP Web Services, which are based on the SOAP protocol and provide a concrete formal description of the web service (2) REST Web Services, which represent a lighter weight approach based on the concept of *resource* and make extensive use of HTTP as a transport protocol. By the end of this tutorial, the reader will be able to (1) understand SOAP and REST Web Services, (2) develop SOAP and REST web services that expose ROS components to the end-users and client applications.

There were a few previous attempts to integrate ROS with Web technologies [6–8]. In [6], the author proposed RoboWeb, a SOAP- based service-oriented architecture that virtualizes robotic hardware and software resources and exposes them as services through the Web. This tutorial paper improves on [6] is that it provides the integration of SOAP and REST Web services with ROS. In [8], the authors proposed `rosbridge` to integrate ROS with Web technologies (not Web services) so that to to use commonly available Internet browsers for non-roboticians users to interact with a ROS-enabled robot. In [7], the authors proposed ROStful by extending rosbridge to support REST Web services and developed a lighweight Web server that exposes ROS topics, services and actions through RESTful Web services. Our work improves on [7] by proposing a unified framework for both SOAP and REST Web services, and is completely independent from rosbridge.

The rest of this tutorial is organized as follow. Section 2 introduces RESTful and SOAP web services. Section 3 introduces the ROSJAVA API, which we use to implemented the ROS web services using Java language. In Sect. 4, we provide the detailed steps of integrating SOAP and RESTful web services with ROS ecosystem.

2 Web Services

Web Services (WS) are software components exposed to the Internet for **machine-to-machine interaction**. They represent the main implementation of the concept of Service Oriented Architecture (SOA), which basically consists in exposing resources as *services* to end-users. They can be seen as online APIs that a developer can remotely use without having to locally add these libraries into his project, but rather, these APIs are remotely *invoked* on the server. The main benefit of web services is to ensure *interoperability* between heterogeneous systems. Web services are also *loosely coupled* systems allowing seamless access, use and transaction between software components, independently of the programming language in which they are written and the (software/hardware) platforms where they are deployed. There are two categories of web services: (1) SOAP web services, and (2) RESTful Web services. In what follows, we present an overview of these two types of web services.

2.1 SOAP Web Services

SOAP web services are defined as an implementation of SOA architecture over the web. It offers communication and exchange of data between heterogeneous, distributed software components. The service provider and the service consumer exchange data in **XML format** using the Simple Object Access Protocol (**SOAP**). The Web Services Description Language (**WSDL**) file is a contract file containing the description of the service. It is published in the registry by the service provider to be discovered and used by the service consumer. Next, we will present the technologies used by SOAP Web Services including SOAP protocol, XML message format and WSDL contract file.

Figure 1 shows the typical execution model of SOA. A *service provider* **registers** its service into a *public registry* so that it can be discovered by clients. A *service consumer* fetches and **finds** services into the registry, and then it **binds** to the service of interest with the service providers. Finally, the service consumer (also referred to as client) **invokes** the methods of interest provided by the web service.

Fig. 1 SOA execution
model

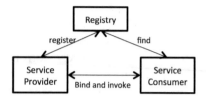

SOAP web services represent the main implementation of previous SOA execution model. It is characterized by the following protocols:

- **SOAP**: It stands for Simple Object Access Protocol. It is a W3C recommendation. It is defined as an XML-based communication protocol which allows distributed and heterogeneous applications to exchange data via service transport protocols such as FTP, SMTP or HTTP. SOAP basically represents a protocol of *message format* that two web services entities (a server and a client) exchange to communicate between each other. The message format is represented through the XML (eXtensible Markup Language) language.
- **WSDL**: It stands for Web Services Description Language. It is a W3C recommendation. It is an XML-based document that describes Web Services. It specifies the location of the service, the functionalities that the service exposes, the communication protocols the service supports and the correct format of exchanged messages. WSDL can be seen as a *contract* that provides clients with all information they need to interact with the web service.

It has to be noted that, in the early days of SOAP web services, the *UDDI* (Universal Description, Discovery and Integration) protocol was proposed as a core protocol of SOAP web services for automatic service discovery and registration. However, it was quickly abandoned as it was not proven to be really useful in real deployments.

In the rest of this subsection, we will show how to implement a SOAP web service in Java using JAX-WS standard specification. The objective is to give the reader a little background on how SOAP web services are implemented. If you are already familiar with SOAP WS implementation in Java, you can skip this subsection.

The following example presents the main steps to create a simple Hello World application, and exposes it as SOAP web service using JAX-WS API. Typically, we consider four classes to organize the implementation of SOAP web services: (1) the HelloWorldInterface.java interface, which is a generic Java interface that defines the web service's methods, (2) the HelloWorldWS.java class which contains the implementation of the web service interface defined above, (3) the HelloWorldWSPublisher.java class that publishes the web service through the web service platform (Glassfish, or Tomcat) and exposes it to client application, and finally, (4) the HelloWorldWSClient.java class, which is a simple Java client that consumes the published web service.

Note: *No prerequisites are required for creating and deploying SOAP Web Services because the Java Platform, Entreprise Edition (Java EE) supports the Java API for XML Web Services (JAX-WS 2.2).*

Class 1: HelloWorldInterface.java

```
1   package com.soap.ws;
2
3   import javax.jws.WebMethod;
4   import javax.jws.WebService;
5   import javax.jws.soap.SOAPBinding;
6   import javax.jws.soap.SOAPBinding.Style;
7
8   @WebService
9   @SOAPBinding(style = Style.RPC)
10  public interface HelloWorldInterface{
11      @WebMethod String sayHello();
12  }
```

Class 1 presents a simple Hello World application interface of a SOAP web service in Java. It contains one method,called, `sayHello(...)`, that can be invoked by a client. It must return a String that will be sent back to the client invoking the Web method. A good programming practice is to separate the interface from the implementation of the Web Service.

From line 3 to line 6, we import required libraries to implement SOAP Web Services. It has to be noted that `javax.jws` package belongs to the JAX-WS API that is inherently available in the Java EE 6 (version 2.0). Implementing a SOAP web service is quite easy. We just need to add the following annotations:

- `@WebService` annotation is required to indicate that the annotated class is a web service (line 8).
- `@SOAPBinding` annotation is optional and used to specify the style of the SOAP message. The default style is *document* (line 9).
- `@WebMethod` annotation is optional and used to indicate that the annotated method is a web service operation (line 11).

Class 2 presents the web service implementation of the `HelloWorldInter face.java` interface and provides the implementation of all methods that will be exposed as web services.

Class n° 2: HelloWorldWS.java

```
1   package com.soap.ws;
2
3   import javax.jws.WebService;
4
5   @WebService(endpointInterface = "com.soap.ws.HelloWorldInterface")
6   public class HelloWorldWS  implements HelloWorldInterface {
7       @Override
8       public String sayHello(){
9           return "hello";
10      }
11  }
```

In line 5, we refer to the `HelloWorldInterface.java` interface that the web service class implements by using the attribute `endpointInterface` in `@WebService` annotation.

Class 3 publishes the web service using the static method `publish` of the class `Endpoint`. Note that it is important to specify the URL that will be used to access the web service and instantiate the web service implementation class.

Class n° 3: HelloWorldWSPublisher.java

```
1   package com.soap.ws;
2
3   import javax.xml.ws.Endpoint;
4
5   public class HelloWorldWSPublisher {
6       public static void main(String[] args) {
7           Endpoint.publish("http://localhost:1239/helloWorld", new ←
                HelloWorldWS());
8       }
9   }
```

To test if the above web service works as expected, we need to run `HelloWorld WSPublisher` class and check the WSDL file by pasting the URL "`http://localhost:1239/helloWorld?wsdl`" in the browser. Note that you must replace localhost and the port number by the IP address of the server machine and the port running the web service, if you make them different.

TEST YOUR CODE
Once the web service is deployed and running in the server, we need to invoke the web service using a client. In what follows, we present a small client program showing how to invoke a SOAP web service.

Class n° 4: HelloWorldWSClient.java

```
1   package com.soap.client;
2
3   import java.net.URL;
4   import javax.xml.namespace.QName;
5   import javax.xml.ws.Service;
6   import com.soap.ws.HelloWorldInterface;
7
8   public class HelloWorldWSClient {
9       public static void main(String[] args) throws Exception{
10          URL url = new URL("http://localhost:1239/helloWorld?←
                wsdl");
11          QName qname = new QName("http://ws.soap.com/", "←
                HelloWorldWSService");
12          Service service = Service.create(url, qname);
13          HelloWorldInterface hello = service.getPort(←
                HelloWorldInterface.class);
14          System.out.println(hello.sayHello());
15      }
16  }
```

The code above is a web service client that accesses the HelloWorld published web service.

First, we create a qualified name structure of the web service using the package name and the web service class implementation as parameters (line 11). Then, we create a generic web service client object using the Service class using the URL of the

Web Service where it is deployed, and the qualified web service name (line 12). Finally, we create a proxy service that will be used to invoke the web method sayHello(...). The execution of the client program above should return the String hello as specified in the implementation of the Web method.

2.2 RESTful Web Services

RESTful web service is an architectural style applied to web services. The design of REST-based web services is based on the concept of resources that will be exposed through the Internet for remote access using the HTTP concepts.

REST was designed by Roy Fielding, who is also one of the authors of HTTP specification, thus, he was inspired a lot by the concepts of HTTP when designing the REST Web Services. In this subsection, we will present the main REST principles and concepts and explain some guidelines for appropriate design of REST web services. The basic principle of HTTP is that each resource (such as a web page) is identified by a distinct Uniform Resource Locator (URL), and manipulated using HTTP methods including GET, POST, PUT and DELETE. In the same way in REST, a web service is defined by a resource that can be invoked through one of the four HTTP methods GET, POST, PUT and DELETE.

We distinguish four HTTP concepts: (1) *Resource Location*, (2) *HTTP Methods*, (3) *Metadata*, and (4) *Content Types*.

- **Resource Location**: a resource in a website can be accessed through it URL, which is composed of three major parts: (1) the schema, which is the protocol used to access the resource namely http://, (2) host which is the server location and (3) URI, which is the path/address to a resource inside the server. So, each resource has a unique identifier (URI) that defines the path/address to access this resource. Just like web pages, RESTful resources are also accessed through their Uniform Resource URIs. The main difference is that RESTful APIs have resource-based addresses; however, websites have action-based addresses. For instance, /*messages*/ is a resource-based URI, however /*getMessages*/ is an action-based URI.
- **HTTP Methods**: HTTP defines four *methods* or *verbs* to deal with resources namely GET, POST, PUT and DELETE. RESTful web service APIs take benefit of these methods to interact with RESTful resources. Unlike SOAP web service requests that are only based on the POST HTTP method, RESTful web service makes use of all HTTP methods.
  Table 1 shows the most commonly used HTTP methods and their descriptions.
- **Metadata**: HTTP response messages are composed of three main parts: (1) message body, (2) response header and (3) status code. Response headers and status code are metadata that contain useful information such as the Content-Type and message status. RESTful Web Services use status code and send it in code format to the client to specify message status.

Table 1 HTTP methods

HTTP methods	Description
GET	Get information from a resource
POST	Create a resource
PUT	Update a resource
DELETE	Remove a resource

- **Content Types**: "Content-Types" is a part of message header that indicates the format of messages exchanged between server side and client side. RESTful web services take benefit of this HTTP concept to indicate the format of messages exchanged. For each method, it is possible to specify a particular format for returning the result to the calling client, namely as a plain text, or JSON format (*application/json*) or XML format (*text/xml*), which provides multiple ways to exploit the data at the client side.

In the rest of this subsection, we will show how to design, implement and expose a RESTful web service using JAX-RS standard specification in Java. We will present the steps to create a simple Hello World example and expose it as RESTful web service using JAX-RS API. Table 2 shows the design of the RESTful web service example. It consists of "Robot" resource accessible through the "/{robotname/}" URI. This resource contains a web method called "sayHello(name)" which can be invoked using HTTP GET method. For that, we will implement three classes: (1) the RobotResource.java class which defines the resource's methods, (2) the RobotWsPublisher.java class that exposes the resource's methods as web services in the web and finally (3) the RobotWSClient.java class which is a simple Java client that consumes the published web service.

Class 1 presents the Java implementation of the RESTful resource presented in Table 2. It consists of a Java class named RobotResource that is accessible through "/robotname" URI by using @Path annotation. It contains one method called sayHello(), which is invoked by clients using the HTTP GET method. This method contains a parameter, which is linked to the resource's path by using @PathParam annotation.

Table 2 Design of RESTful Web Service

Resource:	Robot
URI:	/robotname
HTTP methods supported:	GET: sayHello(name)

Class n° 1: RobotResource.java

```
1   package com.rest.resource;
2
3
4   import javax.ws.rs.GET;
5   import javax.ws.rs.Path;
6   import javax.ws.rs.PathParam;
7   import javax.ws.rs.core.Response;
8
9   @Path("/{robotname}")
10  public class RobotResource {
11        @GET
12        public Response sayHello(@PathParam("robotname") String name)↵
              {
13              String output = "Hello: " + name;
14              return Response.status(200).entity(output).build();
15        }
16  }
```

In Class 2, we create a HTTP server that deploys and publishes the RESTful web service created in Class 1 (line 14) by using the static method `create` of the class `HttpServerFactory`. Note that it is important to specify the URL that will be used to access the web service (`http://localhost:7001/`). You should replace localhost and the port number by the IP address of the server machine and the port running the web service, if they are different.

Class n° 2: RobotWsPublisher.java

```
1   package com.rest.endpoint;
2
3   import java.io.IOException;
4   import java.util.logging.Level;
5   import java.util.logging.Logger;
6   import com.rest.resource.RobotResource;
7   import com.sun.jersey.api.container.httpserver.HttpServerFactory;
8   import com.sun.jersey.api.core.ClassNamesResourceConfig;
9   import com.sun.net.httpserver.HttpServer;
10
11  public class RobotWsPublisher {
12        public static void main(String[] args) {
13              try {
14                    HttpServer create = HttpServerFactory.create("↵
                        http://localhost:7001/", new ↵
                        ClassNamesResourceConfig(RobotResource.class)↵
                        );
15                    create.start();
16              } catch (IllegalArgumentException e) {
17                    e.printStackTrace();
18              } catch (IOException ex) {
19                    Logger.getLogger(RobotWsPublisher.class.getName()↵
                        ).log(Level.SEVERE, null, ex);
20              }
21        }
22  }
```

To test if the REST web service is correctly deployed, we need to run `RobotWs Publisher.java` class and invoke our RESTful web service by pasting the URL `http://localhost:7001/turtlebot` in a web browser. You must get `Hello: turtlebot` as a response displayed in the web browser.

TEST YOUR CODE

Once the REST web service is up and running in the server, we need to invoke the web service using a client. In what follows, we present a small client showing how to invoke a RESTful web service.

Class 3 consists of a RESTful web service client that allows access to our published web service. It creates an instance of a `HttpURLConnection` class with the web service server (lines 15–17), then it sends a `HTTP GET` request to invoke the web method `sayHello()` of our resource by using the URL `http://localhost:7001/turtlebot` (lines 18–20). The execution of the client program above should return `Hello: turtlebot` as a response.

Class n° 3: RobotWSClient.java

```
 1  package com.rest.client;
 2
 3  import java.io.BufferedReader;
 4  import java.io.InputStreamReader;
 5  import java.net.HttpURLConnection;
 6  import java.net.URL;
 7
 8  public class RobotWSClient {
 9      private final String USER_AGENT = "Mozilla/5.0";
10      public static void main(String[] args) throws Exception {
11          RobotWSClient http = new RobotWSClient ();
12          http.sendGet ();
13      }
14      private void sendGet () throws Exception {
15          String url = "http://localhost:7001/turtlebot";
16          URL obj = new URL(url);
17          HttpURLConnection con = (HttpURLConnection) obj.
                openConnection ();
18          con.setRequestMethod("GET");
19          con.setRequestProperty("User-Agent", USER_AGENT);
20          int responseCode = con.getResponseCode();
21          System.out.println("\nSending GET request to URL : " + url)
                ;
22          System.out.println("Response Code : " + responseCode);
23          BufferedReader in = new BufferedReader(new
                InputStreamReader(con.getInputStream()));
24          String inputLine;
25          StringBuffer response = new StringBuffer();
26          while ((inputLine = in.readLine()) != null) {
27              response.append(inputLine);
28          }
29          in.close();
30          System.out.println(response.toString());
31      }
32  }
```

3 ROSJAVA API

ROSJAVA is a Java API that was designed to extend ROS for Java developers. It provides a set of tools and libraries that offer ROS-style communication mechanisms including topic publishing/subscribing, service request/response, and parameters set/get. Using ROSJAVA API, Java developers can implement robotic applications with Java using ROS concepts.

In the rest of this subsection, we will show an illustrative example that implements and runs a ROS node that publishes a message on a ROS topic using ROSJAVA API. This example consists of two Java classes: (1) the "Talker.java" class which implements a ROS node that publishes std_msgs.String messages on the /chatter topic and (2) the TalkerRunner.java class, which executes the ROS node implemented in Talker.java class.

Note: In Sect. 4, we will show how to get and import ROSJAVA APIs and how to prepare all prerequisites to develop a Java application using ROSJAVA to interact with ROS ecosystem.

The code presented in Class 1 is a Java implementation of a ROS publisher node using ROSJAVA API. ROSJAVA package contains an abstract class named AbstractNodeMain that represents the superclass of any ROS node to be created with ROSJAVA. In other words, any ROSJAVA node should extend the ROSJAVA superclass AbstractNodeMain and override the method onStart (ConnectedNode cn) that represents the entry point of the node. Connected Node class contains methods needed to launch the ROS subscribers and publishers and to communicate with the rosmaster. It allows creating publishers and subscribers using newPublisher and newSubscriber methods respectively.

In our example (class 1), we create a new publisher (line 16) that publishes std_msgs.String messages to the /chatter topic using the method new Publisher() of the class ConnectedNode. Then, we create the message to be published by using the method newMessage of Publisher class (line 17). Finally, we publish the message using publish() method of Publisher class (line25).

Class n° 1: Talker.java

```
1  package com.rosjava;
2
3  import org.ros.concurrent.CancellableLoop;
4  import org.ros.namespace.GraphName;
5  import org.ros.node.AbstractNodeMain;
6  import org.ros.node.ConnectedNode;
7  import org.ros.node.topic.Publisher;
8
9  public class Talker extends AbstractNodeMain {
10
11  public GraphName getDefaultNodeName() {
12     return GraphName.of("rosjava/talkerNode");
```

```
13 |  }
14 |
15 |  public void onStart(final ConnectedNode connectedNode) {
16 |      final Publisher<std_msgs.String> publisher = connectedNode.↩
   |          newPublisher("chatter", std_msgs.String._TYPE);
17 |      std_msgs.String str = publisher.newMessage();
18 |      str.setData("Hello world!");
19 |
20 |      try {
21 |          Thread.sleep(300);
22 |          } catch (InterruptedException e) {
23 |              e.printStackTrace();
24 |          }
25 |      publisher.publish(str);
26 |  }
27 |  }
```

In Class 2, we execute the ROS publisher node implemented in Class 1 by using the static method `execute()` of the class `NodeMainExecutor` provided by ROSJAVA API.

Class n° 2: TalkerRunner.java

```
1  |  package com.rosjava;
2  |
3  |  import org.ros.node.DefaultNodeMainExecutor;
4  |  import org.ros.node.NodeConfiguration;
5  |  import org.ros.node.NodeMainExecutor;
6  |  import com.google.common.base.Preconditions;
7  |  import java.net.URI;
8  |
9  |  public class TalkerRunner{
10 |      public static void main(String[] args) {
11 |          Talker talkerNode = new Talker();
12 |          NodeMainExecutor nodeMainExecutor = DefaultNodeMainExecutor↩
   |              .newDefault();
13 |          URI masteruri = URI.create("http://192.168.1.4:11311");
14 |          String host = "192.168.1.4";
15 |          NodeConfiguration talkerNodeConfiguration = ↩
   |              NodeConfiguration.newPublic(host, masteruri);
16 |          Preconditions.checkState(talkerNode != null);
17 |          nodeMainExecutor.execute(talkerNode,talkerNodeConfiguration↩
   |              );
18 |      }
19 |  }
```

TEST YOUR CODE
To test your code, run the rosmaster by running in the terminal:

```
roscore
```

First, check the list of running topics using the following terminal command:

```
rostopic list
```

Observe that two topics should appear as follows:

```
/rosout
/rosout_agg
```

Now, run "`TalkerRunner.java`" class from your IDE like any Java program to execute the publisher ROS node implemented in "`Talker.java`" class:

Now, observe that there are three running topics including the /`chatter` topic.

```
rostopic list
/chatter
/rosout
/rosout_agg
```

This means that the publisher is currently working. You can use the command "`rostopic echo /chatter`" to see the content of the topic published by the ROJAVA node.

4 ROS Web Services

In this section, we will show how to integrate web services (both SOAP and RESTful web services) into ROS using ROSJAVA API to allow client applications especially WEB developers to seamlessly and transparently interact with the ROS-enabled robots. For that purpose, we will implement SOAP and RESTful web services and deploy them on the robot. We will work on turtlesim simulator for illustration purposes, but the code below can be easily generalized to any type of ROS-enabled robot.

Figure 2 shows the architecture of ROS Web Services and explains how to integrate web services into ROS ecosystem. The web services present an intermediate layer between client applications and ROS ecosystem. They provide an abstraction of ROS resources, including topics, services and actions by exposing them as web services to developers who do not have prior background on robots or on ROS. They allow

Fig. 2 ROS web services architecture

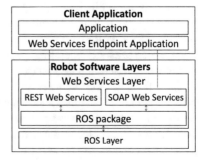

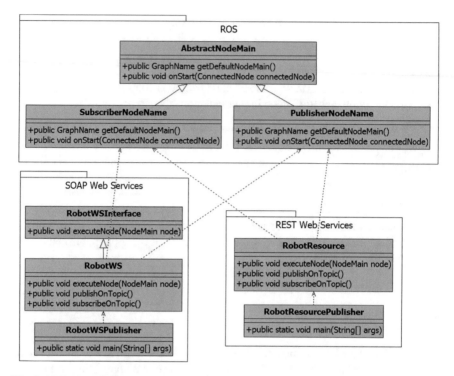

Fig. 3 UML class diagram for Web Services Layer

(i) subscribing to and publishing any ROS topic, action or service (ii) delivering the ROS messages to clients subscribing to a particular topic, and (iii) forwarding messages received from the client acting as publishers of ROS messages to ROS. This means that for any ROS topic, action or service that must be exposed, the Web service should instantiate a ROS subscriber and/or a publisher for the topic or the service of interest. Then, Web methods and interfaces should be defined according to which functionalities the robot would like to expose and execute.

Our architecture comprises two main parts: (1) **Client Application** which uses the web services endpoint to consume web services implemented on the robot and exposed to the public and (2) **Robot Software Layers** which consists of: (a) **Web Services Layer** that implements SOAP and RESTful web services using ROSJAVA API to interact with ROS components and expose robots' capabilities, and (b) **ROS layer** which is a meta-operating system for robots which provides tools and libraries that allow interaction with robots.

Let's focus on the Web Services Layer as we are interested in integrating web services with ROS. Figure 3 shows the UML class diagram that we proposed for Web Services Layer. It can be instantiated for any type of robots. We used the generic term Robot as a prefix in the name of the classes, but it can be replaced with the name of a concrete robot when instantiated for a real robot. Web Services Layer comprises three software packages: (i) `ROS`, (ii) `SOAP Web services`, and (iii) `REST Web services`.

ROS package: The ROS package provides an additional level of abstraction on top of ROS. It exposes the functionalities of ROS system to Java developers by providing ROS-style communication mechanisms including topic publishing/subscribing and service request/response. The ROS package contains classes purely written in Java using ROSJAVA API and does not include any Web service functionality. Classes in this package consist of publishers and subscribers of all ROS topics and ROS services that will be later exposed as web services and used to control and manipulate robots.

SOAP Web services package: The SOAP Web services package, that we call `ros-ws`, provides a unified view of the robot and its environment. It abstracts individual robot capabilities, allowing the robot to be controlled through SOAP web services. This package uses the ROS package to interact with ROS ecosystem and manipulate ROS nodes to publish/subscribe on ROS topics and request ROS services allowing the control of the ROS-enabled robot. A robot is represented with a Java class called with the robot's name suffixed with "'WS'" like `turtlebotWS.java` for turtlebot robot. This class contains web methods that are exposed as SOAP web services. These web methods execute ROS nodes implemented in `ROS package` to publish/subscribe on corresponding ROS topics and request corresponding ROS services in order to manipulate and control the robot.

REST Web services package: The REST Web services package, that we call `ros-rs`, presents the robot's capabilities and exposes them as RESTful web services. A robot consists of a set of devices including wheels, arms, camera, location sensor, temperature sensor and others. Each device is represented as resource in the REST Resource model we propose (Fig. 4). Resources are organized following device categories. For instance, location sensor and temperature sensor belong to sensors category, however wheels and arms belong to actuators category. The REST Web services package defines a class for each particular resource that must be exposed to public. This class implements methods that manipulate the corresponding device by subscribing/publishing on ROS topics and requesting ROS services that control the device by using ROS nodes implemented in `ROS package` using ROSJAVA API.

In the rest of this section, we will explain all prerequisites required to implement such applications including installing ROSJAVA API and RESTful and SOAP web services APIs. Then, we will validate our architecture through an experimental deployment of REST and SOAP web services into Turtlesim.

Fig. 4 REST Resource model

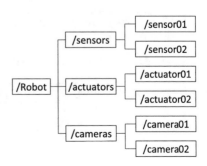

4.1 Prerequisites

Our examples are implemented on ROS Hydro version on Ubuntu 12.04, but they should also work for Indigo (although not tested). It is recommended to use Eclipse as an IDE for development. Installation and configuration steps of Eclipse IDE can be found in http://www.coins-lab.org/psu/cs460/index.php?title=Eclipse.

Make sure to install JDK 1.6 using the command:

```
sudo apt-get install Openjdk-6-jre
```

- **Getting ROSJAVA API**:

  To implement Java programs that manipulate ROS components, we need to import ROSJAVA libraries to our projects. For this purpose, we will install ROSJAVA environment to get all libraries we may need to develop ROS using Java.

  If you do not have ROSJAVA installed, then open a terminal and install it as follows:

```
sudo apt-get install ros-hydro-catkin
ros-hydro-ros ros-hydro-rosjava python-wstool
```

If the installation is successful, you can find the maven repository of all rosjava artifacts in /opt/ros/hydro/share/maven.
In /opt/ros/hydro/share/maven/org/ros, you can search and find all ROSJAVA libraries.
To facilitate getting the basic ROSJAVA libraries, we will create a ROSJAVA project.

Creating ROSJAVA project:

First, we need to create a workspace for ROSJAVA using this command:

```
mkdir ~/rosjava_workspace
```

Then, we create /src folder where to put all ROSJAVA packages. Write this command in the terminal:

```
cd ~/rosjava_workspace
mkdir src
```

Next, we create a ROSJAVA package by running this command:

```
catkin_create_rosjava_pkg rosjava_package
```

Then, we compile workspace to generate /build and /devel folders required to source the workspace

```
cd ~/rosjava_workspace
catkin_make
source devel/setup.bash
```

Next, we create a ROSJAVA project inside the ROSJAVA package we have created by running:

```
cd ~/rosjava_workspace/src/rosjava_package
catkin_create_rosjava_project rosjava_project
```

Then, we compile the project we have created by running:

```
cd ~/rosjava_workspace
catkin_make
```

If everything works fine, we find basic ROSJAVA libraries in the following directory:
~/rosjava_workspace/src/rosjava_package/rosjava_project/build/install/rosjava_project/lib.

- **Web Service APIs**:
  To implement Web Services, we need to import SOAP and REST libraries to our project. For our examples, we are using JDK 1.6 which supports the Java API for XML web services (JAX-WS). Regarding RESTful web services, we need to download "javax.ws.rs-api-2.0.1.jar" library for using JAX-RS API and "http-2.2.1.jar" and "jersey-bundle-1.8.jar" libraries for creating the HTTP server.

- **Adding ROSJAVA and Web Services APIs to Eclipse**:
  To facilitate access to ROSJAVA and REST libraries, we will create two user libraries: (1) the *"rosjava_libraries"* that contains all ROSJAVA libraries and (2) the *"rest_libraries"* that contains all REST libraries. So, we can easily add these libraries to any project when needed by just calling them.

  In a first step, you need to start Eclipse, and choose your workspace where to put Java projects.
  In Eclipse's toolbar, select window->preferences->Java->Build Path->User Libraries. Then, press "New" and type the name of the library "rosjava_Libraries" and click OK. Now, back to "User Libraries", select the library you have created and click "Add External Jars". Then, navigate the location of the basic rosjava libraries:
  ~/rosjava_workspace/src/rosjava_package/rosjava_project/build/install/rosjava_project/lib.

Fig. 5 ROSJAVA libraries

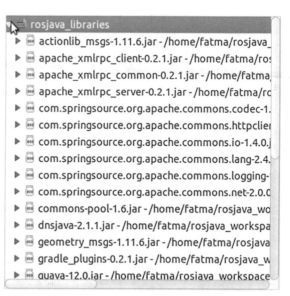

Select all JARs and click validate. Now you have defined your library in Eclipse that contains all ROSJAVA libraries required to implement ROS using Java. Figure 5 shows libraries in the user library "'rosjava_libraries'".

Note: We do the same thing with the user library "'rest_libraries'" by adding REST libraries we have already downloaded.

4.2 Implementation and Experimentation

In this subsection, we will take benefit of technologies described in Sects. 2 and 3 to develop SOAP/REST web services interfaces using the powerful features of Java EE in combination with ROSJAVA API to interact with ROS environment in order to tele-operate the turtlesim by publishing a message of type geometry_msgs.Twist on the ROS topic /turtle1/cmd_vel, and to get its position by subscribing on the ROS topic /turtle1/pose and receiving messages of type turtlesim.Pose.

ROS package: According to our architecture (Fig. 2), we need first to declare and implement all ROS nodes (publishers and subscribers) required to control the robot. For our illustrative example, we implement two Java classes: (1) "motion.java" which consists of a ROS node that publishes a message of type geometry_msgs. Twist on the ROS topic /turtle1/cmd_vel in order to control the movement of the turtlesim, and (2) "Velocity.java" to control the position of turtlesim. It consists of a ROS node that subscribes on the ROS topic /turtle1/pose and receives messages of type turtlesim.Pose.

Class n° 1: Position.java:

```
1   package com.ros.nodes;
2   import org.ros.message.MessageListener;
3   import org.ros.namespace.GraphName;
4   import org.ros.node.AbstractNodeMain;
5   import org.ros.node.ConnectedNode;
6   import org.ros.node.topic.Subscriber;
7
8   public class Position extends AbstractNodeMain{
9       public static turtlesim.Pose turtlesimPose;
10
11      public GraphName getDefaultNodeName() {
12          return GraphName.of("rosjava/turtlesim_listeners");
13      }
14
15  public void onStart(ConnectedNode connectedNode) {
16      //Create a new subscriber on /turtle1/pose topic
17      Subscriber<turtlesim.Pose> poseSubscriber = connectedNode.↵
            newSubscriber("/turtle1/pose", turtlesim.Pose._TYPE);
18      //When a new message is received by /turtle1/pose topic, a ↵
            MessageListener will be called with the incoming message as↵
            an argument to MessageListener.onNewMessage.
19      poseSubscriber.addMessageListener(new MessageListener<turtlesim↵
            .Pose>() {
20          public void onNewMessage(turtlesim.Pose poseMessage) {
21              turtlesimPose = poseMessage;
22              System.out.printf("Pose: (\%.2f, \%.2f, \%s)\n", ↵
                    poseMessage.getX(), poseMessage.getY(), poseMessage↵
                    .getTheta());
23          }
24      });
25  }
26  }
```

Class n° 2: Motion.java:

```
1   package com.ros.nodes;
2
3   import org.ros.concurrent.CancellableLoop;
4   import org.ros.namespace.GraphName;
5   import org.ros.node.AbstractNodeMain;
6   import org.ros.node.ConnectedNode;
7   import org.ros.node.NodeMain;
8   import org.ros.node.topic.Publisher;
9
10  public class Motion extends AbstractNodeMain{
11
12      private String direction;
13      private double speed;
14
15      public Motion(String direction, double speed){
16          this.direction = direction;
17          this.speed =speed;
18      }
19
20      public GraphName getDefaultNodeName() {
21          return GraphName.of("rosjava/turtlesimMotion");
22      }
23        public void onStart(final ConnectedNode connectedNode) {
24            //create new publisher on /turtle1/cmd_vel topic
```

```
25    final Publisher<geometry_msgs.Twist> publisher = ↵
          connectedNode.newPublisher("/turtle1/cmd_vel", ↵
          geometry_msgs.Twist._TYPE);
26    //create new message
27    geometry_msgs.Twist twist = publisher.newMessage();
28    //set the message by indicating the speed and the ↵
          direction
29    if(this.direction=="back"){
30        twist.getLinear().setX(this.speed*(-1));
31        twist.getLinear().setY(0);
32        twist.getLinear().setZ(0);
33
34        twist.getAngular().setX(0);
35        twist.getAngular().setY(0);
36        twist.getAngular().setZ(0);
37    }
38    if(this.direction=="forward"){
39        twist.getLinear().setX(this.speed);
40        twist.getLinear().setY(0);
41        twist.getLinear().setZ(0);
42
43        twist.getAngular().setX(0);
44        twist.getAngular().setY(0);
45        twist.getAngular().setZ(0);
46        }
47    if(this.direction=="left"){
48        twist.getLinear().setX(0);
49        twist.getLinear().setY(0);
50        twist.getLinear().setZ(0);
51
52        twist.getAngular().setX(0);
53        twist.getAngular().setY(0);
54        twist.getAngular().setZ(this.speed);
55    }
56
57    if(this.direction=="right"){
58        twist.getLinear().setX(0);
59        twist.getLinear().setY(0);
60        twist.getLinear().setZ(0);
61
62        twist.getAngular().setX(0);
63        twist.getAngular().setY(0);
64        twist.getAngular().setZ(this.speed*(-1));
65    }
66    try {
67        Thread.sleep(300);
68    } catch (InterruptedException e) {
69        e.printStackTrace();
70    }
71    //publish the message
72    publisher.publish(twist);
73    }
74 }
```

As explained in Sect. 3, we have implemented two ROS nodes in Java using ROS-JAVA API. The "Position.java" class subscribes on the ROS topic /turtle1/pose. Once executed, it should return the position of the robot as a response. In the other hand, the execution of "Motion.java" class should move the turtlesim with the speed and direction already specified by the executor.

Now, the question is how to send these requests through a Web service client. So, the objective is to expose the publisher of the topic /turtle1/cmd_vel and the subscriber on the topic /turtle1/pose as a Web services so that they can be invoked by a Web service client to send a request to the robot for controlling its movement and position respectively.

SOAP Web Services package:
Referring to our architecture (Fig. 2), the SOAP Web Services package implements web methods that expose robot capabilities using ROS package to publish/subscribe on a ROS topic and request ROS service. For our illustrative example, we implement a Java class defined as a web service and called "TurtlesimWS.java" to control the turtlesim. This class contains web methods that execute "Position.java" and "Motion.java" classes implemented in ROS package to publish on the topic /turtle1/cmd_vel and subscribe on the topic /turtle1/pose.

In a first step, we declare "TurtlesimWSInterface.java" which is a Java interface for the "TurtlesimWS.java" web service. This represents a good practice in Web Services design so as to separate the implementation of the Web service from its interface.

Class n° 1: TurtlesimWSInterface.java:

```
1   package com.soap.ws;
2
3   import javax.jws.WebMethod;
4   import javax.jws.WebService;
5   import javax.jws.soap.SOAPBinding;
6   import javax.jws.soap.SOAPBinding.Style;
7
8   @WebService
9   @SOAPBinding(style = Style.RPC)
10  public interface TurtlesimWSInterface {
11
12      @WebMethod
13      public void getPosition();
14
15      @WebMethod
16      public void move(String direction, double speed);
17
18  }
```

The interface contains two Web methods exposed to the Web service clients: (1) The getPosition(...) method is responsible for executing the request for getting the position of turtlesim, and (2) The move(...) method which takes the speed and the direction values as parameters, is responsible for executing the request for moving the turtlesim. It is clear that this interface provides a clear description of the service provided by the service robot, and hides implementation details from the end-users. All what clients applications need to know is the service description, also known as the contract, that defines the methods that can be invoked. They just need to invoke these web methods to control the robot

Now, in the back-end of the Web service layer, we define the implementation of the Web service interface in "TurtlesimWS.java" class.

Class n° 2: TurtlesimWS.java:

```
1   package com.soap.ws;
2
3   import java.net.URI;
4   import javax.jws.WebService;
5   import org.ros.node.DefaultNodeMainExecutor;
6   import org.ros.node.NodeConfiguration;
7   import org.ros.node.NodeMain;
8   import org.ros.node.NodeMainExecutor;
9   import com.google.common.base.Preconditions;
10  import com.ros.nodes.Position;
11  import com.ros.nodes.Motion;
12
13
14  @WebService(endpointInterface = "com.soap.ws.TurtlesimWSInterface")
15  public class TurtlesimWS implements TurtlesimInterface{
16
17      private void executeNode(NodeMain node){
18
19          URI masteruri = URI.create("http://192.168.1.4:11311");
20          String host = "192.168.1.4";
21          NodeMainExecutor nodeMainExecutor = DefaultNodeMainExecutor↩
                .newDefault();
22          NodeConfiguration nodeConfiguration = NodeConfiguration.↩
                newPublic(host, masteruri);
23          Preconditions.checkState(node != null);
24          nodeMainExecutor.execute(node, nodeConfiguration);
25      }
26
27      @Override
28      public void getPosition(){
29          Position position = new Position();
30          executeNode(position);
31      }
32
33      @Override
34      public void move(String direction, double speed){
35          Motion motion = new Motion(direction,speed);
36          executeNode(motion);
37      }
38  }
```

Observe that we define three methods: (1) the executeNode(...) method which is responsible for executing ROS nodes. It takes as a parameter the "NodeMain" object which is a super class of ROS nodes. This method instantiates "NodeMainExecutor" class provided by ROSJAVA API to allow the execution of ROS nodes, (2) move(...) and (3) getPosition(...) web methods which instantiate "executeNode()" method to execute ROS nodes in order to publish on the topic /turtle1/cmd_vel and subscribe on the topic /turtle1/pose respectively.

Now, we need to publish the web service implementation in order to expose defined web methods to the public. We present "TurtlesimWSPublisher.java" which is a publisher of the web service implementation "TurtlesimWS.java;;.

Class n° 1: TurtlesimWSPublisher.java:

```
 1  package com.soap.endpoint;
 2
 3  import javax.xml.ws.Endpoint;
 4
 5  import com.soap.ws.TurtlesimWS;
 6
 7  public class TurtlesimWSPublisher {
 8
 9      public static void main(String[] args) {
10
11          Endpoint.publish("http://localhost:1239/turtlesim", new ↵
                TurtlesimWS());
12      }
13  }
```

Note that you may refer to Sect. 2 for more clarification and explanation of the code.

Now, we define a client web service to consume the SOAP web methods that we have implemented and exposed for Turtlesim. We will just make a simple test by moving the turtlesim forward.

Class n° 1: TurtlesimWSClient.java:

```
 1  package com.soap.client;
 2
 3  import java.net.URL;
 4  import javax.xml.namespace.QName;
 5  import javax.xml.ws.Service;
 6  import com.soap.ws.TurtlesimInterface;
 7
 8  public class TurtlesimWSClient {
 9
10      public static void main(String[] args)  throws Exception{
11
12          URL url = new URL("http://localhost:1239/turtlesim?wsdl");
13          QName qname = new QName("http://ws.soap.com/", "↵
                TurtlesimWSService");
14          Service service = Service.create(url, qname);
15
16          TurtlesimWSInterface turtlesim = service.getPort(↵
                TurtlesimWSInterface.class);
17          turtlesim.move("forward",2);
18      }
19  }
```

We notice that SOAP web services hide all ROS details and provide a new way to interact with robots. The web service client do not need any background on ROS ecosystem or on robotics. He just need to execute a web service in order to interact with the robot. In our example, the web service client sends a request to the move(...) web method in order to move the turtlesim forward with speed of 2 m per second.

TEST YOUR CODE

To test the code, please run the rosmaster by running in the terminal:

```
roscore
```

Then, run turtlesim node by taping in the terminal:

```
rosrun turtlesim turtlesim_node
```

Now, run "`TurtlesimWSPublisher.java`" class and then "`Turtlesim WSClient.java`" class. You should see the turtlesim moving forward.

RESTful Web Services:

As explained in our architecture, RESTful web services expose the robot's capabilities by executing corresponding ROS nodes implemented in the ROS package. For our example, we consider turtlesim as a single resource. Thus, we define "`TurtlesimResource.java`" class that implements three methods: (1) the "`getPosition(...)`" web method which executes "`Position.java`" class to subscribe on `/turtle1/pose` topic and get the position of turtlesim, (2) the "`move(...)`" web method that executes "`Motion.java`" class to publish on `/turtle1/cmd_vel` topic and move the turtlesim and (3) the "`executeNode(...)`" method which executes ROS nodes using ROSJAVA API. Class 1 presents an implementation of the "`TurtlesimResource.java`" class for controlling turtlesim.

Class n° 1: TurtlesimResource.java:

```
1   package com.rest.resource;
2
3   import java.net.URI;
4   import javax.ws.rs.GET;
5   import javax.ws.rs.POST;
6   import javax.ws.rs.Path;
7   import javax.ws.rs.PathParam;
8   import javax.ws.rs.core.Response;
9   import org.ros.node.DefaultNodeMainExecutor;
10  import org.ros.node.NodeConfiguration;
11  import org.ros.node.NodeMain;
12  import org.ros.node.NodeMainExecutor;
13  import com.google.common.base.Preconditions;
14  import com.ros.nodes.Position;
15  import com.ros.nodes.Motion;
16
17  @Path("/turtlesim")
18  public class TurtlesimResource {
19
20      private void executeNode(NodeMain node){
21
22          URI masteruri = URI.create("http://192.168.0.123:11311");
23          String host = "192.168.0.123";
24          NodeMainExecutor nodeMainExecutor = DefaultNodeMainExecutor↩
                  .newDefault();
```

```
25      NodeConfiguration nodeConfiguration = NodeConfiguration.↵
           newPublic(host, masteruri);
26      Preconditions.checkState(node != null);
27      nodeMainExecutor.execute(node, nodeConfiguration);
28  }
29
30  @POST
31  @Path("/position")
32  public void getPosition() {
33      Position position = new Position();
34      executeNode(position);
35  }
36
37  @GET
38  @Path("/{direction}/{speed}")
39  public void move(@PathParam("direction") String direction, ↵
           @PathParam("speed") double speed) {
40      Motion moveForwardTurtlesimNode = new Motion(direction,↵
           speed);
41      executeNode(moveForwardTurtlesimNode);
42  }
43 }
```

In "TurtlesimResource.java" class, the getPosition(...) Web method is executed when the Web service receives an HTTP request using the POST operation on the following URI /turtlesim/position. However, the move(...) web method is executed when an HTTP GET request is received on the URI /turtlesim/, added to this the list of parameters passed to the Web method like /turtlesim/forward/2 to request to move the turtlesim forward with speed of 2 m per second.

After preparing the implementation of RESTful web service for turtlesim, we implement the "TurtlesimResourcePublisher.java" class to publish and expose the web service to the public.

Class n° 1: TurtlesimResourcePublisher.java:

```
1  package com.rest.endpoint;
2
3
4  import java.io.IOException;
5  import java.util.logging.Level;
6  import java.util.logging.Logger;
7  import com.rest.resource.TurtlesimResource;
8  import com.sun.jersey.api.container.httpserver.HttpServerFactory;
9  import com.sun.jersey.api.core.ClassNamesResourceConfig;
10 import com.sun.net.httpserver.HttpServer;
11
12 public class TurtlesimResourcePublisher {
13     public static void main(String[] args) {
14         try {
15             HttpServer create = HttpServerFactory.create(
16             "http://localhost:7001/",new ClassNamesResourceConfig(↵
                   TurtlesimResource.class));
17             create.start();
18         } catch (IllegalArgumentException e) {
19             e.printStackTrace();
20             } catch (IOException ex) {
```

```
21            Logger.getLogger(TurtlesimResourcePublisher.↩
                 class.getName()).log(Level.SEVERE, null, ex↩
                 );
22          }
23       }
24  }
```

Observe that we deploy the RESTful resource "TurtlesimResource" on an HTTP server which is accessible through the URL http://localhost:7001/.

We can test our web service by sending an HTTP GET request on the URI /turtlesim/forward/2 to move the turtlesim forward with speed of 2 m per second.

Now, we implement a client web service to consume the RESTful web service that we have implemented and exposed for Turtlesim.

Class n° 1: TurtlesimResourceClient.java:

```
1   package com.rest.client;
2
3   import java.io.BufferedReader;
4   import java.io.InputStreamReader;
5   import java.net.HttpURLConnection;
6   import java.net.URL;
7
8   public class TurtlesimResourceClient {
9
10      private final String USER_AGENT = "Mozilla/5.0";
11
12      public static void main(String[] args) throws Exception {
13
14          TurtlesimResourceClient http = new TurtlesimResourceClient↩
                 ();
15
16          System.out.println("Testing 1 - Send Http POST request");
17          http.moveForwardTurtlesim();
18
19      }
20
21      // HTTP GET request
22      private void moveForwardTurtlesim() throws Exception {
23
24          String url = "http://localhost:7001/turtlesim/position";
25
26          URL obj = new URL(url);
27          HttpURLConnection con = (HttpURLConnection) obj.↩
                 openConnection();
28
29          con.setRequestMethod("POST");
30
31          //add request header
32          con.setRequestProperty("User-Agent", USER_AGENT);
33
34          int responseCode = con.getResponseCode();
35          System.out.println("\nSending 'POST' request to URL : " + ↩
                 url);
36          System.out.println("Response Code : " + responseCode);
37
38          BufferedReader in = new BufferedReader(
39                  new InputStreamReader(con.getInputStream()));
40          String inputLine;
41          StringBuffer response = new StringBuffer();
42
43          while ((inputLine = in.readLine()) != null) {
```

```
44        response.append(inputLine);
45      }
46      in.close();
47
48      //print result
49      System.out.println(response.toString());
50
51    }
52
53  }
```

Observe that the client application is completely unaware of the ROS ecosystem and the details of the implementation. It mainly consists in invoking the Web method `getPosition(...)` by sending `HTTP POST` request on the URI `/turtlesim/position` in order to get the position of the turtlesim. The developer makes a complete abstraction of ROS and is able to command the robot through available service interfaces.

TEST YOUR CODE

To test the code, please run the rosmaster by running in the terminal:

```
roscore
```

Then, run turtlesim node by taping in the terminal:

```
rosrun turtlesim turtlesim_node
```

Now, run "`TurtlesimResourcePublisher.java`" class and then "`TurtlesimResourceClient.java`" class. You should see the position of the turtlesim

5 Conclusion

In this chapter, we presented a tutorial on how to integrate SOAP and REST Web services into ROS. The objective of this integration is to provide new abstraction layers on top of ROS to make easier the development of robotics applications by non roboticians. We proposed a software architecture for SOAP and REST Web services integration with ROS and we validated it through a real implementation. This integration is one step towards the new paradigm of cloud computing.

Acknowledgments This work is supported by the myBot project entitled "MyBot: A Personal Assistant Robot Case Study for Elderly People Care" [1] under the grant number 34–75 from King AbdulAziz City for Science and Technology (KACST). This work is partially supported by Prince Sultan University.

References

1. Mybot Project, KACST Project Number, (2015), pp. 34–75
2. Video Demonstration of Mybot Service Robot for Courier Delivery Application, (2015). https://www.youtube.com/watch?v=otltmx2-uca
3. A. Koubâa (2015) ROS as a service: web services for robot operating system. J. Software Eng. Robot. 6(1), ISSN:2035-3928
4. C. Pautasso, O. Zimmermann, F. Leymann, Restful Web Services Versus "Big" Web Services: Making the Right Architectural Decision, in *Proceedings of the 17th International Conference on World Wide Web*, ser. WWW '08. (New York, NY, USA: ACM, 2008), pp. 805–814. http://doi.acm.org/10.1145/1367497.1367606
5. A. Koubâa, M. Sriti, H. Bennaceur, A. Ammar, Y. Javed, M. Alajlan, N. Al-Elaiwi, M. Tounsi, E.M. Shakshuki, COROS: A Multi-agent Software Architecture for Cooperative and Autonomous Service Robots, in *Cooperative Robots and Sensor Networks 2015*, (2015), pp. 3–30. http://dx.doi.org/10.1007/978-3-319-18299-5_1
6. A. Koubaa, A Service-Oriented Architecture for Virtualizing Robots in Robot-as-a-Service Clouds, in *Architecture of Computing Systems—ARCS 2014*, (2014)
7. Introducing Rostful: Ros Over Restful Web Services, (2015). http://www.ros.org/news/2014/02/introducing-rostful-ros-over-restful-web-services.html
8. S. Osentoski, G. Jay, C. Crick, B. Pitzer, C. DuHadway, O.C. Jenkins, Robots as Web Services: Reproducible Experimentation and Application Development using Rosjs, in *2011 IEEE International Conference on Robotics and Automation (ICRA)*, (2011)

rapros: A ROS Package for Rapid Prototyping

Luca Cavanini, Gionata Cimini, Alessandro Freddi,
Gianluca Ippoliti and Andrea Monteriù

Abstract ROS framework lacks of an internal tool to design or test control algorithms and therefore developers have to test their algorithms on-line, directly on the robotic platform they are working with. This is not always safe and possible, and a rapid prototyping tool can help during the design phase. Users can develop their algorithms directly on the controller board and safely test them in a simulated scenario. Although some rapid prototyping tools exist in the ROS community, none of them take Simulink® into consideration. In this work the authors provide an open source Rapid Prototyping tool which integrates ROS and Simulink. The proposed package is useful for control designers, who are frequently used to exploit Simulink features for control deployment. The tool can be downloaded from https://github.com/gionatacimini/rapros.

1 Introduction

In the last years Rapid Prototyping (RP) has become very popular in several engineering fields. Generally, it refers to strategies, tools or platforms which help the developers, by reducing the design time phase and increasing the overall efficiency of development process. It is clear that RP is a general concept and can assume

L. Cavanini · G. Cimini (✉) · G. Ippoliti · A. Monteriù
Dipartimento di Ingegneria Dell'Informazione, Università Politecnica delle Marche,
via Brecce Bianche 12, 60131 Ancona, Italy
e-mail: g.cimini@univpm.it

L. Cavanini
e-mail: l.cavanini@univpm.it

G. Ippoliti
e-mail: g.ippoliti@univpm.it

A. Monteriù
e-mail: a.monteriu@univpm.it

A. Freddi
Università degli Studi eCampus, via Isimbardi 10, 22060 Novedrate, CO, Italy
e-mail: alessandro.freddi@uniecampus.it

© Springer International Publishing Switzerland 2016   491
A. Koubaa (ed.), *Robot Operating System (ROS)*, Studies in Computational
Intelligence 625, DOI 10.1007/978-3-319-26054-9_19

different meanings depending on the field it is referring to [14, 15]. In robotics these differences are emphasized due to the several fields involved for the complete developing of a robot, i.e. mechanics, electronics, informatics and automation. Thus frameworks for RP design of robots are very common, [22] as well as tool-chains for the rapid, and possible easier, low-level code embedding, together with communication infrastructure testing, [16, 19, 20, 27]. In this paper we focus on RP for robots' behaviour design, whose potentialities have been widely used and applications detailed in the robotics literature [12, 17, 25]. Whereas the control algorithm design could be include in the ROS framework, this framework lacks of tool for RP. At the moment, the developer has to test on-line the algorithms, directly on the robotic platform. But this is not always safe or possible. The proposed solution is a package which enables controller RP directly in the ROS-based applications. It facilitates the system- and component-level design and the simulation, together with continuous test and validation of the control system. In both academy and industry, one of the most appreciated tool for system modelling and control design is MAT-LAB [13]. In particular, Simulink is very widespread in the control community as it provides a graphical interface and libraries to build dynamical models using block diagrams. This allows quickly and clearly to model a system or design a control, while limiting at the same time the code to write inline. ROS framework lacks of a modelling language or tool for dynamical models, thus Simulink is the perfect candidate to accomplish this task.

Several solutions have been recently proposed to provide an effective ROS and Matlab interaction. From 2015, a Mathworks® official library is available to interface Matlab and Simulink environment and ROS systems; unfortunately it is available only in a payment version. Furthermore independent research groups and universities proposed several solution [1–3]. The main disadvantage of all available tools is that they are tailored for Matlab command-line and consequently they are hard or even impossible to use in Simulink due to the limited functions support in the simulation environment. In addition, most of them are based on external libraries, and their support and integration highly depend on external sources. It is worth noticing that a well known platform for ROS users for robot simulation is Gazebo, but, despite its very good features in terms of simulation, it is not suited for control design and RP [21].

The previous background and motivations inspired the research presented in this work. The novelty and the main contribution is a tool for RP of control algorithms which allows the integration of ROS framework and Simulink. On one side, the control community will benefit from this tool since testing controllers on real hardware will turn to be an easier task thanks to the hardware abstraction of the ROS framework. On the other side, ROS users will benefit from the tool for its RP capability, which is the main contribution of the work. There is a huge spread of Matlab users in the ROS community, therefore the majority of the software's audience is not required to learn a new tool or language. The proposed tool is easy to install and use; it guarantees also an optimal supportability and maintainability as it is completely based on Matlab and ROS native functions, mostly Python-based. The tool is named *rapros* (*ra*pid *p*rototyping *ROS*), and can be downloaded from https://github.com/gionatacimini/rapros.

The main feature provided by the software is the possibility to directly develop algorithms on the controller board and safely test them in a simulated scenario. The proposed package will be useful in the design, test and tuning of control algorithms for robot applications in ROS. The *ros_control* package is available in ROS framework to accomplish control design tasks [4]. It is based on a generic control loop feedback mechanism, typically a PID controller, for driving the output sent to the actuators. It contains a set of plug-in and standard hardware interfaces, and allows to connect to sensors and actuators. However, it is oriented to the development of low-level controller, without allowing the test and prototyping of the designed algorithm. Thus the proposed tool can be used also in combination with the *ros_control* package to extend its feature, allowing a direct communication with MATLAB.

To better understand its potentialities, the authors provide two examples of the possible uses of the tool. The first example regards the Unmanned Aerial Vehicle (UAV) stabilization, while the second is the control of a fan. In both the examples, linear controllers have been used, and the setup scenario for RP, consists of a PC with Matlab/Simulink and a board with ROS framework embedded; both the BeagleBone and the BeagleBoard-xM have been used and tested as target boards.

The paper is organized as follows. Section 2 provides a background detailing the main features of RP. Section 3 describes the steps needed to get the tool working in both ROS and Matlab. Section 4 presents the above mentioned examples where the tool is used for rapid prototype controller design. Finally Sect. 5 describes the package structure and operation.

2 Background

The adopted terminology in the field of RP is confusing, since it is strictly related to the engineering field of application. In the following, we provide some definitions which hold true for the rest of the chapter. They are specifically tailored for the application of interest (i.e. robotics and control) [23, 24]:

- **Rapid Prototyping** (RP): it is the set of procedures which helps to design and to develop controllers Fig. 1 for robotics applications, it is a general concept which includes both Processor in the Loop and Hardware in the Loop;
- **Processor In the Loop** (PIL): it is the RP technique where the real control hardware is involved and tested, and all the other hardware components are modelled on a PC [26];
- **Hardware In the Loop** (HIL): it is the RP technique where the hardware components (or part of them) are real, while the control algorithm runs on a PC, thus it is simulated [11].

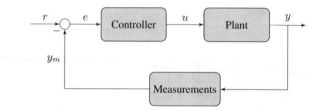

Fig. 1 Closed loop control scheme

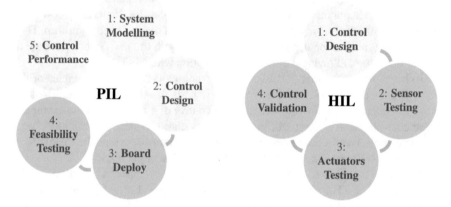

Fig. 2 Rapid prototyping: PIL (*left*) an HIL (*right*) concepts. In *red* the steps on ROS embedded, in *yellow* the steps in MATLAB framework

PIL and HIL are both useful and complementary in the design of robot behaviours, allowing continuous testing and deployment of features. Figure 2 shows the steps in which PIL and HIL can be divided. When speaking about prototyping, PC and real robotic hardware are considered. In the figure the main steps involving PC, namely Simulink, are highlighted in red, while the steps involving real hardware are highlighted in orange.

In PIL techniques, the robot model, or one of its subsystems, is implemented in Simulink, which makes the task east thanks to the block-like programming language. Once the dynamic system of the process is modelled, a control algorithm is designed and tested in simulation (see Fig. 3a). The satisfactory control is then deployed on the real platform. This steps are very easy to accomplish as Simulink provides several toolboxes for the control design, e.g. fuzzy, PID, neural networks etc. Furthermore, the deployment is assisted by the automatic code generation feature which, at least, represents a good starting point for the board deployment. The feasibility testing is the key point of the PIL procedure. Indeed the processor's performances are evaluated, mainly, in terms of computational burden and memory allocation. Embedded platforms provide Micro Controller Units (MCUs) which have poor features respect to the PC where the algorithm is initially tested. Complex algorithms could not fit into the available memory or could take more than a sampling interval to be evaluated.

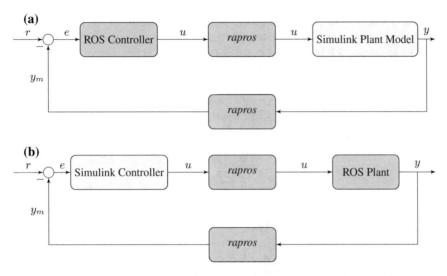

Fig. 3 Block schemes for the possible uses of *rapros*. **a** PIL objective: test the real control hardware on a plant modelled in Simulink. **b** HIL objective: test the real plant hardware (or part of it) on a controller synthesized on a PC

For this reason, the MCU is profiled, and the user decides if it is worth optimizing the code or switching to a more powerful device. Once the feasibility testing is passed, the control performance are evaluated. This is important for some critical aspects which are difficult to predict during the control design test. For example the arithmetic precision of the embedded platform is usually lower (i.e. single precision) with respect to a PC, and this can lead to different behaviours, especially in complex algorithms.

In control design, HIL follows the general steps depicted Fig. 2. HIL is devoted to test the hardware resources of the system, excluding from the processor, thus the control algorithm is designed and runs on a PC (see Fig. 3b). The hardware to be tested consists of sensors and actuators. Both of them have to be characterized, and the availability of a PC based application, facilitates this task. Acquisitions, open-loop tests, changes in the software are easily performed. Moreover, it is common in robotic platform to have multiple peripherals. The key factor for a successful integration of all the peripherals, could be the exploitation of an incremental integration of hardware trough RP. Thus this approach permits to start the design phase with a completely simulated model, and to add, one by one, real hardware components allowing the interfaces and behaviours tests [25].

3 ROS Environment Configuration

The importance of RP in embedded applications has been highlighted in the previous section. For this reason the proposed *rapros* package has been installed ad tested on embedded boards, namely BeagleBone Black and BeagleBoard-xM, running Linux kernels and the ROS framework; these will used as control boars for testing the examples detailed in Sect. 4. The setup procedure on this low-cost boards is divided into three steps: (i) Linux kernel installation, (ii) ROS environment configuration, (iii) *rapros* package setup. It is worth noticing that *rapros* package could be tested also on a single PC, running ROS and Simulink.

The setup and configuration procedure are explained in the following. The guidelines highlight Unix OS commands; obviously the procedure can be performed on Microsoft Windows© systems. On Windows PC, Putty software is required [5]; in this case the correct configuration includes board IP (port 22) and *ssh* connection type.

3.1 BeagleBone Black Setup

BeagleBone Black is a low-cost development platform, which provides a AM335x 1GHz ARM® Cortex-A8 with 512MB DDR3 RAM and 3D acceleration.

It runs Ångström, a Linux distribution developed in order to work ARM units. The Ångström kernel image can be download from [6]. The installation of the operating system is easy following the guidelines in [7]. In the following we provide the main steps to build up the complete system. The board communicates with a PC through a USB port and a *ssh* connection, established with:

```
> sshroot@beaglebone.local
```

Usually username and password are both *root* at the first log-in.

Even if it is not mandatory, it is preferable to have an updated version of Ångström:

```
>opkg update
>opkg updrage
```

The ROS version, suitable for the board, can be downloaded from a *git* project. To directly clone the repository into the system, the last software version must be installed through *opkg*:

```
>opkg install git
>git clone git://github.com/vmayoral/beagle-ros.git
```

To install the ROS tool, an easy to use shell script is provided.

```
>cd beagle-ros/scripts
>bash ./minimal-ros-install-angstrom.sh
```

Once the installation is finished, it is advisable to configure the ROS environment variables in the *.bashrc* file. This is possible by using a software editor, like nano, typing in home directory:

```
>nano .bashrc
```

and adding to the end of the opened file, the following raws:

```
export ROS_HOSTNAME=localhost
export ROS_MASTER_URI=http://localhost:11311
```

The workspace is created during the installation task, but if this does not happen, it is possible to create it following the ROS tutorial available on the reference website [8].

3.2 BeagleBoard-xM Setup

BeagleBoard-xM offers a AM37x 1GHz ARM Cortex-A8 processor, with 512 MB LPDDR RAM and 3D acceleration. The procedure to install ROS on these boards is provided. In this case, the OS selected for the installation is the ARM version of Linux Ubuntu 10.04 LTS. This Linux distribution is more stable than the others Linux OS embedded alternatives. Indeed, Ångström release for this board is not still fully working.

The OS is downloaded and copied on the MicroSD card, and some local parameters must be set:

```
>wget https://rcn-ee.net/rootfs/2015-01-06/microsd/bbxm-ubuntu-14.04.1-
console-armhf-2015-01-06-2gb.img.xz
>unxz bbxm-ubuntu-14.04.1-console-armhf-2015-01-06-2gb.img.xz
>sudo dd if=./bbxm-ubuntu-14.04.1-console-armhf-2015-01-06-2gb.img
of=/dev/sdX
>sudo update-locale LANG=C LANGUAGE=C LC_ALL=C LC_MESSAGES=POSIX
```

In this case the ROS framework can be installed following the standard tutorial for desktop installation [9]. Occasionally, the authors have experienced unwanted

changes in Ethernet MAC address after a reboot. The following lines are sufficient
to avoid the problem:

```
>sudo gedit /etc/network/interfaces
>auto eth0
>iface eth0 inet dhcp
>hwaddress ether 01:02:03:04:05:06 >sudo /etc/init.d/networking restart
```

where the fourth command requires the correct Ethernet MAC address

3.3 rapros Package Installation

The proposed package can be downloaded from [8]. The *rapros* package comes with
two working examples:

- a customized *rapros* node and the related Simulink block to perform a loop-back
  test in Simulink and the ROS environment;
- the node and the Simulink block developed during the PIL test.

To install the package, *rapros* must be placed in the /src folder inside the ROS
workspace. From MATLAB side, the Matlab_Simulink directory inside the selected
example package must be added to the Matlab path. The *rapros* with the Simulink
block will appear in the Matlab palette. Finally, each launch file and each *rapros*
python file, in the launch directory needs the administrator privileges to be launched.
This is possible typing:

```
>sudo chmod +x name_file.launch
>sudo chmod +x name_file.py
```

3.4 rapros Parameters Setup

rapros package requires the correct setup of the connection parameters. These parameters are set on the launch file, for the ROS side, and on the *rapros* block mask, for
the Simulink side. The launch file allows to change the following paramenters:

- **ip_board**: in the IP parameter group, this indicates the IP address of the board
  connected to the personal computer through direct Ethernet connection;
- **ip_pc**: in the IP parameter group, this indicates the IP address of the personal
  computer connected to the board through direct Ethernet connection;
- T_s: indicates the sample time of the control system.

The user is encouraged to run the loop-back test, to verify the correct behaviour of the system. The example is located in *rapros_test*. It can be completely run on a PC without the use of a board, setting the *ip_board* and *ip_pc* to the local host address, namely 127.0.0.1.

In "*rapros* Simulink block", the parameters can be set through the block mask. The following fields can be customized:

- **Input dimension**: it indicates the dimension of the input from the Simulink to the *rapros* block;
- **Output dimension**: it indicates the dimension of the output from the *rapros* block to the Simulink model;
- **Sample time**: it indicates the sample time of the control system, and it must be the same of T_s value set in the launch file;
- **Host address receive**: this parameter indicates the address of the personal computer connected to the board through direct Ethernet connection;
- **Host address send**: it indicates the address of the board connected to the PC through direct Ethernet connection.

Once the configuration parameters are correctly set, the launch file of the needed example package starts the ROS-side package. If everything is working, the following message will be displayed:

```
wait upd packet
```

When the simulation starts in Simulink, the *rapros* node is directly synchronized with the model. If the *rapros* node stops, Simulink model stops the simulation with an error. When Simulink interrupts the simulation, the node returns in wait state. It is worth noticing that after each simulation, the *rapros* node must be restarted, in order to reset the paramter server of ROS Master, and the variables' values inside the node.

4 Starting with a Test

In this chapter we show the *rapros* functionalities into standard control development tasks, exploiting two well known examples in the literature. The first example is about an application of the package to a PIL task. In this example the package is used to verify the correct behaviour of a controller developed in ROS, and implemented on a board, with a model of the real system deployed in Simulink.

The second example is a typical Hardware In the Loop (HIL) task. The real system is controlled by a board connected to hardware sensors and actuators. The control algorithm is directly developed in the Simulink simulation environment and, thanks to the *rapros* package, it controls directly the real system.

The software employed in these tests are:

- Ubuntu 14.04 LTS (PC);
- Matlab/Simulink 2013b;
- Ubuntu Arm 14.04 LTS (Beagleboard-xM);
- Ångström 12.12 (Beagleboard Black);
- ROS Hydro Medusa (Beagleboard Black);
- ROS Indigo Igloo (Beagleboard-xM);
- *rapros* 1.0.

To run the test, the first step is to copy the *rapros* package into the *src* folder of the catkin workspace into the board:

```
>mv -r path/to/local/directory path/to/board/directory
```

The second step is to modify the example launch file, to insert the PC and board IP addresses in the appropriate section. It is possible do that by nano editor software:

```
>cd /catkin_ws/src/rapros/example/launch
>nano rapros_name_example.launch
```

Afterwards , at the computer side, the next step is to set the same IP addresses on the Simulink mask of the *rapros* block present in the Simulink example file.

Finally, in order to start the selected example, first the user has to run the modified launch file on the board:

```
>roslaunch rapros_name_example.launch
```

Then, to obtain the ROS node, the user has to wait that the simulation stars on Simulink environment; this is notified by the following message:

```
>wait udp packet
```

At this point it is possible to start the simulation on Simulink. The proper operation of the example will be notified by the iterative generation of the following messages on the board interface terminal:

Fig. 4 The quadrotor scheme

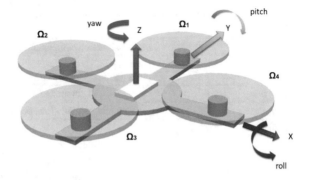

```
>wait udp packet
>received
>pc_ip
>Transmitted to Simulink
```

4.1 A PIL Example: The Quadrotor Stabilization

This example concerns the stabilization of a linearized quadrotor model, as the one depicted in Fig. 4, for which the dynamical model is built in Simulink. For the ease of reading, the notation is clarified in the following.

The parameters characterizing the non-linear and the linearized model [18] of the quadrotor are summarized in Table 1. The control purpose is to stabilize the z position component around the origin. The non-linear simulation model, implemented in Simulink, is:

$$
\begin{cases}
\dot{\Gamma}_{\text{NED}} &= R^T \dot{V}_B \\
\dot{\Phi} &= p + q \sin \Phi \tan \Theta + r \cos \Phi \tan \Theta \\
\dot{\Theta} &= q \cos \Phi - r \sin \Phi \\
\dot{\Psi} &= q \dfrac{\sin \Phi}{\cos \Theta} + r \dfrac{\cos \Phi}{\cos \theta} \\
\dot{u} &= rv - qw - g \sin \Theta \\
\dot{v} &= pw - ru - g \cos \Theta \sin \Phi \\
\dot{w} &= qu - pv + g \cos \Theta \cos \Phi - \dfrac{T}{m} \\
\dot{p} &= \dfrac{lb}{I_x}(\Omega_2{}^2 - \Omega_4{}^2) - qr \dfrac{I_z - I_y}{I_y} \\
\dot{q} &= \dfrac{lb}{I_x}(\Omega_1{}^2 - \Omega_3{}^2) - qr \dfrac{I_x - I_z}{I_y} \\
\dot{r} &= \dfrac{d}{I_z}(\Omega_1{}^2 - \Omega_2{}^2 + \Omega_3{}^2 - \Omega_4{}^2)
\end{cases}
\tag{1}
$$

Table 1 Variables and parameters of the considered quadtorot

Variables		
x, y, z	x, y, z Position components	(m)
Φ, Θ, Ψ	Roll, pitch, yaw angle orientation	(rad)
u, v, w	x, y, z Speed components	(m/s)
p, q, r	Roll, pitch, yaw rotation speed	(rad/s)
Parameters		
g	Gravity acceleration	9.8 (m/s^2)
Ω_{0i}	ith Motor rotation initial value	323 (rad/s)
l_y, l_x	Moment of inertia (roll, pitch)	0.0081 (kg/m^2)
l_z	Moment of inertia (Yaw)	0,0162 (kg/m^2)
m	Vehicle mass	0.85 (kg)
l	Arm length from the center of mass	0.2 (m)
b	Spin coefficient	1.46×10^{-5}

where

$$R^T = \begin{bmatrix} c\Theta c\Psi & s\Phi s\Theta c\Psi - c\Phi s\Psi & c\Phi s\Theta c\Psi + s\Phi s\Psi \\ c\Theta s\Psi & s\Phi s\Theta s\Psi - c\Phi c\Psi & c\Psi s\Theta s\Psi \\ s\Theta & s\Phi c\Theta & c\Phi c\Theta \end{bmatrix} \qquad (2)$$

and $s\alpha, c\alpha$ are equal to $\sin\alpha, \cos\alpha$ respectively.

In order to stabilize the quadrotor model is linearized around the initial condition Ω_{0i}, (see Table 1), and this model has been adopted for control purpose.

To stabilize the plant, a standard Proportional, Integral, Derivative (PID) controller is designed and tuned with the help of Simulink toolbox. In *rapros_quadrotor* node, the Python code for a discrete time PID controller is implemented.

The system control inputs Ω_i are referred to the rotation speed of the single motor. These inputs are mixed to obtain simple control signal. This is made in order to control the movement directions of the quadrotor controlling a single input signal u_i. This mix of signal is realized by a control matrix, presented below.

$$u_1 = \Omega_1 + \Omega_2 + \Omega_3 + \Omega_4 \qquad (3)$$
$$u_2 = \Omega_2 - \Omega_4 \qquad (4)$$
$$u_3 = \Omega_1 - \Omega_3 \qquad (5)$$
$$u_4 = \Omega_1 - \Omega_2 + \Omega_3 - \Omega_4 \qquad (6)$$

The Simulink model contains both models of the quadrotor system stabilized by the ROS controller and the PID block. In this way it is possible to compare the simulation result and appreciate the error introduced by implementing the control algorithm on a board rather than using that implemented in Simulink.

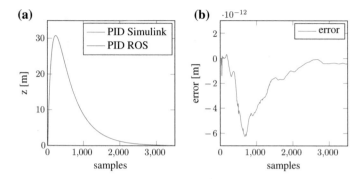

Fig. 5 Control results for quadrotor stabilization. **a** z component stabilization. **b** Error between PC-based and PIL test

The simulation starts running the *rapros_quadrotor* node and then the Simulink model containing the *rapros* block. Figure 5a, b show the control result and the error between PC-based and PIL test, respectively. The results presented are relative to both Beagleboard-xM and Beagleboard Black. The two boards provided exactly the same results and for this reason only one dataset is reported.

The first is the plot of the output signal of the controlled systems: there is not a visible mismatch between them.

The second plot is about the numerical mismatch between the two systems' output signals. This mismatch is computed as the difference between the two signals. This difference is really low, in the order of 10^{-12}, and near to the numerical precision of the computer.

This last result shows as the tool allows to obtain the same results of a native Simulink simulation, working with a different environment, or also with an external computation board.

4.2 An HIL Example: The FAN Control

The HIL example is based on the interaction of the board with a real system (Fig. 6). The selected system to control is a coil, connected to a rotating plane which rotates depending by the fan coil rotation speed. The control system is developed in order to maintain the plane to a constant tilt, as shown in Fig. 7. This HIL example has the objective to show that the proposed ROS tool can be effectively applied to a real control system problem, i.e. to control the tilt angle of the fan system. For this reason the ROS code for the deployment of the control system is here not reported.

The controller for this system is developed in a range of the tilt angle between 20° and 40°, because in this range the plane is characterized by a semi-linear behaviour. For bigger tilt angles, the system assumes a very strong non-linear behaviour due to the decrease of the plane surface hit by the wind. For controller development,

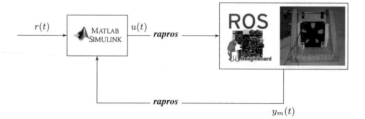

Fig. 6 The fan test control scheme

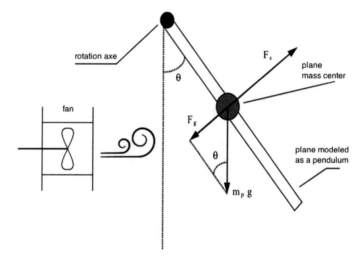

Fig. 7 The controlled system

the mathematical system model has been not employed, and a simple and realistic model was created by means of process identification techniques. These techniques allow to obtain a mathematical representation of the model behaviour through the following transfer function

$$W(s) = \frac{0.296}{1 + 0.7s} \tag{7}$$

The *rapros* allows to establish a direct connection from PC-running Simulink to the ROS embedded board (see Fig. 6). The board is directly connected with the low level fan control system, and to the fan tilt measurement sensor. It allows to generate directly a PWM control signal, read measurements from the sensor and routing data. In this case, the plant is actually made of the fan system, tilt sensor and the Beagleboard, on which ROS is installed.

The results presented are related to the fan controlled output signal. In Fig. 8 the fan tilt is controlled to follow a step signal of 39°. The system is controlled by a

Fig. 8 Controlled system
response to step signal

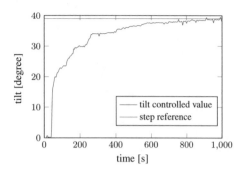

PID controller, developed in Simulink, which directly controls the real system. This result validates the controller development on the system model, with the tilt signal stabilized to the step reference signal value.

5 Package Description

The *rapros* package is composed by:

- *rapros.slx* Simulink block;
- *rapros.py* Python node.

These two components are detailed in the following sections. Depending on how the tool is used, it allows both PIL and HIL applications. Figures 9 and 10 represent the block diagram of PIL and HIL operation, respectively.

5.1 *rapros Simulink Block*

The Simulink block is configurable through a user-friendly mask (see Fig. 11) and allows to set the connection parameters to the ROS. The communication is imple-

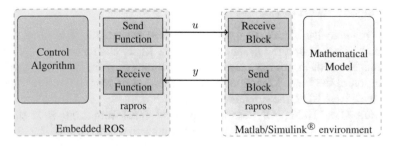

Fig. 9 PIL block scheme

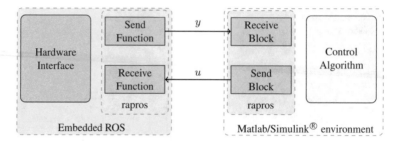

Fig. 10 HIL block scheme

Fig. 11 The *rapros*
Simulink block mask

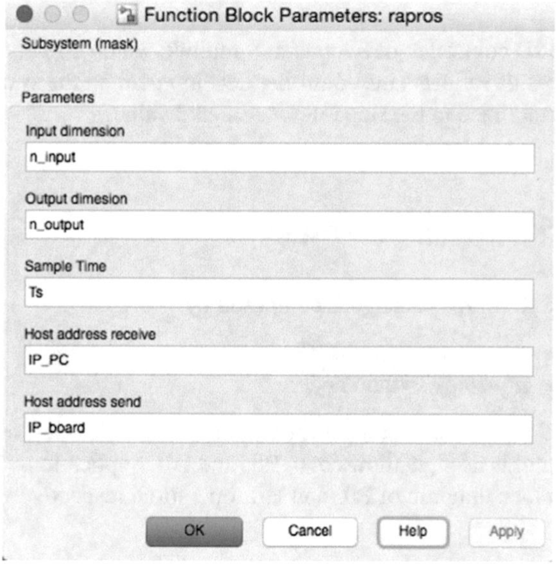

mented with UDP protocol, which does not manage the packets re-sending. Protocols
with re-sending features can destroy the correct order of data sequence which is of
utmost concern in this application. The blocks check also the communication status. When data are not received within a fixed timeout, the simulation stops and a
communication error is generated.

The block contains two sub-blocks:

- UDP send/receive blocks, that realize the UDP connection. The simulation is
  correctly blocked and the synchronization for RP is maintained by this block,
  through the built-in priority feature. Indeed, Simulink simulation runs until the
  execution of the "send" function, which causes a wait state. After receiving the
  data back from ROS, the simulation restarts;
- The Zero Older Hold (ZOH) block samples the signal from and to ROS environment, with a sample time T_s which is configurable in the mask of the block.

5.2 *rapros.py* Node

In Python, *rapros.py* node contains the code to interface with the Simulink block, in particular send and receive functions are implemented. The synchronization, as previously explained, is completely managed from Simulink side; thus, Python node remains always in a wait state. The node unpacks the UDP data and saves it in a "Float32MultiArray" of the ROS standard messages library. "Float32MultiArray" is a particular message which permits to define variables and topics messages of dynamic dimensions. This message is used to create automatically a data array of appropriate dimensions respect to the application, avoiding to hard-coding the node code. Dimensions are automatically calculated by the node for each UDP packet: this permits to handle a variable number of inputs and outputs. The node is defined by default publisher (to the "U" topic) and subscriber (to the "Y" topic) that allow to route data, of any dimension, from and to ROS environment. The proposed package is working with any release of ROS, from Groovy to the newest ones. To run the *rapros* also on previous ROS versions, the user only has to convert manifest and CMakeList files to Fuerte (and previous) format [10].

References

1. https://github.com/nmichael/ipc_bridge
2. https://code.google.com/p/mplab-ros-pkg/wiki/java_matlab_bridge
3. https://github.com/mozcelikors/Matlab-Ros-Interface
4. http://wiki.ros.org/ros_control
5. http://www.putty.org
6. http://www.angstrom-distribution.org/building-angstrom
7. http://downloads.angstrom-distribution.org/demo/beaglebone/
8. https://github.com/gionatacimini/rapros
9. http://wiki.ros.org/indigo/Installation/UbuntuARM
10. http://wiki.ros.org/catkin_or_rosbuild
11. Hardware-in-the-loop simulation
12. B. Bona, M. Indri, N. Smaldone, Rapid prototyping of a model-based control with friction compensation for a direct-drive robot. IEEE/ASME Trans. Mechatron. **11**(5), 576–584 (2006)
13. R. Bucher, S. Balemi, Rapid controller prototyping with Matlab/Simulink and Linux. Control Eng. Pract. **14**(2), 185–192 (2006)
14. G. Calisse, G. Cimini, L. Colombo, A. Freddi, G. Ippoliti, A. Monteriu, M. Pirro, Development of a smart led lighting system: rapid prototyping scenario, in *2014 11th International Multi-Conference on Systems, Signals Devices (SSD)*, pp. 1–6, Feb 2014
15. G. Cimini, M.L. Corradini, G. Ippoliti, G. Orlando, M. Pirro, A rapid prototyping scenario for power factor control in permanent magnet synchronous motor drives: control solutions for interleaved boost converters. Electr. Power Compon. Syst. **42**(6), 639–649 (2014)
16. G. Cimini, G. Ippoliti, G. Orlando, M. Pirro, PMSM control with power factor correction: rapid prototyping scenario. In *2013 Fourth International Conference on Power Engineering, Energy and Electrical Drives (POWERENG)*, pp. 688–693, May 2013
17. J. de Carufel, E. Martin, J.-C. Piedboeuf, Control strategies for hardware-in-the-loop simulation of flexible space robots. IEE Proc. Control Theory Appl. **147**(6), 569–579 (2000)

18. A. Freddi, S. Longhi, A. Monteriù, Actuator fault detection system for a mini-quadrotor, in *IEEE International Symposium on Industrial Electronics*, pp. 2055–2060, Bari, Italy, 4–7 July 2010

19. J.O. Hamblen, G.M.E. van Bekkum, An embedded systems laboratory to support rapid prototyping of robotics and the internet of things. IEEE Trans. Educ. **56**(1), 121–128 (2013)

20. K.-S. Hwang, W.-H. Hsiao, G.-T. Shing, K.-J. Chen, Rapid prototyping platform for robotics applications. IEEE Trans. Educ. **54**(2), 236–246 (2011)

21. N. Koenig, A. Howard, Design and use paradigms for Gazebo, an open-source multi-robot simulator, in *IEEE/RSJ International Conference on Proceedings Intelligent Robots and Systems, 2004. (IROS 2004)*, vol. 3, pp. 2149–2154, Sep 2004

22. T. Laliberte, C.M. Gosselin, G. Cote, Practical prototyping. IEEE Robot. Autom. Mag. **8**(3), 43–52 (2001)

23. H. Li, M. Steurer, K.L. Shi, S. Woodruff, D. Zhang, Development of a unified design, test, and research platform for wind energy systems based on hardware-in-the-loop real-time simulation. IEEE Trans. Ind. Electron. **53**(4), 1144–1151 (2006)

24. B. Lu, X. Wu, H. Figueroa, A. Monti, A low-cost real-time hardware-in-the-loop testing approach of power electronics controls. IEEE Trans. Ind. Electron. **54**(2), 919–931 (2007)

25. A. Martin, M.R. Emami, Dynamic load emulation in hardware-in-the-loop simulation of robot manipulators. IEEE Trans. Ind. Electron. **58**(7), 2980–2987 (2011)

26. B. Murphy, A. Wakefield, J. Friedman, *Best practices for verification, validation, and test in model-based design* (Technical report, SAE Technical Paper, 2008)

27. L. Rai, S.-J. Kang, Knowledge-based integration between virtual and physical prototyping for identifying behavioral constraints of embedded real-time systems. IEEE Trans. Syst. Man Cybern. Part A: Syst. Humans **39**(4), 754–769 (2009)

HyperFlex: A Model Driven Toolchain for Designing and Configuring Software Control Systems for Autonomous Robots

Davide Brugali and Luca Gherardi

Abstract A huge corpus of open source robotic software libraries is available on ROS repositories that can be reused to develop a large variety of robot control systems. The difficult challenge consists in selecting and integrating a coherent set of components that provide the required functionality taking into account their mutual dependencies and architectural mismatches. The HyperFlex approach presented in this chapter enables the explicit representation of robot system architectures, functional variability, and application requirements as software models that can be manipulated by a system configuration engine.

Keywords Model driven engineering · Software variability · Robotics

1 Introduction

The Robot Operating System (ROS) favors a software development approach that consists in designing fine-grain components, which implement common robotic functionalities. This approach is embodied by a repeated mantra among ROS developers [12]: "We don't wrap your main". The strength of this approach is the possibility to develop a large variety of different control systems by composing in multiple ways reusable software building blocks. Its weakness is the lack of support to the reuse of effective solutions to recurrent architectural design problems. Consequently, application developers and system integrators have to solve the difficult architectural design problems always from scratch.

The difficult challenge consists in selecting, integrating, and configuring a coherent set of components that provide the required functionality taking into account their mutual dependencies and architectural mismatches.

D. Brugali (✉)
School of Engineering, DIGIP, University of Bergamo, Bergamo, Italy
e-mail: brugali@unibg.it

L. Gherardi
Institute for Dynamic Systems and Control, ETH Zurich, Zurich, Switzerland
e-mail: lucagh@ethz.ch

© Springer International Publishing Switzerland 2016   509
A. Koubaa (ed.), *Robot Operating System (ROS)*, Studies in Computational
Intelligence 625, DOI 10.1007/978-3-319-26054-9_20

This challenge is exacerbated by the peculiarity of the robotics domain: robots can have many purposes, many forms, and many functions. Consequently, each robotic systems has to be configured with a specific mix of functionalities and that strongly depends on the robot mechanical structure (a rover with zero or multiple arms), the task to be performed (cleaning a floor, rescuing people after a disaster), and the environmental conditions (indoor, outdoor, underground).

In various application domains, software product line (SPL) development has proven to be the most effective approach to face this kind of challenges. A SPL is a family of applications (products) that share many (structural, behavioral, etc.) commonalities and together address a particular domain [11].

The core of an SPL is a stable software architecture that clearly separates common features from the variations reflected in the products and prescribes how software components can be assembled to derive individual products. Each new application is built by configuring the SPL, i.e. by selecting the variants (e.g. functionalities, software resources) that meet specific application requirements.

This chapter aims at presenting the *HyperFlex* approach and toolchain for the development of SPLs for autonomous robots based on robotic component frameworks, such as ROS and Orocos.

HyperFlex is a Model-driven engineering (MDE) environment [23], whose development started in the context of the EU FP7 BRICS project [7].

The chapter is organized as follows. Section 2 provides an overview of the proposed approach. It introduces the concept of software variability modelling using a simple example and illustrates the HyperFlex toolchain. Section 3 presents the ROS models and metamodels at the basis of the HyperFlex tools. Section 4 illustrates a realistic case study. Section 5 discusses the related works. The relevant conclusions are presented in Sect. 6.

2 Software Product Lines Development with HyperFlex

2.1 Modeling Stable Architectures

The architecture of a SPL plays the role of reference architecture [21] for a family of products. While reference architectures have been defined in other embedded systems domains (e.g. the AUTOSAR standard [3]), in Robotics, the goal of defining a reference architecture for every robotic application is elusive, because robots can have many purposes, many forms, and many functions.

Nevertheless, we argue that stable architectures can be defined for functional systems that provide common robotic capabilities: navigation, dexterous manipulation, perception, planning, and control are enduring business themes that represent the essence of the robotics domain. In this context, stable means that a significant variety of robot control systems can be developed by configuring (i.e. resolving the robotic variability) and integrating functional systems.

A functional system is a composition of elemental components and/or subsystems that together provide a specific robot capability (e.g. motion planning). Elemental components are basic modeling entities that are implemented as ROS Nodes or Orocos components.

It should be noted that, at runtime, a functional system is not distinguishable from the rest of the system as a self-contained composition unit. Indeed, ROS Nodes interact with each other by exchanging messages through a common bus according to the peer-to-peer architectural style. Nodes are equally privileged, equipotent participants in the application.

At design time, *HyperFlex* allows to explicitly model the architecture of functional systems in terms of components, interfaces, connectors, and components wiring. Models of functional systems can be reused as a building blocks and hierarchically composed to build more complex systems and applications.

As an example, Fig. 1a represents the architecture of a *3D Object Detector* system, which is made of two elemental components. The *Blob Detector* component receives as input a RGB image and computes the 3D position of objects of a specific color. A property called *BlobColor* is used for configuring the detector. The *Filter* component processes the input image for noise reduction (e.g. deblurring). This component is **optional** and is needed only when the images are acquired by a moving camera. Figure 1b represents a variant of the same architecture, where the *Filter* component is not present. In both variants of the architecture, the *Blob Detector* component is **mandatory**.

Components define provided and required interfaces (depicted as yellow and cyan squares, respectively), which can be connected by means of registers (green rectangles) according to the topic-based publisher/subscriber paradigm. Similarly, the *3D Object Detector* system exposes provided and required interfaces, which allows the composition with other systems. Figure 1c shows the composite architecture of a

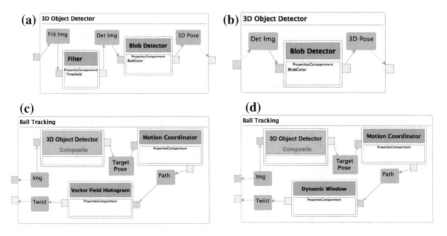

Fig. 1 The architectural models of the Ball Tracking SPL

functional system for ball tracking. It is composed of the *3D Object Detector* system and two ROS elemental components for motion coordination and obstacle avoidance. Figure 1d shows the same architecture but with an **alternative** implementation of the obstacle avoidance functionality.

2.2 Modeling Robotic Requirements

Each functional system is designed as a software product line, which specifies variation points and variants (e.g. alternative components, optional components).

HyperFlex uses the Feature Models formalism [19] to symbolically represent the variability of a functional system and the constraints between variation points and variants that limit the set of valid configurations.

Figure 2a represents the Feature Model of the *Object Detector* functional system. The link with the white circle specifies that the filtering functionality is optional, while the link with the black circle indicates that the blob color is a mandatory property. Two variants of blob color are supported, i.e. Red and Green. The cardinality indicates that exactly one color should be selected.

An instance of Feature Model is the collection of selected features, which are marked in green. As such, the instance represented in Fig. 2a corresponds to the configuration of the *Object Detector* functional system depicted in Fig. 1b, where the *Filter* component is not present.

Feature Models can be hierarchically composed to reflect the composition of functional systems. At each level the feature names abstract the relevant concepts of the corresponding system composition level. Low-level names represent functional and technical terms while high level names are closer to the application requirements. This approach ensures that the terminology is well known by the system integrators that operates on a specific level.

For example, the Feature Model depicted in Fig. 2b represent the variability of the *Ball Tracking* system depicted in Fig. 1d. Here, the feature *Object Detector* is a link to the root feature of the Feature Model depicted in Fig. 2a. The Feature *Ceiling* indicates that the camera is mounted in a fixed position, thus the *Object Detector* subsystem does not need to include the *Filter* component.

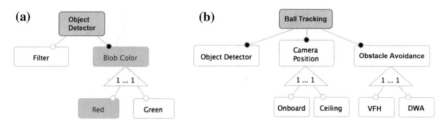

Fig. 2 The feature models of the object detector and ball tracking systems

During the variability resolution the *System Integrator* operates on the highest level Feature Model and the selected features trigger the automatic selection of the mapped features in the subsystem Feature Models. If necessary, the *System Integrator* can select features of lower-level Feature Models for specific configuration properties. The idea is similar to the concept of *configuration links* proposed in [22]. The approach is not limited to two levels but can be hierarchically extended according to the specific needs. Systems made of subsystems can be further composed in order to design more complex systems.

2.3 System Configuration

System configuration is a crucial phase in the robot application development process. It requires to select, integrate, and fine tune the functionalities of the autonomous robots according to the environment conditions (often completely beyond the control of the system integrator), the task to be performed (often continuativelly for long periods without human intervention), and the available resources (subject to failures and malfunctioning). When maintenance is needed, re-configuration has to be performed quickly to limit service disruption. Typically, system integrators are experts in a specific application domain (e.g. autonomous robots for inventory management), but are not specifically trained in software engineering. System integrators could be students who participate to a robocup competition, or professionals who bring together robotic subsystems in a manufacturing scenario and ensure that they function properly.

According to the SPL approach, a system configuration corresponds to a specific instance of the SPL architecture, where the software variability has been completely resolved by selecting appropriate variants for each variation point. *HyperFlex* allows to specify criteria for system configuration as model-to-model transformations associated to the features of the SPL feature model.

Figure 3 represents the Resolution Model for the *3D Object Tracking* system. In this example, a transformation is associated to the feature *Filter* and is executed when the feature is not selected. This transformation connects the *Blob Detector* input image with the *Object Detector* input port and specifies that the *Filter* component is not required.

The ROS framework provides a command-line application (i.e. the *roslaunch* tool) that parses an xml-based launch file (i.e. ROS launch file) and activates the required components.

Once the architectural model has been configured according to the feature selection, a model to text transformation allows to generate the corresponding ROS launch file. The transformation recursively inspects the nested subsystems in the configured model in order to identify all the atomic ROS components. For each component a *node tag* is added to the ROS launch file. Subsequently, for each Node the transformation remaps the topic names (by adding *remap tags* to the ROS launch file)

Fig. 3 The resolution model
of the object tracking
subsystem

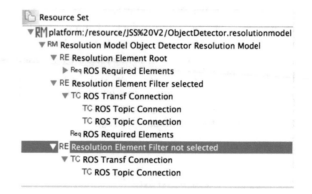

according to the connections specified in the architectural model of each functional system. Finally the transformation creates a *param tag* for each node's parameter.

Listing 1.1 illustrates an example of ROS launch file, which corresponds to the subsystem depicted in Fig. 1c. The file contains a *node tag* for each component and a few *param tags* for their parameters. In addition the *remap tag* changes the name of the topic on which the publisher of the node *Blob Detector* publishes the results of its computation.

```
<launch>
  <node name="Filter" pkg="object_detection" >
    <param name="threshold" value="0.5" />
  </node>
  <node name="BlobDetector" pkg="object_detection" >
    <param name="blob_color" value="5" />
    <remap from="3D_Pose" to="Target_Pose"/>
  </node>
  <node name="MotionCoordination" pkg="ball_tracking" />
  <node name="VectorFieldHistogram" pkg="ball_tracking" />
</launch>
```

Listing 1.1. An example of ROS launch file

2.4 The HyperFlex Toolchain

The *HyperFlex* toolchain is a collection of Eclipse plugins implemented by means of the Eclipse Modeling Project [1]. It is available open source on GitHub [2]. The toolchain includes four graphical editors and two model transformation engines as depicted in Fig. 4 for modeling, composing, and configuring robotic SPLs.

The *Architecture Editor* is used to design the Component&Connector Architectural Model of functional systems and applications.

The *Feature Editor* is used to model the application requirements and the variability of functional systems as Feature Models.

Fig. 4 Tools and models of the HyperFlex toolchain

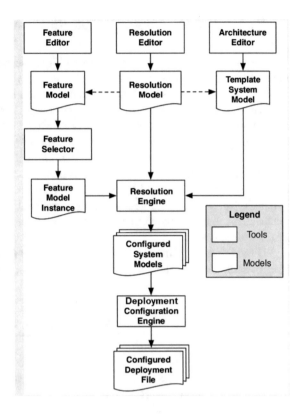

The *Resolution Editor* is used to define the Resolution Model as set of transformation rules that specify how the functional variability in the software architecture should be resolved when the features of the Feature Model are selected.

The *Feature Selector* is used to create instances of Feature Models that correspond to specific system configurations.

The *System Resolution Engine* receives as input a collection of Feature Model instances (one for each functional system), executes the transformation rules defined in the corresponding Resolution Models, and generates the variability-free instances of the Architectural Models.

The *Deployment Configuration Engine* transforms the variability-free Architectural Models into configuration files for the deployment of the functional systems (e.g. the ROS launch files). In [17] we describe how configuration files include all the *machine tags* required to deploy the different components on different computational nodes taking into account their available resources.

3 Variability Modeling, Composition, and Resolution

The core aspect of our approach is the possibility to design robotic control systems as hierarchical composition of functional subsystems, whose internal variability can be modeled and resolved using Feature Models and model-to-model transformations.

As depicted in Fig. 5, models of functional systems are structured in two levels of abstraction (Meta-Models (M2) and concrete Models (M1)) and organized in three categories (*Architecture variability*, *Symbolic variability*, and *Variability resolution*).

Each robotic component framework has its own specific M2 meta-model for modeling the architecture of a functional system. We have defined component meta-models for Orocos [16] and ROS (Sects. 3.1 and 3.2). The M2 meta-models for variability resolution are organized in two levels: abstract (*M2A*), i.e. framework-independent, and framework-specific (*M2S*). We have also defined a meta-model for Feature Models [15].

M1 models are organized in two sub-levels. The *M1-TPL* level refers to the template of a functional system, i.e. the software architecture of a functional system that

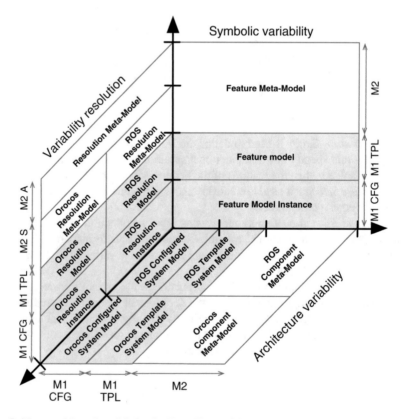

Fig. 5 Meta-models and models for the HyperFlex toolchain

needs to be configured by resolving its internal variability. The *M1-CFG* level refers to a configured functional system, i.e. a the software architecture of a functional system that can be deployed on a specific robotic system.

Template System Models are framework-specific, i.e. defined according to the corresponding component model (e.g. ROS or OROCOS), and support four variability mechanisms:

- separation of a component interface from its implementation and the consequent possibility to replace the implementation of a given component;
- definition of component properties, i.e. attributes that are used in the component implementation;
- explicit representation of connections between components and the possibility to add or remove them;
- explicit representation of optional components and the possibility to include or exclude them in the configured system.

For example, a *ROS Template System Model* defines the architectural model of a ROS-based functional system SPL, while a *ROS Configured System Model* defines the architectural model of a specific system derived from it. A *ROS Resolution Model* defines the model-to-model transformations.

The functional variability of a *Template System Model* is symbolically represented in a corresponding *Feature Model (M1-TPL)*. A *Feature Model Instance (M1-CFG)* represents the relevant features selected by a system integrator to configure a functional system.

HyperFlex supports the interconnection of heterogeneous functional systems (e.g. based on ROS and Orocos) to form heterogeneous composite systems, which in turn can be components of higher-level systems. For this purpose, we have defined (see Sects. 3.3 and 3.4) the *Abstract Heterogeneous Composition meta-model*, which is specialized for Orocos-ROS heterogeneous systems, and the *Heterogeneous Composition Resolution meta-model*.

3.1 ROS Component Meta-Model

ROS is a communication infrastructure supporting the integration of independently developed software components, called ROS nodes. Nodes are blocks of functional code and are implemented as classes that wrap robotic software libraries and provide access to the communication mechanisms of the underlying infrastructure (the ROS core). Nodes are compiled as stand-alone executables.

The pair *node package/node type* has to be unique in the file system and identifies the implementation of a node. Multiple instances of the same node are possible at runtime and are distinguished by the *node name*.

A ROS system is a computation graph consisting of a set of nodes communicating with one another by exchanging typed messages asynchronously according to the publish/subscribe communication paradigm and/or services according to the client/server paradigm.

Messages are organized by topics, which correspond to information subjects that allow subscribers to recognize the events they are interested in. When a node receives a message belonging to a subscribed topic, a message handler performs some computation on the message payload data and possibly generates a new message of a given topic to publish the computation results. A Node may also define *services* that are invoked by other nodes according to the client-server paradigm.

Figure 6 depicts the ROS Component meta-model, which has been split in three parts (a, b, c) for sake of readability.

Figure 6a depicts the entities used to model the mechanisms for message-based publish-subscribe communication. Nodes may have several provided and required interfaces (`NodeMsgProducer` and `NodeMsgConsumer`, respectively), which are connected to `Topic` entities. Topics represent message queues implemented in the ROS infrastructure and are distinguished by a name that represents a category of messages. A ROS `System` contains an instance of `Composite`, which hierarchically aggregates `Node` components according to the Composite Design Pattern [14]. `Composite` is only a modeling entity that does not correspond to any implementation entity. A composite may expose the topics of the messages

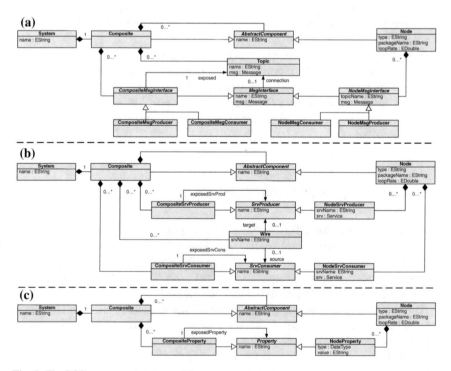

Fig. 6 The ROS component meta-model

produced or consumed by the internal nodes. For this purpose, a component may have a `CompositeMsgProducer` and/or a `CompositeMsgConsumer` interface for each exposed topic. An OCL constraint ensures that all the producers and consumers associated with the same topic conform to the same message type.

In order to make the communication between a producer and a consumer possible, they both have to refer to a topic defined by the same name, which is actually hard-coded in the nodes implementation. When this condition is not true, for example because the components have been implemented by different developers, a mapping is required between the names of the topic defined in each node implementation and a common logical name for the same topic. The logical name can be defined at design time, when the nodes are composed into composites, while the mapping is performed at deployment time by the ROS infrastructure.

Figure 6b depicts the entities used to model the mechanisms for service-based client-server communication. A node defines provided and required service interfaces (`NodeSrvProducer` and `NodeSrvConsumer`, respectively). They are typed by a `Service`, which defines the interface in terms of request and response messages and can be promoted at the level of composite interfaces (`CompositeSrv Producer` and `CompositeSrvConsumer`, respectively). Provided and required interfaces can be connected by means of `Wire` entities. OCL constraints ensure that wire connections are created between a pair of children of the same composite the interfaces of which conform to the same service.

Figure 6c depicts the mechanisms for the definition of properties. The `Property` entity provides an interface for setting the value of a parameter defined in the component implementation of a ROS node (e.g. the task execution period or an algorithm parameter). Properties can be promoted at the composite level.

3.2 ROS Resolution Meta-Model

The upper part of Fig. 7 depicts the *Abstract Resolution meta-model*, which is independent from any robotic software framework.

Class `ResolutionModel` encapsulates a set of `RMResolutionElements`, which associate instances of class `Feature` defined in the *Feature Model* with a set of model-to-model transformations (the `RMTransformations`) and/or to a set of architectural elements (`RMRequiredElements`).

Each `RMResolutionElement` has the boolean attribute `activeIf Selected`, which specify when the resolution element is active: if `true` the resolution is active when the associated feature is selected, if `false` the resolution is active when the associated feature is not selected.

`RMRequiredElements` reflects a collection of architectural elements that are defined in the Template System Model and that have to be present in the Configured System Model when the resolution element is active. *RMTransformations* are instead actions to be performed on the architectural elements defined in the Template System Model and are organized in three types.

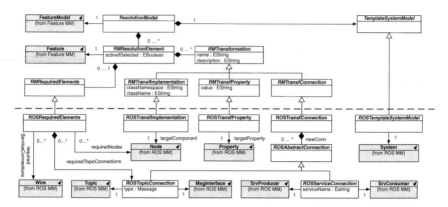

Fig. 7 The ROS resolution meta-model (framework-independent classes in the upper part, framework-specific classes in the bottom part.)

- **Implementation Transformation**: it specifies a link to a given component in the Template System Model and the implementation that has to be associated with it (in ROS it is identified by a package name and a node type, which are modeled by the attributes `classNamespace` and `className`).
- **Property Transformation**: it specifies a link to a property of a given component and the value that has to be assigned to it.
- **Connection Transformation**: it specifies a set of new connections that have to be created between pairs of components.

The lower part of Fig. 7 depicts the *ROS Resolution Meta-Model*, which is specific for the *ROS Template System Model*. Class `TemplateSystemModel` has to be extended with classes that hold a reference to the Template System Model of each specific software framework. Similarly, the abstract classes representing the different model-to-model transformations and the required architectural elements have to be extended in order to encapsulate references to framework-specific classes.

This design allows the definition of uniform resolutions models for every possible architectural model that support the four variability mechanisms described in Sect. 3, i.e. implementation, property, connection, and component variability. Following this approach, we have defined the resolution meta-model also for Orocos and the Service Component Architecture (SCA).

In particular, class `ROSRequiredElement` has references to ROS nodes, and wires and specifies which topic should be associated with message interfaces (see the ROS component meta-model in Fig. 6).

The `ROSTransfConnection` encapsulates a collection of `ROSAbstract Connections`, which specify the topic and service connections that have to be created between ROS nodes. If an existing connection needs to be removed, it will not be included among the `ROSRequiredElements`.

3.3 Architecture Composition Meta-Models

Figure 8 depicts the *Abstract Heterogeneous Composition meta-model*, which provides the mechanisms for modeling the architecture of complex systems by hierarchically composing heterogeneous subsystems designed with different software frameworks.

A CompositionModel is defined by a name and a System, which is a composition of Composite entities. The abstract class Composite is a wrapper to a functional system and needs to be specialized with a concrete class (i.e. ROSComposite in Fig. 9), which holds a reference to an instance of class Composite defined in the meta-model of each specific component model. Composite entities define CompositeProvidedInterfaces, CompositeRequiredInterfaces and CompositeProperties, which expose at composition level the interfaces and the properties of the concrete functional subsystem.

The class System also aggregates instances of class Connection, which connects provided and required interfaces of the encapsulated functional systems. The abstract class Connection needs to be specialized in order to define concrete connections between pairs of functional systems according to their component models. For example the ROSOrocosConnection in Fig. 9 allows the interconnection of a ROS system with an Orocos system.

Class System represents a heterogeneous system that may be composed of a set of homogeneous subsystems developed according to several component models (i.e. ROS and Orocos functional systems). It should be noted that class System might also encapsulate references to instances of other heterogeneous systems modeled by separated CompositionModels. For this purpose the class SystemComposite (Fig. 8) specializes the class Composite while the class System defines specific

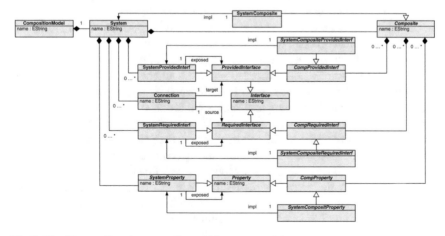

Fig. 8 The Abstract Heterogeneous Composition meta-model

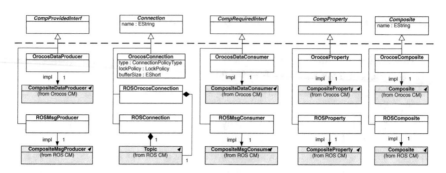

Fig. 9 The Heterogeneous Composition meta-model for ROS and Orocos (framework-independent classes are depicted above the *dashed line*)

provided and required interfaces and properties which expose the interfaces and properties of the encapsulated subsystems.

Due to limited space, Fig. 9 does not report classes for service interfaces and connection, which are however defined in the meta-model.

3.4   Resolution Composition Meta-Models

The *Heterogeneous System Resolution Model* represents the bridge between the features defined in the *Heterogeneous System Feature Model* and the configuration on the system architectures (i.e. *Heterogeneous Template System Models*). Figure 10 is structured in three parts. The central part depicts the entities of the *Abstract Resolution Meta-Model* already described in Sect. 3.2. The other entities define the concrete classes of the *Heterogeneous Composition Resolution Meta-Model*.

The class CompResolutionModel specializes the ResolutionModel. It is the root of the meta-model, which holds references to the ResolutionModels of the composed subsystems, to the FeatureModel of the heterogeneous system, and to the CompTemplateSystemModel (it should be noted that the references to the Resolution Models of the composed subsystems, implicitly provide references to their Feature Models and Template Subsystem Models).

CompRequiredElements defines a set of Composites and Connections, which reflect required subsystems and required connections between them. In addition, CompRequiredElements allow the specification of a set of required Features belonging to the Feature Models of the functional subsystems. This association defines the mapping between a feature of the Feature Model that abstracts the functional variability of the composite system and a set of features in the Feature Models of the functional subsystems. The transformation CompTransfImplementation allows the replacement of a composite, with another composite, which can even conform to a different meta-model, but must conform to the same set of interfaces. This is guaranteed during the definition of the

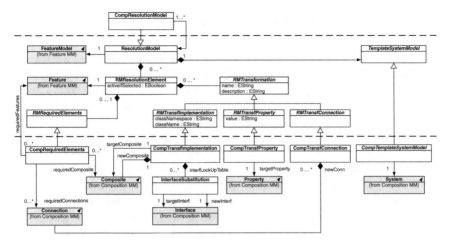

Fig. 10 The Heterogeneous Composition Resolution meta-model (framework-independent classes in the central part, framework-specific classes in the bottom and upper parts)

transformation thanks to a set of constraints. The transformation specifies the composite to be replaced (`targetComposite`), the replacing composite (`newComposite`), and a lookup table that defines a set of `InterfaceSubstitutions` (interface to be replaced and replacing interface). When this transformation is applied, all the connections that regard the replaced interfaces are updated.

In the simplest case, when the replaced and replacing composites conform to the same meta-model, only the fields `target` or `source` in the class `Connection` need to be changed. In a more complex case, it might be necessary to change the connection type. For example if the original connection was a `ROSConnection` and the new interface is an Orocos interface, the transformation will replace the old connection with a `ROSOrocosConnection`. Finally, `CompTransfProperty` and `CompTransfConnection` are used for setting the value of composite properties and for creating new connections between the subsystems.

4 Case Study: Autonomous Logistics

This section illustrates the HyperFlex approach to the development of a SPL for logistic applications. The case study is split in two parts: (a) the development of a ROS-based functional system for robot navigation, and (b) the composition of ROS and OROCOS functional subsystems for building a family of autonomous logistic applications.

For sake of simplicity, we illustrate the modeling process from requirements analysis to architectural design only for the navigation system.

4.1 The Robot Navigation Functional System

Modern warehouse logistics employ autonomous robotic systems with advanced robotic capabilities to move parts and objects in the warehouse. A major concern is the robot ability to precisely follow a given trajectory, which requires the ability to self-localize in the operational environment. While a large variety of algorithms exists for these capabilities (e.g. visual marker localization, laser-based mapping and localization), their mix and integration strongly depend on the requirements of the specific application scenario. For example, installing visual markers on a large surface area with hundreds of pick and place positions is cost efficient when a large number of robots (e.g. hundreds in the Amazon warehouse) operate in the same environment. In this case, the robots are equipped with simple sensors (e.g. low-cost camera) and functionality that rely on the regularity of the environment. On the contrary, for single robot scenarios, it is more convenient to equip the autonomous robot with a more complex sensory system (e.g. stereo vision, laser scanners, and inertial measurement unit) and more sophisticated algorithms for map-based localization and navigation.

Typically, an engineering company wants to achieve customer value through large commercial diversity of its products (e.g. different control systems for the above scenarios) with a minimum of technical diversity at minimal cost.

This requires to analyze the functional variability in the proposed application scenario in order to identify those aspects that are stable within the domain, and those aspects that are more likely to be affected by the evolution of the application domain (e.g. new sensors, new algorithms, new usage scenarios). Here stability can be defined as a system's resilience to changes in the original requirements specification.

Robot navigation involves several functionalities, such as *Localization*, *Path Planning*, *Trajectory Generation*, *Trajectory Adaption*, *Trajectory Following*.

A large variety of algorithms for these functionalities are available as open source libraries and components. While ROS favors the definition of standard messages for data exchange among components (i.e. the *Twist* and *Odometry* messages for rover drivers), the interoperability of different components providing the same functionality is often limited by context dependencies that are hidden in their implementations (e.g. the specific rover kinematic model, the specific type of sensor, etc.). It is therefore necessary to explicitly model the dependencies and constraints among reusable software components that limit the variability of a functional system.

Table 1 summarizes some examples of the variability that characterizes robot navigation. Column *Variation point* exemplifies three typical variability concerns Column *Variants* enumerates the possible variants for each variation point. Column *Type of Variant* specifies three type of variants:

- *Algorithm* indicates that the variants are provided by different algorithms for which may exists even multiple implementations.
- *Architecture* indicates that the variants correspond to different sets and arrangements of functionalities for which different algorithms may be used.
- *Attribute* indicates that the variants correspond to different values of some algorithms parameters.

Table 1 Variation points in robot navigation

Variation point	Variants	Variant type	Functionality
Rover kinematics	Omnidirectional differential	Algorithm	Trj-Generation, Trj-Adapter, Trj-Following
Sensor	3D Camera, laser	Algorithm	Localization, Trj-Adapter
Environment	Static, dynamic	Algorithm	Trj-Adapter
Environment	Structured, unstructured	Architecture	Path planning, localization
Behaviour	Performance, safety	Attributes	Trj-Generator, Trj-Adapter

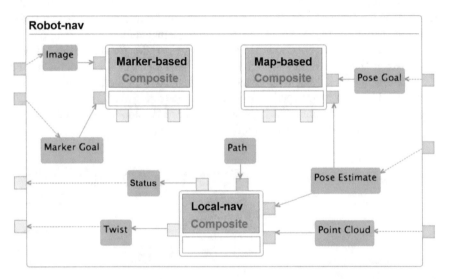

Fig. 11 The Robot Navigation composite

The architectural model of the Robot Navigation SPL is depicted in the screenshot of Fig. 11. It is a composition of three homogeneous functional subsystems, i.e. developed using the same robotic framework (ROS in this example).

The *Local Navigation* composite is a stable part of a Robust Navigation architecture. Its internal architecture defines the relationships between the basic functionalities to drive a robot between two waypoints of a path and to avoid unexpected obstacles detected by the sensors. Each functionality is provided by a specific component, whose interface is stable, while its implementation can be replaced. The path between a start position and a goal position can be generated according to two strategies based on the nature of the environment: marker-based and map-based.

The *Map-based Navigation* composite provides the functionality to plan a geometric path in the free space given a map of the environment. This functionality requires the robot to estimate its current position with respect to the map reference frame accurately.

The *Marker-based Navigation* composite provides the functionality to follow a path defined by a specific sequence of visual markers placed on the floor or on the walls. In this case, the robot needs only to estimate its relative position with respect to the next visual marker.

The *Map-based Navigation* composite and the *Marker-based Navigation* are optional components of the Robot Navigation SPL. At least one should be included in the configured system by adding a connection to the *Local Navigation* composite by means of the *Path* topic and publishing messages on the *Status* topic.

The Feature Model depicted in Fig. 12 captures the functional variability of the Robot Navigation functional system. For sake of simplicity, the figure does not represent the variability regarding all the different implementations of the components.

The selected features correspond to a sytem configuration for an *Omnidirectional* rover, which should exhibit a *Reactive* motion behaviour (i.e. high values for the acceleration limits). Obstacle avoidance is performed with the *Dynamic Window Approach*. The rover is equipped with a camera for detecting markers placed in *Fixed* positions in the environment (i.e. only horizontally on the floor or only vertically on the walls). *ARTK +* is the algorithm for marker localization.

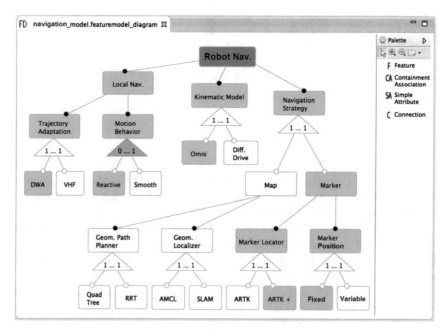

Fig. 12 A screenshot that shows the feature model of the Robot Navigation system (*green* features correspond to selected features)

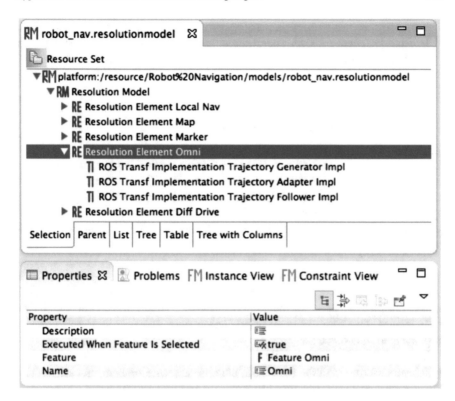

Fig. 13 The resolution model of the Robot Navigation system

The Resolution Model for the Robot Navigation functional system defines a set of resolution elements associated with the features presented above.

Figure 13 depicts an extract of the Robot Navigation Resolution Model (the name of the resolution elements reflect the associated features). The lower part shows the properties of the *Omni* resolution element, which refers to the feature *Omni*. For the feature selection depicted in Fig. 12 the following resolution elements are executed by the resolution engine:

- *Local Navigation*: specifies that all the components and the connections reported in the Local Navigation composite are required elements.
- *Marker*: specifies that the components of the Marker Based Navigation system and the connections between them are Required Elements. It also creates the connections between the *Local Navigation* composite and the *Marker-based* composite.
- *Omni*: specifies a set of implementation transformations, which set the implementations of algorithms for *Trajectory Generator*, *Trajectory Adapter* and *Trajectory Follower* that generate velocities and accelerations compatible with an omnidirectional kinematic model.

- *Reactive*: specifies a set of property transformations, which set the values of the *Trajectory Generator*, *Trajectory Adapter* and *Trajectory Follower* properties (i.e. acceleration limits).

4.2   The Autonomous Logistics SPL

Figure 14 depicts the architecture of the application layer. It has been designed with HyperFlex as a Heterogeneous System Model (see Sect. 3.3). *Driver* and *Task Manager* are mandatory functional systems, while *Robust-nav* and *Manipulation* are optional functional systems. This flexibility allows the configuration of control systems for simple manipulation tasks, simple navigation tasks, or composite mobile manipulation tasks.

The *Driver* functional system groups the components that provide access to the physical devices, such as the rover driver, which subscribes to twist messages and provides pose estimates, the laser scanner driver, which provides point clouds containing information on the surrounding obstacles, and the RGB camera, which captures images. Due to typical real-time requirements, this functional subsystem is implemented in OROCOS.

The *Task Manager* functional system groups the ROS components that, given a high level description of the robot task, plan the sequence of elemental actions that activate the system functionality (e.g. plan and execute a path, detect and grasp an object, etc.)

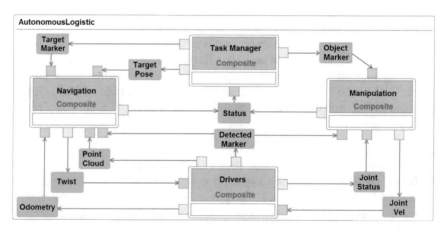

Fig. 14 Heterogenous template system model of the application layer for the *Autonomous Logistic* SPL

The *Robot Navigation* functional system (described in the previous section) is modeled here as a building block of the Autonomous Logistic SPL. It defines a set of interfaces, which can be used for interfacing the robot navigation system with the other functional systems.

The *Manipulation* functional system is structured as a composition of functional subsystems based on the OROCOS framework. As an example, Fig. 15 (copied from [16]) shows the model of the *Arm Trajectory Follower*. It should be noted that this subsystem is modeled according to the OROCOS component model, which does not use topics for inter-component communication.

The interaction of heterogeneous subsystems requires the use of software libraries that address the architectural mismatches between different software frameworks. These libraries are available for most of the robotic frameworks (e.g. ROS-Orocos [24], SCA-Orocos [9], SCA-ROS [8]). The HyperFlex *Architecture Editor* automatically identifies the connection type according to the type of the connected systems and associates with the connection the information required to configure the interconnection library.

The feature model of the Robot Navigation functional system represents concepts that are relevant for an expert in robot navigation, who knows which algorithm is most appropriate for specific operational conditions. For example, the path planning algorithm has to be chosen taking into account the structure of the environment, which could be open space or cluttered by thin obstacles. The selection of the obstacle avoidance algorithm depends on the dynamic of the environment, where only static or also moving obstacles are present. Robot localization can be performed using an RGB camera to detect visual markers, if the environment can be structured, or using a depth sensor, if a geometric map of the environment is available.

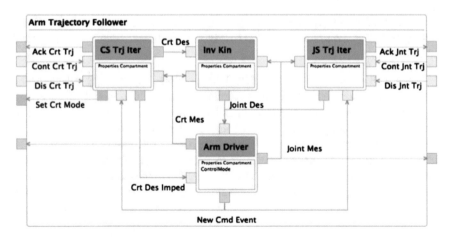

Fig. 15 Template System Model of an OROCOS-based functional system

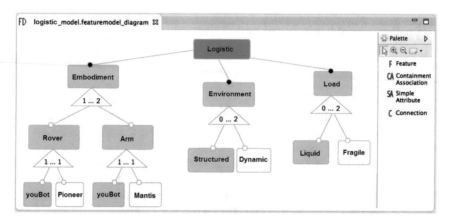

Fig. 16 The feature model of the *Autonomous Logistic* SPL

This knowledge can be captured by a higher level feature model, such as the *Logistic SPL* Feature Model depicted in Fig. 16. At this level of abstraction, the features represent concepts that are relevant to the system integrator. Three variation points have been defined (for sake of simplicity, only a limited number of variants are represented).

- *Embodiment* specifies the type of rover (*youBot* or *Pioneer*) and the type of arm (*youBot* or *Mantis*).
- The *Environment* can be either unstructured or structured, dynamic or static. More features may describe other properties, such as empty space or cluttered space, etc.
- The types of objects (*Load*) that can be manipulated and/or transported might have properties (e.g. *Liquid* and *Fragile*) that require particular attention (e.g. limited velocity).

When a feature of the *Logistic SPL* Feature Model is manually selected, one or several features are automatically selected in the Feature Model of the *Robot Navigation* SPL and of the other functional systems. For example, the selection of the feature *Rover* in the *Logistic SPL* Feature Model activate the *Robot Navigation* feature and consequently the use of the corresponding subsystem. The selection of the feature *youBot* from the *Rover* variation point selects the feature *Omni* from the *Robot Kinematic Model* variation point. The selection of the feature *Liquid* activates a resolution element, which selects the feature *Smooth* in the Feature Model of the *Robot Navigation* SPL.

5 Related Works

The following subsections illustrates related works on MDE approaches in two areas: (a) MDE approaches for software variability management, (b) Robotics-specific MDE approaches.

5.1 MDE for Software Variability Management

In GenArch [10] the variability model and the configuration model are represented using the same meta-model, while in OMG CVL [4] the variability model and the resolution model are not explicitly separated. In our approach, the first three models are completely orthogonal, i.e. they can vary independently, while the configuration model is an instance of the variability model and can be hierarchically composed as described in the previous sections.

The *Compositional Variability* [5] approach supports the hierarchical composition of architectural models and feature models. The associations between a high-level feature model and a low level feature models are defined by means of the so called *Configuration Links*, which are similar to the feature dependencies defined in HyperFlex. Differently from HyperFlex, this approach defines an abstract component model and does not provide the capabilities for modeling domain-specific and heterogeneous component-based systems.

In order to maximize the reuse between product families in consumer electronics, Philips' researchers defined the product population approach. The architectural variability among the products is modeled by means of the Koala component model, which provides variability mechanisms such as parameterization (for configuring components) and switches (for changing connections) [25]. Differently from HyperFlex, this approach does not model functional variability explicitly.

A model driven approach for the design of embedded component based systems is presented in [18]. The approach is based on the Flex-eWare component model, which defines an architectural model in terms of components, composites, interfaces, and connections as in HyperFlex, and additional concepts for modeling computational nodes (e.g. micro-controllers) and quality of service properties. The Flex-eWare component model allows the generation of source code for three different target platforms (eC3M, Fractal and OASIS). Differently from HyperFlex, Flex-eWare models software variability only at architectural level.

5.2 Robotics-Specific MDE approaches

The real-time software framework OpenRTM [6] consists of a component meta-model, a component-manager (deployment infrastructure) and a set of tools. The component meta-model is divided in *Component profile* and *System profile*. Similar to the component model presented in this chapter, RT-Middleware provides the primitives for modeling software component and hierarchically composing them. However, in OpenRTM the composite is used for coordinating the execution of its components, while in HyperFlex the composite is a modeling entity and the coordination is managed by one of its internal components. Therefore, the HyperFlex approach is more general and can model component systems based on software frameworks that do not support hierarchical composition at implementation level.

The Proteus project [13] has developed an approach to modeling robotics control systems based on heterogeneous component models (e.g. ROS and Orocos). Proteus defines a framework-independent component meta-model to design the system architecture. It provides tools for code generation of component skeletons for different component frameworks but not for explicitly modeling robotic software variability.

The Smartsoft project [20] has developed a toolchain that supports variability modeling in task sequencing. Software models define sequences of actions with variation points that are bound at runtime according to the execution context and policies for task fulfillment. For example, the order of the actions that the robot executes for cleaning a table changes according to the objects that are on the table. In contrast, HyperFlex models the functional variability of a control system independently from the specific task that the robot has to execute.

6 Conclusions and Future Works

MDE is often considered to be synonymous with code generation and, frequently, research and development efforts in MDE for Robotics are motivated by the objective of simplifying application development by robotic experts with limited software engineering skills.

Interestingly, a study on the state of practice in MDE in industry reports that productivity gains due to automatic code generation are not considered significant enough to drive an MDE adoption effort, due to increased training costs and substantial organizational changes [26]. It turns out that the main advantages are in the support that MDE provides in defining the architecture of a software system.

For this reason, the HyperFlex approach builds upon the concepts of stable software architecture and variability modeling. The HyperFlex approach is supported by a set of MDE tools that allow software developers of robotic control systems to:

- Model the software architecture and the functional variability of a family of similar control systems (SPL) developed for a given application domain (e.g. autonomous logistics).
- Model the requirements of a specific application (product) with a modeling language that uses the vocabulary of robotic experts and system integrators.

- Automatically derive a configured system (product) according to the user requirements without the need for the system integrator to be an expert in robot functionality and software architectures.

This chapter has presented the HyperFlex approach in the context of the development of ROS-based robotic systems. For this reason, it has focused on ROS models and meta-models and on ROS-specific tools, such as the automatic generation of launch files from the models of configured systems.

Ongoing works aim at extending the ROS meta-models with the entities for modeling additional configuration mechanisms, such as the *pluginlib* package for writing and dynamically loading plugins using the ROS infrastructure. This extension will enable the possibility to use HyperFlex for the development of dynamically adaptive robotic systems. Robotic engineers will define several variation points (resources, algorithms, control strategies, coordination policies, etc.). Depending on the context, the control system dynamically chooses suitable variants to realize those variation points. These variants may provide better quality of service (QoS), offer new services that did not make sense in the previous context, or discard some services that are no longer useful.

Acknowledgments The research leading to these results has received funding from the European Community's Seventh Framework Programme (2007–2013) under grant agreement no. FP7-ICT-231940-BRICS (Best Practice in Robotics). The authors would like to thank all the partners of the BRICS project for their valuable comments.

References

1. The Eclipse Modeling Project (2013), http://www.eclipse.org/modeling
2. The HyperFlex Toolchain (2014), http://robotics.unibg.it/hyperflex/
3. AUTomotive Open System ARchitecture (2015), http://www.autosar.org/
4. Common Variability Language (2015), http://www.omgwiki.org/variability
5. A. Abele, H. Lönn, M.-O. Reiser, M. Weber, H. Glathe, EPM: a prototype tool for variability management in component hierarchies, in *Proceedings of the 16th International Software Product Line Conference* vol. 2 (ACM, 2012), pp. 246–249
6. N. Ando, S. Kurihara, G. Biggs, T. Sakamoto, H. Nakamoto, T. Kotoku, Software deployment infrastructure for component based RT-systems. J. Robot. Mechatron. **23**(3), 350–359 (2011)
7. R. Bischoff, et al., Brics-best practice in robotics, in *Robotics (ISR), 2010 41st International Symposium on and 2010 6th German Conference on Robotics (ROBOTIK)* (VDE, 2010), pp. 1–8
8. D. Brugali, A. Fernandes da Fonseca, A. Luzzana, Y. Maccarana, Developing service oriented robot control system, in *8th International Symposium on Service-Oriented System Engineering*, Oxford, UK, 7–10 Apr 2014. IEEE
9. D. Brugali, L. Gherardi, M. Klotzbuecher, H. Bruyninckx, Service component architecture in robotics: the SCA-Orocos integration, in *International Symposium on Leveraging Applications of Formal Methods, Verification and Validation* (Springer, 2011), pp. 46–60
10. E. Cirilo, U. Kulesza, C. Lucena, A product derivation tool based on model-driven techniques and annotations. J. Univ. Comput. Sci. **14**(8), 1344–1367 (2008)
11. P.C. Clements, L. Northrop. *Software Product Lines: Practices and Patterns*. SEI Series in Software Engineering (Addison-Wesley, Boston, 2001)

12. S. Cousins, B. Gerkey, K. Conley, W. Garage, Sharing software with ROS [ROS topics]. IEEE Robot. Autom. Mag. **17**(2), 12–14 (2010)
13. S. Dhouib, S. Kchir, S. Stinckwich, T. Ziadi, M. Ziane. RobotML, a domain-specific language to design, simulate and deploy robotic applications, in *Simulation, Modeling, and Programming for Autonomous Robots* (Springer, 2012), pp. 149–160
14. E. Gamma, R. Helm, R. Johnson, J. Vlissides, *Design Patterns: Abstraction and Reuse of Object-Oriented Design* (Springer, New York, 1995)
15. L. Gherardi, D. Brugali, An eclipse-based feature models toolchain, in *6th Italian Workshop on Eclipse Technologies (EclipseIT 2011)*, 2011
16. L. Gherardi, D. Brugali, Modeling and reusing robotic software architectures: the HyperFlex toolchain, in *IEEE International Conference on Robotics and Automation (ICRA 2014)*, Hong Kong, China, 31 May–5 June 2014. IEEE
17. N. Hochgeschwender, L. Gherardi, A. Shakhirmardanov, G. Kraetzschmar, D. Brugali, H. Bruyninckx, A model-based approach to software deployment in robotics, in *26th IEEE/RSJ International Conference on Intelligent Robots and Systems (IROS 2013)*, Tokyo, Japan, 3–7 Nov 2013. IEEE/RJS
18. M. Jan, et al., Flex-eware: a flexible model driven solution for designing and implementing embedded distributed systems. Softw. Pract. Exper. **42**(12) (2012)
19. K. Kang, Feature-oriented domain analysis (FODA) feasibility study (Technical report, DTIC Document, 1990)
20. A. Lotz, et al., Managing run-time variability in robotics software by modeling functional and non-functional behavior, in *Enterprise, Business-Process and Information Systems Modeling* (Springer, New York, 2013), pp. 441–455
21. E.Y. Nakagawa, P.O. Antonino, M. Becker, Reference architecture and product line architecture: a subtle but critical difference, in *Proceedings of the 5th European Conference on Software Architecture*, ECSA'11 (Springer, Berlin, Heidelberg, 2011), pp. 207–211
22. M.-O. Reiser, R.T. Kolagari, M. Weber, Compositional variability-concepts and patterns, in *42nd Hawaii International Conference on System Sciences, 2009. HICSS'09* (IEEE, 2009), pp. 1–10
23. D. Schmidt, Guest editor's introduction: model-driven engineering. Computer **39**(2), 25–31 (2006)
24. R. Smits, H. Bruyninckx, Composition of complex robot applications via data flow integration, in *2011 IEEE International Conference on Robotics and Automation (ICRA)* (IEEE, 2011), pp. 5576–5580
25. R. van Ommering, Mechanisms for handling diversity in a product population, in *Fourth International Software Architecture Workshop*, Citeseer, 2000
26. J. Whittle, J. Hutchinson, M. Rouncefield, The state of practice in model-driven engineering. IEEE Softw. **31**(3), 79–85 (2014). May

Integration and Usage of a ROS-Based Whole Body Control Software Framework

Chien-Liang Fok and Luis Sentis

Abstract Controllt! is a ROS-based high performance feedback control framework that enables Whole Body Control (WBC) algorithms to be implemented, instantiated, and integrated into ROS applications. It operates above individual joint controllers but below planners and takes a holistic view of the robot to achieve multiple simultaneous objectives. Such capabilities are particularly useful for highly redundant and multi-branched robots like humanoids where the large number of degrees of freedom (DOFs) and intrinsic multi-tasking like reaching for an object while maintaining balance requires advanced feedback control strategies. Controllt! provides two software abstractions, a compound task and a constraint set, that enables users to configure, use, and integrate whole body controllers. The compound task consists of prioritized tasks with controllers that operate in a relatively low dimensional space compared to the number of joints. The constraint set specifies physical limits of the robot like points of contact with the environment and mechanical couplings between joints. Controllt! comes with an implementation of the Whole Body Operational Space control (WBOSC) algorithm, one of the original WBC algorithms. Through prioritized null-space projection, WBOSC achieves each tasks' objectives subjected to limitations from higher priority tasks and the constraint set. Using tasks and constraints, users can make high-DOF multi-branched robots execute sophisticated multi-objective and adaptive behaviors. This chapter presents Controllt! and provides examples of advanced whole body behaviors it enables.

Keywords Controllt! ROS framework WBC WBOSC

C.-L. Fok (✉) · L. Sentis
Department of Mechanical Engineering, University of Texas at Austin, Austin, USA
e-mail: liangfok@utexas.edu
URL: http://www.me.utexas.edu/~hcrl/

L. Sentis
e-mail: lsentis@austin.utexas.edu

© Springer International Publishing Switzerland 2016
A. Koubaa (ed.), *Robot Operating System (ROS)*, Studies in Computational
Intelligence 625, DOI 10.1007/978-3-319-26054-9_21

1 Introduction

Whole Body Control (WBC) strategies are particularly useful for multi-branched highly redundant robots like humanoids due to their ability to achieve multiple control objectives and incorporate equality and inequality constraints into the control problem. This allows control strategies to deal with expected changes in interactions with the environment such as contact transitions [1]. Unlike traditional controllers that work at the single joint level or whole-body planners that operate off-line or infrequently relative to the WBC servo frequency, which is currently around 0.5–2 kHz, whole body controllers take a holistic view of the entire robot and use every joint to achieve the user-specified objectives via a real-time feedback control process. This real-time feedback control enables WBC-enabled robots to be more adaptive and responsive to unexpected contextual changes relative to systems that rely entirely on open-loop planners for coordinating whole body behaviors. There are many forms of whole body controllers including inverse dynamics controllers [2] and optimal controllers [3–5]. While these types of Multi-Input-Multi-Output (MIMO) controllers may be supported by ControlIt! in the future, for now ControlIt! comes with one type of whole body controller based on the Whole Body Operational Space Control (WBOSC) algorithm [6–9].

WBOSC is one of the first WBC algorithms developed. It enables unified motion/force control of multiple prioritized operational space objectives. Example objectives include end effector position and orientation, center of pressure locations, and internal force distributions within the robot. The whole body controller attempts to achieve these operational space objectives while deterministically handling joint redundancies through a lower priority posture specification and adhering to physical constraints. ControlIt! is a ROS-based framework that provides a state-of-the-art open source implementation of WBOSC and is designed to be extensible via plugins.

ControlIt! was originally developed and tested on Valkyrie, NASA's first humanoid robot, as shown in Fig. 1a. In the run-up to the DARPA Robotics Challenge (DRC) Trials in December 2013, it was successfully used to accomplish several tasks mandated by the DRC including industrial valve turning, door opening, and power tool manipulation [10]. After the DRC Trials, ControlIt! was integrated and tested on Dreamer, a humanoid upper body built by Meka Robotics (now owned by Google) and shown in Fig. 1b. Using ControlIt!, Dreamer was able to execute a product disassembly task [11] and various human-robot interactions like waving, shaking hands, and making University of Texas' "Hook-em Horns" gesture [12]. While ControlIt! is currently only tested with two robots, its architecture is designed to be robot-independent. The process for integrating ControlIt! with a new robot consists of developing two plugins that enable ControlIt! to access to the robot hardware and real-time clock capabilities, and specifying a whole body controller configuration similar to that shown in Fig. 2. More details will be described later in this chapter.

ControlIt! currently works with ROS Hydro and ROS Indigo. Dependencies include Eigen 3.0.5 [13] and the Rigid Body Dynamics Library (RBDL) 2.3.2 [14]. To enable testing in simulation, ControlIt! includes a plugin for Gazebo [15], an open

(a)   **(b)**

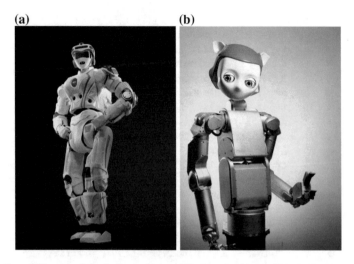

Fig. 1 ControlIt! was originally integrated and evaluated with NASA's Valkyrie and Meka's Dreamer humanoid robots. Both robots have a large number of DOFs, are multi-branched, and contain torque controlled series elastic actuators. **a** Valkyrie. **b** Dreamer

source robot dynamics simulator, that enables whole body controllers to control a simulated robot via a shared memory communication link [16].

As the provider of a whole body controller, ControlIt! is just one of many software components within a ROS-based system. Its placement in the overall software stack is shown in Fig. 3. Components that logically reside below ControlIt! include robot hardware (e.g., sensor and actuator) drivers, joint controllers, and joint controller managers like `ros_control` [17]. Components that logically reside above ControlIt! include planners and trajectory generators like MoveIt! [18], behavior sequencers like SMACH [19], cognitive processes, application logic, and user interfaces. ROS provides an infrastructure that spans the component hierarchy. Components further down the stack operate at higher update frequencies enabling more responsive feedback control.

There is a strong synergistic relationship between ControlIt! and ROS. ControlIt! makes use of ROS' infrastructure for supporting software development, process execution, code organization, parameter management, data logging, data visualization, and component based software architecture. In return, ControlIt! enables other ROS nodes to make use of a whole body controller to achieve sophisticated multi-objective and multi-constrained behaviors in robots with joint redundancies. ControlIt! runs as a node within a ROS network that communicates with other nodes via ROS topics [20] and ROS services [21]. Despite being a single node, extensive use of ROS' `pluginlib` infrastructure [22] enables a high degree of extensibility. The configuration of ControlIt! is initially done through the ROS parameter server [23], but can be dynamically changed at run time via ROS topics and services. Details will be described later in this chapter.

```
1   tasks:
2    - name: RightHandPosition
3      type: controlit/CartesianPositionTask
4    - name: LeftHandPosition
5      type: controlit/CartesianPositionTask
6    - name: RightHandOrientation
7      type: controlit/3DOrientionTask
8    - name: LeftHandOrientation
9      type: controlit/3DOrientionTask
10   - name: Posture
11     type: controlit/JointPositionTask
12  compound_task:
13    name: DreamerCompoundTask
14    task_list:
15      - {name: RightHandPosition, priority: 0}
16      - {name: LeftHandPosition, priority: 0}
17      - {name: RightHandOrientation, priority: 1}
18      - {name: LeftHandOrientation, priority: 1}
19      - {name: Posture, priority: 2}
20  constraints:
21   - name: ContactConstraint
22     type: controlit/FlatContactConstraint
23   - name: TorsoTransmission
24     type: controlit/TransmissionConstraint
25  constraint_set:
26    name: My Constraint Set
27    active_constraints:
28      - {name: ContactConstraint}
29      - {name: TorsoTransmission}
```

Fig. 2 An example configuration file that specifies a whole body controller for Dreamer. Control points include the Cartesian position and orientation of Dreamer's wrists and the overall posture. Task parameter details are omitted

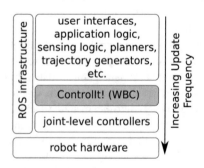

Fig. 3 ControlIt!'s relationship with ROS and other ROS nodes

The remainder of this chapter is organized as follows. Section 2 provides an overview of WBOSC's mathematical foundations. Section 3 presents ControlIt!'s software architecture. Section 4 describes the plugin libraries included with ControlIt!, which include the tasks and constraints that constitute the primitives for configuring whole body controllers. Section 5 provides examples of how ControlIt! was used on actual robots for a variety of applications. Section 6 describes the installation process. Section 7 details how to run various demos in simulation. The chapter ends with conclusions in Sect. 8.

2   Overview of Whole Body Operational Space Control

The mathematical foundations of WBOSC are detailed in previous publications [6–9] and those interested in the full details should refer to them. This section only provides an overview of the mathematics that underpins WBOSC.

WBOSC provides a servo loop that operates as a kinematics and dynamics calculation module. It cycles at a user-specified servo frequency, which is limited by the speed of the processor and is typically in the range of 0.5–2 kHz. During each cycle of the servo loop, WBOSC takes as input the robot's current state, tasks, and constraints, and outputs the desired joint torques that joint-level controllers must achieve. Over time, assuming the robot is able to perform WBOSC's commands, the desired multi-objective whole body behavior emerges.

To enable support for mobile robots, the total state consists of both the robot's joint states and world state, as shown in Fig. 4. The world state is the robot's position and orientation in the world, i.e. the robot's global pose, in addition to the reaction forces with respect to the environment. Let n_{joints} be the number of actual DOFs in the robot. The robot's joint positions are represented by the vector q_{actual} as shown by the following equation.

$$q_{actual} = < q_1 \ldots q_{n_{joints}} > \tag{1}$$

The robot's global Cartesian position and orientation are represented by a 6-dimensional floating virtual joint that connects the robot's base link to the world, i.e., three rotational and three prismatic virtual joints. It is denoted by vector $q_{base} \in \mathbb{R}^6$. The two partial state vectors, q_{actual} and q_{base}, are concatenated into a single state vector $q_{full} = q_{actual} \cup q_{base}$. This combination of real and virtual joints into a single vector is called the *generalized* joint position vector. Let n_{dofs} be the number real and virtual DOFs in the model that is used by WBOSC. Thus, $q_{full} \in \mathbb{R}^{6+n_{joints}} = \mathbb{R}^{n_{dofs}}$.

The underactuation matrix $U \in \mathbb{R}^{n_{joints} \times n_{dofs}}$ defines the relationship between the actuated joint vector and the full joint state vector as shown by the following equation.

$$q_{actual} = U q_{full} \tag{2}$$

The total state that is provided to the whole body controller consists of the full joint position vector q_{full} and the full joint velocity vector $\dot{q}_{full}$.

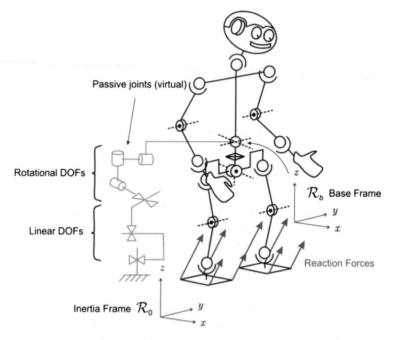

Fig. 4 Floating base dynamic model

Let A be the robot's generalized joint space inertia matrix, B be the generalized joint space Coriolis and centrifugal force vector, G be the generalized joint space gravity force vector, J_c be the contact Jacobian matrix that maps from generalized joint velocity to the velocity of the constraint space dimensions, λ_c be the co-state of the constraint space reaction forces, and $\tau_{command}$ be the desired force/torque joint command vector that is sent to the robot's joint-level controllers. The robot dynamics can be described by a single linear second order differential equation shown by the following equation.

$$A\begin{pmatrix} \ddot{q}_{base} \\ \ddot{q}_{actual} \end{pmatrix} + B + G + J_c^{\mathrm{T}}\lambda_c = \begin{pmatrix} 0_{6\times 1} \\ \tau_{command} \end{pmatrix} \qquad (3)$$

Constraints are formulated as follows. Let $\dot{p}_c$ be the velocity of the constrained dimensions, which we approximate as being completely rigid and therefore yielding zero velocity at the contact points, as shown by the following equation.

$$\dot{p}_c = J_c\begin{pmatrix} \dot{q}_{base} \\ \dot{q}_{actual} \end{pmatrix} = 0 \qquad (4)$$

Tasks are formulated as follows. Let $\dot{p}_t$ be the desired velocity of the task, J_t be the Jacobian matrix of task t that maps from generalized joint velocity to the velocity of the task space dimensions, and N_c be the generalized null-space of the constraint set. Furthermore, let J_t^* be the contact consistent reduced Jacobian matrix of task t, i.e., it is consistent with U and N_c. The definition of $\dot{p}_t$ is given by the following equation where operator $\overline{arg}$ is the dynamically consistent generalized inverse of arg [7].

$$\dot{p}_t = J_t \begin{pmatrix} \dot{q}_{base} \\ \dot{q}_{actual} \end{pmatrix} = J_t \overline{U N_c} \dot{q}_{actual}$$
$$= J_t^* \dot{q}_{actual} \tag{5}$$

Let Λ_t^* be the contact-consistent task-space inertia matrix for task t, $\ddot{p}_{t,ref}$ be the reference, i.e., desired, task-space acceleration for task t, β_t^* be the contact-consistent task-space Coriolis and centrifugal force vector for task t, and γ_t^* be the contact-consistent task space gravity force vector for task t. The force/torque command of task t, denoted F_t, is given by the following equation.

$$F_t = \Lambda_t^* \ddot{p}_{t,ref} + \beta_t^* + \gamma_t^* \tag{6}$$

To achieve multi-priority control, let $J_{t|prev}^*$ be the Jacobian matrix of task t that is consistent with U, N_c, and all higher priority tasks. The equation for $\tau_{command}$ is the sum of all of the individual task commands multiplied by the corresponding $J_{t|prev}^*$ matrix as shown by the following equation.

$$\tau_{command} = \sum_t J_{t|prev}^{*T} F_t \tag{7}$$

Finally, when a robot has more than one point of contact with the environment, there are internal tensions within the robot as shown in Fig. 5. By definition, these "internal forces" are orthogonal to joint accelerations, i.e., they result in no net movement of the robot. The control structures like the multicontact/grasp matrix that are used to control these internal forces are documented in previous publications [8]. Let L^* be the nullspace of $(U N_c)$ and $\tau_{internal}$ be the reference (i.e., desired) internal forces vector. The contribution of the internal forces can thus be added to Eq. 7 as shown by the following equation.

$$\tau_{command} = \sum_t \left(J_{t|prev}^{*T} F_t \right) + L^{*T} \tau_{internal} \tag{8}$$

This concludes the overview of WBOSC's mathematical foundation. WBOSC is a WBC algorithm that supports constraints, prioritized tasks, and internal tensions. Successive null-space projections are used to enforce priority semantics. When multiple contact points with the environment exists, a separate structure is used to control internal tensions. This is possible since internal tensions do not result in joint accelerations and thus are orthogonal to the tasks and constraints.

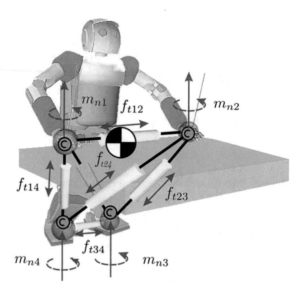

Fig. 5 Internal tension model

3 Software Architecture

ControlIt!'s software architecture is shown in Fig. 6. It is divided into three levels: configuration, WBC, and a Hardware Abstraction Layer (HAL). Configuration
classes parameterize the whole body controller and include the compound task, constraint set, and robot model. The WBC layer consists of a coordinator that implements
the servo loop and the whole body controller that implements the whole body control
logic. The HAL consists of a robot interface and a clock. The robot interface enables
ControlIt! to work with a wide variety of robots while the clock implements the servo
loop's thread and enables support for different real-time frameworks like RTAI [24]
and RT-Preempt [25].

ControlIt! is designed to be highly extensible via ROS plugins [22]. The elements
that are extensible include tasks, constraints, the whole body controller, the clock,
and the robot interface. They are shown in gray in Fig. 6. In the future, the robot
model may also be a plugin.

The remainder of this section is structured as follows. Section 3.1 discusses
the architecture's core classes. Section 3.2 discusses how parameters are handled
and bound to ROS topics. Finally, Sect. 3.3 presents ControlIt!'s multi-threaded
architecture.

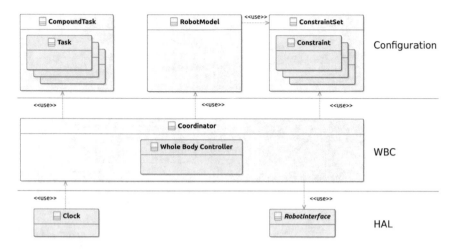

Fig. 6 ControlIt!'s software architecture

3.1 Core Classes

Robot Model. WBOSC is a model-based controller meaning it relies on a software model that specifies the kinematic and dynamic properties of the robot being controlled. Figure 7 includes a UML diagram of ControlIt!'s `RobotModel` class. Internally, `RobotModel` uses the `Model` class that is provided by RBDL [14]. This library includes algorithms for computing forward and inverse kinematics, forward and inverse dynamics, frame transformations, the inertia matrix, Coriolis and centrifugal forces, and the gravity vector. In other words, the RBDL model is used to derive the variables A, B, and G, and J_c in Eq. 3. Method `init()` initializes a `RobotModel` by instantiating a ROS node handle and getting the relevant parameters off the ROS parameter server [23]. This includes a Universal Robot Description Format (URDF) description of the robot that is used to instantiate a RBDL model, and a YAML-specification of the constraint set that is used to instantiate the constraint set. The robot model uses the constraint set to determine which of the real joint are actuated. This is necessary because some robots like Dreamer have co-actuated joints where one actuator controls multiple joints. In Dreamer's case, the two torso pitch joints are co-actuated and thus always have the same state. ControlIt! models this via a transmission constraint that makes one joint a slave of the other joint, as will be discussed.

During each cycle of the servo loop, the real-time servo thread within the coordinator obtains the latest joint state from the robot interface and passes this state to `RobotModel::setJointState()`, which saves the information in member variable `RobotModel::jointState`. Another thread that's dedicated to updating the model periodically calls `RobotModel::update()`, which updates `RobotModel::model` and `RobotModel::inertiaMatrix`, i.e., variable A in Eq. 3. By using a separate thread to update the model, we offload a significant

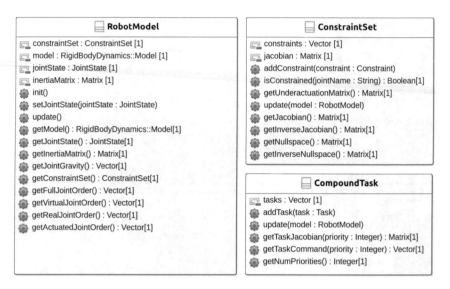

Fig. 7 UML diagrams of ControlIt!'s configuration classes

amount of computation from the real-time servo thread enabling it to achieve higher servo frequencies, which is often needed for increasing closed loop stability.[1]

Note that the robot model is usually incorrect necessitating the use of a whole body feedback controller. Future work includes the integration of system identification algorithms that adjust the model at run-time to reduce model inaccuracies. This should enable increasingly higher feedback controller gains and thus higher performance behaviors to be achieved over time.

Constraint Set. The constraint set contains constraints that specify the natural physical limits of the robot. During initialization, the constraint set's configuration is determined by a YAML specification stored on the ROS parameter server under `/[controller name]/config/constraint_set`. Figure 7 contains a UML diagram of class `ConstraintSet`. The constraint set computes a Jacobian matrix that is the vertical concatenation of the J_c matrices belonging to the constraints as defined in Eq. 4. It also computes U in Eq. 2, $\overline{U N_c}$ in Eq. 5, and whether each joint is constrained. The coordinator passes this information to the whole body controller, which uses it to ensure the commands reside within the constraint set's nullspace.

Figure 8 contains a UML diagram of class `Constraint`. All constraints are named, specify the number of constrained DOFs, and provide a Jacobian matrix J_c. There are two types of constraints: contact and transmission. Contact constraints specify how a robot contacts its environment. It is parameterized by the link and the

---

[1] Increasing feedback controller gains too much is not desirable since doing so may lead to saturation of the robot's actuators and instability. By using a multi-threaded architecture, ControlIt! simply provides the user with the option to increase the servo frequency higher than otherwise possible.

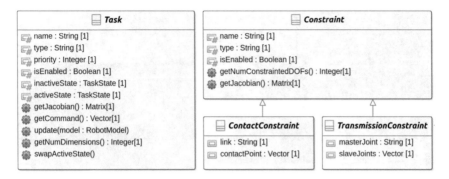

Fig. 8 UML diagrams of ControlIt!'s `Task` and `Constraint` classes

point on the link where the contact is modeled to occur, e.g., it can be a contact region's mid-point, center of pressure, or zero moment point. Transmission constraints specify dependences between joints due to co-actuation. It is parameterized by a specification of which joint is the master and which is the slave. The slave joint's behavior is dependent on the master joint's behavior.

Compound Task. The compound task contains a set of prioritized tasks, each of which specifies an operational or postural objective for the whole body controller to achieve. During initialization, the configuration of the compound task is determined by a YAML specification stored on the ROS parameter server under `/[controller name]/config/compound_task`. The compound task is the key software abstraction through which users can configure a whole body controller. Figure 7 contains a UML diagram of class `CompoundTask`. For each priority level, the compound task vertically concatenates the Jacobians and commands belonging to the tasks at the priority level. The coordinator takes this information and passes it to the whole body controller. WBOSC uses these concatenated Jacobian matrices and command vectors to enforce task prioritization and multiple tasks at the same priority level, while adhering to the constraint set, as defined by Eq. 7.

Figure 8 contains a UML diagram of class `Task`. All tasks are named, have a priority level, can be enabled and disabled, provide a task-space command vector and a Jacobian matrix that converts the command to joint space, and maintains two sets of states, active and inactive. The active state is used by the real-time servo thread while the inactive state is updated by a separate thread. Like the process of updating the robot model, the purpose of using a separate thread to update the task states is to offload the amount of computations that need to be performed by the real-time servo thread and thereby increase the maximum achievable servo frequency. The real-time servo thread periodically calls `Task::swapActiveState` that checks if an update is available and, if so, swaps the active and inactive states.

The `Task` and `Constraint` classes are abstract; concrete implementations are included through plugins. Both have names and types for easy identification and can be enabled or disabled based on context. In the future, support for dynamically adding and removing tasks and constraints (not just enabling/disabling) will be added. Currently-provided tasks and constraints are described in Sect. 4.

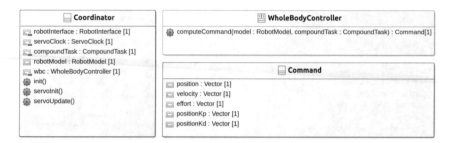

Fig. 9 UML diagrams of ControlIt!'s WBC classes

Whole Body Controller. The whole body controller implements the actual WBC algorithm. Figure 9 shows its UML diagram. Since ControlIt! is designed to be extensible, the whole body controller is actually an interface definition. Concrete implementations are provided via dynamically loadable plugins and will be discussed in Sect. 4. The interface `WholeBodyController` defines a single method named `computeCommand()`. Inputs to this method are a `RobotModel` and a `CompoundTask`. Using these input parameters, the method performs the WBC computations that generate a command for each joint under its control and returns the commands within an object of type `Command`. Figure 9 contains a UML diagram of `Command`. As shown in the figure, the command contains the desired position, velocity, and effort (i.e., force or torque) values for each joint in the robot, along with a couple parameters for the joint-level controllers. Note that depending on the type of whole body controller and joint-level controllers employed, not all of the variables within a `Command` object are used. For example, Dreamer only uses the effort command because its joints are torque controlled whereas Valkyrie used all of the parameters because its joints are impedance controlled.

Coordinator. As shown in Fig. 6, the coordinator is a central component in ControlIt!'s architecture. It contains a whole body controller and uses the configuration objects and the robot interface. It implements the servo loop that is shown in Fig. 12, which is periodically executed by the servo clock. The coordinator is implemented by class `Coordinator` whose UML class diagram is given in Fig. 9. As shown in the figure, `Coordinator` contains a robot interface, clock, compound task, robot model, and whole body controller as member variables. These variables are instantiated when the `init()` method is called. The coordinator also implements methods `servoInit()` and `servoUpdate()`. Method `servoInit()` initializes the robot model by reading the latest robot joint state from the robot interface and passing this information to the robot model. This is necessary because some robot interfaces contain data structures that are only accessible to the real-time thread that's provided by the clock. Once initialized, the clock periodically calls method `servoUpdate()`, which implements the servo loop.

RobotInterface. The robot interface decouples the rest of ControlIt! from robot-specific software. This enables ControlIt! to support different robots without major software changes. Figure 10 contains a UML class diagram of the robot

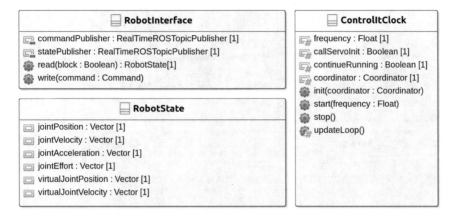

Fig. 10 UML diagrams of ControlIt!'s HAL classes

interface. Recall that `RobotInterface` is an abstract class. Concrete implementations are introduced via plugins that will be described in Sect. 4. The robot interface provides two methods: `read()`, which obtains the latest robot joint state, and `write()`, which sends a command to the robot. For diagnostic purposes, it has two real-time ROS topic publishers for revealing the states and commands. `RealtimeROSTopicPublisher` uses a thread-pool to offload the publishing process from the servo thread.

Clock. The clock instantiates the real-time servo thread and contains a reference to `Coordinator`. It calls `Coordinator::servoInit()` once upon startup and then `Coordinator::servoUpdate()` periodically. It is also an abstract class with concrete implementations made available via plugins.

3.2 Parameter Binding

ROS provides a component-based architecture consisting of multiple communicating software processes called nodes one of which is ControlIt!. A parameter binding mechanism is provided to integrate ControlIt! with other nodes. Figure 11 contains UML diagrams of the relevant classes. `Parameter` stores information about a parameter like its name, value, and bindings. It also provides method `set()`, which updates the value and the bindings. `ParameterReflection` is the parent-class of all classes that contain parameters. It allows child classes to declare and access parameters and emit events. `Event` contains a logical expression over the parameters within a `ParameterReflection` object. When this logical expression turns true, a message containing the event's name is published onto ROS topic `/[controller name]/events`. This enables event-triggered behaviors. Events are continuously evaluated by the servo loop as indicated in Fig. 12. `Binding` contains a `BindingConfig` and a `Parameter`. `BindingConfig` stores details

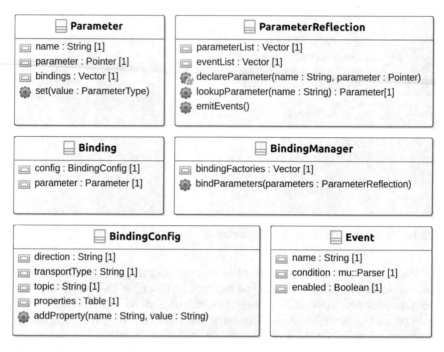

Fig. 11 UML class diagrams related to parameters and parameter bindings

about a binding like which transport protocol to use, in which direction, and transport protocol-specific parameters. `BindingManager` creates and stores the bindings. To support extensibility in terms of transport protocols, `Binding` is an abstract class. Concrete transport layer-specific instances are provided via plugins.

3.3 Multi-threaded Architecture

To increase the servo frequency, ControlIt! uses a multi-threaded architecture where computationally-intensive updates that do not need to occur every cycle of the servo loop are done by child threads. This is possible since some state like the robot model and task Jacobian matrices typically do not significantly change from one cycle of the servo loop to the next. Figure 12 shows the finite state machines of the threads used in ControlIt!. As shown in the figure, there are three threads: (1) a real-time servo thread, (2) a task updater thread, and (3) a model update thread. To prevent race conditions between the threads, two robot models are maintained: an active one that is used by the real-time servo thread, and an inactive one that is updated by the model update thread. Likewise, tasks maintain active and inactive states where the active ones are used by the servo thread and inactive one is used by the task update thread. Since the servo thread is real-time, it should never be blocked by either the

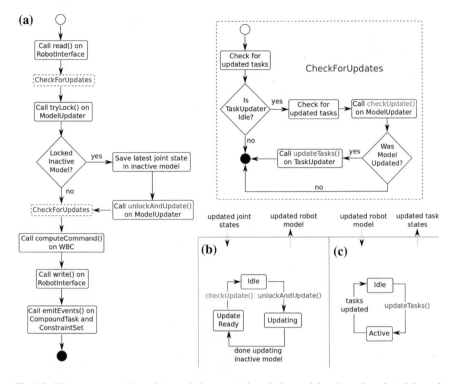

Fig. 12 Finite state machines of the real-time servo thread, the model updater thread, and the task updater thread

task updater thread or model updater thread. This is done by having the servo thread swap the inactive and active states at certain points in the servo loop. Using this multi-threaded architecture, the controller's execution frequency is stable as shown by Fig. 13.

4 Plugin Libraries

As a framework, ControlIt! is designed to work with a wide variety of robots and applications. This is achieved by enabling key aspects of ControlIt! to be extended via dynamically loadable plugins based on ROS pluginlib [22]. To provide a robust base set of functionalities, ControlIt! comes with numerous plugins that are organized into libraries. Specifically, ControlIt! comes with a task library, constraint library, whole body controller library, clock library, and robot interface library. Each of these libraries contain plugins that can be added to ControlIt!. New plugins can be developed for general use or specific applications, and for hardware platforms that are not covered by existing plugins. We now describe each of these libraries.

Fig. 13 A histogram of ControlIt!'s servo frequency when running on Dreamer hardware for 70 s. The desired frequency was 1 kHz

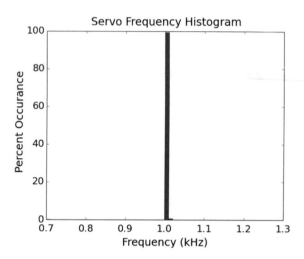

Table 1 The task library

Name	Key parameters
JointPositionTask	Desired joint position
CartesianPositionTask	Control frame, control point
	Desired Cartesian position,
2DOrientationTask	Control frame, control vector,
	Desired frame, desired vector
3DOrientationTask	Control frame, desired Quaternion
COMTask	Control frame, control point,
	Desired COM location
COPTask	Control frame, desired COP location
InternalForcesTask	Desired internal forces

Task Library. The plugins in the task library are shown in Table 1. There are currently seven tasks in the library. JointPositionTask controls the position and velocity of every joint in the robot. It is typically the lowest priority task in a compound task, specifies the robot's overall posture, and is needed to handle redundant joints in high DOF robots. CartesianPositionTask controls the world position of a point on the robot. 2DOrientationTask and 3DOrientationTask control a robot link's two or three orientation dimensions. 2D orientation is useful when one dimension is constrained like in a mobile wheeled platform. COMTask controls the robot's Center Of Mass (COM). COPTask controls the location of a robot link's Center Of Pressure (COP) when it is in contact with the environment. Finally, InternalForcesTask specifies the desired internal forces within the robot. The current task implementations use PID controllers. In the future, the controllers within tasks may be plugins enabling other Single-Input-Single-Output (SISO) and Multi-Input-Multi-Output (MIMO) controllers to be used.

Table 2 The constraint library

Name	Key parameters
FlatContactConstraint	Constrained link, contact normal, contact point
PointContactConstraint	Constrained link, contact point
OmniWheelConstraint	Constrained link, wheel axis, contact point, normal axis
CoactuationConstraint	Master joint, slave joint, transmission ratio

The task library represents "WBC primitives." Combinations of these primitives can be configured for a wide range of applications and robots. Their capabilities directly impact the whole body behaviors that can be achieved. The integration of ControlIt! into ROS applications is done by binding the task parameters to ROS Topics and other transport protocols.

Constraint Library. The plugins in the constraint library are shown in Table 2. There are currently four constraints in the library. FlatContact-Constraint is used when a link is unable to translate or rotate due to contact with the environment. PointContactConstraint is used when a link can rotate but not translate. OmniWheelConstraint restricts one rotational DOF and one translational DOF based on the current orientation of the wheel. CoactuationConstraint enables ControlIt! to handle robots with co-actuated joints, like Dreamer's torso pitch joints. The transmission ratio specifies how much the slave joint moves relative to the master joint.

Multiple instances of the same constraint may exist in the constraint set if they are for different parts of the robot. For example, a biped robot like Valkyrie would have a FlatContactConstraint for each foot. Additional contact constraints can be added if, for example, the robot's arms contact the environment.

WBC Library. The WBC library currently includes two implementations of WBOSC as shown in Table 3. The first implementation, available via the WBOSC plugin, implements the WBC algorithm described in Sect. 2. It takes the constraint set, compound task, and robot model, and outputs an effort (i.e., force/torque) command vector that minimizes the tasks errors subjected to constraint and task priority specifications. The output of WBOSC is then sent to an effort-controlled robot like Dreamer.

The WBOSC_Impedance plugin works with Impedance-controlled robots, which require commands containing the desired positions, velocities, and optionally gravity compensation torques. The main benefit is higher impedance due to the ability to place the damping portion of the joint feedback controller closer to the control

Table 3 The WBC library

Name	Application
WBOSC	Effort-controlled robots
WBOSC_Impedance	Impedance-controlled robots

Table 4 The robot interface library

Name	Transport protocol
`RobotInterfaceROSTopic`	ROS topics
`RobotInterfaceSharedMemory`	Shared memory
`RobotInterfaceUDP`	UDP
`RobotInterfaceTCP`	TCP

plant, which results in lower communication latencies [26]. The implementation of `WBOSC_Impedance` actually extends `WBOSC` with an internal model that uses the effort command generated by `WBOSC` to derive the expected joint positions and velocities.

Robot Interface Library. The plugins in the robot interface library are shown in Table 4. The robot interfaces differ in the type of transport protocol supported, which vary in their latency, bandwidth, reliability, level of abstraction, and whether they enable a distributed architecture where ControlIt! runs on a different machine than the robot hardware drivers. Shared memory [16] has the lowest latency and highest bandwidth but does not support distributed operation, which all others support. The difference between `RobotInterfaceROSTopic` and `RobotInterfaceTCP` is the level of software abstraction since ROS Topics by default use TCP. Whereas TCP packets are defined using raw bytes, ROS topic messages are defined by ROS' message description language, which includes higher level data types [27]. We provide `RobotInterface` plugins that are not based on ROS topics for robots that cannot run ROS. In addition to the above, specialized robot interfaces for Dreamer and Valkyrie exist but are not part of the library since they are robot specific.

To support simulation testing, ControlIt! includes a corresponding Gazebo [15] plugin for each of the robot interfaces in the library. This enables developers to quickly switch between evaluating an application based on ControlIt! in simulation and on real hardware.

Clock Library. The plugins in the clock library are shown in Table 5. They support clocks based on RT- Preempt [25], ROS time [28], and C++'s `std::chrono` library [29]. In addition, a separate `ClockRTAI` is included in a separate package that enables use of the Real-Time Application Interface (RTAI) for real-time operation [24]. In the future, this RTAI-based clock may be included with the ControlIt! Clock Library by using conditional compilation and RTAI's LXRT mode, which will enable the library to be compilable even on non-RTAI platforms.

Table 5 The clock library

Name	Clock type
`ClockRTPreempt`	RT-Preempt
`ClockROS`	ROS Time
`ClockChrono`	C++'s `std::chrono` library

Name	Transport protocol
InputBindingROS	ROS topic
OutputBindingROS	ROS topic
InputBindingSM	Shared memory
OutputBindingSM	Shared memory

Table 6 The parameter binding library

Parameter Binding Library. The plugins in the parameter binding library are shown in Table 6. As shown in the table, ControlIt! currently provides bindings for ROS topics and shared memory transport layers. Two types of bindings are provided for each transport layer, one input and one output. Input bindings enable other nodes to change the values of parameters within ControlIt!. Output bindings enable other nodes to monitor the values of ControlIt! parameters.

5 Example Whole Body Control Configurations

The software architecture presented in Sect. 3 and the plugin libraries described in Sect. 4 provide sufficient flexibility and expressiveness to control numerous multi-branched mobile robots with a large number of joints, like humanoids, and make them do general tasks. This section describes several whole body controller configurations used on actual robot hardware, specifically Valkyrie and Dreamer.

Towards the end of September 2013, Valkyrie hardware and embedded system development reached a point where a whole body controller could be tested on the full robot. Till now, ControlIt! was only tested with Valkyrie in simulation. For this test, a total of 29 joints were controlled by the whole body controller. They include two six-DOF legs, a 3-DOF waist, and two 7-DOF arms. The neck and finger joints were controlled by separate ROS nodes. To reduce complexity and increase the probability of success, a relatively simple whole body controller was used. Specifically, the constraint set consisted of two `FlatContactConstraint` constraints, one for each foot, and the compound task consisted of an `InternalForcesTask` and a `JointPositionTask`. Using this configuration, ControlIt! was able to make the robot stand, as shown in Fig. 14a. It was even able to withstand some light disturbances like gently pushing it from behind or the side. At this time, the joint-level controllers implemented torque controllers, so `WBOSC` was used as the whole body controller.

At the time, Valkyrie's immediate objective was to compete in the DARPA Robotics Challenge Trials in December 2013, which required that Valkyrie perform various locomotion and manipulation tasks. Given the tight deadline and to enable problems with the lower body to be resolved in parallel with upper body development (e.g., the ankle and knee joints tended to overheat), ControlIt! was configured to work with Valkyrie's 14-DOF upper body to practice some of the manipulation tasks. Figure 14b shows ControlIt! controlling Valkyrie's upper body to turn an indus-

(a)   **(b)**

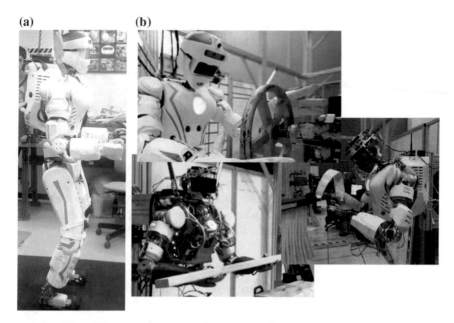

Fig. 14 Two whole body controller configurations used on NASA's Valkyrie robot. **a** ControlIt! is applied to Valkyrie's full body and is configured with a `FlatContactConstraint` for each foot, an `InternalForcesTask`, and a `JointPositionTask` to make the robot stand. **b** ControlIt! is applied to Valkyrie's upperbody and is configured with a `FlatContactConstraint` at the hip, high priority `CartesianPositionTask` and `3DOrientationTask` for each wrist, and a low priority `JointPositionTask` to make the robot turn an industrial valve, grab a fire hose, and lift debris

trial valve, manipulate a fire hose, and pick up debris. Since the upper body was mounted on a fixed platform for these tests, the constraint set consisted of a single flat contact constraint assigned to the robot's hip. To facilitate manipulation capabilities, a more sophisticated compound task was used. It consisted of two priority levels. The high priority level contained four tasks: a `CartesianPositionTask` and a `3DOrientationTask` for each of the two wrists. The lower priority level contained a `JointPositionTask` that defined the robot's overall posture and prevented nondeterministic behavior due to joint redundancy. For these tests, the joint-level controllers were modified to be Impedance controllers, meaning `WBOSC_Impedance` was used as the whole body controller.

ControlIt! has also been integrated with Dreamer, a 16-DOF humanoid upper body with series elastic joints (two 7 DOF arms and a 2-DOF torso). The torso yaw joint was broken at the time of testing and thus disabled. The joints in the right fingers, left gripper, and head were controlled by separate controllers in different ROS nodes that use ROS topics to access the robot hardware via ControlIt!'s Dreamer-specific `RobotInterface`. As shown in Fig. 15, ControlIt! was able to make Dreamer perform a variety of operations including a complex product disassembly task that requires coordination of both end effectors, a University of Texas hook'em

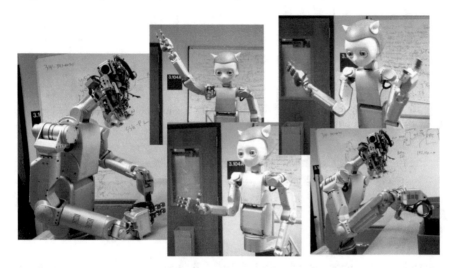

Fig. 15 ControlIt! is used on Dreamer. The constraint set consisted of a `FlatContact Constraint` on the torso's base and a `CoactuationConstraint` on the two torso pitch joints, which are physically linked together in a 1:1 ratio. The compound task consisted of a high priority `CartesianPositionTask` and `3DOrientationTask` for each wrist, and a low priority `JointPositionTask` to make the robot disassemble a product, perform a University of Texas Hook'em Horns gesture, shake hands, wave, and store an object in a container

horns gesture, shake hands, wave, and place a product in a container. All of these behaviors were accomplished using the ControlIt! configuration shown in Fig. 2. Specifically, the constraint set consists of a `FlatContactConstraint` on the torso's base and a `TransmissionConstraint` on the two torso pitch joints, which are physically linked together in a 1:1 ratio. The compound task consists of a high priority `CartesianPositionTask` and `3DOrientationTask` for each wrist, and a low priority `JointPositionTask`.

The system architecture used to achieve the manipulation behaviors on Valkyrie and Dreamer is shown in Fig. 16. From highest to lowest levels, the components consist of a user interface, application logic, planners and trajectory generators, ControlIt!, and finally the robot itself. The user interface is the component that the user directly interacts with. For Valkyrie, the user interface consisted of RViz [30] and Robot Task Commander [31]. For Dreamer, the user interface consisted of RViz and a command line terminal. The application logic determines which behavior to perform. It does this by providing coarse-granularity task-space (e.g., Cartesian space) waypoints. Planners and trajectory generators take these coarse waypoints and generate fine-grained task-space waypoints. For Valkyrie, the Reflexxes [32, 33] motion library was used. For Dreamer, cubic-spline was used. The fine-grained trajectories are then passed to ControlIt!, which issues the appropriate joint-level commands to the robot to achieve the desired behavior. State feedback from the robot is used by the other components to detect and adjust for anomalies. ControlIt! can handle small disturbances by adjusting the joint effort commands. Larger disturbances can be handled

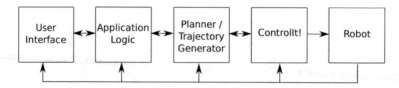

Fig. 16 The overall architecture used to implement the manipulation behaviors on Valkyrie and Dreamer using ControlIt!

through replanning. Extreme disturbances can be handled by the application logic or user intervention through the user interface.

6 Installation

ControlIt! is open sourced under a LPGLv2.1 license. Currently it must be downloaded as source code and manually compiled. By providing the source code and compilation instructions, users have the flexibility to modify ControlIt! to work in other Linux distributions and versions of library dependencies. For those who do not need to modify ControlIt! and can work with Ubuntu and ROS, work is underway to enable automated installation via Debian packages. For the latest installation instructions, consult ControlIt!'s website, http://robotcontrolit.com [34]. The following instructions are for the source-based installation.

The package management and build system used by ControlIt! is catkin [35]. Installing and compiling ControlIt! consists of setting up a ROS workspace, adding the relevant git repositories, updating the workspace (this automatically downloads the source code), installing RBDL, and then compiling the source code.

Before proceeding, ensure the following dependencies are met. First, the target computer needs to run Ubuntu 12.04 or 14.04 and have ROS Hydro or Indigo. If simulation testing is desired, Gazebo [15] should be installed. Finally, Ubuntu 14.04 systems need to install yaml-cpp 0.3.0 [36] since its API is incompatible with the default yaml-cpp 0.5.0.

Once the above-mentioned dependencies are met, ControlIt! can be installed and compiled. Follow the installation instructions on ControlIt's website at http://robotcontrolit.com/installation. After installing ControlIt!, compile it by executing the following commands:

```
$ roscd; cd ..
$ rm -rf build devel
$ cakin_make
```

Many of ControlIt!'s demos use shared memory to communicate with the simulation. To prevent needing to allocate this shared memory each time you restart your computer (and having to type your sudo password), permanently allocate sufficient shared memory by executing the command below.

```
$ rosrun shared_memory_interface \
  set_shared_memory_size_persistent 536870912
```

This concludes the installation of ControlIt!. Instruction on how to use ControlIt! is covered in the next section.

7 Usage

To demonstrate how to use ControlIt! and integrate it into ROS applications, several robot models and sets of configuration files are available. This section describes how to run some examples using these files.

When testing a new WBC algorithm, configuration, or behavior, it is often necessary to start simple and then gradually increase complexity. For example, ControlIt! was made to work with Dreamer by adding one joint at a time. Each time a new joint was added, ControlIt! was thoroughly re-tested and the feedback control gains were hand-tuned to ensure continuation of desired controller behavior. Note that, in the future, automatic gain tuning tools can be developed and used. For instance, gain tuning rules were recently developed for series elastic actuators [37], which could be generalized for multi-input multi-output systems. When integrating ControlIt! with Valkyrie, each limb was physically detached from the rest of the robot and tested separately before combining them into a full humanoid.

To support incremental testing, `controlit_robot_models` comes with a set of primitive shape-based models that span a wide range of complexity from the lower half of one leg to a full bipedal humanoid as shown in Fig. 17. Primitive shapes are simpler than mesh-based models and thus help maintain reasonably fast simulation times, which is helpful when debugging a new whole body control algorithm or configuration.

The simplest model is shown in Fig. 17a and is called *stickbot_lowerleg_3dof*. To use ControlIt! with this robot model in simulation, execute the following commands:

(a)   **(b)**   **(c)**   **(d)**   **(e)**

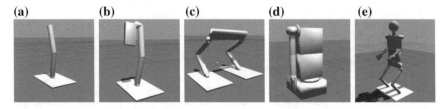

Fig. 17 Primitive shape-based robot models used to test ControlIt! that span a wide range of complexity. **a** A 3-DOF lower leg, **b** a 6-DOF leg, **c** a 12-DOF biped, **d** a 10-DOF upper body, and **e** a 32-DOF full humanoid. Incrementally increasing complexity is useful when testing new WBC algorithms, configurations, and behaviors

```
$ roscd stickbot_lowerleg_3dof_controlit/models
$ ./generate_stickbot_lowerleg_3dof_controlit_urdfs.sh
$ roslaunch stickbot_lowerleg_3dof_controlit simulate_jpos.launch
```

The first command changes the current working directory. The second command generates the Universal Robot Description Format (URDF) [38] file, i.e., the robot model, used by Gazebo. The models used by ControlIt! and RViz [30] are generated automatically when executing the third command. After executing the third command, Gazebo's GUI appears with the robot loaded but in a paused state, and another visualization of the robot in RViz also appears. Click on the start button within Gazebo to start the simulation, and observe the robot go into the configuration shown in Fig. 17a. The whole body controller has a `FlatContactConstraint` assigned to the robot's foot and a `JointPositionTask` with target joint angles that enable the robot to remain upright. This constitutes the simplest example of how to use ControlIt!. Similar commands exist for the more sophisticated robot models shown in Fig. 17. Full details are available on ControlIt!'s website [34].

ControlIt!'s website also contains examples of how to use ControlIt! to achieve advanced whole body behaviors. One particularly useful example is the integration of MoveIt! [18] with ControlIt!. MoveIt! provides many useful functions including planners based on the Open Motion Planning Library (OMPL) [39, 40]

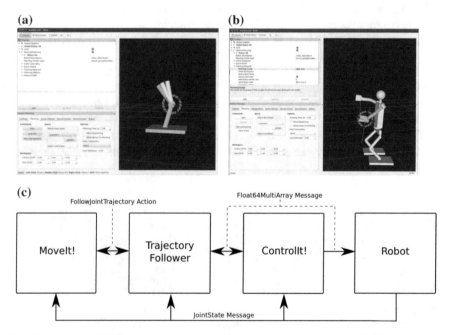

Fig. 18 MoveIt!'s GUI can be used to plan and issue motion trajectories for ControlIt! to follow. **a** Joint position control of `stickbot_lowerleg_3dof`. **b** Cartesian position control of `stickbot_humanoid_32dof`. **c** The architecture for integrating ControlIt! and MoveIt!

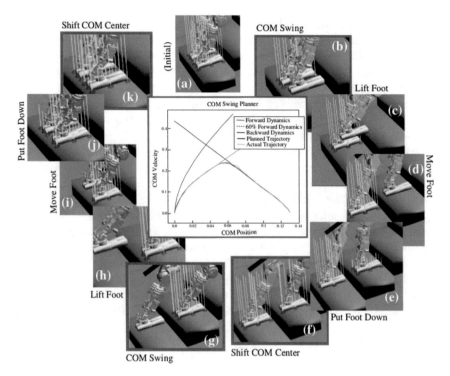

Fig. 19 An example of using a phase-space locomotion planner in conjunction with ControlIt! to make a bipedal robot walk up a flight of stairs in simulation. Due to space constraints, only the feet of the bipedal robot are shown

and a GUI for enabling users to specify goals using 6-DOF interactive markers [41]. Figure 18 shows how MoveIt! can be used to control the joint positions of `stickbot_lowerleg_3dof` and the Cartesian wrist positions of the 32-DOF humanoid shown in Fig. 17e, which is called `stickbot_humanoid_32dof`. As shown in Fig. 18c, the integration is done by introducing an adapter node called `TrajectoryFollower` that provides a ROS action server of type `control_msgs::FollowJointTrajectoryAction` and communicates with ControlIt! via ROS topics. It accepts action requests from MoveIt!, generates a trajectory from the robot's current state to the requested state using a spline algorithm, and transmits the points along this trajectory to ControlIt!. It monitors ControlIt!'s progress via ROS topics and updates MoveIt! using the ROS `actionlib` communication interface [42].

Another example of using ControlIt! is shown in Fig. 19. This figure shows how the integration of a phase-space locomotion planner [43] with ControlIt! enables an early model of Valkyrie to walk up a flight of stairs in simulation. The phase-space locomotion planner uses an inverted pendulum model to generate a rough estimate of the robot's dynamics when it swings its Center Of Mass (COM) sideways as it takes a step. As shown in the figure, the planner implements a finite state machine consisting of eleven states. The first five states swings the COM and moves one foot forward. The second six states swing the COM in the opposite direction and moves the other foot to be alongside the first foot. ControlIt! is configured with a flat contact constraint on each foot that can be enabled and disabled based on whether that foot is in contact with the ground. The compound task consists of a high priority `COMTask` and a lower priority `JointPositionTask`. The locomotion planner communicates with ControlIt! via ROS topics.

ControlIt!'s website contains many additional examples of how ControlIt! can be used to enable ROS applications to achieve advanced whole body behaviors on high-DOF multi-branched robots. Figure 20 shows some of these examples. Due to space constraints, full details are omitted but are available on-line. As shown in the figure, ControlIt! works with numerous robot models including various versions of Dreamer, Valkyrie, and Atlas. Atlas is a hydraulically-actuated humanoid made by Boston Dynamics (now owned by Google) and was provided by the US government for the DARPA Robotics Challenge. In preparation for this challenge, ControlIt! was used to make these robot models perform useful tasks like stand, locomote, vehicle ingress, pick up debris, open a door, manipulate tabletop items, climb a ladder, hook up a hose, and use a hand drill. These are only a subset of the behaviors enabled by ControlIt!.

8 Conclusions

ControlIt! is a high performance and highly flexible ROS-based framework that enables whole body controllers, specifically those based on the Whole Body Operational Space Control (WBSOC) formulation, to be integrated into a ROS application. It defines a software architecture and set of software abstractions for instantiating and configuring whole body controllers, and integrating them into a wide range of robots and applications. High performance with controller servo frequencies in the range of 0.5–2 kHz is achieved by using multiple threads to offload the amount of computations within the servo loop. This high frequency feedback control enables real-time adaptation to unmodeled disturbances that cannot be achieved by whole body planners. Software flexibility is achieved by extensive use of dynamically loadable plugins. These plugins enable new whole body control programming primitives like tasks and constraints to be introduced into the system. Combinations of these primitives are structured into compound tasks and constraint sets, resulting in levels of expressiveness that are sufficient to achieve a wide range of whole body robot behaviors. The whole body controller itself is a plugin, and to date, two forms of WBOSC are provided, one for torque-controlled robots like Dreamer and another for impedance-controlled robots like Valkyrie. Platform independence is achieved through robot interface and clock plugins, which enables ControlIt! to work with a variety of robot hardware platforms and real-time frameworks like RTAI and RT-Preempt, respectively. To date, ControlIt! was tested on two hardware platforms,

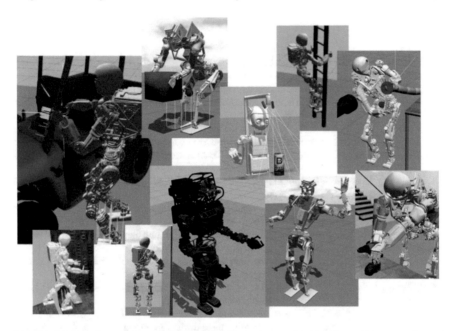

Fig. 20 Additional examples of advanced WBC behaviors enabled by ControlIt! on a variety of robot models including various versions of Dreamer, Valkyrie, and Atlas. Most were obtained in preparation for the DARPA Robotics Challenge Trials. All images were taken from the Gazebo dynamics simulator

Valkyrie and Dreamer. It was successfully used in combination with various planners and user interfaces to perform numerous manipulation tasks. In simulation, ControlIt! was demonstrated to perform even more advanced behaviors like locomotion on numerous additional robot models. In the future, we will work on further improving ControlIt! and integrating it with exteroceptive sensing capabilities that will enable, for example, visual servoing [44]. We will also integrate ControlIt! with software processes that enable both greater autonomy (i.e., via human behavior modeling and decision-making processes [45]), and ease of programming (e.g., via demonstration and reinforcement-based learning [46, 47]).

Acknowledgments We thank Gwendolyn Johnson and John D. Yamokoski for helping with ControlIt!'s initial implementation and the creation the example robot models. We thank Polam Liu and Joshua James for integrating MoveIt! with ControlIt! and developing the shared memory transport layer.
We thank the entire NASA JSC DARPA Robotics Challenge team for their help on integrating and using ControlIt! with Valkyrie. Finally, we thank Advanced Automation Accumulator Limited, the US National Robotics Initiative, and the Texas ETF for sponsoring our research.

References

1. IEEE Robotics and Automation Society, Whole body control technical committee (2015), http://www.ieee-ras.org/whole-body-control. Accessed 13 Feb 2015
2. M. Mistry, J. Buchli, S. Schaal, Inverse dynamics control of floating base systems using orthogonal decomposition," in *2010 IEEE International Conference on Robotics and Automation (ICRA)*, May 2010, pp. 3406–3412
3. W. Hyun, I.-H. Suh, J. Lim, Resolved motion control of redundant robot manipulators by neural optimization networks, in *Intelligent Robots and Systems'90. 'Towards a New Frontier of Applications', Proceedings. IROS'90. IEEE International Workshop on*, July 1990, pp. 627–634 vol.2
4. A. Herzog, L. Righetti, F. Grimminger, P. Pastor, S. Schaal, Momentum-based balance control for torque-controlled humanoids (2013), arXiv:1305.2042
5. A. Escande, N. Mansard, P.-B. Wieber, Hierarchical quadratic programming: fast online humanoid-robot motion generation. Int. J. Robot. Res. **33**(7), 1006–1028 (2014)
6. L. Sentis, O. Khatib, Synthesis of whole-body behaviors through hierarchical control of behavioral primitives. Int. J. Humanoid Robot. **2**(4), 505–518 (2005)
7. L. Sentis, Synthesis and control of whole-body behaviors in humanoid systems, Ph.D. dissertation, Stanford University (2007), supervised by Oussama Khatib
8. L. Sentis, J. Park, O. Khatib, Compliant control of multicontact and center-of-mass behaviors in humanoid robots. IEEE Trans. Robot. **26**(4), 483–501 (2010)
9. L. Sentis, J. Peterson, R. Philippsen, Implementation and stability analysis of prioritized whole-body compliant controllers on a wheeled humanoid robot in uneven terrains. Auton. Robots **35**(4), 301–319 (2013)
10. N.A. Radford, P. Strawser, K. Hambuchen, J.S. Mehling, W.K. Verdeyen, S. Donnan, J. Holley, J. Sanchez, V. Nguyen, L. Bridgwater, R. Berka, R. Ambrose, C. McQuin, J.D. Yamokoski, S. Hart, R. Guo, A. Parsons, B. Wightman, P. Dinh, B. Ames, C. Blakely, C. Edmonson, B. Sommers, R. Rea, C. Tobler, H. Bibby, B. Howard, L. Nui, A. Lee, M. Conover, L. Truong, D. Chesney, R. P. Jr., G. Johnson, C.-L. Fok, N. Paine, L. Sentis, E. Cousineau, R. Sinnet, J. Lack, M. Powell, B. Morris, A. Ames, Valkyrie: NASA's first bipedal humanoid robot. J. Field Robot. **10** (2014)
11. C.-L. Fok, Dreamer product disassembly using ControlIt! (2015), https://youtu.be/I3OCZW7lpGU. Accessed 27 Mar 2015
12. C.-L. Fok, Human robot interactions using ControlIt! (2015), https://youtu.be/uagk5brDXWw. Accessed 27 Mar 2015
13. B. Jacob, G. Guennebaud, The eigen project (2015), http://eigen.tuxfamily.org/. Accessed 13 Feb 2015
14. Martin Felis, Rigid body dynamics library (2015), http://rbdl.bitbucket.org/. Accessed 13 Feb 2015
15. Open Source Robotics Foundation, Gazebo simulator website (2015), http://gazebosim.org/. Accessed 13 Feb 2015
16. Robot Operating System. ROS shared memory interface (2015), https://bitbucket.org/jraipxg/ros_shared_memory_interface. Accessed 13 Feb 2015
17. Robot Operating System, ROS control (2015), http://wiki.ros.org/ros_control. Accessed 13 Feb 2015
18. I.A. Sucan, S. Chitta, Moveit! (2015), http://moveit.ros.org/. Accessed 13 Feb 2015
19. Robot Operating System, ROS SMACH task-level architecture (2015), http://wiki.ros.org/smach. Accessed 26 Mar 2015
20. Robot Operating System, ROS topic (2015), http://wiki.ros.org/Topics. Accessed 27 Mar 2015
21. Robot Operating System, ROS service (2015), http://wiki.ros.org/Services. Accessed 27 Mar 2015
22. Robot Operating System, ROS pluginlib (2015), http://wiki.ros.org/pluginlib. Accessed 13 Feb 2015

23. Robot Operating System, ROSParam (2015), http://wiki.ros.org/rosparam. Accessed 13 Feb 2015
24. Dipartimento Di Scienze e Tecnologie Aerospaziali del Politecnico di Milano, Real-time application interface (2015), https://www.rtai.org/. Accessed 13 Feb 2015
25. L. Fu, R. Schwebel, Rt-preempt (2015), https://rt.wiki.kernel.org/index.php/RT_PREEMPT_HOWTO. Accessed 29 Mar 2015
26. Y. Zhao, N. Paine, K. Kim, L. Sentis, Stability and performance limits of latency-prone distributed feedback controllers. IEEE Trans. Ind. Electron. **PP**(99), 1–1 (2015)
27. Robot Operating System. ROS msg (2014), http://wiki.ros.org/msg. Accessed 29 Jun 2015
28. Robot Operating System, ROS Time (2015), http://wiki.ros.org/roscpp/Overview/Time. Accessed 02 Apr 2015
29. CPP Reference, Date and time utilities (2015), http://en.cppreference.com/w/cpp/chrono. Accessed 2 Apr 2015
30. Robot Operating System, ROS RViz (2015), http://wiki.ros.org/rviz. Accessed 3 Apr 2015
31. S. Hart, P. Dinh, J. Yamokoski, B. Wightman, N. Radford, Robot task commander: a framework and IDE for robot application development, in *2014 IEEE/RSJ International Conference on Intelligent Robots and Systems (IROS 2014)*, Sept 2014, pp. 1547–1554
32. T. Kröger, *On-Line Trajectory Generation in Robotic Systems*, Springer Tracts in Advanced Robotics, vol. 58. Springer, Berlin (2010)
33. T. Kroeger, Rigid body dynamics library (2015), http://www.reflexxes.com/. Accessed 03 Apr 2015
34. C.-L. Fok, Controlit! website (2015), https://robotcontrolit.com/. Accessed 13 Feb 2015
35. Robot Operating System, Catkin (2015), http://wiki.ros.org/rosbuild. Accessed 2 Apr 2015
36. J. Beder, yaml-cpp (2015), https://github.com/jbeder/yaml-cpp. Accessed 2 Apr 2015
37. Y. Zhao, N. Paine, L. Sentis, Feedback parameter selection for impedance control of series elastic actuators, in *2014 14th IEEE-RAS International Conference on Humanoid Robots (Humanoids)*, Nov 2014, pp. 999–1006
38. Robot Operating System, URDF (2014), http://wiki.ros.org/urdf. Accessed 14 Feb 2015
39. Kavraki Laboratory, Open motion planning library (2015), http://ompl.kavrakilab.org/. Accessed 4 Apr 2015
40. I.A. Şucan, M. Moll, L.E. Kavraki, The open motion planning library. IEEE Robot. Autom. Mag. **19**(4), 72–82 (2012), http://ompl.kavrakilab.org
41. Robot Operating System, ROS interactive markers (2015), http://wiki.ros.org/rviz/Tutorials/Interactive%20Markers%3A%20Getting%20Started. Accessed 3 Apr 2015
42. Robot Operating System, ROS actionlib (2015), http://wiki.ros.org/actionlib. Accessed 4 Apr 2015
43. D. Kim, Y. Zhao, G. Thomas, L. Sentis, Accessing whole-body operational space control in a point-foot series elastic biped: balance on split terrain and undirected walking (2015), arXiv:1501.02855
44. S. Hutchinson, G. Hager, P. Corke, A tutorial on visual servo control. IEEE Trans. Robot. Autom. **12**(5), 651–670 (1996)
45. J.G. Trafton, L.M. Hiatt, A.M. Harrison, F. Tamborello, S.S. Khemlani, A.C. Schultz, ACT-R/E: an embodied cognitive architecture for human robot interaction. J. Human-Robot Interact. **2**, 30–55 (2013)
46. M. Cakmak, A.L. Thomaz, Eliciting good teaching from humans for machine learners. Artif. Intell. **217**, 198–215 (2014)
47. B. Akgun, M. Cakmak, K. Jiang, A. Thomaz, Keyframe-based learning from demonstration. Int. J. Soc. Robot. **4**(4), 343–355 (2012)

Part VII
ROS Simulation Frameworks

Simulation of Closed Kinematic Chains in Realistic Environments Using *Gazebo*

Michael Bailey, Krystian Gebis and Miloš Žefran

Abstract Simulation is an integral part of the robo design process; it allows the designer to verify that the mechanical structure, sensors and software work together as intended. It can also serve as a collaboration platform for a team. *Gazebo* is a particularly attractive simulation platform as the physical behavior of the robot can be simulated in parallel with the ROS software that controls it. A lesser known feature of *Gazebo* is its ability to simulate closed kinematic chains. This is partly due to a lack of a well-established procedure for creating such simulations. This chapter describes in detail how robots with closed kinematic chains can be simulated in *Gazebo*. It explains how a robot model created with a computer-aided design (CAD) program such as *SolidWorks* can be exported to *Gazebo* so that closed kinematic chains are properly modeled, and how a realistic simulation environment can be generated. We provide detailed step-by-step examples that can be used by the reader to easily create new simulations using *Gazebo* and *SolidWorks*. *SolidWorks* was chosen as the CAD tool because it can partially export kinematic structures. Closed kinematic chains can then be relatively easily added to these exported structures so they can be used in *Gazebo*.

Both M. Bailey and K. Gebis contributed equally to this work.

M. Bailey (✉)
Department of Mechanical and Industrial Engineering, University of Illinois
at Chicago, Champaign, USA
e-mail: mbaile20@uic.edu
URL: http://www.uic.edu

K. Gebis · M. Žefran
Department of Electrical and Computer Engineering, University of Illinois
at Chicago, Champaign, USA
e-mail: kgebis2@uic.edu

M. Žefran
e-mail: mzefran@uic.edu

© Springer International Publishing Switzerland 2016
A. Koubaa (ed.), *Robot Operating System (ROS)*, Studies in Computational
Intelligence 625, DOI 10.1007/978-3-319-26054-9_22

1 Introduction

Simulation is a key tool for the design of robotic systems and robot simulators are an active area of development. While the behavior of simple systems in simple environments can be easily predicted, as systems and environments become more complex, so do their interactions. The ability to readily simulate such interactions can help the designers make informed design decisions, gain qualitative understanding of the system's challenges, compare design alternatives, and develop algorithms for robot control.

The main advantage of *Gazebo* is that it easily interfaces with ROS, the infrastructure for robot control. ROS allows the robot to be controlled using a hierarchy of networked processes (nodes). The lowest level in the hierarchy abstracts the robot hardware; the rest of the system operates on this abstraction. A robot controlled through ROS can be easily simulated in a dynamic environment using *Gazebo*. During the simulation, *Gazebo* is treated as a node describing the hardware layer. ROS then communicates with these simulated hardware nodes. This approach allows for a physical system to be simulated in conjunction with the robot control algorithms [1].

Gazebo can simulate multi-body dynamics, including interactions between bodies such as impact, using several existing physics engines [1]. In *Gazebo*, both the robot model and environment model must be defined. However, since *Gazebo* can easily import models from other design tools, this approach can be made quite modular. For example, a robot model created in *SolidWorks*, a popular computer-aided design (CAD) package used for mechanical design, can be exported for use in *Gazebo* using an appropriate plugin. And the robot model as well as the environment can be realistically rendered by using 3D packages such as Autodesk 3DStudioMax for texturing. This allows for robots to be simulated directly using the models employed for their design, and for environments to be created and adjusted in a modular fashion so that a robot's ability to perform tasks can be tested in a wide range of scenarios.

Simulating open kinematic chains has been fairly well established in *Gazebo*. However, simulating closed chain mechanisms presents some technical challenges. Closed kinematic chains consist of a series of rigid bodies connected by joints, where a child link connects back to the parent link. They are typically used to generate a desired output motion or force at one link from the input at another link, and are the basis of many mechanisms [2, 3].

Unified Robot Description Format (URDF) is an XML representation used in ROS to describe the robot model. A `sw_urdf_exporter` plugin can be used in *SolidWorks* to export a model as an URDF file. However, URDF relies on a tree-like structure to describe robot kinematics, making it incapable of representing the more general graph topology associated with closed kinematic chains. On the other hand, *Gazebo* uses Simulation Description Format (SDF), another XML specification for describing robot models, which can deal with the graph representation of the kinematics. Fortunately, it is not too difficult to convert a URDF file to an SDF file and add the missing connections so that models from *SolidWorks* can be accurately reproduced.

Through the use of examples this chapter aims to create a tutorial exposition for the creation of robot simulations in complex environments, as well as a procedural approach for simulating robots featuring closed kinematic chains (kinematic loops). All the files and code used in these examples are publicly accessible [4]. We will start with simple robot systems and test environments and then increase complexity with each additional example. A brief discussions of the kinematics will be provided to provide the necessary background.

2 Preliminaries

Robots consist of multiple interconnected subsystems. A useful analogy to visualize how these systems work together is thinking of the robot as a system containing a skeleton, muscles, eyes and brain. The mechanical design of a system constitutes its skeleton. Actuators, sensors, and algorithms, can be considered the muscles, eyes and brain of the robot, respectively [5]. To simulate the robot in *Gazebo*, each subsystem needs to be described in an appropriate format. In particular, the mechanical structure is described by specifying links and joints. The actuators and the sensors can then be added by using appropriate ROS plugins. Finally, the algorithms are defined through the network of interconnected ROS nodes that use the mechanical structure, actuators and sensors. To give context to this discussion, we present a brief introduction to kinematics as well as a discussion on the two different description formats, SDF and URDF.

2.1 *Kinematic Linkages*

The robot's mechanical structure is defined through links and joints. A link is typically assumed to be a rigid body (incapable of deformation), and it can move relative to other links. A joint connects two links and determines how they can move relative to each other; joints are also called kinematic pairs [6]. The most commonly used joints are so-called lower pairs, shown in Table 1. More complicated joints can be typically obtained by combining the six lower pairs.

The following definitions are given to provide a contextual basis for the examples discussed in this chapter (Fig. 1).

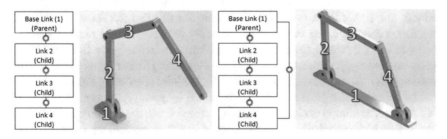

Fig. 1 Tree structured open chain and graph structured closed chain

Table 1 Lower pair joints

	Degrees of Freedom	Type of Motion
Revolute	1	Planar Rotation
Prismatic	1	Planar Translation
Helical	1	Screw Motion
Cylindrical	2	Independent Rotation and Translation
Spherical	3	Spatial Rotation
Planar	3	Spatial Translation

Parent Link: A link to which another link is connected via a joint and whose motion moves the links attached to it.

Child Link: A link which connects to a parent link.

Kinematic Chain: An assembly of links and joints which are connected in a way such that an output motion is created in response to a supplied input motion [6].

Open Kinematic Chain: A series of rigid bodies (links) connected by joints where child links are not connected to any of their ancestors [6].

Closed Kinematic Chain (Kinematic Loop): A series of rigid bodies (links) connected by joints where a child link connects back to one of its ancestors [6].

2.1.1 Closed Kinematic Chains

This section provides a brief discussion of closed kinematic chains and motivates the need for a procedural approach for simulating such linkages. In closed kinematic chains (loops), the motion between the links is restricted through additional constraints due to the kinematic closure. They are typically designed to either generate a desired trajectory of one link in the loop (output link) as a result of the motion of another link (input link), or for force transmission [2]. Figure 2 shows an example of a four-bar linkage, one of the most frequently used closed loop mechanisms. As the input link (2) rotates, the output link (4) oscillates. In the figure, the black circle represents the range of positions of the joint connecting links (2) and (3), while the blue curve represents the range of positions for the joint connecting links (3) and (4).

The behavior of a closed chain is determined by joint types, joint location and link ratios. The natural choice for designing closed chain mechanisms is CAD programs. In addition to their usual capability of creating the geometries of the links and calculating their physical properties, CAD packages such as *SolidWorks* also allow the designer to design and test the input/output relation of the mechanism.

Fig. 2 A crank rocker four bar linkage mechanism evaluated at 30° intervals

Unfortunately, CAD models can not be used directly in *Gazebo* so a procedure needs to be established for converting a CAD model to an appropriate simulation model. In particular, the CAD model needs to provide the necessary link and joint parameters for simulation.

There are several prominent examples of closed kinematic chain mechanisms with applications in robotics. Robot grippers often use kinematic loops in their designs. A human knee is a complex joint which is often modeled as a four-bar mechanism to achieve an analogous output motion [7]. Closed chains are also used in wheeled mobile robots to create passive suspension systems and increase climbing capacity [8, 9]. A four-bar linkage is an integral part of Ackerman steering. Tracks and roller chains are examples of closed kinematic chains used for power transmission. All these examples are challenging to simulate.

2.1.2 Virtual Links

In order to represent complex rigid links that are composed of several parts it is useful to create virtual links and connect them through fixed joints. With a fixed joint, all the degrees of freedom are restrained so the two bodies effectively behave as one link. Such assemblies do not change the kinematics of the robot, but they can be used for texturing or to create a reference for various plugins.

2.2 Description Formats

There are two primary description formats used with *Gazebo* and *ROS*, URDF and SDF. Both file formats can describe the kinematics and dynamics of a robot and additional plugins are added to this structure as sources of feedback or to utility to the robot [10, 11]. In this way these two formats fill similar roles. However, the nuances of these two formats have distinct differences that the creator of a simulation must consider.

URDF file type is older and has been historically used in ROS. One notable advantage of this description format is that there has more development that utilizes URDF. For example, it is possible to export SimMechanics models (a popular 3d simulation environment which is part of *Matlab*), and to export *SolidWorks* CAD models to URDF [12, 13] . One potential disadvantage is that URDF has a tree structure where elements branch outwards from a base. This has a few implications; one of the more notable ones is robots created in URDF can only use open kinematic chains (see Fig. 1). However, this tree structure combined with URDFs historical use can be advantageous as well. The popular package robot_state_publisher uses the URDF tree structure to compute the position of orientation of links offering an efficient method of computing transforms for feedback units back to an reference frame [14], which cannot be used by SDF as it has a "graph structure".

SDF was developed to fix some of the shortcomings of URDF, such as URDF's improper use of XML syntax [11]. SDF improves upon URDF by offering more features such as additional joint descriptions, offering better control of the physical descriptions and features of a robot and environment items [10, 11]. Importantly, with its graph structure is capable of describing closed kinematic chain structures (Fig. 1). However, because it is newer and features a different structure there is less SDF specific development.

It is fairly direct to convert from URDF and SDF as part of the libsdformat library. However, converting SDF back to URDF is less developed [15]. This is useful because it allows for some of the tools, such as exporting to URDF, to be used indirectly to create an SDF. Before deciding on a description format one should look at the specifics of the robot and perform a trade off analysis to determine which format would be optimal for the application. Understanding how the restrictions and benefits provided by these two different formats affect the simulation is important. Simulations are by their nature simplifications of a real world scenario. By using SDF the robot's physical response can be simulated more holistically in parallel with the software architecture. For certain applications this can be necessary or advantageous. However, if it is possible to create an acceptable simplification of the system that works within the restrictions of URDF, one might facilitate more rapid development by allowing for a wider use of ROS packages. In these two formats current climate it is advantageous to utilize URDF if the simulated robot does not require the additional descriptions offered by SDF. Early examples in this chapter are examples where the additional features of SDF are not necessary and a URDF can be used. The final example discussed features a closed kinematic chain which cannot be simplified requiring the use of SDF.

3 Creating a Basic Robot Model with *Gazebo*

To begin the discussion on simulation, it will be useful to describe the process of creating a basic robot and environment within *Gazebo*. In this example, we will create a simple skid steer robot. We will then create a sample environment using the *Gazebo*'s library of objects. Before beginning creation of a simulation model, one must know the parameters of the model's individual components. For this example, the parameters that must be known are the location and geometry of the robot's individual links, joints, joint types, and how the robot moves.

3.1 Basic Skid Steer Robot

The model below will be used to explain how the various aspects of creating a simple simulation model. A basic four wheeled model can be described with five different parts in a tree-like structure. Figure 3 shows a block diagram of the robot's links.

The central chassis of the robot will become the base frame. As shown in Fig. 4, the yellow rectangular box will be the base link. Since we want our wheels to be able to rotate, this requires us to create separate links for each individual wheel on the robot; these can be described as child links.

With the geometry of links created, now the link location, joint type, and joint axis must be defined. A link location can be classified by position and orientation, while a joint type can be classified as a revolute, continuous, fixed, or prismatic. A joint axis must be specified to define joint position and orientation. From Fig. 4, we see the location of joints, and from our knowledge of the system we know that the joints connecting the wheel to the chassis are revolute (continuous in URDF) joints.

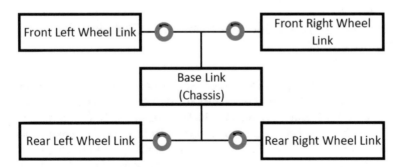

Fig. 3 Basic skid steer robot link block diagram

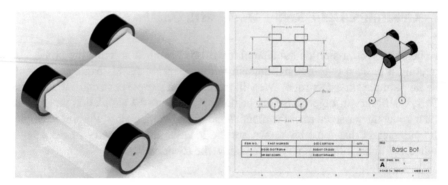

Fig. 4 Basic skid steer robot

3.1.1 Synthesizing Model with ROS

With the kinematics and physical parameters of the robot defined and created the ROS plugins can be added to the model [16]. For this basic example only the skid steer plugin will be used. This will allow for the robot to be driven as a skid steer system. Skid steering is one of the more simple methods of maneuvering, where the robot turns by rotating the wheels on either side of the robot at varying speeds, dragging (skidding) the wheels to turn. To drive this model, add the skid steer plugin at the end of the SDF or URDF file. For this plugin to work, the robot must have four wheel joints to define as wheel locations. Additionally, the wheel base and wheel diameter must be defined [16]. The final SDF or URDF file should now include descriptions of link geometries, physical properties, joint locations, joint descriptions, and the skid steer plugin.

4 Creating a Simulation of a Robot in a Complex Test Environment

In this section, we will simulate a robot, Scipio, created by Chicago Engineering Design Team at the University of Illinois at Chicago in its intended environment: the course for the Intelligent Ground Vehicle Competition. In this fashion, we will demonstrate how a robot's ability to achieve tasks in a complex environment can be tested. This robot is designed for autonomous navigation utilizing four wheel skid steer for mobility. The feedback units, which inform Scipio's path planning algorithms are a GPS unit, inertial measurement unit, wheel encoders, a sweeping laser range finder, and a stereo camera (Fig. 5).

Fig. 5 Scipio autonomous skid steer robot

4.1 Creation of Scipio Model: Skid Steer Robot

The process for creating the model for this robot is fairly straightforward. Its link formulation is similar to the Simple Skid Steer Bot introduced in Sect. 3. We want to maintain the physical parameters of the robot so the *Gazebo* model will be created directly from the CAD model, which was used to design the robot. Before exporting, it is useful to simplify the model; CAD models used for the mechanical design of the system will often have extraneous parts not needed for simulation model. Details like bolts, nuts, and other internal components do not need to be included; unnecessary components will make the simulation more computationally expensive. However, before removing these components, it is important to calculate parameters such as the moment of inertia and mass in order to generate an accurate model of the robot. Before we start the exporting process, we first must think about the overall structure of our robot; what parts of our robot move, what parts are grouped together, and what are the most important aspects of the robot that need to be simulated. The revelant files to this discussion by be found in our repository linked in Sect. 1 [4] (Fig. 6).

Now that we have established the grounds on which we want to export the robot, we can begin exporting the model to URDF. When you open the CAD model in *SolidWorks*, start the *sw_urdf_exporter* by going to File > Export URDF. Then, the URDF Exporter window will appear. The export process begins by defining an initial link called base_link; this link will signify the highest parent link. We will want anything relating to the base chassis of our robot to be a part of the base_link. To tell the exporter which components to included in the base_link, first expand the entire assembly by clicking on the plus sign to the right of the question mark in the

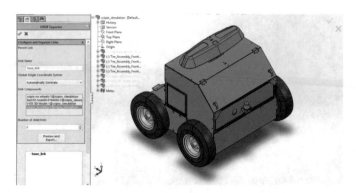

Fig. 6 Export procedure: base link

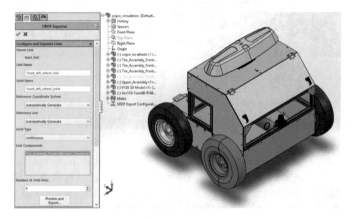

Fig. 7 Export procedure: wheel link

top panel of the exporter. This way all subassemblies and parts of the CAD model can be selected. Then choose which sub assemblies to include in the link; in our case, Scipio has two main chassis sub-assemblies, as well as the laser scanner and GPS sub-assemblies.

Next, we will want to add the wheels to our export tree. This can be done in two ways; either increase the number of childs links in the respective textbox, or right click on a link (base_link) in the exporter tree and choose add child. Create a separate child for each individual wheel: front_left_wheel (Fig. 7).

For each of these child links created, specify a similar corresponding joint name. For example, for the front_left_wheel_link, name the joint similarly to say front_left_wheel_joint. This will be the name of the joint that joins the base_link, and the front_ left_wheel_link. Once the joint name and joined links have been specified, it is good practice to define the type of joint connecting the two links this will deter-mine the type interaction between the base_link and front_left_wheel_link. Special consideration should be made as to how the assembly was built. Going back to Fig. 6,

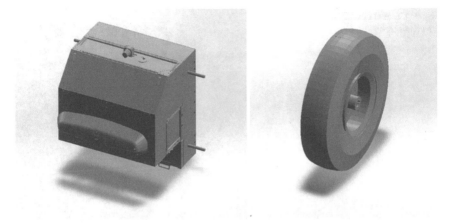

Fig. 8 Created STL files from export

we can notice that the shafts for the wheels are included in the base_link. For the URDF exporter to recognize a revolute pair the two links must have a concentric relationship.

After the export is complete a folder will be created containing the familiar launch, meshes, robots, and textures folders, and the manifests XML file. The STLs created by the URDF exporter, which at this moment constitute the visual and collision mesh for the robot, will not have any texturing. Figure 8 displays the robot mesh files directly after export.

In order to create a realistic model, a visual model for the robot must be textured. It is important to note that the position of the mesh in its local coordinate system should be maintained when texturing models, or the new mesh will not correspond to the intended location. In order to texture the elements of the robot, we need to create a new visual mesh. In addition to being able to preserve physical parameters of the robot, one of the benefits of the robot model from a CAD model is that the textures from the CAD model can be used to create a visual mesh. A Collada file will be used to create the final visual mesh. To reach this point, first the links of the robot must be exported to a file type that can be opened by a 3d packages such Autodesk 3ds Max or Blender. The most straightforward way of doing this is to import the created STLs from the export process. Another option is to export the grouping of assemblies and parts which constitute a defined link as IGES. Now the mesh can either be textured manually, or if exported in IGES format, the textures should be fairly well-maintained and no additional texturing should be needed. With the models imported into a 3d package and the level of desired detail satisfied these new files can now be exported as Collada files (.DAE) (Fig. 9).

Often adding visuals to the robot itself does not improve the performance of the simulation. However, it could become impactful in multi-robot simulations and creates a strong presentation tool and increases the aesthetic value of the simulation.

Fig. 9 Created collada files
for visual mesh

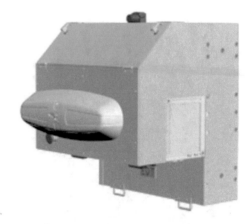

Fig. 10 Created scipio
model in *Gazebo*

In a qualitative way this provides an improved experience by visual feedback for
users of a simulation (Fig. 10).

4.1.1 Synthesizing Model with ROS

In order to simulate the robot with of *Gazebo*, we will need to add various *Gazebo*
plugins which will allow us to control the robot using ROS topics. When you install
the full ROS desktop package, *Gazebo* will be installed as well as a few of the ROS-
Gazebo plugins to allow for interfacing. Although most of the plugins we will be
using already come with the full ROS desktop download, we have decided to use
some of the *Gazebo*-ROS plugins from the *hector_gazebo_plugins* ROS package
[17]. Here is a list of the plugins we will be including in the model of Scipio:

- Skid Steer Motor Driver Plugin (libgazebo_ros_skid_steer_drive.so)
- Inertial Measurement Unit (IMU) Plugin (libhector_gazebo_ros_imu.so)
- GPS Plugin (libhector_gazebo_ros_gps.so)
- 2D Laser Scanner Plugin (libgazebo_ros_laser.so)
- Bumblebee (libgazebo_ros_multicamera.so)

The bumblebee camera plugin is also a package that does not come with the native ROS desktop install; this package is known as *lcsr_camera_models* [18]. Although this package will be included in our Github repository for this book chapter, you can find the standalone package by using this link: https://github.com/jhu-lcsr/lcsr_camera_models .

The *hector_gazebo_plugins* package can easily be installed by entering the following command into your terminal:

sudo apt-get install ros-indigo-hector-gazebo-plugins

Now that we have the plugin packages, we add them into the robot. We can now start to fill in the parameters of each individual plugin. We will look into the skid steer and 2D Laser Scanner Plugin as examples. All these plugins can be found in our robot description by visiting our repository under Scipio simulation [4].

Plugins have parameters that the user must define. For example, the skid steer plugin (Fig. 11) references the wheel joints created in the export to inform system behavior (red); this is an example of how the skeleton informs the muscles. Another parameter we will need to input is wheel separation and wheel diameter (orange). This is another advantage to utilizing a CAD package as a tool for creating the simulation because these values can be measured directly from this model than inputted. The torque parameter (yellow) is going to be taken from the specifications of the drive motor used in the system. Again, the final lines (blue) describe ROS topic names that the skid steer plugin (as a ROS node) publishes or subscribes to.

The laser range finder plugin (Fig. 12) is an example of a plugin that represents a feedback unit, or the eyes of the system with our analogy. In this plugin, the user must

```
<plugin name="skid_steer_drive_controller"
        filename="libgazebo_ros_skid_steer_drive.so">
<updateRate>35.0</updateRate>
    <robotNamespace>/</robotNamespace>
    <leftFrontJoint>back_right_wheel_joint</leftFrontJoint>
    <rightFrontJoint>back_left_wheel_joint</rightFrontJoint>
    <leftRearJoint>front_right_wheel_joint</leftRearJoint>
    <rightRearJoint>front_left_wheel_joint</rightRearJoint>
    <wheelSeparation>0.854431</wheelSeparation>
    <wheelDiameter>0.379476</wheelDiameter>
    <robotBaseFrame>base_footprint</robotBaseFrame>
    <torque>40</torque>
    <commandTopic>cmd_vel</commandTopic>
    <odometryTopic>raw_odom</odometryTopic>
    <odometryFrame>odom</odometryFrame>
</plugin>
```

Fig. 11 Skid steer motor driver plugin

```
<sensor type="ray" name="laser">
        <pose>0 0 0 0 0 0</pose>
        <visualize>true</visualize>
        <update_rate>15</update_rate>
        <ray>
            <scan>
            <horizontal>
            <samples>720</samples>
                <resolution>1</resolution>
                <min_angle>-1.570796</min_angle>
                <max_angle>1.570796</max_angle>
            </horizontal>
            </scan>
            <range>
                <min>0.08</min>
                <max>10.0</max>
                <resolution>0.05</resolution>
            </range>
        <noise>                      <type>gaussian</type>
            <mean>0.0</mean>
            <stddev>0.000</stddev>
        </noise>
        </ray>
        <plugin name="gazebo_ros_laser"
            filename="libgazebo_ros_laser.so">
            <topicName>scan</topicName>
            <frameName>laser</frameName>
        </plugin>
</sensor>
```

Fig. 12 2D laser scanner plugin

specify the pose, position, and orientation (red). The update rate (orange) determines the maximum number of times per second the model will attempt to record new data. Parameters that are associated with the laser's field of operation are denoted yellow. Min and max angle describe the angular scan range. The min and max range describes the minimum and maximum distance to which the laser operates. This information is dependent on the type of laser used and can be taken from a specific laser's data sheet. Users may add noise (green) to the feedback unit to increase the realism in their simulation. This information may be provided by the manufacturer; other times, one must calibrate the laser through testing.

4.2 Creation of Environment

The test environment for this robot will be a simulated course of the Intelligent Ground Vehicle Competition [19]. Within this environment, Scipio can test its ability to navigate a course autonomously and simulate the competition as closely as possible. All simulations must be simplifications of real systems. In order to create an effective simulation environment, denote the important features that the simulation environment should encapsulate.

In this simulation, the course size must be to scale, lines must indicate the boundaries, and the surface texture must provide an uneven surface and be visually similar to grass. Course objects such as construction barrels and different types of fencing will be present. *Gazebo* provides a library of objects that can be added into an envi-

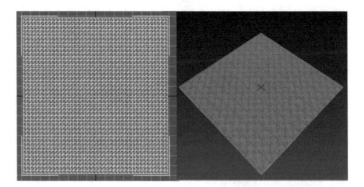

Fig. 13 Ground plane mesh pre-texture and textured Collada file

ronment. If the library is not sufficient, new objects can be created by creating a URDF model for that object. In this scenario the two primary parameters which can describe most of these features are object color and geometry. First, we need to create a ground plane. An object mesh is required in order to create a ground plane for a robot to drive upon. In this example, this was achieved first by creating a plane in Autodesk 3ds Max; however, this plane could be created in any general 3d package. We then dimensioned the plane to appropriately fit the elements of the course. From this point a grass visual texture can be done by applying a bitmap texture to this object. It is a good practice to select an isotropic (identical in all directions) image so that the tiling of the image does not appear unnatural (Fig. 13).

With the ground plane appropriately sized and with a visual texture, terrain elevations can be added to add realism to the simulation. In this scenario the robot is intended to operate on grass, the unstructured terrain elevations created by an uneven surface impacts sensor measurements and the robots performance. Therefore the created environment should reflect the waviness and roughness present in this terrain. To achieve this the mesh which describes the grass ground plane should reflect these parameters in its geometry. There are several ways to achieve this depending on which tool is being used to create this object. However, the height deviations should correspond to something logical, such as a a road roughness index, in this example. In this scenario, we used the displacement modifier to create height variations based on the grass bitmap. Export the file as a collada file (.DAE). This DAE file will represent the collision and visual mesh of the ground plane.

To create the lines for the environment, create an appropriately scaled bitmap image to texture another object, which will represent the lines in the simulation (Fig. 14). This image file can be created in an image editing software, such as Photoshop. In this example a sample course image was used to generate the bitmap image which maps the visual texture of the lines_link. In a similar fashion to before map this image to a plane and create a Collada file of this mesh.

Ground Plane and Lines Model

```xml
<?xml version="1.0" ?>
<sdf version="1.4">
  <model name="grass">
    <static>true</static>
    <link name="link">
      <pose>0 0 0 0 0 1.57</pose>
      <must_be_base_link>1</must_be_base_link>
      <collision name="collision">
        <geometry>
          <mesh>
            <scale>1 1 1</scale>
            <uri>model://grass/meshes/grass.DAE</uri>
          </mesh>
        </geometry>
        <surface>
          <friction>
            <ode>
              <mu>100</mu>
              <mu2>50</mu2>
            </ode>
          </friction>
        </surface>
      </collision>
      <visual name="visual">
        <cast_shadows>false</cast_shadows>
        <geometry>
          <mesh>
            <scale>1 1 1</scale>
            <uri>model://grass/meshes/grass.DAE</uri>
          </mesh>
        </geometry>
      </visual>
    </link>
    <link name="lines_link">
      <pose>0 0 0.075 0 0 0</pose>
      <must_be_base_link>1</must_be_base_link>
      <visual name="lines_visual">
        <cast_shadows>false</cast_shadows>
        <geometry>
          <box>
            <size>38 38 0.0001</size>
          </box>
        </geometry>
        <material>
          <script>
            <uri>model://grass/materials/scripts</uri>
            <uri>model://grass/materials/textures</uri>
            <name>grass/lines</name>
          </script>
        </material>
      </visual>
    </link>
  </model>
</sdf>
```

The SDF file contains both the grass link (green) and the lines (blue) in shown in the code segment above. This model represents the ground and the line markers. It is denoted static so that it cannot move (orange). Visual meshes are outlined in red while collision meshes are outlined in gray. The collision model for the grass is present but the collision model lines was removed so that the contour of the ground is determined purely by the created grass.DAE mesh. The lines visuals are simply maped onto a standard SDF geometry and the texture is defined in the materials section (outlined in red). This model is a component of the complete test environment, the final world contains this created grass .SDF as well as representations of barrels, fencing and

Fig. 14 Course layout

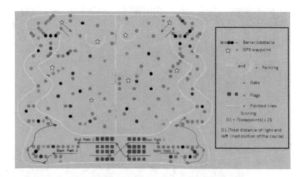

Fig. 15 Robot in simulatation environment

other object, which were added using objects from the *Gazebo* library, as shown in Fig. 15.

4.2.1 Example Use of Simulation

To use examples of Scipio, clone the Github repository into your catkin workspace (inside of the catkin_ws/src folder) [4]. Compile your workspace, so that all the necessary components get installed into the proper directories, and also make sure you have sourced your workspace (source < path_to_your_workspace > /devel/setup.bash). Once you have compiled and sourced your workspace, you can simply run the following command to bring up the entire Scipio simulation setup.

roslaunch scipio_simulation gazebo.launch

This should bring up two windows; the *Gazebo* and RViz GUI's (Fig. 16).

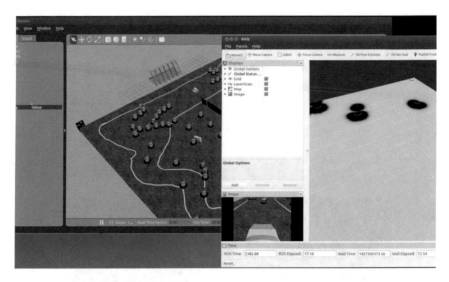

Fig. 16 Simulation GUI upon launch

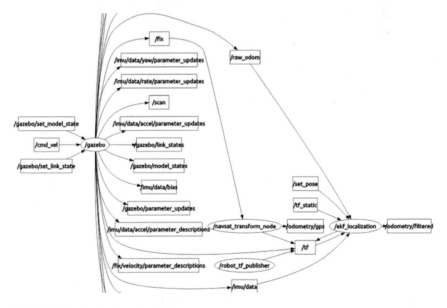

Fig. 17 ROS node structure for Scipio

If you would like to use the RViz display configuration provided in the github repo, in RViz just click File > Open Config, and navigate to where you clone your repo and select the scipio_nav.rviz file. You can now begin to implement your various algorithms, and switch your work in between your real life robot, and your *Gazebo* simulation.

Figure 17 shows the primary interest topics that would be published out from the *Gazebo* node. Do not get confused by this arrangement, however, as many people may wonder where the nodes for IMU, GPS, laser, and bumblebee are on the ROS node graph. The actual node is the *Gazebo* ROS api plugin, which allows for communication between ROS and *Gazebo*. The *Gazebo* node then loads up various plugins, which do tasks that normal ROS nodes would do; publish data and subscribe to data. In the node graph, a instance of navsat_transform_node and ekf_localization is also being executed; these are nodes that are part of the Extended Kalman Filter package called robot_localization. A launch file for Scipio's configuration is provided on the Github Repo.

5 Simulating a Robot Featuring Kinematic Loops in a Complex Test Environment

As mentioned previously a passive suspension system for mobile robot configuration is one application of kinematic loops. In this section we will create a simulation model and environment for a robot Surus created by Chicago Engineering Design Team at the University of Illinois at Chicago. This robot was designed as a robotic excavator to compete in the NASA Robotic Mining Competition. This robot features a passive suspension system which is a derivative of the system presented in RCL-E exomars breadboard chassis [8, 20, 21]. It features two parallel longitudinal bogies, each of which feature a five bar closed kinematic chain and one rear transverse bogie. This example will act as a tutorial exposition for creating closed kinematic chains in robotic simulations using *Gazebo* and simulating this type of robot in a complex environment. The importance of including the kinematic loop in the simulation will be demonstrated by presenting experimental results, where two chassis configurations are compared: One which is articulated, featuring the kinematic loop, and one where the wheels links are connected directly to a rigid chassis. The files used in this discussion can be found on our repository [4] and an accompanying video discussion by be viewed at: https://youtu.be/Vpfxjh4mPXk (Fig. 18).

5.1 Creation of Robot Model Featuring Kinematic Loops

As in previous examples certain parameters of the robot such as geometry of links, joint locations, mass and inertial properties are defined through the *SolidWorks* model. From here again using the plugin ***sw_urdf_exporter*** [13], the CAD design of the robot made in *SolidWorks* can be exported into Universal Robot Description Format (URDF) [22]. URDF defines joints and links along with their respective collision and geometric models. The technical challenge arises because URDF is only capable of

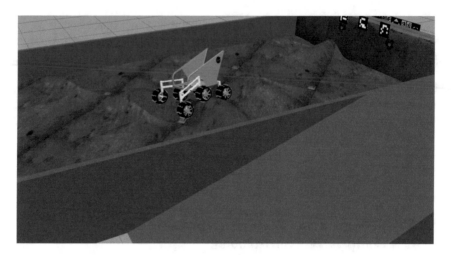

Fig. 18 Simulation environment for robot featuring articulated suspension system

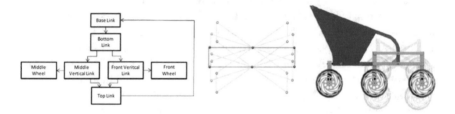

Fig. 19 Longitudinal bogie a closed kinematic chain

handing "tree-like" kinematic structures so it can not be used if the robot includes closed kinematic chains (Fig. 19).

In order to close the chain, first the model must be exported into URDF as a series of open chains. The series is then converted to SDF (Simulation Description Format) [10], where the additional joints will be added in order to close the chain. The locations of these joints can either be deduced from the location of the other joints, or by examining the original assembly file. Again, the first chosen link will be the base link for the entire export. This means that any link created after the first one will have to be a child to that parent link or a child of an existing child of the parent link, resulting in a tree-like structure. It is important that a reference configuration for the assembly is chosen so that joint locations can be obtained easily, as these distances will be needed later (Fig. 20).

At this point the top link of the mechanism does not have any children; they will have to be added manually as additional joints to links (4) and (5) to create a closed kinematic chain and obtain a "graph-like" structure. Note that since the system consists of two parallel front longitudinal bogies, this procedure can be repeated for the other side of the robot. The last chain which makes up the rover is the rear

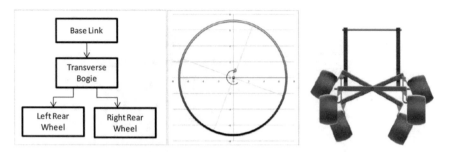

Fig. 20 Transverse bogie an open kinematic chain

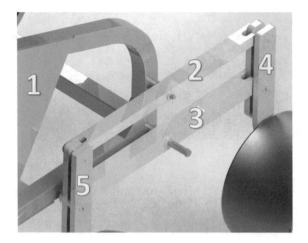

Fig. 21 Base link and the longitudinal bogie assembly

transverse bogie. This chain is open so it can be exported directly. Each longitudinal bogie consists of five links, excluding the wheels, as shown in Fig. 21. The base link will have five direct children links in total, four from the front bogies and one from the rear transverse bogie. Figure 22 shows the complete tree structure with all the links, including the wheels, named (the numbers refer to Fig. 21 and have been added for clarity, they are not part of the description).

Once all of the links and the joints have been created and named, the URDF model is ready for export. As the export process happens the plugin will create origins and axis for the joints. During the export, the plugin will create several folders, among them one called *robots*. Inside of the *robots* folder there is a URDF file that can be converted to a SDF file.

Convert the file from URDF to SDF with the *Gazebo* using the *gzsdf* command-line tool. For example, to convert a URDF file *exported-robot.urdf* to a SDF file *exported-robot.sdf* under Linux, one uses the command:

$$\textit{gzsdf print exported-robot.urdf} > \textit{exported-robot.sdf}$$

Fig. 22 Initial exported tree structure

```
<joint name='left_top_back_joint' type='revolute'>
    <child>left_bot_back_link</child>
    <parent>left_top_link</parent>
    <pose>0 0.0762 0 0 0 0</pose>
    <axis>
        <xyz>-1 0 0</xyz>
        <limit>
            <lower>-1e+16</lower>
            <upper>1e+16</upper>
        </limit>
    </axis>
</joint>

<joint name='left_top_front_joint' type='revolute'>
    <child>left_bot_front_link</child>
    <parent>left_top_link</parent>
    <pose> 0 0.0762 0 0 0 0 </pose>
    <axis>
        <xyz>-1 0 0</xyz>
        <limit>
            <lower>-1e+16</lower>
            <upper>1e+16</upper>
        </limit>
    </axis>
</joint>
```

Fig. 23 Code fragments that define joints connecting the front and back links of the longitudinal bogie to the top link

Once the SDF file is generated, we will have to manually connect (both *left_* and *right_*) *bot_front_link* and *bot_back_link* back to the top link through rotational joints to create the closed kinematic chains. In the newly created *exported-robot.sdf* file, two new joints can be created using the code fragments shown in Fig. 23. For each manually added joint, the child and the parent link needs to be specified (blue and green), as well as joint type (orange), the position and of the joint relative to the child, the joint limits, and the direction of the joint axis (outlined in red) [10].

5.1.1 Synthesizing Model with ROS

Inorder to simulate the complete robot system in *Gazebo*, we have used the gazebo_ros_pkgs wrapper, which allows ROS messages, services, and dynamic reconfiguration parameters to interface with the *Gazebo* simulator. This allows us to take advantage of various existing ROS plugins such as six wheel differential drive (*DiffDrive6W*) and IMU (*GazeboRosImu*) from [17], and monocular camera (*Camera*) [16]. For example, by specifying the wheel joints to the *DiffDrive6W* plugin, we are able to send a velocity ROS message command, which can be used to navigate the robot through the environment. All of these hardware plug-ins can be configured to closely match the manufacturer specifications and additional realism can be achieved by adjusting parameters like Gaussian error, motor torque, as well as camera focal length, distortion and field of view. This allows for physical system, control algorithms and sensors to all be simulated and tested at same time.

5.2 Creation of Environment

Environments can be designed in a similar procedure to the robots discussed in this chapter. Designing these components is generally more straightforward depending on the environment. This step is important because it provides the surroundings that the robot system will interact with. The first thing to consider is what parameters are important to the creation of the environment and what can be neglected. Remember to be cognizant of the assumptions used to create the environment and how these assumptions can affect the outcome of a simulation when compared to the real life case. In this application the most important features that were included in the creation of the environment for this example were geometry of the terrain, and a visual feedback. In order to study the mobility of this articulated system the geometry of the terrain is important and visual feedback is important for camera sensor data.

The simulated environment created in this environment was made to mimic the conditions of the NASA Robotic Mining Competition which aims to simulate a planetary excavation mission. The test arena for this competition features a chaotic terrain environment in which the ground varies from minimum and maximum elevations (± 0.3 m) [23]. One way to mimic these conditions is to create a description format for an environment item which will contain the geometry that the robot will drive on. The mesh for the rigid body can be designed using a computer automated design package, which will define the geometry of this body. In this case we created the environment using *SolidWorks* using the freeform tool to vary the elevations to the aforementioned predefined elevations and exported it using the ***sw_urdf_exporter*** plugin. However, because the terrain was modeled as a single body there is not a major disadvantage to creating the mesh in any general CAD or 3d package. Because environment objects are generally simple URDF format is sufficient. The URDF file could be converted to SDF format for consistency, but that is not necessary.

Similar to the discussion presented in Sect. 4.2, the terrain can be textured, by taking the STL mesh, adjusting it in a 3d package, such as *Autodesk 3ds Max* or

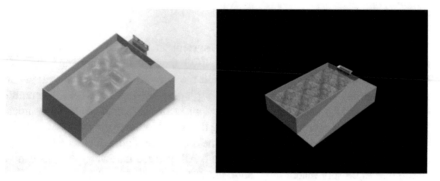

Fig. 24 LunArena environment before and after texturing

Blender, and exporting this augmented mesh as a collada format. With this the camera can have feedback. In this way the ground can have a texture and objects, such as ArUrco markers, can be colored, so that the robot can "see" these items similar to how it would in real life. The collada file type allows for textures created by images to be applied to the model. When exporting a file in collada (.DAE) format, be sure to check that the images used for texturing are included.

As mentioned earlier, it is important to take note of assumptions made in creation of the environment. In this case the ground was modeled as a singular rigid body. However, in reality the robot would be driving on soil. Both URDF and SDF formats allow for the physical parameters of the environment to be defined and tuned to varying degrees. For example using SDF format world physics parameters as static friction, dynamic friction, and surface contact layer to add additional realism (Fig. 24).

5.2.1 Example Use of Simulation

With the previous steps completed, the robot model and environment can be used for various robot simulations. By including the kinematic loops in the simulation the robot's the simulated suspension mechanism's response is integrated with the robot's hardware and software architecture. This allows for the algorithms used to govern the robot to be tested in parallel with the robot's response to the simulated chaotic terrain, yielding a more holistic simulation where the robot's ability to perform tasks can be evaluated.

By simulating a robot in this manner, the simulated results can give engineers insights into how a system operates. One useful application of this type of simulation is observing how different robot systems respond to an environment. In this example two chassis are compared one which features the closed chain system documented above and one where the articulated system is not included.

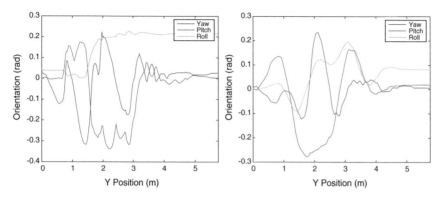

Fig. 25 Simulation of the non-articulated (*left*) and articulated (*right*) rover driving forward

During simulation, the complete state of the robot (*Gazebo*_model_state) was recorded so that position and orientations of the two rovers could be compared. In addition, data was collected from a simulated inertia measurement unit (IMU) as well as from the simulated camera.

By collecting this data, for example, we can investigate to how each configuration's response to the terrain would affect a robot's trajectory and the sensor measurements that are used for higher-level algorithms. These measurements can be recorded and saved as a delimited .txt file. This data can be then imported to software used for analysis such as *Excel* or *Matlab*. In Fig. 25 the *Gazebo*_model_state was imported and graphed in *Matlab* to make comparisons. The same procedure was repeated for driving in reverse.

In the simulations, the Y axis of the robot points forward while Z points up. From the data collected from the simulation we can see that the rigid (non-articulated) chassis (left panel of Figs. 25 and 26) shows oscillations that are not present in the trajectories of the articulated chassis (right panel of Figs. 25 and 26). This shows how ROS and *Gazebo* offer a convenient and powerful platform for robotic simulations. Mechanisms developed in CAD package such as *SolidWorks*, can be integrated into robotic simulations, where the system can be simulated as a whole in various environments. Not only can algorithms and the ROS architecture be tested using this type of simulation, but engineers and roboticists can also use this formulation to collect simulated data to make quantitative assessments of how a system behaves.

6 Lessons Learned

Closed kinematic chains are the basis for many mechanisms; mechanisms provide a robot with the ability to perform tasks so including these kinematic structures in simulation is important. URDF is a common historical description format with considerable technology development; however, it has a tree structure making it

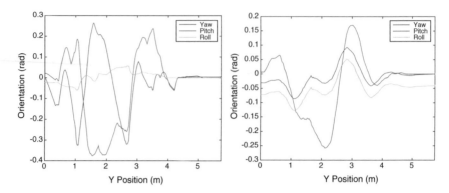

Fig. 26 Simulation of the non-articulated (*left*)) and articulated (*left*) rover driving in reverse

unable to simulated closed chains. SDF is a newer more complete format which utilizes a graph structure allowing it to encompass these closed chains. Creating a description format for a robot can become arduous. By utilizing tools such as *sw_urdf_exporter* plugin an accurate simulation model for a robot can be created easily. However, it must be exported as an open chain. To close the chain this model must be then converted to SDF.

When simulating a robot perform a trade-off analysis based on the advantages and disadvantages of each of these formats to determine which description format should be used. Environment objects can be created in a similar fashion to a robot utilizing a description format. Simulations can be used to test the ROS architecture of a system as well as an evaluation tool for how a mechanical element synthesizes with the system. One should be conscious of how simplifications in a simulation may impact results. This work aims to formalize a procedure for creating closed chains in *Gazebo* simulations, providing two benefits. First, a greater range of robot configurations can be simulated in *Gazebo*. Second, it allows for a robot and its ROS architecture to be simulated in parallel with its physical or kinematic response more holistically in *Gazebo*.

References

1. N. Koenig, A. Howard, Design and use paradigms for gazebo, an open-source multi-robot simulator, in *2004 IEEE/RSJ International Conference on Proceedings Intelligent Robots and Systems (IROS 2004)*, vol. 3 (IEEE, 2004), pp. 2149–2154. http://ieeexplore.ieee.org/xpls/abs_all.jsp?arnumber=1389727
2. A. Erdman, G. Sandor, S. Kota, *Mechanism Design: Analysis and Synthesis*, vol. 1 (Prentice Hall, Upper Saddle River, 2001)
3. F. Freudenstein, Design of four-link mechanisms, Ph.D. thesis, Columbia University, New York, United States, 1954. http://search.proquest.com/dissertations/docview/301956609/citation?accountid=14552

4. l0g1x/SpringerROS_gazebo2015		GitHub,		https://github.com/l0g1x/SpringerROS_Gazebo2015
5. S. Cetinkunt, *Mechatronics with Experiments* (Wiley, New York, 2015)
6. R. Norton, *Design of Machinery: An Introduction to the Synthesis and Analysis of Mechanisms and Machines*, McGraw-Hill series in Mechanical Engineering (McGraw-Hill, New York, 2012)
7. C. Radcliffe, Four-bar linkage prosthetic knee mechanisms: kinematics, alignment and prescription criteria. Prosthet. Orthot. Int. **18**(3), 159–173 (1994)
8. A. Seeni, B. Schafer, B. Rebele, N. Tolyarenko, Robot mobility concepts for extraterrestrial surface exploration, in *Aerospace Conference* (IEEE, 2008), pp. 1–14. http://ieeexplore.ieee.org/xpls/abs_all.jsp?arnumber=4526237
9. T. Thueer, P. Lamon, A. Krebs, R.Y. Siegwart, CRAB-Exploration rover with advanced obstacle negotiation capabilities, in *9th ESA Workshop on Advanced Space Technologies for Robotics and Automation (ASTRA)* (Noordwijk, Netherlands, 2006), http://robotics.estec.esa.int/ASTRA/Astra2006/Papers/ASTRA2006-2.2.1.03.pdf
10. osrf/sdformat—Bitbucket. https://bitbucket.org/osrf/sdformat
11. Gazebo : Tutorial : URDF in Gazebo, http://gazebosim.org/tutorials/?tut=ros_urdf
12. simmechanics_to_urdf—ROS Wiki, http://wiki.ros.org/simmechanics_to_urdf
13. Modeling for Gazebo A Design Guide for Proper Exporting from Solidworks for Gazebo Simulation, http://blogs.solidworks.com/teacher/wp-content/uploads/sites/3/WPI-Robotics-SolidWorks-to-Gazebo.pdf
14. robot_state_publisher—ROS Wiki, http://wiki.ros.org/robot_state_publisher
15. osrf / sdformat / Pull request #165: urdf format support in sdf âĂŤ Bitbucket, https://bitbucket.org/osrf/sdformat/pull-request/165/urdf-format-support-in-sdf/diff
16. Gazebo : Tutorial : Gazebo plugins in ROS, http://gazebosim.org/tutorials?tut=ros_gzplugins
17. hector_gazebo_plugins—ROS	Wiki,	http://wiki.ros.org/hector_gazebo_plugins?distro=indigo
18. jhu-lcsr/rtt_gazebo Âů GitHub, https://github.com/jhu-lcsr/rtt_gazebo
19. IGVC—Intelligent Ground Vehicle Competition, http://www.igvc.org/
20. V. Kucherenko, A. Bogatchev, M. Van Winnendael, Chassis concepts for the ExoMars rover, in *8th ESA Workshop on Advanced Space Technologies for Robotics and Automation (ASTRA)* (Noordwijk, Netherlands, 2004), http://robotics.estec.esa.int/ASTRA/Astra2004/Papers/astra2004_D-05.pdf
21. C.G.Y. Lee, J. Dalcolmo, S. Klinkner, L. Richter, G. Terrien, A. Krebs, R.Y. Siegwart, L. Waugh, C. Draper, Design and manufacture of a full size breadboard exomars rover chassis, in *6th ESA Workshop on Advanced Space Technologies for Robotics and Automation (ASTRA)* (Noordwijk, Netherlands, 2006), http://robotics.estec.esa.int/ASTRA/Astra2006/Papers/ASTRA2006-2.1.1.03.pdf
22. urdf—ROS Wiki, http://wiki.ros.org/urdf
23. Robotic Mining Competition|NASA, http://www.nasa.gov/offices/education/centers/kennedy/technology/nasarmc.html#.VPd-3S5RJQn

RotorS—A Modular Gazebo MAV Simulator Framework

Fadri Furrer, Michael Burri, Markus Achtelik and Roland Siegwart

Abstract In this chapter we present a modular Micro Aerial Vehicle (MAV) simulation framework, which enables a quick start to perform research on MAVs. After reading this chapter, the reader will have a ready to use MAV simulator, including control and state estimation. The simulator was designed in a modular way, such that different controllers and state estimators can be used interchangeably, while incorporating new MAVs is reduced to a few steps. The provided controllers can be adapted to a custom vehicle by only changing a parameter file. Different controllers and state estimators can be compared with the provided evaluation framework. The simulation framework is a good starting point to tackle higher level tasks, such as collision avoidance, path planning, and vision based problems, like Simultaneous Localization and Mapping (SLAM), on MAVs. All components were designed to be analogous to its real world counterparts. This allows the usage of the same controllers and state estimators, including their parameters, in the simulation as on the real MAV.

Keywords ROS · Gazebo · Micro Aerial Vehicles · Benchmarking

F. Furrer (✉) · M. Burri · M. Achtelik · R. Siegwart
ETH Zurich, Autonomous Systems Lab, Leonhardstrasse 21,
8092 Zurich, Switzerland
e-mail: fadri.furrer@mavt.ethz.ch
URL: http://www.asl.ethz.ch

M. Burri
e-mail: michael.burri@mavt.ethz.ch

M. Achtelik
e-mail: markus.achtelik@mavt.ethz.ch

R. Siegwart
e-mail: rsiegwart@mavt.ethz.ch

© Springer International Publishing Switzerland 2016
A. Koubaa (ed.), *Robot Operating System (ROS)*, Studies in Computational
Intelligence 625, DOI 10.1007/978-3-319-26054-9_23

1 Introduction

To test algorithms on MAVs, one needs access to expensive hardware and field tests usually consume a considerable amount of time and require a trained safety-pilot. Most of the errors, occurring on real platforms, are hard to reproduce, and often result in damaging the MAV. The RotorS simulation framework was developed to reduce field testing times and to separate problems for testing, making debugging easier, and finally reducing crashes of real MAVs. This is also convenient for student projects, where access to an expensive and complex real platform cannot always be granted. To solve higher-level tasks such as path-planning, the provided simulated MAVs can be used without any additional modification to the models. Besides a simulation of MAVs, the framework also includes a position controller and a state estimator, that work with the provided models.

The focus of this chapter will be to describe in detail the steps required to set up the RotorS simulator, depicted in Fig. 1, including the Robot Operating System (ROS) and Gazebo, and to present an evaluation framework. Once this chapter has been read, the reader will be able to set up the simulator, attach basic sensors to a MAV, and get it to fly autonomously in the simulation. The reader will also be able to compare different (custom) algorithms using the evaluation scripts. All the aspects learned and methods developed in this chapter can then be applied to a real MAV.

An overview of the main components of the RotorS simulator are shown in Fig. 2. In this chapter, the focus is put on the simulation part, shown on the left side in Fig. 2, but a lot of effort was put into keeping the structure of the simulator analogous to

Fig. 1 A screenshot of the RotorS simulator. The scene is built up from Gazebo default models and a Firefly hex-rotor helicopter

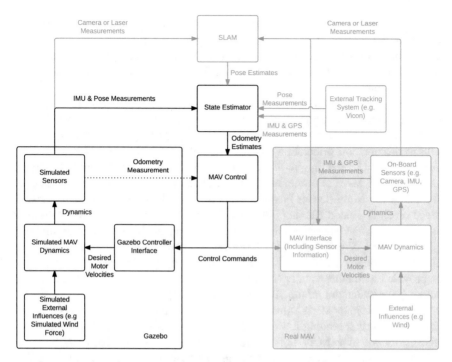

Fig. 2 Necessary building blocks to get an MAV airborne. This book chapter has a focus on the simulation part on the *left side*. All parts in *black* are covered in this book chapter and are available open source. The structure of the simulator is designed to match the real MAVs as close as possible and can be seen by comparing to the *right*

the real system. Ideally, all components used in the simulated environment can be run on the real platform without any changes.

All components, found on real MAVs, are simulated by Gazebo plugins and by the Gazebo physics engine. In the simulator we created a modular way of assembling MAVs. A MAV consists of a body, a fixed number of rotors, which can be placed at user specified locations, and some sensors attached to the body. Each rotor has motor dynamics, and accounts for the most dominant aerodynamic effects according to [7]. The parameters were identified on a real MAV, a "Firefly" from Ascending Technology[1] using recorded flight data. Several sensors, such as an Inertial Measurement Unit (IMU), a generic odometry sensor and a visual inertial sensor, consisting of a stereo camera and an IMU, and sensors developed by the user can be attached to the body. To simulate realistic conditions, we implemented noise models for the applied sensors. All the simulation data can be logged using rosbags [2], by having direct access to simulation data. Hence, ground truth data is available for each sensor, which eases debugging and evaluation of new algorithms.

---

[1] Ascending Technologies http://www.asctec.de/.

To facilitate the development of different control strategies, a simple interface is provided. We present an implementation of a geometric controller [5], which gives access on various levels of commands, such as angular rates, attitude, or position control. Additionally, this acts as an ideal starting point to implement more advanced control strategies.

The last building block is the state estimation, used to obtain information about the state of the MAV at a high rate. While state estimation is crucial on real MAVs, in the simulation this part can be replaced by a generic (ideal) odometry sensor. This means position, orientation, linear and angular velocity of the MAV are directly provided by a Gazebo plugin. In the first tutorial section, Sect. 3.3, we are showing how to get the MAV airborne, using this ideal sensor. In the following section, we show how the Multi Sensor Fusion (MSF) framework, an open source framework based on an Extended Kalman Filter (EKF) for 6 Degrees of Freedom (DoF) pose estimation [6], can be used to handle noisy, low-rate, and high-latency sensor data.

Once the MAV is flying, higher level tasks can be tackled and tested in the simulation environment, such as obstacle avoidance and path planning. The reader will be provided with representative simulation environments including the widely used octomap representation of the 3D occupancy grid [4].

This book chapter starts with a short summary of the underlying theory and references for further reading in Sect. 2. In particular it covers modeling, control and state estimation of an MAV. In the following section, Sect. 3, the basics of the simulator are explained, which are needed to get the MAV hovering. Also, the available sensors are explained in detail and how to add them to the Gazebo simulator. In Sect. 4 advanced topics are covered. This includes a deeper insight in the XML Macros (Xacro) scripting language, with an example of developing a custom asymmetric quad-rotor helicopter and a controller to enable hovering. Additionally, we show examples for solving higher level tasks using the simulator, such as collision avoidance or path planning. Finally, thoughts into the transition to real MAVs are given.

2   Background

In this section, we are presenting the theoretical background, needed to understand the following tutorials and how the simulator works. Throughout this chapter we refer to MAVs as multi-rotor helicopters, while many of the concepts are not limited to multi-rotor helicopters only.

The most important part is how to model the dynamics of an MAV. This allows to develop control strategies on different command levels such as attitude or position, which is explained in the second part of this section. Furthermore, we give a short introduction to state estimation, which is used to provide state information to the controller. For higher level tasks, such as path planning or local collision avoidance, we give an overview of the octree representation of a 3D occupancy grid.

Fig. 3 Forces and moments acting on the center of a single rotor

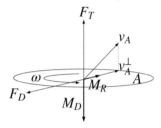

2.1 MAV Modeling

The forces and moments that are acting on an MAV can be split up into the forces and moments acting on each rotor, and the gravitational force acting on the Center of Gravity (CoG) of an MAV. All these forces combined together describe the full dynamics of an MAV.

2.1.1 Single Rotor Forces and Moments

As depicted in Fig. 3, we will analyze the thrust force F_T, the drag force F_D, the rolling moment M_R, and the moment originating from the drag of a rotor blade, M_D, from [7].

$$F_T = \omega^2 C_T \cdot e_{z_B} \tag{1}$$

$$F_D = -\omega C_D \cdot v_A^\perp \tag{2}$$

$$M_R = \omega C_R \cdot v_A^\perp \tag{3}$$

$$M_D = -\epsilon C_M \cdot F_T \tag{4}$$

where ω is the positive angular velocity of the rotor blade, C_T is the rotor thrust constant, C_D is the rotor drag constant, C_R is the rolling moment constant, and C_M is the rotor moment constant. All of these constants are positive. ϵ denotes the turning direction of the rotor, namely $+1$ (counter clockwise) or -1 (clockwise). e_{z_B} is the unit vector pointing in the z-direction in the rotor's body frame. ($u^\perp$ denotes the projection of a vector u onto the rotor plane as can be seen in Fig. 3. It can be calculated as:

$$u^\perp = e_{z_B} \times (u \times e_{z_B}) = u - (u \cdot e_{z_B}) \cdot e_{z_B} \tag{5}$$

2.1.2 MAV Dynamics

Figure 4 shows a quad-rotor helicopter with numbered rotors and the forces acting on it. Looking at the whole MAV, we can write down the equations of motion from Newton's law, and Euler's equation as:

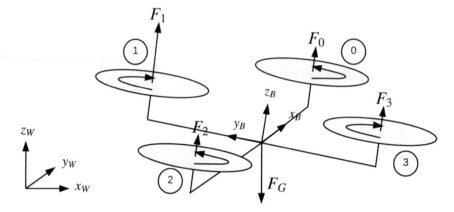

Fig. 4 Sketch of a quadrotor with the body centered body frame B and the global world frame W. The main forces from the individual rotors F_i and gravity F_G are acting on the main body

$$F = m \cdot a \tag{6}$$
$$\tau = J \cdot \dot{\omega} + \omega \times J \cdot \omega \tag{7}$$

with m, the mass of the MAV, a its acceleration, J its inertia matrix, and ω its angular velocities. The linear part in (6) is expressed in the world coordinate system, while the rotational part in (7) is expressed in the rotating body frame.

For a multi-rotor helicopter with n motors, (6) and (7) can be written as:

$$\sum_{i=0}^{n-1}(R_{WB}\underbrace{(F_{T,i} + F_{D,i}))}_{F_i} + F_G = m \cdot a \tag{8}$$

$$\sum_{i=0}^{n-1}(M_{R,i} + M_{D,i} + F_i \times r_i) = J \cdot \dot{\omega} + \omega \times J \cdot \omega \tag{9}$$

where R_{WB} is the rotation matrix rotating vectors from the body frame B to the world frame W, and r_i denotes the vector from the CoG of the MAV to the center of the ith rotor.

2.2 Control

In this section, we only cover the control of multi-rotor helicopters. Most of the current multi-rotor helicopters have rotor configurations, where the rotor axes normals point in the z-axis of the body frame z_B, therefore we will look at these configuration in particular.

In order to control an MAV, we first need to find a mapping between the system's output, which is the resulting thrust T, the accumulated thrust of all rotors, and torque τ acting on the CoG of the helicopter, and the system's input, which are the angular velocities of each rotor ω_i. We are only considering the thrust forces of each rotor, its resulting moments, and the drag moments for now. Hence we can formulate the following equation:

$$\begin{pmatrix} T \\ \tau \end{pmatrix} = A \cdot \begin{pmatrix} \omega_0^2 \\ \omega_1^2 \\ \vdots \\ \omega_n^2 \end{pmatrix} \tag{10}$$

We refer to this mapping matrix A as the allocation matrix, which is of size $A \in \mathbb{R}^{4 \times n}$. For a quad-rotor helicopter, as in Fig. 4, such an allocation matrix is given by:

$$A = \begin{pmatrix} C_T & C_T & C_T & C_T \\ 0 & lC_T & 0 & -lC_T \\ -lC_T & 0 & lC_T & 0 \\ -C_T C_M & C_T C_M & -C_T C_M & C_T C_M \end{pmatrix} \tag{11}$$

By looking at a multi-rotor helicopter with all rotor axes pointing in the same direction, we can only generate a thrust T pointing in the direction of the rotor blade's normal vector, we assume here, that this coincides with z_B. Hence, only the thrust T, and the moments around all three body axes x_B, y_B, and z_B can be controlled directly. To be able to navigate in 3D space, the vehicle has to be tilted towards a setpoint. Therefore we want to control the overall thrust of the vehicle, the direction of z_B (by controlling the roll- and pitch angle), and the yaw rate ω_z. This is usually referred to as the attitude controller. Because the dynamics of the attitude are usually much faster than the translational dynamics, often a cascaded control approach is chosen. The attitude control loop usually runs onboard a micro controller at high rate, while the position loop, which calculates the desired attitude and thrust, runs at a lower rate. This yields a control structure as depicted in Fig. 5. In this simulator, we are using the geometric controller proposed in [5], which has the same structure, but directly calculates the resulting thrust and moments at the same rate.

2.3 State Estimation

One of the key elements for enabling stable and robust MAV flights is accurate knowledge of the state of the MAV. In this section, we give a brief overview of how to fuse IMU measurements with a generic 6 DoF pose measurement in order to obtain estimates of the states that are commonly needed for a position controller for an MAV. Such a 6 DoF pose measurement could be obtained for instance from visual-

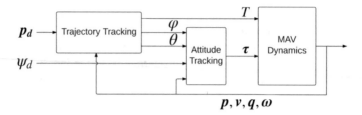

Fig. 5 Controller sketch, with the desired position p_d and the desired yaw angle ψ_d. Usually, position control is split into two parts: An outer trajectory tracking controller calculates attitude and thrust references, that are tracked by an inner attitude tracking controller

odometry or visual SLAM systems, laser rangefinder-based techniques or marker-based localization. Another very common technique to get 6 DoF pose measurements is to use an external tracking system, which provides highly accurate measurements at high rate. This method however has the disadvantage of being tied to external infrastructure. There are many other sensor types available, like pressure, optical flow, or ultrasonic sensors, but an in-depth explanation would be beyond the scope of this chapter. A good reference and overview of the properties of these sensors can be found in [11].

While IMU measurements are available on all of today's MAVs, the pose can be acquired by different methods as outlined above. IMU measurements and pose measurements have very complementary properties: Measurements from IMUs, typically used onboard MAVs, are available at high rate and with low delay, but are corrupted by noise and a time-varying bias. As a result, solely time-discrete integration (dead-reckoning) of these sensors makes a steadily accurate estimation of the absolute pose of the vehicle nearly impossible. In contrast, the methods for estimation of the 6 DoF pose usually exhibit no drift or only very low drift, but their measurements usually arrive at a much lower rate, and with high delay, due to their computational complexity. Combining both types of measurements yields a (almost) drift-free estimate of the state, at high rate and with low delay. We describe the essentials of how to do this using an EKF formulation, as shown below, and refer to [1] for a detailed description.

2.3.1 IMU Sensor Model

A common IMU model is given by

$$\omega_m = \omega + b_\omega + n_\omega \tag{12}$$
$$a_m = a + b_a + n_a \tag{13}$$

where the measured quantities are denoted with subscript m. b_ω and b_a are biases on the measured angular velocities, and accelerations respectively. These biases are modeled as random walk, having zero mean white Gaussian noise as time-derivative:

$$\dot{b}_\omega = n_{b_\omega} \tag{14}$$

$$\dot{b}_a = n_{b_a} \tag{15}$$

where n_{b_ω} and n_{b_a} are noise levels.

2.3.2 State Representation

Most controllers are separated into a position loop and an attitude loop. This works well under the assumption, that the rotational dynamics are faster than the translational motion. For the outer loop, the position p and the velocity v are needed in world coordinates. The inner attitude loop needs the orientation $\bar{q}$ and the angular velocity ω, which is provided by the IMU. Together with the bias states from the IMU model, this leads to the following state vector:

$$x = \begin{bmatrix} p^T & v^T & \bar{q}^T & b_a^T & b_\omega^T \end{bmatrix}^T \tag{16}$$

Thanks to the low complexity of time-discrete integration of IMU measurements, IMU measurements are commonly used as input for the time-update phase ('prediction') of the EKF. This has also the advantage of being independent from specific vehicle dynamics and their model parameters, and furthermore avoids having the angular rate in the state. This leads to the following dynamic model:

$$\dot{p} = v$$
$$\dot{v} = C \cdot (a_m - b_a - n_a) + g$$
$$\dot{\bar{q}} = \frac{1}{2}\bar{q} \otimes \begin{bmatrix} 0 \\ \omega_m - b_\omega - n_\omega \end{bmatrix}$$
$$\dot{b}_a = n_{b_a}$$
$$\dot{b}_\omega = n_{b_\omega}$$

For the derivation of the time-discrete integration, the error-state dynamics, the system propagation matrix, and the noise covariance matrix for this particular problem, we refer to [1]. A very good tutorial on the background is given in [10].[2]

2.3.3 Measurement Model

The measurement equations can be separated into position p_m and attitude measurements $\bar{q}_m$, which express the measured pose of the IMU with respect to the world frame.

---

[2]Be aware that in [10], the JPL quaternion notation is used, while [1] and many libraries, including ROS and Eigen, use the Hamilton notation.

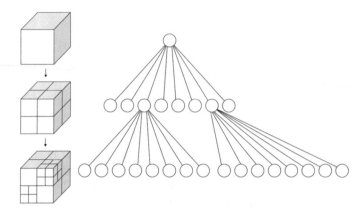

Fig. 6 *Left* Recursive subdivision of a cube into octants. *Right* The corresponding octree

$$p_m = p + n_p \tag{17}$$
$$\bar{q}_m = \bar{q} \otimes \delta\bar{q}_n \tag{18}$$

where $\delta\bar{q}_n$ represents a small error rotation and n_p is zero mean, white Gaussian noise. This simple model assumes that the origin of the pose-sensor coincides with the IMU, which is usually not the case. Also, the frame of reference of the pose-sensor may not align with the world frame. However, one can compensate for these misalignments, since they are often observable, and thus do not need to be calibrated beforehand. For the derivations and an observability analysis, we refer to [1, 11].

2.4 Octree Representation

For 3D collision avoidance and path planning, it is important to have an efficient representation of obstacles (necessary for collision checking). In recent work, octree representations are widely used for that purpose. An octree is a tree where every node has exactly eight children, which makes it well suited for efficient storage in memory. It is often used as a representation of whether a 3D space is occupied or not. Every node represents a certain part of a 3D space, this part can get subdivided into eight parts (of equal size), denoted as octants, of this subspace. This can recursively be applied until a leave has the desired resolution of the represented space, see Fig. 6. If every octant of a node has the same value, the node value can be set to this value, and the octant nodes can be omitted in the tree.[3]

---

[3]Octree2 by WhiteTimberwolf, PNG version: Nü—Own work. Licensed under CC BY-SA 3.0 via Wikimedia Commons—http://commons.wikimedia.org/wiki/File:Octree2.svg#/media/File:Octree2.svg.

3 Tutorials

This section explains how to use the RotorS simulator with its main components. It demonstrates how to get the MAV into hovering mode, and how to attach sensors to it. The setup of the simulator, for OS X and Ubuntu, is shown in Sect. 3.1. An overview of the different components of the simulator is given in Sect. 3.2. Then, we explain how to use our controllers to get the MAV into hovering mode, by assuming that we have a sensor available, which delivers a full *Odometry* message[4] in Sect. 3.3. Followed by Sect. 3.4, which covers the same scenario, but with a setup closer to real world applications, where only a pose sensor and an IMU are available. The state of the MAV is estimated from these measurements. We explain how sensors can be added in Sect. 3.5, and how to use our evaluation scripts in Sect. 3.6. For more advanced tutorials on how to build a custom model, controller, or sensor plugin, see Sect. 4. Directions on how to solve high level tasks with the simulator are given in Sect. 5.

3.1 Simulator Setup

Before installing the MAV simulator RotorS, the following steps are necessary: First, install ROS according to the official wiki-page http://wiki.ros.org/indigo/Installation. Second, the following external packages are needed.

3.1.1 Ubuntu

Is the recommended OS to run ROS and the package manager should be used to install the necessary dependencies.

```
$ sudo apt–get install ros–indigo–joy ros–indigo–octomap–ros python–wstool python–
    catkin–tools
```

Now, to install the RotorS simulator packages use the following command.

```
$ sudo apt–get install ros–indigo–rotors–simulator
```

The RotorS packages can also be compiled from source, as described in Sect. 3.1.3.

3.1.2 OS X

Can run ROS native, using homebrew, and we recommend using pip to install the necessary dependencies. Here, the RotorS simulator packages need to be installed from source, as described in Sect. 3.1.3.

---

[4]As described on: http://docs.ros.org/api/nav_msgs/html/msg/Odometry.html.

```
$ pip install wstool
$ pip install catkin-tools
```

3.1.3 Installing RotorS from Source

Is independent of the operating system and can be done using wstool, assuming
the catkin workspace is located at ~/catkin_ws/src. Make sure that wstool is
initialized before running the following commands.

```
$ cd ~/catkin_ws/src
$ wstool init
$ wstool set rotors_simulator https://github.com/ethz-asl/rotors_simulator.git —
    git -y
$ wstool set mav_comm https://github.com/ethz-asl/mav_comm.git —git -y
$ wstool update
$ cd ~/catkin_ws/
$ catkin build
$ source ~/catkin_ws/devel/setup.bash
```

If you want use the demos with an external state estimator, the following additional
packages are needed:

```
$ cd ~/catkin_ws/src
$ wstool set ethzasl_msf https://github.com/ethz-asl/ethzasl_msf.git —git -y
$ wstool set rotors_simulator_demos https://github.com/ethz-asl/
    rotors_simulator_demos.git —git -y
$ wstool set glog_catkin https://github.com/ethz-asl/glog_catkin.git —git -y
$ wstool set catkin_simple https://github.com/catkin/catkin_simple.git —git -y
$ wstool update
$ cd ~/catkin_ws/
$ catkin build
$ source ~/catkin_ws/devel/setup.bash
```

3.2 Simulator Overview

The RotorS simulator is split up in various packages as shown in Fig. 7. The general
way to incorporate an already existing robot into the simulation is by describing
its geometry, and kinematic properties. This procedure is outlined in Sect. 4.3. As
a second part, an arbitrary number of sensors can be added, which is described in
Sect. 3.5. Once at this stage, the robot can be moved, in our case the MAV will be
put into hovering mode, as described in Sect. 3.3 without state estimation, and in
Sect. 3.4 with a state estimator. The whole procedure is sketched below:

1. Select your model

   (a) Pick one of the models provided by RotorS, or
   (b) Build your own model as described below in Sect. 4.3

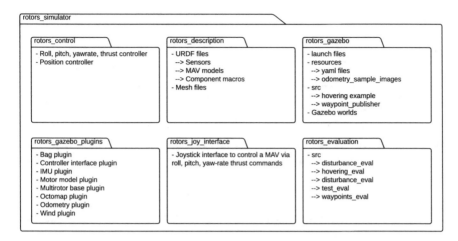

Fig. 7 Structure of the packages contained in the RotorS simulator

2. Attach sensors to the MAV

    (a) Use one of the Xacro for the sensors shipped with RotorS, or
    (b) Create your own sensor (look at our plugins in `rotors_gazebo_plugins` for reference[5]), and attach them directly, or create a Xacro, analogously to the Xacros in `component_snippets.xacro`

3. Add a controller to your MAV

    (a) Start one of the controllers shipped with RotorS, or
    (b) Write your own controller, and launch it

4. Use a state estimator

    (a) Do not run a state estimator, and use the output of the ideal odometry sensor directly as an input to your controller, or
    (b) Use MSF, see Sect. 3.4 on how to use it in the simulator, or
    (c) Use your own state estimator

3.3 Hovering Example

To check if the setup is working, we want to start with a short example. We will run the simulator with an AscTec Firefly hex-rotor helicopter, running an implementation of a position controller by Lee et al. [5]. Here, we assume that we have a sensor

---

[5]The official Gazebo sensor tutorials can be accessed on: http://gazebosim.org/tutorials?cat= sensors.

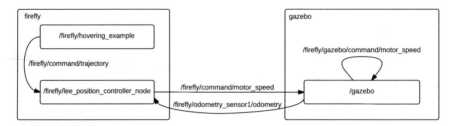

Fig. 8 Graph of ROS nodes and topics of the minimal hovering example with a Firefly. This example consists of a node that sends the initial waypoint, a controller node and the Gazebo simulator (including plugins)

available, which publishes odometry data that can directly be used by a controller. To start the simulation run:

```
$ roslaunch rotors_gazebo mav_hovering_example.launch
```

You will see the hex-rotor helicopter taking off after 5 s, and flying to the point $p = \begin{pmatrix} 0 & 0 & 1 \end{pmatrix}^T$.

In Fig. 8 you can see a re-drawn output of `rqt_graph`. This tool gives an overview of all ROS nodes that are running, and the topics on which the nodes are communicating, which is very helpful for debugging.

Gazebo is only shown as one ROS node, but internally all the Gazebo plugins are running, such as IMU and individual motors that are mounted on the frame. In this example, a generic odometry sensor is mounted on the Firefly, which publishes the following messages in the `/firefly/odometry_sensor1` namespace:

- *nav_msgs/Odometry* message on the `odometry` topic.
- *geometry_msgs/PoseStamped* message on the `pose` topic.
- *geometry_msgs/PoseWithCovarianceStamped* message on the `pose_with_covariance` topic.
- *geometry_msgs/PointStamped* message on the `position` topic.
- *geometry_msgs/TransformStamped* message on the `transform` topic.

The sensor was added to the Firefly model in the `firefly_generic_odometry_sensor.xacro`-file, which describes the model and gets assigned to the `robot_description` parameter in the launch file.

All the states needed by the position controller, depicted in Fig. 5, are contained in the *Odometry* message. Namely, these (states) are the position, orientation, and linear and angular velocity of the MAV. Hence, the position controller running alongside the simulation can directly subscribe to the *Odometry* message. The controller publishes *Actuators* messages, which are read by the Gazebo controller interface and forwarded to the individual motor model plugins.

Commands for the controller are read from *MultiDOFJointTrajectory* messages, which get published by the `hovering_example` node. These messages contain references of poses, and its derivatives. In this example only the position and yaw values of the messages are set. The `hovering_example` node publishes such a message to initiate the take-off maneuver. In addition this node un-pauses the Gazebo physics. *MultiDOFJointTrajectory* messages are used to be compatible with planners, such as Movelt! [9]. But waypoints can also be published as *PoseStamped* messages.

You can change the position reference of the MAV by sending a *MultiDOFJoint-Trajectory* message. Usually this is done by a planner or waypoint publisher. For this tutorial, we implemented a simple ROS node, that reads the position and yaw from the command line, translates it into a *MultiDOFJointTrajectory* message and sends it to the controller.

```
$ rosrun rotors_gazebo waypoint_publisher 1 0 1 0 _ _ns:=firefly
```

The first three parameters are the position, followed by the yaw angle in degrees. Because all our nodes are running in the `firefly` namespace, we also need to run the waypoint publisher in that namespace, which is done by the last argument.[6]

All the topics that are published in this configuration can be listed with:

```
$ rostopic list
```

and to look at the published data:

```
$ rostopic echo <the_desired_topic>
```

You can exchange the vehicle by setting the `mav_name`-argument, for example, you could launch the simulation with a *Pelican* quad-rotor helicopter.

```
$ roslaunch rotors_gazebo mav_hovering_example.launch mav_name:=pelican
```

Currently, these MAVs are available: `asymmetric_quadrotor`, `firefly`, `hummingbird`, and `pelican`. All the control parameters are set in the `lee_controller_<mav_name>.yaml`, and the vehicle parameters used by the controller in `<mav_name>.yaml`.

Note 1 These values do not have to be identical with the values in the model's description. On real systems, the vehicle parameters are usually unknown, and only approximate values can be identified.

Both files are located in the `rotors_gazebo/resources` folder.

3.4 Hovering with State Estimation

On real MAVs there is usually no direct odometry sensor, that gives information about all the states. Instead there is a broad variety of sensors, like GPS and magnetometer, cameras or lasers to do SLAM, or external tracking systems that provide a full 6 DOF pose. In this section, we want to show how to use the MSF package [6] to get the full state from a pose sensor and the IMU.

---

[6]More on namespaces can be found on: http://wiki.ros.org/Names.

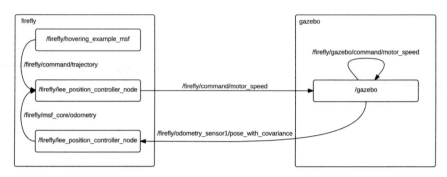

Fig. 9 Graph with ROS nodes and topics of the hovering example with state estimation. This example consists of a node that sends the initial waypoint, a state estimator node, a controller node and the Gazebo simulator (including plugins)

To run this example you need the `rotors_simulator_demos` package and its dependencies, see Sect. 3.1. The following command will start all the needed ROS nodes.

```
$ roslaunch rotors_simulator_demos mav_hovering_example_msf.launch
```

This time, the MAV is not as stable as in the previous example, and has a small offset at the beginning. The wobbling comes from the simulated noise on the pose sensor and the offset from the IMU biases. After a while the offset will disappear, because the EKF estimates the biases correctly.

Having a look at the re-drawn `rqt_graph` in Fig. 9, the following changes from the previous example can be observed: First, the Gazebo odometry plugin only publishes the pose of the MAV and not the full odometry message as before. On real systems, this could be an external tracking system. Second, an additional node is started, the MSF `pose_sensor`, but more on this node later. The controller node now subscribes to the odometry topic from the MSF instead of the odometry from Gazebo. This is done in the launch file, by a topic remapping with the following line:

```
<remap from="odometry" to="msf_core/odometry" />
```

As explained in the previous section, the following node can be used to move the MAV.

```
$ rosrun rotors_gazebo waypoint_publisher <x> <y> <z> <yaw> _ _ns:=firefly
```

The MSF package consists of a core that performs the state propagation based on IMU measurements. For the state update many predefined sensor combinations are available, from which we use the `pose_sensor` in this example. The parameters are loaded from the `msf_simulator.yaml` located in the `rotors_simulator_demos/resources` folder.

3.5 Mount Sensors

This section explains the different sensors, and how these are mounted on the MAVs. Any sensor can be attached to the vehicle in the corresponding Xacro files. This is done by calling one of the macros, located in the `rotors_description/urdf/component_snippets.xacro` file. The `mav_hoverin_example.launch` file, for instance loads the robot description from the `firefly_generic_odometry_sensor.gazebo` file, which adds an odometry sensor to the Firefly, by calling:

```
<xacro:odometry_plugin_macro (here go the macro properties)>
  <inertia (with its properties) />
  <origin (with its properties) />
</xacro:odometry_plugin_macro>
```

The macro properties are explained in the `component_snippets.xacro`-file. Properties in macros that are preceded by a * have to be set as tag-blocks within the opening and closing tags of the macro. The `inertia` and `origin` properties of the odometry macro are such blocks. An inertia block is written as:

```
<inertia ixx="1.0" ixy="0.0" ixz="0.0" iyy="1.0" iyz="0.0" izz="1.0" />
```

Properties present in all of RotorS' sensor macros are:

namespace assigns a ROS namespace to this sensor
parent_link the sensor gets attached to this link (with an offset, relative to this link, specified in the `origin`-block)
some_topic topic name on which the sensor publishes messages

Macro properties can be set, as illustrated for the `parent_link` property below:

```
<xacro:odometry_plugin_macro
    ...
    parent_link="base_link"
    ...
  >
  ...
</xacro:odometry_plugin_macro>
```

Note 2 All values of properties are passed as strings, hence, they must be enclosed by quotation marks.

Note 3 In Gazebo links must have a certain weight (at least 10^{-5} kg), otherwise they get omitted by the physics engine. This will result in not finding the links by the plugins.

An overview of the sensors currently used in RotorS, for which Xacros are provided, is given in Table 1.[7] In Fig. 10, a VI-Sensor was added to the Firefly hexacopter.

---

[7]Additional Gazebo plugins are listed on: http://gazebosim.org/tutorials?tut=ros_gzplugins.

Table 1 This table gives an overview of the different sensor Xacros provided in the RotorS simulator

Sensor	Description	Xacro-name Gazebo plugin(s)
Camera	A configurable standard ROS camera	`camera_macro` `libgazebo_ros_camera.so`
IMU	This is an IMU implementation, with zero mean white Gaussian noise and a random walk on all measurements as described in (12) and (13)	`imu_plugin_macro` `librotors_gazebo_imu_plugin.so`
Odometry	Simulate position, pose, and odometry sensors. A binary image can be attached to specify where the sensor works in the world, in order to simulate sensor outages. White pixel: sensor works, black: sensor does not work. It mimics any generic on- or off-board tracking system such as Global Positioning System (GPS), Vicon, etc.	`odometry_plugin_macro` `librotors_gazebo_odometry_plugin.so`

VI-Sensor	This is a stereo camera with an IMU, featuring embedded synchronisation and time-stamping developed at Autonomous Systems Lab (ASL). In simulation, one can choose to enable images, depth images and point clouds, and ground truth odometry	`vi_sensor_macro`
		`libgazebo_ros_camera.so`
		`librotors_gazebo_odometry_plugin.so`
		`librotors_gazebo_imu_plugin.so`
		`libgazebo_ros_openni_kinect.so`

*For more information about the VI-sensor: http://www.skybotix.com/

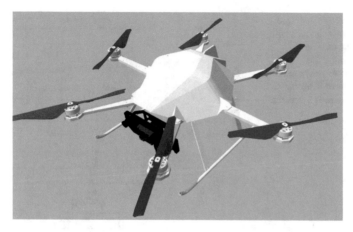

Fig. 10 The Firefly hexacopter with a VI-Sensor mounted on it pointing to the *front*, and slightly downwards

3.6 Evaluation

We are particularly interested in tracking the MAV's location to compare it with a user specified position, and how it reacts to disturbances, such as a wind gust. Great interest lies also in checking how quickly an MAV can fly to a certain location, and again how accurately it stays around the specified setpoint. As a first test, we can generate a bag file of a flight with a *Firefly*, by running the hovering example with logging:

```
$ roslaunch rotors_gazebo mav_hovering_example.launch enable_logging:=true
```

The bag files will be stored in the .ros-folder in the user's home directory by default. To run the evaluation of the controller:

```
$ rosrun rotors_evaluation hovering_eval.py –b ~/.ros/<your_firefly_bag_file>.bag
    —save_plots True —mav_name firefly
```

Note 4 If there is an error about an unindexed bag file, reindex it with:

```
rosbag reindex <bagfile>
```

We get the following output by the script (the evaluation measures the differences to a hovering setpoint of $p = (0\ 0\ 1)^T$, from 10 to 20 s):

```
Position RMS error: 0.040 m
Angular velocity RMS error: 0.000 rad/s
```

Additionally, the script will produce three plots showing the position, the position error, and the angular velocities. We evaluate angular velocities to expose overly

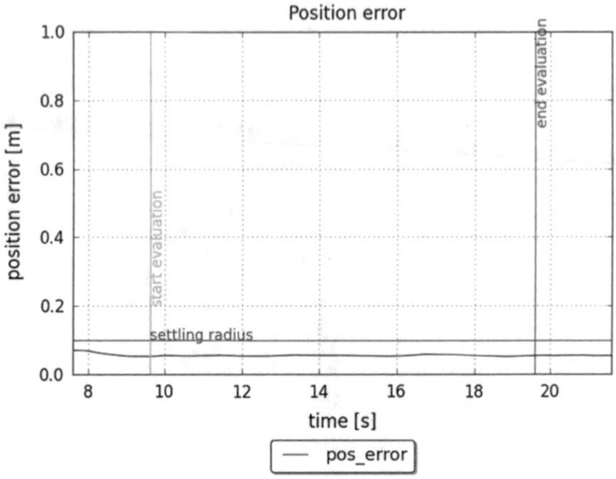

Fig. 11 Position error plot
of a hovering sequence with
noise on the odometry
sensor. Because the
controller does not contain
an integrator or disturbance
estimator, there is a constant
offset

Fig. 12 Angular velocities
plot of a hovering sequence
with noise on the odometry
sensor

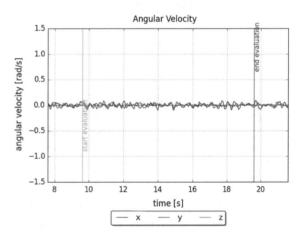

aggressive-tuned controllers. The plots show that we have a static offset in the z-axis.
This is the case since there is no integral part in the controller, and the mass value
in the controller is not exactly set to the correct value. As there is no noise on the
sensors, the angular velocity RMS error is zero. More realistic conditions would
be achieved by adding some noise on the odometry sensor, and then runningh the
simulation and evaluation again. The noise parameters of the odometry sensor can
be set in the `firefly_generic_odometry_sensor.gazebo`-file. You will
get plots like the ones shown in Fig. 11 for the position error, and for the angular
velocities in Fig. 12 (these plots might differ, depending on the magnitude of the
noise that was set). There are evaluation scripts available for multiple waypoints:
`waypoint_eval.py`, and for evaluating the reaction to external disturbances:
`disturbance_eval.py`.

4 Advanced Tutorials

Up to this point we learned how to use the the the components provided by RotorS. If the reader is not only interested in using this set of MAVs with some standard sensors and controllers, this section will give more insight on how to develop a custom controller in Sect. 4.1. In Sect. 4.3, it is described how to integrate a new MAV, how to write new sensors, and how to work on a state estimator.

4.1 Developing a Custom Controller

Here, we show how you build a controller for an arbitrary MAV. The message passing is handled by the `gazebo_controller_interface`, and the motor dynamics are incorporated in `gazebo_motor_model`. Hence, the task of developing a controller can be reduced to subscribing to sensor or state estimator messages, reference commands, and publishing *Actuators* messages on the `command/motor_speed` topic. Another option is to run the *RollPitchYawrate-ThrustController* and build a position controller that publishes *MultiDOFJointTrajectory* messages.

Note 5 Our position controller listens to *MultiDOFJointTrajectory* messages as control input, whereas the roll-pitch-yawrate-thrust controller listens to *RollPitchYawrateThrust* messages. Both listen to odometry messages either from a sensor directly, or from a state estimator.

There are two parts in our design of the controller, the first part handles the parameters and the message passing, this is an executable, a ROS-node. The second part of the controller is a library, which gets loaded by the executable, and does all the computations. We encourage to use one of our controller ROS-nodes as a template, located in the `rotors_control/src/nodes`-folder. It reads the controller parameters from the ROS-parameter server. The parameters are usually set in `yaml`-files, and split up into controller specific parameters and vehicle specific parameters. Through a *launch*-file, the parameters from a `yaml`-file are passed to the ROS-parameter server, by adding the following line between your `<node>`-tags.

```
<rosparam command="load" file="$(find rotors_gazebo)/resource/your_parameters_file.
     yaml" />
```

The vehicle parameter file includes the mass of the MAV, its inertia and rotor configuration. Controller gains are specified in the controller parameters file.

With all the parameters in your controller and the callback methods for the *Odometry* and *MultiDOFJointTrajectory* messages, we can now start writing the actual controller, the controller library. These libraries work without ROS, and hence would be able to run on an MAV directly.

Our controller libraries are located in the `rotors_control/src/library`-folder. Again, we encourage to use one of the libraries as a template to develop your controller.

We have a method, `CalculateRotorVelocities`, which gets called every control iteration, that calculates the required rotor velocities ω from the controller's input, and information about the current MAV state. As mentioned before, the reference consists of the desired position p_d and its derivatives, and the yaw angle ψ and its derivatives. The input is the content of a *MultiDOFJointTrajectory* message. In our implementation of [5], this method calls other methods, which do the calculations of the desired accelerations $\ddot{x}$, and the angular accelerations $\dot{\omega}$. To wrap it up, if you want to implement your own controller, all that needs to be done is to compute the rotor velocities in the `CalculateRotorVelocities` method, based on the state information of the MAV. Or use cascaded controllers, that is, run our roll-pitch-yawrate-thrust controller and publish *RollPitchYawrateThrust* messages in your controller.

4.2   Tutorial on Xacros

This section should give a short overview on what Xacro [3] is, it can be skipped if the reader is already familiar with the concepts of Xacro. Xacro is an Extensible Markup Language (XML) macro language, which is used to generate more readable and often shorter XML code. In our `rotors_description`-package we use it extensively. First, a very quick introduction to XML, which is built out of tags. A tag is dividing parts of XML code in blocks. These block have opening tags `<tagName>` and closing tags `</tagName>`. Each tag can have any number of properties, which are set as:

```
<someTag property1="value1" property2="all values are passed as strings" property3
    ="4.0" />
```

As seen in the line above, if there is no further content between the opening and closing tag, a tag can be closed directly as `<tag/>`.

Since XML is solely text based, there are a number of reasons, as can be seen below, for using Xacro. The `xmlns:xacro="http://ros.org/wiki/xacro"` (tag) property needs to be set to the first XML tag, such that the converter recognizes that the current file is of Xacro type. Here we will start with a minimal example on how to set and use Xacro properties. These are particularly useful for repeating values, such as constants or parameters.

```
<foo xmlns:xacro="http://ros.org/wiki/xacro">
  <xacro:property name="pi" value="3.14159265359" />
  <xacro:property name="moment_constant" value="0.016" />
  <bar property1="${moment_constant}">${pi}</bar>
</foo>
```

The code above can be stored to a text file, `test.xacro`, for instance. Then it can be converted to a regular XML file by:

```
$ rosrun xacro xacro.py test.xacro
```

Properties surrounded by $-brackets ($ { }) get replaced in this step, as you can see in the output of the line above. This is usually written to a file, by appending `-o` `test.xml` to the example above, or piped to the ROS parameter server in a launch file:

```
<param name="robot_description" command="
  $(find xacro)/xacro.py 'test.xacro'"
/>
```

Below the remaining basic Xacro functionalities are explained:

Property Blocks If there is content between the opening and the closing tag of Xacro properties, we call it property blocks. It can be inserted as shown below in the macros part.

```
<xacro:property name="an_origin">
  <origin xyz="0.0 0.0 0.05" rpy="0 0 0" />
</xacro:property>
```

Math Expressions can be used to do basic arithmetic operations, such as: $ { (3 0 + 1) * pi/180}.

Conditional Blocks can be used to conditionally include code in the output:

```
<xacro:if value="${add_camera}"><...some extra code...></xacro:if>
<xacro:unless value="${moment_constant}"><...some extra code...></xacro:unless>
```

Rospack commands such as `find`, can be used within $-parentheses `$(find rotors_gazebo)`.

Macros specify your own snippets such as the one shown below to calculate the inertia of a box:

```
<xacro:macro name="box_inertial" params="x y z mass *origin">
  <inertial>
    <mass value="${mass_box}" />
    <xacro:insert_block name="origin" />
    <inertia ixx="${0.0833333 * mass * (y*y + z*z)}" ixy="0.0" ixz="0.0"
      iyy="${0.0833333 * mass * (x*x + z*z)}" iyz="0.0"
      izz="${0.0833333 * mass * (x*x + y*y)}" />
  </inertial>
</xacro:macro>
```

To define a macro, the `<xacro:macro>` tag is used, where you specify its name and an arbitrary number of parameters. Parameters starting with an asterisk (*), denote property blocks, which are read as the content between the opening and closing tag of the macro. The above just defines a macro, the code gets only placed in your output file if you call the macro with its parameters as:

```
<xacro:box_inertial x="0.2" y="0.4" z="0.1">
  <origin xyz="0.1 0.2 0.05" rpy="0 0 0" />
</xacro:box_inertial>
```

Include other Xacro files with:

```
<xacro:include filename="../other.xacro" />
```

4.3   Assembling a Model

As you are able to fly with the MAVs that are shipped with RotorS, we want to encourage you to bring your own MAV into RotorS. In this section, we describe the procedure of how to get the geometry of your robot, and the locations of the actuators into RotorS. In the next section, we describe how custom sensors can be written as a Gazebo plugin. For the description of the robot we are using the Unified Robot Description Format (URDF), with Xacro, which is briefly explained in the previous section.

Note 6 Gazebo has its own format to describe robots, objects, and the environment, called Simulation Description Format (SDF). We are sticking with URDF here, as this format can be displayed in the ROS 3D Robot Visualizer (rviz), and all the SDF-specific properties can be added by putting them in a `<gazebo>`-block. Internally Gazebo converts the URDF-files to SDF-files.

If you have a specific robot that you want to use in your simulation, this procedure is quite straight forward. You have to check which parts of your robot are fixed, or rigid bodies, in URDF these parts are called *links*, and connect these links; connections are called *joints*. In Fig. 13, the different links and joints of an example quad-rotor helicopter are shown. There are a number of different joints as can be seen on http://wiki.ros.org/urdf/XML/joint. Each joint has a *parent* and a *child* link. For our application we only use three types:

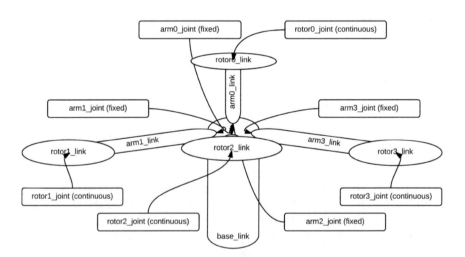

Fig. 13 A draft of a multirotor helicopter with four non-symmetrical aligned rotors. All the joints and links are named to get an overview of the necessary components in the Xacro files

continuous joints are hinge joints without any mechanical limits, such as our rotors.

fixed joints are connecting two links rigidly, such that there is no movement possible. These joints get removed by Gazebo's physics engine, that means, no information can be gathered about these joints and its child links in Gazebo (plugins).

revolute joints are the same as continuous joints, but have a lower and upper limit, which limits their turning angle. We are using these joints with both limits set to zero, as fixed joints, if the child link needs to be accessible from a plugin.

As we go on with this section, we want to develop a model of a conceptual quad-rotor helicopter, as depicted in Fig. 13. This quad-rotor helicopter model has four rotors, of which three are distributed equally on the edge of a circle with radius l, while the remaining rotor is placed in the back of the vehicle with an arm of length $l/3$. All the rotors are of the same dimension.

We can now start to build our model in a URDF file. Every robot that is built with URDF needs a `<robot>`-tag in the beginning of the file, with a unique name. Within the opening- and the closing tag, the robot will be assembled. We start by adding the base of the robot, there is a macro for doing this.

```
<robot name="asymmetric_quadrotor" xmlns:xacro="http://ros.org/wiki/xacro">
  <xacro:multirotor_base_macro
    robot_namespace="${namespace}"
    mass="${mass}"
    body_width="${body_width}"
    body_height="${body_height}"
    use_mesh_file="${use_mesh_file}"
    mesh_file="">
    <xacro:insert_block name="body_inertia" />
  </xacro:multirotor_base_macro>
</robot>
```

The listing above needs some Xacro-properties, such as `namespace` and `mass` to be set, this is explained in Sect. 4.2.

One would continue adding the arms and the rotors to the base link by connecting them with joints—we will show this for one arm and rotor. If you have a mesh file of your multi-rotor helicopter with arms, and know its inertia (from the body including the arms), you can omit this step, and attach the rotors directly to the base link. For the other MAVs that are provided with RotorS, the rotors are directly attached to the base. The reader is encouraged to use the macros provided in the `component_snippets.xacro` and `multirotor_base.xacro` files.

```
<link name="arm1_link">
  <xacro:box_inertial x="${arm_length}" y="0.03" z="0.01" mass_box="${mass_arm}" />
  <visual>
    <origin xyz="${arm_length/2} 0 0" rpy="0 0 0" />
    <geometry>
      <box size="${arm_length} 0.03 0.01" />
    </geometry>
  </visual>
  <collision>
```

```
      <origin xyz="${arm_length/2} 0 0" rpy="0 0 0" />
      <geometry>
        <box size="${arm_length} 0.03 0.01" />
      </geometry>
    </collision>
  </link>
  <joint name="arm1_joint" type="fixed">
    <origin xyz="0 0 0" rpy="0 0 ${2*pi/3}" />
    <parent link="base_link" />
    <child link="arm1_link" />
  </joint>
  <xacro:vertical_rotor
    robot_namespace="${namespace}"
    suffix="1"
    direction="cw"
    motor_constant="${motor_constant}"
    moment_constant="${moment_constant}"
    parent="arm1_link"
    mass_rotor="${mass_rotor}"
    radius_rotor="${radius_rotor}"
    time_constant_up="${time_constant_up}"
    time_constant_down="${time_constant_down}"
    max_rot_velocity="${max_rot_velocity}"
    motor_number="1"
    rotor_drag_coefficient="${rotor_drag_coefficient}"
    rolling_moment_coefficient="${rolling_moment_coefficient}"
    color="Blue"
    use_own_mesh="false"
    mesh="">
    <origin xyz="${arm_length} 0 ${rotor_offset_top}" rpy="0 0 0" />
    <xacro:insert_block name="rotor_inertia" />
  </xacro:vertical_rotor>
```

Note 7 Since you are using a very similar snippet for all four arms, it is a good idea to create a macro, which could be called `arm_with_rotor`. It would take the arm length, an angle, the motor number, and the mass as parameters.

You can test your current configuration by either just looking at the output of:

```
$ rosrun xacro xacro.py <xacro_file> > <urdf_file>
$ check_urdf <urdf_file> # You need liburdfdom–tools installed for this.
```

or by running our `mav_empty_world.launch` file, which spawns your robot in an empty Gazebo world:

```
$ roslaunch rotors_gazebo mav_empty_world.launch mav_name:=asymmetric_quadrotor
```

Note 8 To use this command your file should be called `asymmetric_quadrotor`
`.xacro` and be placed in the `rotors_description/urdf`-folder. Additionally you need to have a file called `asymmetric_quadrotor_base.xacro`, which includes the above file. We seperated the geometrical properties from the sensors and controllers. Sensors and controllers should be placed in the second file.

4.4 Creating Custom Sensors

To make your robot perceive the environment and its own motion with respect to the inertial frame, you need to add sensors to your model. We already discussed how to add available sensors in Sect. 3.5. In this section, we focus on how to write a Gazebo plugin for a new sensor.

Note 9 Before you create your own plugin, you should make sure that there is no sensor available, which satisfies your requirements. For instance, a laser sensor can be used to mimic an ultrasonic sensor, by just using one ray. This of course is not very accurate, as ultrasonic sensors usually measure the shortest distance to an object in a cone-like shape. But, depending on your application, a measurement along one single ray might also be enough.

In the remainder of this section, we show how one would conceptionally proceed with building a wind sensor. A wind sensor usually measures the airspeed v_{air} on an MAV. This airspeed is the difference between the wind speed v_{wind} and the current velocity v of the MAV.

$$v_{air} = v_{wind} - v \tag{19}$$

To bring this sensor into the simulation, you would start by creating a *Model-Plugin*, as described on:
http://gazebosim.org/tutorials?tut=plugins_model&cat=write_plugin.
You would create a subscriber to the `wind`-topic, and store the pointer to the link, to which the plugin got attached to. Additionally, you would create a publisher on your desired topic, `air_speed`, for instance. All of this can be done in the `Load` method of the plugin. In the `OnUpdate` method, which gets called every single simulation iteration, you are then getting the link's velocity in the world frame by

`link_->GetWorldLinearVel()`

and perform the calculation described in (19). You can either publish the calculated value directly, or add some additional noise. Take a look at the *GazeboImuPlugin* and the *GazeboOdometryPlugin* to get an idea how to add noise to your calculated value. Of course, you can also reduce the publishing frequency.

Note 10 Since on a lot of aircraft, this sensor is measuring the difference between the static pressure and the stagnation pressure due to inflowing air in a pitot tube [12], it is arguable if the above sensor design makes sense. To make this airspeed sensor more realistic, the example above could be modified by, for example, only taking the wind and MAV velocities in a single direction in the MAV's body frame, and then transforming them back into the world frame.

5   Using RotorS for Higher Level Tasks

One of the biggest advantages of the RotorS simulator is, that it comes with a fully functional trajectory tracking controller. This allows implementing higher level tasks like collision avoidance or path planning without having to implement state estimation or control first. Also in simulation, you have access to perfect ground truth, which is usually hard to get on real systems. Once the algorithms are working in simulation, the changes necessary when transitioning to real world systems are typically small. In this section we want to give some ideas, how two example problems can be solved using the simulator.

5.1   Collision Avoidance

A commonly used strategy to solve collision avoidance on MAVs is to down-project the environment onto a 2D ground plane and use the ros navigation stack. To tackle this problem, an appropriate sensor needs to be mounted on the MAV, to get an estimate of the surrounding. Gazebo provides plugins for 2D laser scanners like the Hokuyo, that could potentially be mounted on a real MAV. This approach has the big disadvantage of reducing the operating space of the MAV to a plane at a constant height.

For full 3D collision avoidance, front looking depth cameras are a good starting point, like the Kinect-sensor which is already implemented in Gazebo. These sensors are light-weight and give rich information about the surrounding. Another possibility would be to mount cameras and use structure from motion in the monocular case, or a stereo camera to calculate a disparity image. One possible approach to perform 3D collision avoidance is to do it directly on the disparity images as proposed in [8].

5.2   Path Planning

In the RotorS simulator, we prepared an example environment with a power-plant and a waypoint publisher to test different planning algorithms. The waypoints are chosen such that the straight line solution is in collision, and the MAV is going to crash into the wall without a planner. A file with the octree representation, as described in Sect. 2.4, of the power-plant can be found in the `rotors_gazebo/resources` folder and can be used with the `octomap_server` package. The octree representation allows for efficient collision checking in the planner and is well suited for 3D environments. To run the example, use the following command.

```
$ roslaunch rotors_gazebo mav_powerplant_with_waypoint_publisher.launch
```

ROS provides a big collection of planning algorithms in the MoveIt! package [9], that can be adapted to 3D path planning in static environments.

6 Transfer to a Real MAV

Having a simulation of an MAV is of course nice, but it raises the question, how well it represents the real world and how easy the transition from simulation to real MAVs is. During the development of this simulator, a lot of effort was put into keeping the structure of the simulator as close as possible to the real system. In the best case, this means just switching the simulation environment with a ROS node that communicates with the hardware. The simulator should be a tool that enables the development of algorithms, to be deployed on a real MAV later on. To make the transition of the code from the simulator to the code running on the actual hardware as simple as possible, the interface is designed in a way which mimics most of the interfaces on the real systems. Not all messages are replicated in the simulator, but one can easily extend the simulator with models of a battery or other parts which are currently not implemented. One of the biggest challenges on real platforms are the changing delays and the resulting non deterministic order of measurements. This leads to complex message queues, tedious time delay estimation and delay compensation.

To give an impression on how well the simulator replicates the real MAV, we are using the same controller gains in the simulation as on the real Hummingbird and Firefly MAVs.[8] Here we have to mention that in general we send attitude commands to the low level controller of the real helicopters, and use the manufacturer's low level attitude controller to compute motor commands.

7 Conclusion

In this chapter, we presented an overview of the necessary background to understand the basics for autonomous flights of MAVs. We then explained step-by-step how to run the simulation with existing models, with ideal conditions first, and then how to use it with more realistic sensors, and a state estimator. As advanced tutorials, we presented instructions on how to implement a custom quad-rotor helicopter, including guidance to create own controllers and sensors. We believe that this framework will be very helpful for both research and teaching, as it gives an easy start into the topic. Interfaces are compatible, and the simulated dynamics are close to the real platforms, such that an easy transfer from the simulation to the real world is possible.

---

[8]A video of a Firefly following a path in real world and in the RotorS simulator can be seen on http://youtu.be/3cGFmssjNy8.

References

1. M.W. Achtelik, Advanced closed loop visual navigation for micro aerial vehicles. Ph.D. thesis, ETH Zurich, 2014
2. T. Field, J. Leibs, J. Bowman, Rosbag (2015), http://wiki.ros.org/rosbag. Accessed 27 Mar 2015
3. S. Glaser, W. Woodall, Xacro (2015), http://wiki.ros.org/xacro. Accessed 27 Mar 2015
4. A. Hornung, K.M. Wurm, M. Bennewitz, C. Stachniss, W. Burgard, OctoMap: An efficient probabilistic 3D mapping framework based on octrees. Auton. Robots (2013). http://octomap.github.com
5. T. Lee, M. Leoky, N.H. McClamroch, Geometric tracking control of a quadrotor UAV on SE(3), in *2010 49th IEEE Conference on Decision and Control (CDC)* (IEEE, 2010), pp. 5420–5425
6. S. Lynen, M.W. Achtelik, S. Weiss, M. Chli, R. Siegwart, A robust and modular multi-sensor fusion approach applied to mav navigation, in *2013 IEEE/RSJ International Conference on Intelligent Robots and Systems (IROS)* (IEEE, 2013), pp. 3923–3929
7. P. Martin, E. Salaun, The true role of accelerometer feedback in quadrotor control, in *2010 IEEE International Conference on Robotics and Automation (ICRA)*, pp. 1623–1629, May 2010
8. L. Matthies, R. Brockers, Y. Kuwata, S. Weiss, Stereo vision-based obstacle avoidance for micro air vehicles using disparity space, in *2014 IEEE International Conference on Robotics and Automation (ICRA)* (IEEE, 2014), pp. 3242–3249
9. S.A. Sucan, S. Chitta, Moveit! (2015), http://moveit.ros.org. Accessed 10 Aug 2015
10. N. Trawny, S.I. Roumeliotis, Indirect Kalman filter for 3D attitude estimation. Technical Report 2005-002, University of Minnesota, Department of Computer Science and Engineering, 2005
11. S. Weiss, Vision based navigation for micro helicopters. Ph.D. thesis, ETH Zurich, 2012
12. Wikipedia, Airspeed indicator (2015), http://en.wikipedia.org/wiki/Airspeed_indicator. Accessed 27 Mar 2015

Authors' Biography

Fadri Furrer is a Ph.D. student at the ASL at ETH Zurich since 2015. He received his MSc degree in electrical engineering from the Swiss Federal Institute of Technology, Zurich, in 2011. His research interests are in simulation, object reconstruction and recognition from images and point clouds.

Michael Burri is a Ph.D. student at the ASL at ETH Zurich since 2013. He received his MSc degree in robotics and control from the Swiss Federal Institute of Technology, Zurich, in 2011. His research interests are in control, state estimation, and planning for micro aerial vehicles.

Markus Achtelik received his Diploma degree in Electrical Engineering and Information Technology from TU München in 2009. He finished his Ph.D. in 2014 at the ASL at ETH Zurich, and currently works as Postdoc at ASL on control, state estimation and planning, with the goal of enabling autonomous maneuvers for MAVs, using an IMU and onboard camera(s) as main sensors. Since 2014, he is challenge leader of the "plant servicing and inspection challenge" within the European Robotics Challenges.

Roland Siegwart (born in 1959) is a professor for autonomous mobile robots at ETH Zurich. He studied mechanical engineering at ETH, brought up a spin-off company, spent 10 years as professor at EPFL, was vice president of ETH Zurich and held visiting positions at Stanford University and NASA Ames. He is and was the coordinator of multiple European projects and co-founder of half a dozen spin-off companies. He is recipient of the IEEE RAS Inaba Technical Award, IEEE

Fellow and officer of the International Federation of Robotics Research (IFRR). He is in the editorial board of multiple journals in robotics and was a general chair of several conferences in robotics including IROS 2002, AIM 2007, FSR 2007 and ISRR 2009. His interests are in the design and navigation of wheeled, walking and flying robots operating in complex and highly dynamical environments.

Part VIII
Advanced Tools for ROS

The ROS Multimaster Extension for Simplified Deployment of Multi-Robot Systems

Alexander Tiderko, Frank Hoeller and Timo Röhling

Abstract This tutorial chapter describes how to set up a multi-robot system in ROS with the *multimaster_fkie* package. The package adds ROS support for multiple hosts, which can be added and removed from the network at any time without affecting the remaining nodes. The presented multi-master extension works with the unmodified ROS master and does not change the way ROS nodes communicate or establish connections with each other. Thus, the multi-robot system remains fully compatible with a single-master ROS system. It is easy to set up and execute the ROS masters independently on each robot. The multi-master extension takes care of synchronization and merges the masters into a unified network view. For better usability, the package includes a graphical user interface for monitoring, configuration and control of the ROS components. The latest version can be downloaded from https://github.com/fkie/multimaster_fkie. You can also install the package from http://packages.ros.org. The *multimaster_fkie* package works with all ROS versions since *groovy*.

Keywords Multi-master · Node manager · Dynamic network · Multi-robot system · Synchronization · Configuration management

1 Introduction

In a traditional ROS network, a single master node acts as network hub. All ROS nodes which wish to participate in the network have to register their publishers and sub-

A. Tiderko (✉) · F. Hoeller · T. Röhling
Fraunhofer Institute for Communication, Information Processing and Ergonomics FKIE,
Fraunhoferstr. 20, 53343 Wachtberg, Germany
e-mail: alexander.tiderko@fkie.fraunhofer.de
URL: http://www.fkie.fraunhofer.de

F. Hoeller
e-mail: frank.hoeller@fkie.fraunhofer.de

T. Röhling
e-mail: timo.roehling@fkie.fraunhofer.de

© Springer International Publishing Switzerland 2016
A. Koubaa (ed.), *Robot Operating System (ROS)*, Studies in Computational
Intelligence 625, DOI 10.1007/978-3-319-26054-9_24

scribers with the master. Even though communication in ROS is strictly peer-to-peer, the ROS master acts as the central directory for nodes to look up their communication counterparts, and provides a global registry for configuration values. The location of the ROS master is exposed by the environment variable ROS_MASTER_URI.

This centralized network configuration is prone to failures, in particular if the ROS network spans over multiple hosts with unreliable communication links. As the ROS master is assumed to be reachable by all nodes at all times, it becomes challenging or even impossible to choose a suitable host for the ROS master. This is especially true for a multi-robot system, where each robot basically runs a self-contained ROS network, but still needs to communicate with other robots from time to time.

The multi-master extension has been designed with these challenges in mind and provides an noninvasive way to merge multiple ROS masters into a unified "super-network". Additionally, the *multimaster_fkie* packages provide a graphical user interface (GUI) to monitor and manage ROS nodes, topics, services, parameters and launch files in a ROS network. The key features of the multi-master extension are:

- All ROS components run unmodified and need not specifically be setup for the multi-master extension.
- The behavior of ROS nodes will not be changed, no special API calls are required and no extra libraries need to be linked.
- Robots and control stations can be designed as independent single-master ROS networks and be connected as required. In fact, the multi-master extension can be enabled and disabled without the need to restart any nodes.
- Each ROS master retains a locally consistent network view even if parts of the network are removed or lost due to communication outages.
- Existing launch files and configurations can be used without modifications.
- The extension has a fully functional default configuration, there is no need to tweak any parameters.

The multi-master extension is developed in *Python* and is hosted on Github,[1] where bug reports and feature requests should be posted. The pre-compiled packages are also available from the ROS APT repository.[2] The API documentation and a description of supported ROS parameters can be found in the ROS Wiki.[3] A detailed feature description is also available as Github Page.[4] The remainder of this chapter is organized as follows:

- In Sect. 2, we introduce the methods used for automated ROS master discovery and synchronization.
- In Sect. 3, we explain how to download, install and use the multi-master extension to synchronize two hosts and setup a multi-robot system.

---

[1] https://github.com/fkie/multimaster_fkie.

[2] http://packages.ros.org/ros/ubuntu.

[3] http://wiki.ros.org/multimaster_fkie.

[4] http://fkie.github.io/multimaster_fkie.

- In Sect. 4, we present the *Node Manager* and its features, and we describe how to use it to simplify the setup of a multi-robot system.
- Lastly, we present further features of the multi-master extension, namely the *Capability View* and the *Default Configuration* tool. With these, you can load ROS launch files and execute them remotely on demand or automatically on network startup.

2   Background

In this section, we explain the underlying concepts of the multi-master extension and describe the techniques we use for discovery, synchronization, and the detection of state changes in the ROS master nodes. We also discuss the limitations of our approach and give rationales for our design choices.

As motivated in Sect. 1, the multi-master extension merges the state of multiple ROS master nodes into a unified network view. This goal is achieved by running two additional nodes (`master_discovery` and `master_sync`) alongside each ROS master. These nodes interact with ROS masters using the *ROS Master API*[5] and with other ROS nodes using the *ROS Slave API*.[6] The discovery and synchronization is shown in Fig. 1 and is performed as follows:

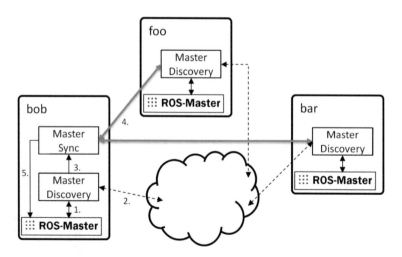

Fig. 1 Conceptual overview for discovering and synchronization

---

[5]http://wiki.ros.org/ROS/Master_API.
[6]http://wiki.ros.org/ROS/Slave_API.

- Each `master_discovery` node connects to its local ROS master and continually monitors for changes (1.). The current state of the ROS master can be queried via an XML-RPC interface.
- The `master_discovery` node broadcasts the time stamp of the last change to the network while simultaneously aggregating a list of the other available `master_discovery` nodes and their time stamps (2.). This list is exposed via a local ROS topic to the `master_sync` node (3.).
- The `master_sync` node connects to all known `master_discovery` nodes and queries the current state of their ROS master whenever the time stamp has changed (4.). Then, it registers or unregisters the remotely available topics and services with the local ROS master (5.).

We will give a more in-depth explanation for each of these three subtasks in the following subsections. However, we want to emphasize first that both the `master_discovery` and `master_sync` nodes access their local ROS master only. This ensures that the multi-master extension will continue to function properly if the network topology changes. Also, in principle, the `master_sync` node only needs to run on hosts where topics from remote ROS masters are to be subscribed. In most cases, this would be a control station monitoring one or more robots. However, some nodes communicate with each other both as publishers and subscribers. In these cases, the `master_sync` node must be run on all participating hosts. It is generally safe to do so anyway.

Furthermore, as the multi-master extension is designed to be as little intrusive as possible, all requirements and restrictions for running a ROS network with a single master also apply to multiple masters. In particular, all ROS nodes still communicate in a peer-to-peer fashion, meaning all participating host names must be resolvable and all IP addresses be reachable.

2.1 ROS Master Change Detection

In order to keep the multi-master network in a consistent state, new topics and services running on a host (or old ones being terminated) have to be detected in a timely manner. Since the ROS master does not support push notifications, the `master_discovery` node has to poll the ROS master for changes periodically. The update rate can be configured with the ~`rosmaster_hz` parameter and defaults to once per second. The rate is automatically reduced if the CPU load exceeds 20 %. As all polling queries are made via the loopback interface on the same host, they are free in terms of network bandwidth utilization.

The ROS Master API returns most required information via the RPC call `getSystemState()`. To satisfy the requirements of the multi-master extension, this state is extended with the XML-RPC URI of each node, the type information for the registered topics and services, and the process ID of all locally running nodes. The node URI is required to detect nodes which have been stopped and restarted in between RPC queries.[7] The type information and the process IDs are utilized by the *Node Manager* GUI for advanced system introspection and modification.

2.2 Discovery

Aside from collecting status data from the ROS master, the `master_discovery` node sends and receives broadcasts to compile a list of all reachable instances (and by extension, their ROS masters). Currently, three approaches have been implemented: the Apple *Zeroconf* protocol, multicast UDP, and unicast UDP. The multicast UDP protocol is the recommended choice, where all running `master_discovery` nodes join a fixed multicast group to exchange heartbeat messages.

A heartbeat message has a payload size of 21 octets (see Table 1). Additional data, e.g. the ROS master URI or node names, are requested via the XML-RPC interface of the `master_discovery` node.

The heartbeat message is sent to the multicast group whenever any field is updated, but at least once every 50 s. As an additional feature, the heartbeat messages can be used to determine the link quality of the communication channel. This feature is enabled by setting the parameter ~`heartbeat_hz` to a value between 0.1 and 25.5 Hz. The `master_discovery` node will then estimate the link quality using the package loss, i.e.

$$\text{quality} = \frac{\text{number of received heartbeats}}{\text{number of expected heartbeats}} \cdot 100$$

and publish the values to the topic ~`linkstats`.

Table 1 Heartbeat message format

Field type	Octets	Description
char	1	Identification character 'R'
unsigned char	1	Version of the heartbeat message (currently: 3)
unsigned char	1	Connection health message rate in units of $\frac{1}{10}$ Hz
int	4	Unix time stamp of the last state update (integral part in seconds)
int	4	Unix time stamp of the last state update (fractional part in nanoseconds)
unsigned short	2	XML-RPC port number of the sending `master_discovery` node
int	4	Unix time stamp of the last change not caused by the multi-master extension (integral part in seconds)
int	4	Unix time stamp of the last change not caused by the multi-master extension (fractional part in nanoseconds)

---

[7]The URI is allocated with a random port number, so it will likely change on restarts.

You may notice that the heartbeat message actually contains two distinct time stamps. The first time stamp is updated whenever the ROS master state changes in any way. The second time stamp, also called *lc* time stamp, will be updated only if the state change was not initiated by the multi-master extension itself, i.e. an externally triggered change published by the local ROS master. This distinction prevents spurious RPC queries from the `master_sync` node (see Sect. 2.3). The regular time stamp is utilized by the *Node Manager* to update its user interface.

For increased robustness, the `master_discovery` node will explicitly request a heartbeat message if none has been received for some time (60 s by default). After five unsuccessful tries, the state of the remote master is set to *Offline*. After an additional timeout (300 s by default), the non-responsive instance is removed from the list of known masters. All changes to the list of known masters will be published to the topic ~`changes`.

We found in our experiments that multicast communication can be somewhat fragile at times. There are a few long-standing bugs in the Linux multicast implementation, and certain commercial, IP-based radio link devices lack multicast support altogether. For this reason, the multi-master extension allows for static configuration of hosts with the parameter ~`robot_hosts`, which accepts IP addresses and resolvable host names. These hosts will be discovered using unicast heartbeats, with the negative side-effect of increased bandwidth usage and the need to configure the network topology in advance. The Apple *Zeroconf* protocol, while being a widespread official standard, has certain quirks in its Avahi open source implementation, and does not support link quality estimation nor some other teaming features.

Many aspects of the system can be customized to suit specific needs. The complete list of available parameters for the `master_discovery` node and their descriptions can be found in the default launch file of the `master_discovery` package. If you use the *Node Manager* to launch the `master_discovery` node, some frequently used parameters can be set there without the need to modify the configuration files (see Sect. 4).

2.3 ROS Master Synchronization

Synchronization is the term we use for propagating information about running nodes, topics, and services through the multi-master network. When all ROS masters are in agreement, the network is in a consistent state. Each time a ROS master changes its state, the change is detected by the local `master_discovery` node, annotated with a time stamp, and published to the network. The change is then picked up by the other `master_discovery` nodes and published to their local topic ~`changes`, which triggers the `master_sync` node to query the current state from the original `master_discovery` node[8] and update its local ROS master. Note that the parameter server is not synchronized for the following reasons:

---

[8]Unless the change is a ROS master being marked *Offline*.

- Most parameters are in the private name space of their respective nodes and never accessed by anyone else.
- Most parameters are set by *roslaunch* immediately prior to node startup, accessed by their node once for initialization and then never again. Thus, most parameters have already outlived their usefulness by the time the multi-master extension has had any chance to propagate them.
- The *dynamic_reconfigure* package already provides a way to change parameters on-the-fly using services, which *will* be synchronized.
- Some parameters such as robot_description tend to be huge in size (>1 MB), further reducing the cost to benefit ratio.

The update itself straight-forward as the master_sync merely registers and unregisters topics and services on behalf of the respective node. However, a few detail have to be taken into account to ensure efficient synchronization.

Mitigation of the Thundering Herd Problem: When a change propagates through the network, each master_sync node has to query the originating host for the changes. We cannot piggyback the change on the heartbeat message because the state is usually larger than a single UDP frame would permit. However, if all nodes sent their query at once, it would create a huge spike in network load (akin to a grazing herd suddenly startling, hence the name). Therefore, each node waits for a random period before querying. Also, the *lc* time stamp enables the master_sync node to detect changes which merely propagate from other ROS masters, preventing spurious queries.

Minimization of latencies: Since each master_discovery node preemptively collects all data required by the master_sync nodes, a single query per node is sufficient for the state update. Additionally, each update is forked into its own thread to prevent communication delays from affecting other updates.

Name space collisions: Each master_sync node only modifies data on its local ROS master. If a node name collision occurs, the previously running node is automatically terminated. ROS internal nodes such as /rosout and the multi-master extension nodes are exempt from synchronization. The same goes for topics which are only meaningful on the local host, such as the bond/Status type which is used for nodelet management.

Network topology changes: The discovery process will pick up all remote masters which become reachable on its own. Remote masters which are no longer reachable will be detected as such by the local master_discovery node.

3 Installation and Usage

3.1 Installation

master_discovery and master_sync have no additional dependencies beyond ROS, unless you wish to use *Zeroconf* for discovery, which depends on *python-avahi* and *avahi-daemon*. This tutorial will set up the multi-master extension without *Zeroconf*.

For full *node_manager* functionality, you need to install the following packages (replace *indigo* with your ROS version, backslashes mark line wraps):

```
sudo apt-get install ros-indigo-rqt-gui \
  ros-indigo-rqt-reconfigure \
  ros-indigo-python-qt-binding \
  python-paramiko python-docutils screen
```

Alternatively, *multimaster_ fkie* can be cloned from the Github repository. Change to the folder `src` in your ROS workspace and call

```
git clone https://github.com/fkie/multimaster_fkie.git
```

In order to generate the required message and service types, build the workspace with

```
catkin_make
```

Now the multi-master extension is ready to use.

3.2   Usage Example with One Host

In our first example, we will start two ROS masters on the same host using different `ROS_MASTER_URI`s. Then we enable synchronization to let `rostopic pub` and `rostopic echo` communicate across the two masters. For the sake of simplicity, we do not use launch files.

First, we need two terminal windows with a command shell. In the first shell, we run a ROS master and an example publisher, which publishes *Hello World* to the topic `/test/topic`:

```
export ROS_MASTER_URI=http://localhost:11311
roscore >/dev/null &
rostopic pub /test/topic std_msgs/String 'Hello World' -r 1
```

In the second shell, we start a second ROS master and a `rostopic echo` to receive the *Hello World* messages:

```
export ROS_MASTER_URI=http://localhost:11312
roscore --port 11312 >/dev/null &
rostopic echo /test/topic
```

Note that the `rostopic echo` command does not receive any messages yet. Now, we open two more terminal windows. In these shells, we will execute the `master_discovery` and `master_sync` nodes. Run the following commands in the third window:

```
export ROS_MASTER_URI=http://localhost:11311
rosrun master_discovery_fkie master_discovery &
rosrun master_sync_fkie master_sync
```

In the forth window, we do the same, but we will connect to the second ROS master:

```
export ROS_MASTER_URI=http://localhost:11312
rosrun master_discovery_fkie master_discovery &
rosrun master_sync_fkie master_sync
```

The synchronization is now running and `rostopic echo` starts printing the *Hello World* messages. If synchronization does not work for you, this is most likely caused by a misconfigured network. Please refer to Sect. 3.4 for a list of the most common configuration problems.

3.3 Remarks to Understand Synchronization

Before we continue with another example, we want to focus on a few peculiarities of the synchronization mechanism. In our first example, each ROS master had its own `master_sync`. Often it is also sufficient to run `master_sync` on one of the hosts. Our example is suitable to show when this is the case: It will work if we start `master_sync` merely for the second ROS master on port 11312, but it will not work with the first one.

The reason for this strange behavior lies in the way some ROS nodes handle publishers and subscribers. `rostopic echo` does not actually create a subscriber until a publisher for the topic becomes available. But if the `master_sync` runs on the publisher's side only, the subscriber will never connect, as the `master_sync` node will never learn of the subscriber's intention to subscribe. Only if the `master_sync` runs on the same ROS master as `rostopic echo`, the synchronization will register the publisher's topic with the second ROS master, and the subscription attempt proceeds as intended. With this knowledge, you can run the `master_sync` only on those hosts which have subscribers waiting for remote topics and it will generally work just fine. If you are unsure, however, its is best to run the `master_sync` node on every host. The resource consumption is low enough so that it won't matter.

3.4 Network Setup

As mentioned before, the setup requirements of the multi-master extension largely mirror the requirements of the ROS ecosystem itself. In this section, we will list a few common causes of failed communication.

The network interface is not configured for multicast: In this case, the `master_discovery` exits without unregistering from the ROS master, so you can see the corresponding warning in the *Node Manager* and find the error description in the log file.

A host name cannot be resolved: `master_discovery` prints a warning to its logfile if a host is discovered whose name cannot be resolved. You can fix this by either setting the `ROS_IP` environment variable or adding an appropriate entry to the `/etc/hosts` file on *all* hosts.

A host does not know its name: The multi-master extension does not provide any warnings for this case, but it is a prerequisite for ROS and will cause unexplained communication failures. You can fix it by either setting the `ROS_HOSTNAME` environment variable or adding an appropriate entry to the host's `/etc/hosts` file. Note that you should list the host with its *network* IP address, not `127.0.0.1`.

The reverse lookup of an IP address does not match the host: This is closely related to the previous problem. Make sure that all hosts are listed with their correct IP address in all `/etc/hosts` files.

Multiple Network interfaces: If your host has more than one network address assigned or multiple active network adapters (such as LAN and Wireless), make sure that ROS uses the correct one. You can force the correct interface by setting the URI in the `ROS_MASTER_URI` environment variable to the host name resolving to the correct IP address (or use the IP address directly).

Clock skew issues: This is particularly frustrating with time-sensitive topics such as the *tf* package. Use tools like *chrony* to synchronize the clocks of all hosts in the network.

Further information on configuration and debugging network problems is available on http://wiki.ros.org/ROS/NetworkSetup.

3.5 Synchronization of Multiple Hosts

For our second example, make sure that on both computers, `ROS_MASTER_URI` refers to the respective local host and the ROS master is running. Start the nodes `master_discovery` and `master_sync` on each host, either in a terminal:

```
rosrun master_discovery_fkie master_discovery &
rosrun master_sync_fkie master_sync
```

or from your launch file:

```
<node name="master_discovery"
      pkg="master_discovery_fkie"type="master_discovery"/>
<node name="master_sync"
      pkg="master_sync_fkie"type="master_sync"/>
```

In this manner, you can add more hosts to the multi-master system. You can also use the *Node Manager* to start a multi-master system as described in Sect. 4.1.

3.6 Special Synchronization Parameters

For advanced usage, there are a number of parameters which can be modified. Most of them are fairly self-explanatory, but we want to highlight two slightly obscure ones: sync_topics_on_demand and sync_remote_nodes.

sync_remote_nodes: By default, the synchronization assumes that each participating host runs its own ROS master, so each node belongs to the ROS master of the host it is running on. Therefore, only these nodes are actually synchronized from each multi-master instance. In certain configurations, this assumption might not hold true. In this case, you can set this parameter to *true*.

sync_topics_on_demand: With this parameter, you can delay the synchronization of topics and services until they are actually requested by a local node. As we have seen before, however, some ROS tools (*rviz* also among them) will not work properly because they wait for the topic to become available before they actually subscribe.

3.7 Default Configuration

Sometimes it is useful to start all ROS nodes at startup, or you need to start some nodes on demand with a fixed pre-configured launch file. As part of the multi-master configuration management, we provide the default_cfg tool. It loads a given ROS launch file and provides services to start and stop the defined nodes. It also provides autostart capabilities, including delayed starts, conditional starts, or exempt nodes which shall not autostart at all. These functions are also used by the *Node Manager* described in Sect. 4. In this subsection, we show how to configure default_cfg to use its autostart capabilities.

To enable the autostart of a launch file the parameter autostart must be set to *true*.

```
<node name="default_cfg"pkg="default_cfg_fkie"
                  type="default_cfg">
    <param name="autostart"value="true"/>
    <param name="package"value="node_manager_fkie"/>
    <param name="launch_file"value="demo_bar.launch"/>
```

```
</node>
```

Now all nodes listed in the launch file `demo_bar.launch` will be started if the `default_cfg` node comes up. You can modify the autostart behavior of each node by adding the corresponding `autostart` parameter to the node. To exclude a node from autostart you have to add an exclusion like this:

```
<node name="bar"pkg="foo_package"type="bar_node">
    <param name="default_cfg/autostart/exclude"value="true"/>
</node>
```

You can also use `default_cfg/autostart/delay` to delay the start of a node or `default_cfg/autostart/required/publisher` to start a node once a specific publisher becomes available.

The multi-master extension does not contain any scripts for starting nodes at system boot time. If you are using Ubuntu, we recommend you take a look at the ROS package *robot_upstart* for this.

4 Node Manager

ROS comes with many monitoring and introspection tools. However, these tools are designed to operate with a single ROS master system. The multi-master extension includes a graphical user interface to manage and operate a multi-master system. After installing the multi-master extension, you can start the *Node Manager* by calling `node_manager` in your console. During the start procedure of the *Node Manager*, the ROS master will be launched automatically if it is not already running. The *Node Manager* provides the following features:

- Listing of discovered hosts in the multi-master system
- Online status overview of each ROS master: lists of registered nodes, topics, services and parameter
- Launch file editor with syntax highlighting and context-sensitive opening of included files
- Navigate through the ROS packages and load launch files
- Run or stop single nodes from a launch file
- Quick access for current screen output or log file of each node started by *Node Manager*
- View extended information about each node, topic, or service
- Edit parameters on the parameter server
- Various filter capabilities
- Additional grouping and visualization of nodes for a quick access
- Visual feedback of configuration errors

An example screenshot of the *Node Manager* is shown in Fig. 2.

Fig. 2 A screenshot of the Node Manager

4.1 ROS Network

The ROS Network docking widget shows all discovered ROS masters. In order to detect other ROS masters, the `master_discovery` node must be running. Otherwise, only the local ROS master will be shown (see Fig. 3).

The connection icon next to each host name shows whether that host is reachable. The icon color indicates the connection quality as determined by the `master_discovery` node.

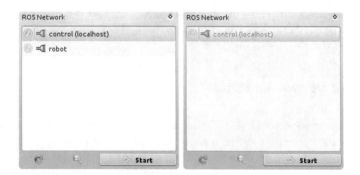

Fig. 3 ROS-Network overview with `master_discovery` running and not

Fig. 4 Start discovery dialog

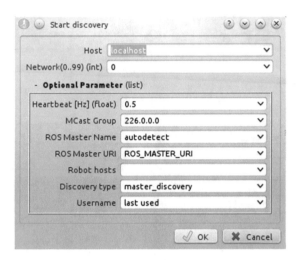

The sync icon at the beginning of each row indicates whether master_sync is running. You can toggle the ROS master synchronization by clicking on this icon. A synchronization dialog with advanced options is available via the synchronization button on the Host Description Panel described in Sect. 4.2. Note that the multi-master capabilities are disabled if the local ROS_MASTER_URI refers to a remote host.

You can initiate the discovery process with the *Start* button on the ROS Network widget. The discovery node can also be started on a remote host. Click on the *Start* button and enter the host name or IP address in the dialog shown in Fig. 4. If the remote ROS master is not running already, it will be started automatically via SSH. Additionally, you can enter a network number (0..99) to create your multiple separated networks (technically, this creates distinct multicast group for the discovery process). This can be useful if two groups of robots shall remain separated although they share the same physical network segment. Note that if no multicast communication is available, you need to set the robot_hosts parameter to the comma-separated list of participating hosts. Also note that this will increase the network load quadratically (doubling the number of hosts quadruples the traffic)!

4.2 Host Description Panel

The host description panel (Fig. 5) displays the name and the time of the last update for the currently selected host. If the *Node Manager* finds a PNG image in node_manager_fkie/images (the path is configurable in the Settings tab) with the same name as the host, it is used as icon. The icon can also be changed by double-clicking on it or by setting the ROS parameter /robot_icon to an image file. The host description panel contains additional buttons to update the state of the

Fig. 5 Host description panel

host or run ROS nodes and tools, which are connected to the selected host, independently from the synchronization state. A warning will be displayed if */use_sim_time* is set to true, as this parameter tends to be forgotten when switching from simulation trials to real world experiments.

4.3   ROS Nodes View and Control

The *Nodes* tab lists all nodes which are either running or available for launch in the current configuration. It can be seen in Fig. 6. Running nodes have a green icon. They are monitored using their process ID, which is updated by the local `master_discovery`. You have to run a `master_discovery` on each host to ensure that information about all nodes is available in the network. If a node is registered with the master but has no associated process ID, the node is considered crashed and marked with warning triangle. Note that only local nodes are pinged. Nodes which are not running on the same host as the ROS master are marked by a green icon with a question mark. Crashes of these nodes cannot be detected. As of now, the only reliable way to detect running nodes is by their registered topics and services. Thus, if a node has neither, it will never be shown as running.

In order to launch nodes, a configuration must be loaded. You can open roslaunch files. If a node with the same name is listed in multiple launch files, you will be asked to choose a configuration when you launch the node. Alternatively, you can use the `default_cfg` (described in Sect. 3.7) to serve launch files to the local network. Nodes which are available through this mechanism are marked by a cloud. If you select a node additional information for this node are shown in description dock. It is described later in Sect. 4.8.

Control Nodes started by the *Node Manager* are each running in a dedicated terminal (using the **screen** tool). Thus you have access to the output of each node. Further, this output is stored in a file that you can access from the control bar on the right side. Nodes are started on remote host with SSH. The default user can be changed in the *Settings* dock or for each host in in the *Host Description Panel*. Since a non-interactive SSH connection is established, you have to ensure that the ROS environment is initialized, e.g. by preparing the `.bashrc` accordingly. You can select multiple nodes to be started with one click; already running nodes are not restarted.

Note: while starting a node only private parameter are retransmitted to remote ROS parameter server. The global parameters are transmitted only once on launch file load. You can force the reload of the global parameters using the *Description*

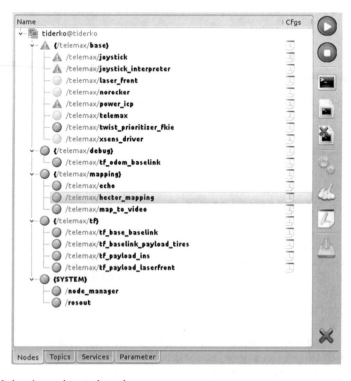

Fig. 6 Nodes view and control panel

Dock, see the Sect. 4.8. When using *default_cfg* all parameter are loaded once at startup and never retransmitted.

The nodes are stopped using the shutdown method of the ROS Slave API.[9] Therefore, unreachable nodes cannot be stopped. In this case, you can use links in the *Description Dock* to send a SIGKILL signal to the node process or force the ROS master to unregister the services and topics of this node. The nodes of the multimaster extension can only be terminated if they are selected individually.

You can also change the parameters of a selected node. There are three methods: launch file, ROS parameter server or dynamic reconfiguration GUI (if available). The *Control Bar* provides three configuration buttons, which are enabled if the corresponding method is available. The *Dynamic Reconfiguration* is only available if the node in question supports it and is running. The configuration of the ROS parameter server is always available, but the displayed parameters depend on the running state of the node. Some parameters are (re-)loaded on start of the node. You can also change the parameter in the loaded launch file. In this case the loaded file will be open and the cursor jumps to the first occurrence of the node name. After the launch file was changed (it is detected automatically), you have to reload this file.

---

[9]http://www.ros.org/wiki/ROS/Slave_API.

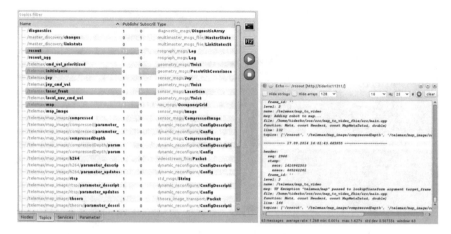

Fig. 7 Topics view and topic echo dialog

4.4 ROS Topics View

The *Topics* Tab (Fig. 7) shows all topics which are registered at the ROS master. The number of publishers and subscribers is shown in the second and third columns. The type of the ROS-topic is also displayed. If a node is selected in the nodes-view, the corresponding topics will be selected. Additional information to the topic is displayed in the *Description Dock*. The control bar in the *Topics* tab provides buttons to display incoming messages. The echo dialog runs in its own process. To display the content of the topic the message type must be compiled. If the type is not available, because it is compiled only on the remote host, the dialog tries to establish a ssh-connection and subscribe the topic on remote host directly.

You can also publish to a selected topic. If no topic is selected, a new one will be created.

4.5 ROS-Services and Parameter View

Figure 8 shows the *Services* and *Parameter* tab. It provides functions to call services or edit parameters. The parameters are not updated automatically; download has to be triggered manually using the top button on the right.

Fig. 8 Services and parameter view

4.6 Launch Dock

The *Launch* Dock (Fig. 9) can be used to find and load or edit the launch files. The lowest folder level contains the paths listed in the ROS_PACKAGE_PATH environment variable. In addition, only subfolders which contain *.launch or *.yaml files are shown. The extention list can be changed in the *Settings* Dock.

On load, the launch file itself and all included files are loaded. If one of these files changes, you will be prompted to reload the launch file. If changes for a node are detected you will be asked if the corresponding nodes should be restarted. At the lowest folder level, the last five loaded launch files are listed for convenience. When needed, the selected history files can be removed by pressing the *Delete*-key on the keyboard.

You can also search for a package folder with this dialog. If you are in a package folder or its subfolder, you can copy and paste files.

All changes to files concern only the local host. You can transfer the modified files (and optionally their includes) to a remote host by clicking the blue arrow button.

4.7 Launch Editor

Launch editor is a simple text editor with *roslaunch XML*[10] syntax highlighting.

Figure 10 shows the syntax highlighting. In addition, you can open the included files in a new tab by *Ctrl+(left mouse click)*. If the file does not exists, it will be created. You can also include files by Drag and Drop.

---

[10]http://www.ros.org/wiki/roslaunch/XML.

Fig. 9 Launch dock for navigation through launch files

Fig. 10 Launch editor with syntax highlighting

It is also possible to add a template of a launch tag by using the *Add tag* button.
Or you can press *Alt+Space* to get a context menu with possible tags or attributes.
    The launch editor also provides a recursive search to find text in included files.

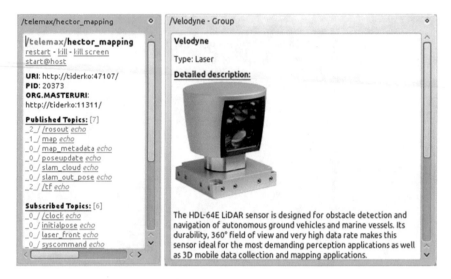

Fig. 11 Examples of description docks

4.8 Description Dock

The *Description Dock* shows additional information about selected nodes, topics or services. It can also contain more control options. Some examples are shown in Fig. 11.

4.9 Capabilities and Additional Description

Since the *Node Manager* tries to improve the overview and control of available ROS nodes on a robot, we added some parameter to allow grouping nodes to capabilities. Figure 6 already showed an example for these groups. To define a group you need to add a `capability_group` parameter to a node.

```
<node name="hector_mapping"pkg="..."type="...">
    <param name="capability_group"value="mapping"/>
</node>
```

If no local `capability_group` parameter is found for a node, the search is expanded to the enclosing namespaces. If a `capability_group` parameter is found there, it will be assigned to the node. In this manner it is possible to assign a group to multiple nodes with just one parameter declaration. It is also possible to add the same node to multiple groups using a prefix for `capability_group`, e.g.:

```
<node name="node_manager"pkg="..."type="...">
    <param name="capability_group"value="System"/>
    <param name="1.capability_group"value="Management"/>
</node>
```

The description of a capability group is stored in a global `capabilities` parameter. This parameter is defined as *rosparam* and contains a list with all group descriptions. The group description itself is a list which consists of *group name*, *group type*, *image* and *description*. *image* is a relative path to the *node_manager_fkie* package, or *$(find PACKAGE)* can be used. The description can be coded as *reStructered-Text*[11] and also contain image references. Since the XML parser will renormalize white spaces, you must use \n to force a line break.

```
<rosparam param="capabilities">
  [["System",
    "core",
    "$(find some_package)/images/system_icon.png",
    "The ''System'' group provides nodes needed to detect and
    synchronize other robots in the ROS network. These are:\n
    \n- Node Manager\n- Master Discovery
    \n- Master Synchronization"
  ]]
</rosparam>
```

The capability group does **not** change the name space of the included ROS nodes.

The parameter `robots` describes the robots in the same manner. The list must contain *host name*, *robot type*, *displayed name*, *image* and *description*.

4.10 Capability View

The *Capability View* (Fig. 12) tabularly shows the discovered hosts and their capabilities. The view is created based on running `default_cfg` nodes. Each available capability group provides buttons to start or stop all nodes of this group.

4.11 Auto Update

By default the *Node Manager* updates the status of changed hosts automatically. The state of the local host is updated periodically to detect communication problems to the local `master_discovery`. Also some of the parameters, like `/use_sim_time`, of selected hosts are updated periodically. This is helpful in high-throughput

---

[11] http://docutils.sourceforge.net/docs/ref/rst/restructuredtext.html.

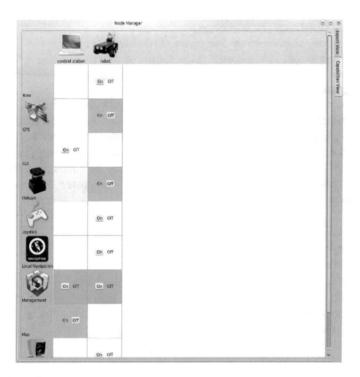

Fig. 12 Capability view

networks, but it can also cause problems in networks with low bandwidth. To disable the auto update feature, the *Settings Dock* provides an `autoupdate` parameter. If you set this parameter to *false* you have to use the reload buttons to update the state of each host, but you can avoid high bandwidth usage after the startup phase.

Authors' Biography

Alexander Tiderko obtained his Dipl.-Inform. degree in computer science from the University of Bonn, Germany in 2006. At present he is a scientist at the Fraunhofer Institute for Communication, Information Processing and Ergonomics FKIE in Wachtberg Germany. His primary research interests include networked multi-robot systems and robot interoperability.

Frank Hoeller obtained his Dipl.-Inform. degree in computer science from the University of Bonn, Germany in 2006. At present he is a scientist at the Fraunhofer Institute for Communication, Information Processing and Ergonomics FKIE in Wachtberg Germany. His primary research interests include mobile robot navigation and robot system structure.

Timo Röhling obtained his Dipl.-Inform. degree in computer science from the University of Bonn, Germany in 2008. At present he is a scientist at the Fraunhofer Institute for Communication, Information Processing and Ergonomics FKIE in Wachtberg Germany. His primary research interests include sensor data processing, mobile robot navigation, and robot system structure.

Advanced ROS Network Introspection (ARNI)

Andreas Bihlmaier, Matthias Hadlich and Heinz Wörn

Abstract This tutorial chapter gives an introduction to Advanced ROS Network Introspection (ARNI), which was released as a solution for monitoring large ROS-based robotic installations. In the spirit of infrastructure monitoring (like Nagios), we generate metadata about all hosts, nodes, topics and connections, in order to monitor and specify the state of distributed robot software based on ROS. ARNI provides a more in-depth view of what is going on within the ROS computation graph out of the box. Any existing ROS node and host can be introspected without prior modification or recompilation. This extends from live network properties to host and node specific ones by running an additional node on each host of the ROS network. Furthermore, it is possible to define reference values for the state of all ROS components based on their metadata attributes. Subsequently, ARNI provides a mechanism to take countermeasures on detection of a violated specification. All features are modular and can be used without modifying existing ROS software. ARNI was written for ROS Indigo and this tutorial has been tested on Ubuntu Trusty (14.04). A link to the source code repository together with complementary information is available at http://wiki.ros.org/arni.

Keywords Introspection · Safety · Reliability · Monitoring

1 Introduction

One advantage of ROS, seen as a middleware, is the flexibility of its publish-subscribe mechanism. The flow of data between nodes is not determined at compile time, rather name remapping allows to specify it at startup time. In addition, nodes and connections can be added and removed during the runtime of other nodes. The downside of

A. Bihlmaier (✉) · M. Hadlich · H. Wörn
Institute for Anthropomatics and Robotics (IAR), Intelligent Process Control
and Robotics Lab (IPR), Karlsruhe Institute of Technology (KIT), 76131 Karlsruhe, Germany
e-mail: andreas.bihlmaier@kit.edu

H. Wörn
e-mail: woern@kit.edu

© Springer International Publishing Switzerland 2016
A. Koubaa (ed.), *Robot Operating System (ROS)*, Studies in Computational
Intelligence 625, DOI 10.1007/978-3-319-26054-9_25

this high flexibility is the difficult to predict the dynamic behavior of the complete system. For example, even adding a new host to the ROS network and executing a node—e.g. providing a human interface to some data stream in the ROS graph— might degrade the performance charateristics of already running node subgraphs. Also, due to the fast changing nature of the ROS graph, it is hard to narrow down the cause of erroneous behavior related to interrupted connections or bottlenecks occuring in the system during runtime.

The Advanced ROS Network Introspection (ARNI) [1] package tries to overcome some of these problems. ARNI consists of components (Fig. 5) that acquire metadata about hosts, nodes, topics and connections in a decentralized manner. As opposed to the actual data, it is possible to aggregate and evaluate all the metadata at a central point in the ROS network. The metadata can then be used to introspect and monitor the entire ROS graph both live and retrospectively. Rather than artificially restricting the important flexibility of ROS, the ARNI package allows to handle the resulting complexity, i.e. diversity and dynamics of the computing graph. Furthermore, ARNI enables to define invariants for certain properties of the ROS network. For example that the robot joint topic should always carry messages with a frequency between 950 and 1050 Hz or that the network bandwidth of each host interface should be below 90 % of its maximum. Since ARNI is exclusively based on existing ROS mechanisms, such as messages, topics and services, it enables certain self-X properties of the robotics software. One example is the automatic running of countermeasures on violation of an invariant, like the restart of a node or the emergency stop of the robot.

The remainder of the chapter consists of the following topics:

- First, a summary of the ROS publish-subscribe mechanism is provided as essential background information for the understanding of ARNI.
- Second, starting from a clean install of ROS Indigo all required steps to install ARNI are detailed. At the end a stand-alone test is provided to verify the correct setup of the ARNI components.
- Third, the architecture of ARNI, all it's components and their interactions are described.
- Fourth to Sixth, step by step instructions are provided to introspect, monitor and finally automatically counteract deviations in the ROS network. The examples used throughout are taken from a real system and cover on the one hand a custom robot control stack and on the other a common community-provided stack for camera sensors.

2 Background

ROS topics are based on the publish-subscribe mechanism with the ROS Master serving as a well-known entry point for naming and registration. Each ROS node advertises the topics it publishes or subscribes to the ROS Master. If a publication and subscription exist for the same topic, a direct connection is created between the

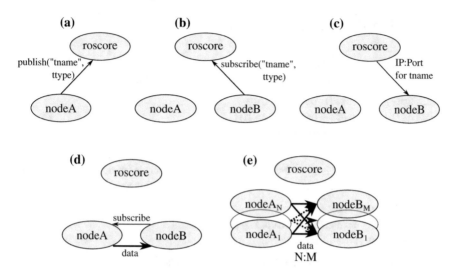

Fig. 1 Overview of the ROS topic mechanism. When a publisher (**a**) and subscriber (**b**) are registered to the same topic, the subscriber receives the network address and port of all publishers (**c**). The subscriber continues by directly contacting each publisher, which in return starts sending data directly to the subscriber (**d**). Many nodes can publish and subscribe to the same topic resulting in a $N : M$ relation (**e**). On the network layer there are $N \cdot M$ connections, one for each (publisher, subscriber) tuple. The nodes can be distributed over any number of hosts within the ROS network

publishing and subscribing node(s), as shown in Fig. 1. To summarize and clearly define the terms used throughout the rest of the chapter, there are four important entities in the ROS computing graph:

- **Host**: A computer within the ROS network, identified by its IP address.
- **Node**: Any process using the ROS client API, identified by its graph resource name.[1]
- **Topic**: A named communication channel as used in the publish-subscribe mechanism, identified by its graph resource name.
- **Connection**: A connection between a (publisher, subscriber) tuple carrying the data of a specific topic (cf. Fig. 1), identified by the tuple (subscriber-node, topic, publisher-node).

Services, synchronous remote procedure calls in ROS, as the second basic communication mechanism of the ROS middleware are also used within ARNI. However, due to their volatile nature, they are not monitored by ARNI and thus not particularly relevant to the scope of this chapter.

---

[1]Cf. http://wiki.ros.org/Names.

3 ROS Environment Configuration

We assume a standard installation[2] of ROS Indigo—on Ubuntu Trusty (14.04). First, up to four further system dependencies need to be installed:

```
sudo apt–get install python–pip
pip install —user —upgrade psutil
pip install —user pysensors
pip install —user pyqtgraph
```

The installation itself is done by compiling the source code using catkin. The path to the catkin workspace has to be adapted to other environments if necessary.

```
cd ~/catkin_ws/src
git clone https://github.com/ROS − PSE/arni.git
cd ~/catkin_ws; catkin_make
```

Now ARNI should be correctly setup and can be started.

3.1 ARNI: Testing of the Setup

In order to insure the correct installation of the Python modules, first try to run

```
rosrun arni_nodeinterface arni_nodeinterface
```

This command should not terminate and should not print any errors about missing or outdated modules. If it is working correctly, abort the program at this point.

Before testing the complete setup one short reminder: If any of the following steps fail—saying "file not found" or the Plugins are not shown in the GUI—try to repeat the installation and make sure to follow all previous steps. Start

```
roscore
```

and execute

```
roslaunch arni_core test_1_steady.launch
```

which will create two nodes communication with each other on a topic as shown in Fig. 2.

Check whether this succeeded with the rqt Graph plugin or run

```
rostopic info/forest
```

If this is the case, then start

```
rqt
```

Go to Plugins, Introspection and open Arni-Detail, the result should look similar to Fig. 3.

---

[2]A desktop-full installation according to http://wiki.ros.org/indigo/Installation/Ubuntu.

Fig. 2 steady_tree publishes with 100 Hz on/forest, ninja_turtle listens to forest

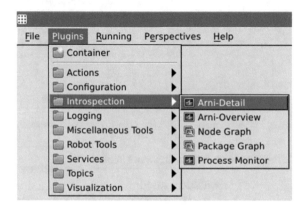

Fig. 3 The location of the Arni-Plugins in rqt

Fig. 4 The GUI showing that the state of the topic/forest is ok

If there is no Arni-Detail under Plugins try to reinstall ARNI. Make sure to uncheck the field "Errorneous Only". Use the filter box to search for or look for forest or manually look for the topic t!/forest. Here the results of monitoring are shown as "state". In this example setup the state of the topic should always be "ok" as shown in Fig. 4. It is continuously monitored by the processing node as specified in the launch file. If instead the state is given as "unknown", the processing node is not working properly. In this case, make sure to check the log for errors and try to fix these.

4 ARNI: Overview

ARNI extends the introspection capabilities of ROS by five elements (cf. Fig. 5): First, distributed measurement and publication of communication and host meta-data. Second, a YAML-based and parameter server compliant format to specify reference states for a ROS subgraph. Third, a processing node, which is responsible

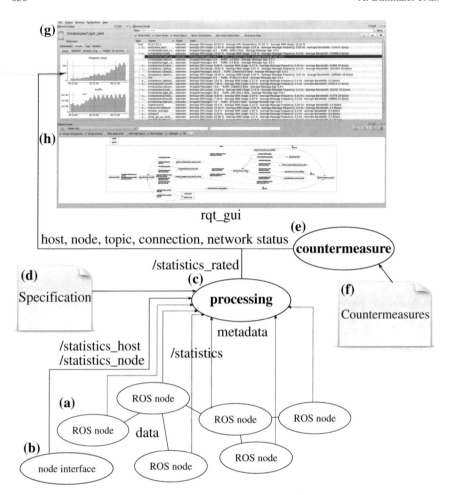

Fig. 5 Overview of the Advanced ROS Network Introspection components: **a** Publishing metadata for topics and connections (/statistics). **b** One node per host to publish metadata for hosts (/statistics_host) and nodes (/statistics_node). **c** The processing node monitors and compares actual values to a YAML specification (**d**) and publishes the result of the comparison (/statistics_rated). **e** The countermeasure node can automatically act on the rated data, given a YAML specification of the countermeasures (**f**). **g** The ARNI rqt_gui to visualize current state of ROS network. **h** The rqt_gui Node Graph plugin can also show/statistics data

for collecting all metadata, comparing it to the reference, if one is available, and publishing the results of the comparison. Fourth, a countermeasure node that runs predefined actions in case a deviation is reported by the monitoring infrastructure. Fifth, an extensive rqt GUI plugin that provides an out-of-the-box overview of hosts, nodes, topics and connections within the ROS network. In case monitoring is active, clearly visible feedback about the overall and detail state of the computation graph is provided.

No new means of communication are introduced by ARNI, instead all components rely solely on features already built into ROS. Most notable of these are topics, parameters and services. Each component will be described in more detail in the remainder of this section. The following sections utilize all these components in two concrete examples.

4.1 Metadata Acquisition

In order to provide live information about the whole ROS graph, it is inevitable to generate this data at its origin. Thus ARNI extends the /statistics information, which can be generated by all ROS—Indigo or later—nodes that use the C++ (roscpp) or Python (rospy) client API. The computation and network overhead is negligible, since no matter how much bandwidth is utilized to transfer data, the amount of metadata to describe it is constant and consists of a single floating point value. The most important metadata for connections and topics is the utilized network bandwidth, message frequency, jitter, latency and dropped messages. ARNI also provides a /statistics_host and /statistics_node topic, which requires to run one additional interface node on each host. Again, the additional CPU and memory usage for this node is very low. The provided information includes per host and per node CPU, memory and raw network utilization as well as hardware health attributes, such as CPU and GPU temperatures. To uniquely identify each component of the ROS graph (cf. 2), each one can be refered to by its SEUID (Statistics Entity Unique IDentifier):

- **Host**: h!*IP*, e.g. h!127.0.0.1
- **Node**: n!*node-graph-name*, e.g. n!/camera_driver
- **Topic**: t!*topic-graph-name*, e.g. t!/cam_left/image_raw
- **Connection**: c!*subscriber-graph-name*!*topic-graph-name*!*publisher-graph-name*, e.g. c!/rqt_gui!/cam_left/image_raw!/camera_driver

Having all this information available live and in a central place enables a much deeper understanding of how the robot software works in terms of its inner communication patterns. This is useful to detect (unwanted) crosstalk, e.g. due to overloaded network equipment or hosts, and soon becomes an indispensable tool to debug non-local errors in the distributed system.

4.2 ARNI rqt GUI

Two rqt plugins provide a graphical overview of all the host, node, topic and connection metrics of the ROS graph. The first plugin, ARNI-Overview (Fig. 9), aggregates all metadata into a single view, which can be seen as a network health dashboard. The second plugin, ARNI-Detail (Figs. 6 and 7), allows to view and select all metadata in a hierarchical manner. Furthermore, plots are available for the metadata, which provide a live view and enable to introspect a moving window history of each metadata item (Fig. 8). If reference states are specified for an item, the conformity to these or deviations from them are graphically indicated. ARNI-Detail is furthermore a tool to gather more information about which key indicator deviates from the reference.

4.3   Reference State Specification

Providing the ROS programmer with live information about all components of his robot is a valuable tool. However, once it is clear what invariants are supposed to apply for a specific part of the whole system, the next step is to specify these invariants and have them automatically be checked by a component of the system. From a system point of view, this corresponds to the property of self-protection. Since flexibility and runtime reconfigurability are highly valued features of ROS, the specifications must also be modular and orthogonal. For example, if a specification is provided for a particular sensor subsystem, it should not affect any other subsystem at all. In addition, it must be possible to only specify constraints for values which are part of the invariant without imposing any constraints on those values which are not. The format will be best understood through the examples provided in the following sections. The general structure can be summarized as being a list of SEUIDs under the parameter namespace `/arni/specifications`, each one containing boundaries for a subset of the available metadata.

4.4   Countermeasures Specification

Based on the evaluation of the current metadata against the specification, the `/statistics_rated` topic is published by the processing node. Rated statistics exist for each item in the reference specification and the provided ratings are: Value unknown, value too low, value within bounds or value too high. The countermeasure node (see Fig. 5e) can use this information together with a list of conditional actions, given as parameter list under `/arni/countermeasure`, to automatically counteract a detected deviation. This corresponds to the property of self-healing. Essentially each countermeasure consists of a condition on rated statistics items together with an action that is executed if the condition is true. Again, the following example should illustrate how these conditional actions work in detail.

5 ARNI: Advanced Introspection

The following three sections will introduce ARNI in detail by showing two exemplary applications in the context of our ROS-based lab setup for cognitive surgical robotics. We begin by starting the ROS network, namely

roscore

and initializing parameters to generate statistics, which must happen *before starting any other ROS node*

roslaunch arni_core init_params.launch

We then start up our ROS network, consisting of various nodes distributed across several machines, as we would when not using ARNI. Now, we can start rqt and open the Arni-Detail plugin. Without any additional action, we get an overview of the communication within the ROS network (see Fig. 6). At this time hosts and nodes do not provide any kind of information, only topics and connections are updated. This is because at the moment the host interface node is not yet started which will be done later. Arni-Detail consists of two Widgets, the Treeview that shows the whole network as a Qt tree widget and the Detail Widget which shows information about the currently selected item in the Treeview (see Fig. 6). The Treeview widget can filter the information in a number of different ways. Any of the four layers host, node, topic and connections can be shown or hidden with checkboxes. There is also a switch for subscribers. Since the Treeview does not show these in the default setup, the subscribers can be recognized by their modified name which contain an additional "–sub" suffix. This enables to search for errors in the ROS network and to see if all nodes listen to the topics they should. The recording bar can be ignored for the moment, it will be used later on Sect. 6.

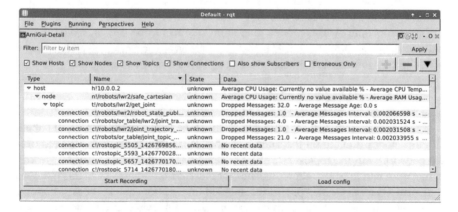

Fig. 6 The Arni-Detail Treeview showing parts of the example network. Since the nodeinterface is not running only topics and connections contain actual data. Those connections displaying "no recent data" have published no information for an extended period of time

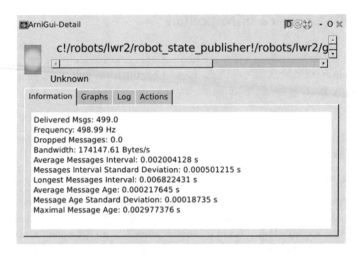

Fig. 7 The Detail widget of Arni-Detail showing information about a connection

The second Widget of Arni-Detail shows information about the currently selected item. As can be seen in Figs. 7 and 8, Arni-Detail shows information for hosts as well as nodes, topics and connections. It consists of four independent parts namely Information, Graphs, Log and Actions. Information shows any kind of information the system has about the item. What this information is for a connection in particular can be seen in Fig. 7. Graphs show a subset of the properties in information and plot them over a history of up to 60 s. If the processing node is running and has a valid specification for the currently selected item, the log will show whenever the items state changes. The following chapter will deal with specifications and explain how to use them. The Actions tab is only activated on nodes and can be used to stop or restart them, this feature is further explained in Sect. 7.

Before opening Arni-Overview one more step has to be executed otherwise it will not show any information. The host interface has to be started on all hosts from which further information should be collected. This done by running

```
rosrun arni_nodeinterface arni_nodeinterface
```

once per host.

The interface node gathers information about the host it is running on and about all other nodes running on the same machine. This information is distributed over the network with the topics /statistics_host and /statistics_node. Arni-Detail will now also show statistics for these hosts and nodes. At this point we start Arni-Overview which provides combined information about all hosts. Figure 9 shows what it looks on our example system. Arni-Overview also contains a large colored button, which is used in analogy to a traffic light. It will be green as long as no specification is loaded and at least one host is online. When specifications are defined and any item deviates from its specification, the light will turn red. As soon

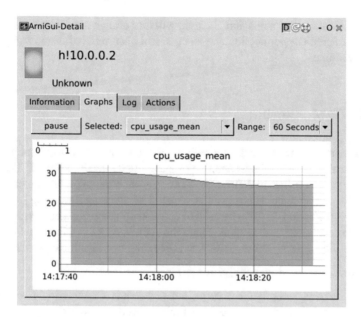

Fig. 8 The Detail widget showing the mean CPU usage plot of a host. The metadata property to plot can be selected as well as the interval over which it is plotted

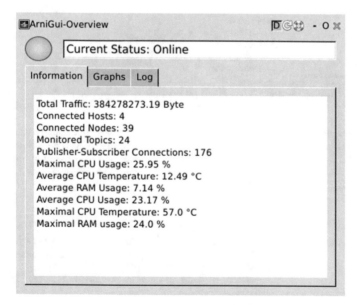

Fig. 9 The Arni-Overview summarizes the information of all hosts. It has the three tabs Information, Graphs and Log which behave very similar to those in Arni-Detail

as all specifications are met again, the light will turn orange for a few seconds before returning to green as notification that an error had occurred, but that it was resolved in the meantime.

6 ARNI: Monitoring

The two parts of the network that will be focussed on are a ROS subgraph for robot control as seen in Fig. 10 and one of a camera depicted in Fig. 11. The ARNI monitoring node enables to define specifications for certain metadata properties and to notify whenever such a constraint is violated by showing a warning in the GUI. The same information can also be used to define automated countermeasures which will be treated in the next section. Since the API is very flexible at this point any kind of notification can be executed. The processing node itself can be started with

rosrun arni_processing arni_processing.

Defining and using a config file is very simple since it is a YAML document that can be loaded onto the parameter server. To simplify the creation of a config file, the GUI provides a tool for "recording". First data will be recorded, then, we will look at the

Fig. 10 This figure shows the relevant actuator ROS subgraph used as one example throughout the chapter. The *ros2fri* node directly translates between the proprietary KUKA FRI protocol and ROS topics. In the *lwr_safe_cartesian* node inverse kinematics and collision detection is implemented. Therefore, it publishes and subscribes a cartesian and joint topic. The *trajectory_action2topic* node provides a ROS action interface to the robot

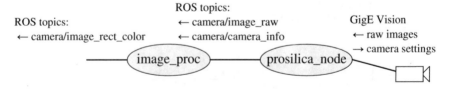

Fig. 11 The figure represents the camera sensor subgraph used as second example. We use the *prosilica_node* (The prosilica_node is provided by the http://wiki.ros.org/prosilica_camera package.) to map the camera's GigE Vision interface to ROS image messages. Furthermore, the *image_proc* (The image_proc node is provided by the http://wiki.ros.org/image_proc package.) node does image rectification in case the camera was intrinsically calibrated

Listing 1.1. Example of recording with rqt GUI

```
[...]
 −t !/sometopic
    frequency: [50,100]
    # other properties
    # further entries like host, node, topic or connection,
    # are all on one level in the YAML file
```

resulting file and its structure. Recording means that all the data of the selected items is recorded for a certain amount of time and a specification is generated from the data. For example, recording one topic for a minute, which has a minimum frequency of 50 Hz and a maximum frequency of 100 Hz, the configuration file will contain an entry as shown in Lst. 1.1.

Loading this YAML file to the parameter server can be done with the GUI or on the command line. In the GUI approach one has to click on the "load config" button, choose the file and open it. It will automatically be uploaded and the processing node is notified so it can update its configuration. If this appears to change nothing, check rqt log which should print errors and the processing node log which should print the amount of loaded entries (one for each host, node, topic and connection).

Trying to change an existing specification by loading a modified file repeatedly with the GUI will not result in the desired behaviour. This is because the GUI pushes every specification into another namespace. The issue can be overcome either by using the command line and the same namespace or by removing the entries from the parameter server. The following command can be executed as given and it will remove all specifications or it can be modified to remove only certain specifications:

rosparam delete/arni/specifications

This leads to the second path of adding specifications, the command line. Specifications can be loaded into a specific namespace, which is important if this is not already done in the specification file. In this manner the same file can be imported multiple times and the specification node will always use the latest specifications, assuming the file is always pushed to the same namespace. First we load the specification to the server (cf. Lst. 1.2) and then we call a service to inform the processing node about the changes:

rosparam load actuator.yaml/arni/specifications
rosservice call/monitoring_node/reload_specifications

As already mentioned, the specification format is a simple YAML structure which will now be further explained. Also note that all valid variants of the YAML syntax are allowed. For example, one may use the "intended" notation for lists or the "[...,...]" notation. If one does not want to type the namespace on every load, it is possible to add them to the YAML file:

```
arni:
  specifications:
    #actual content
```

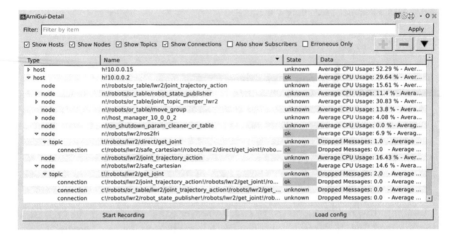

Fig. 12 After loading specifications the system checks if these are fulfilled and displays the result in the GUI

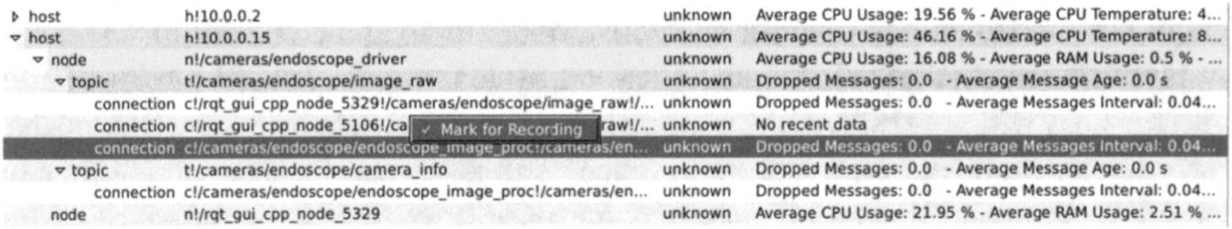

Fig. 13 Items marked for recording appear in the Treeview with a *light orange* color. Items can marked and unmarked via the right-clicking context menu

Remember to use correct indents. The first example pertaining to Fig. 10, acutator.yaml, was generated by hand. We looked at the nodes and considered all properties for which fixed boundaries are known and that should be monitored for safety and reliability. In case of the nodes this means that they should not exceed a certain computation time. The connections have to send constantly at almost exactly 500 Hz. Figure 12 shows how this looks on a properly functioning system which returns "ok" for all items.

If the specification does not work at the first try, have a look at the output of the processing node. The number of specifications loaded is printed on call of the reload_specification service. If none or not all are loaded, check namespaces, indents and whitespaces.

The second—more easy—possibility to start with reference specifications is usage of the GUI recording tool. Items that should be monitored are marked as shown in Fig. 13. Before proceeding, one has to make sure the system is in its target state, i.e. the relevant subsystem is working as intended. This is important because the system will take the maximum and minimum value over the recording period and set it as the margins of the constraint. If the recording is done in an error state, where for example the load is unexpectedly high or low, this will lead to erroneous specification files,

Listing 1.2. actuator.yaml

```
- n!/robots/lwr2/safe_cartesian:
    node_cpu_usage_max: [0, 30]
    node_cpu_usage_mean: [10, 30]
    node_ramusage_mean: [0, 5]
- c!/robots/lwr2/joint_trajectory_action!/robots/lwr2/get_joint!/robots/lwr2/
    ↪ safe_cartesian:
    frequency: [495, 505]
- h!10.0.0.2:
    cpu_temp_max: [0, 80]
    cpu_temp_mean: [0, 70]
    cpu_usage_mean: [0, 90]
    ram_usage_mean: [0, 90]
- c!/robots/lwr2/safe_cartesian!/robots/lwr2/direct/get_joint!/robots/lwr2/ros2fri
    ↪ :
    frequency: [495, 505]
- n!/robots/lwr2/ros2fri:
    node_cpu_usage_max: [0, 30]
    node_cpu_usage_mean: [3, 30]
    node_ramusage_mean: [0, 5]
```

whose values have to be corrected manually. After marking all to be monitored items the "Start Recording" button must be pressed. Recording should be done for at least a few seconds to get meaningful minimum and maximum values. When it is deemed that enough data has been collected, "Stop Recording" can be clicked. The resulting specification file can now be saved and directly loaded as shown above.

The recording feature was used to define specifications for the second subsystem used as example (see Fig. 11). Three items in the camera subsystem were marked and recorded with the GUI. Afterwards we modified the file by removing any metadata item that we either considered to be uncritical or for which no fixed constraint applies. Although not strictly necessary, we advise always to have a close look at the recorded values and often to also manually increase the margin by a few percent and round the values for improved readability.

Applying these specifications on our system we found out that the items sometimes do not match the specifications. In Fig. 14 we see in the Treeview that the topic and the node have problems. In Fig. 15 we further investigated the problem by clicking on the erroneous node item to see which part of the specification is violated. We actually found a previously unnoticed bottleneck in our system, which is due to

Fig. 14 The *orange* color of the node shows that a problem had occurred a short while ago, but that the current state is within the specification. When the topic is marked *red* this means it is currently not working as intended, i.e. violating one of the specifications

Listing 1.3. cam.yaml

```
– c!/cameras/endoscope/endoscope_image_proc!/cameras/endoscope/image_raw!/cameras/
     ↪ endoscope_driver:
  dropped_msgs: [0, 2]
  period_max: [0.033, 0.066]
  period_mean: [0.033, 0.034]
  stamp_age_max: [0.0019, 0.0045]
  stamp_age_mean: [0.0017, 0.0019]
– t!/cameras/endoscope/image_raw:
  bandwidth: [48000000, 63000000]
  dropped_msgs: [0, 2]
  frequency: [24, 35]
  period_max: [0.033, 0.066]
  stamp_age_max: [2.0e−12, 4.0e−12]
– n!/cameras/endoscope_driver:
  node_bandwidth_max: [5000000, 75000000]
  node_bandwidth_mean: [48000000, 65000000]
  node_cpu_usage_max: [0.0, 25.0]
  node_cpu_usage_mean: [0.0, 20.0]
  node_cpu_usage_stddev: [0.0, 3]
  node_message_frequency_max: [0.02, 0.05]
  node_message_frequency_mean: [0.02, 0.05]
  node_ramusage_max: [0.40, 0.7]
  node_ramusage_mean: [0.2, 0.6]
  node_read_max: [0.0, 0.0]
  node_read_mean: [0.0, 0.0]
  node_write_max: [0.0, 0.0]
  node_write_mean: [0.0, 0.0]
```

heavy network load on a different network segment in combination with a central Ethernet switch that contains a backplane with limited bandwidth.

7 ARNI: Countermeasures

In this chapter it is assumed that the previous parts are understood and host interface nodes as well as the processing node are running. Processing needs to have valid constraints since countermeasures can only react to violations of these constraints. To understand how this might be helpful, we like to give a short example. If one looks at the maximum temperature of a server it is clear that it should never exceed a certain limit. Thus, we define a constraint that permanently monitors the CPU temperature. Along with the constraint, we define an action taken on violation of the constraint that guarantees that it does not occur without it being noticed. A good countermeasure in this case would be to email the server administrator or to shutdown the host.

ARNI provides two ways for countermeasures. The first is an interactive comfort function that uses the same mechanisms as the second technique described below. When clicking on a node in Arni-Detail and looking at the detail window, a tab called

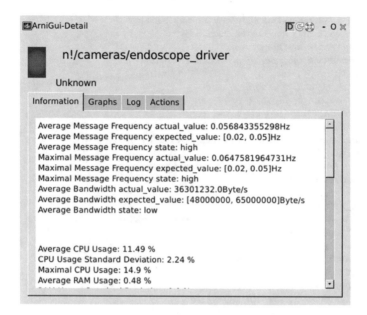

Fig. 15 The node is in an error state since several specifications are violated. For example, the bandwidth is one below the specified one. In this case a network switch was overloaded and thus delayed or dropped packets

"Actions" is available. The node can be stopped or restarted from there. In the latter case the node will be restarted on the same host and with the same command line parameters. Nonetheless this will not work for all nodes and one should always make sure that the node works correctly after this countermeasure.

The second countermeasure technique is much more powerful, although its usage and use cases are highly application dependent. In a similar way to specifications, countermeasures can be created and pushed to the parameter server. We first start the countermeasure node:

rosrun arni_countermeasure arni_countermeasure

Next we load a countermeasure file:

rosparam load counter.yaml/arni/countermeasure

And finally tell the countermeasure node to reload all countermeasures:

rosservice call/countermeasure/reload_constraints

The countermasure node will output the number of loaded countermeasures, which can again be used to detect user errors more easily. Whenever a countermeasure is executed it will also show a short notification. Similar to specifications, multiple countermeasure files can be loaded. The countermeasure files have the format shown in Lst. 1.4.

Listing 1.4. Description of countermeasure specification format

```
# all countermeasures have to be loaded to
# /arni/countermeasure/constraints
some_constraint: # the name of an actual constraint
  constraint:
  -or: # one of: and, or, not
    # property can be high/low or normal
    name_of_the_item: {property: low}
  # the minimum time the constraint has to be
  # violated before the countermeasures is executed
  min_reaction_interval: 1
  # timeout until the action can be executed again
  # this exists so that e.g. a node has sufficient
  # time to restart
  reaction_timeout: 20
  reactions: # one can list as many of these as required
    some_reaction_name: {
      action: restart, # or stop, run, publish
      # specifies the node where the action is run
      node: /cameras/endoscope_driver
      # publish requires "message" to be defined,
      # it is shown in the GUI logging tools
      # run requires "command" to be defined, this
      # shell command will be executed on node's host
  }
```

The possibility to define an autonomy_level will be ignored in this tutorial. It enables the advanced user to define a global countermeasure level and an action will only be executed if its level is below the current global level. This constitutes a more intelligent approach to countermeasure since the system should, for example, take automated countermeasures in development mode but never when used for production—or the other way round. As before, the concrete use of countermeasures is shown (Lst. 1.5) for our two examples. It is assumed that the two previously defined specifications cam.yaml (Lst. 1.3) and actuator.yaml (Lst. 1.2) have been loaded.

A brief explanation of the countermeasures for our example system follow. The first constraint (lwr2_joint_topic) ensures that two connection frequencies within the robot control system are always on the level defined in the specifications. As shown in Lst. 1.2 this frequency value is 495–505 Hz. If this is not the case, something is definitely wrong in this subsystem and the control node is shut down, which in turn immediately leads to an emergency stop of the robot. The second constraint makes sure that the main server's CPU never gets too hot. The server is located in a rack, which experienced such problems in the past, which led to issues that were difficult to debug. The third constraint validates whether the camera sends images with at least 24 Hz. If it falls below this frequency, the node will be automatically restarted—given that we know from experience that reinitialization of the camera often solves this problem.

Listing 1.5. counter.yaml

```
lwr2_joint_topic:
  constraint:
  − or:
      c!/robots/lwr2/safe_cartesian!/robots/lwr2/direct/get_joint!/robots/lwr2/
          ↪ ros2fri: {frequency: low}
      c!/robots/lwr2/joint_trajectory_action!/robots/lwr2/get_joint!/robots/lwr2/
          ↪ safe_cartesian: {frequency: low}
  min_reaction_interval: 1
  reaction_timeout: 20
  reactions:
    stop_node: {action: stop, autonomy_level: 1,
      node: /robots/lwr2/ros2fri}
robotcontrol_host_temp:
  constraint:
  − and:
      h!10.0.0.2: {cpu_temp_max: high}
  min_reaction_interval: 10
  reaction_timeout: 3600
  reactions:
    send_email: {action: run,
      command: mail −s "Warning CPU too hot" admin@example.com,
      autonomy_level: 1,
      node: /robots/lwr2/ros2fri}
endoscope_camera_topic:
  constraint:
  − and:
      t!/cameras/endoscope/image_raw: {frequency: low}
  min_reaction_interval: 1
  reaction_timeout: 20
  reactions:
    restart_node: {action: restart, autonomy_level: 1, node: /cameras/
        ↪ endoscope_driver}
config:
  reaction_autonomy_level: 1
```

8 Conclusion

We hope the examples in this chapter have shown, how easy it is to define specifications and countermeasures for real world systems. ARNI provides powerful advanced tools, but also its very basic functionality can be an important aid in governing big ROS networks. Furthermore, we found that always having the ARNI rqt GUI visible on dedicated lab monitor provides a good overview about the overall state of the ROS system at a glance.

Reference

1. A. Bihlmaier, H. Wörn, Increasing ROS reliability and safety through advanced introspection capabilities. Proc. INFORMATIK **2014**, 1319–1326 (2014)

Authors' Biography

Andreas Bihlmaier Dipl.-Inform., obtained his Diploma in computer science from the Karlsruhe Institute of Technology (KIT). He is a Ph.D. candidate working in the Transregional Collaborative Research Centre (TCRC) "Cognition-Guided Surgery" and is leader of the Cognitive Medical Technologies group in the Institute for Anthropomatics and Robotics—Intelligent Process Control and Robotics Lab (IAR-IPR) at the KIT. His research focuses on cognitive surgical robotics for minimally-invasive surgery, such as a knowledge-based endoscope guidance robot.

Matthias Hadlich is an undergraduate at the KIT studying computer science. He is a research assistant at the High Perfomance Humanoid Technologies Lab (H2T). He is also part of the "Software Engineering Practice" team that created ARNI and is its current maintainer.

Heinz Wörn Prof. Dr.-Ing., studied electronic engineering at the University of Stuttgart. He did his Phd thesis on "Multi Processor Control Systems". He is an expert on robotics and automation with 18 years of industrial experience. In 1997 he became professor at the University of Karlsruhe, now the KIT, for "Complex Systems in Automation and Robotics" and also head of the Institute for Process Control and Robotics (IPR). Prof. Wörn performs research in the fields of industrial, swarm, service and medical robotics.

Implementation of Real-Time Joint Controllers

Walter Fetter Lages

Abstract This tutorial chapter explains the implementation of controllers in the Robot Operating System. The inner working of the ROS real-time loop is explained with discussion of the classes used to implement it. Contrariwise to most available examples of implementation of controllers in ROS, which show the use of single input, single output controllers using the proportional-integral-derivative control law, here controllers are approached in a more general sense, so that any control law can be used. A complete example of implementation of a MIMO nonlinear controller is presented using the computed torque control law. The real-time aspects of the problem are also considered and the controller is ready for running in hard-real-time with the `PREEMPT_RT` kernel patch. The source code of examples are available at public repositories to enable readers to experiment with the examples and adapt them to their robots.

Keywords Controller · Real-time loop · Non-linear controller · Parameter identification · Computed torque · MIMO controller

1 Introduction

Despite ROS being a widely used framework nowadays, its documentation and examples covering low-level controllers are poor and almost all existing tutorials are based on single input, single output (SISO) controllers using the classical Proportional, Integral, Derivative (PID) control law. Most references and textbooks on ROS [9, 18, 21, 23] do not even cover the implementation of controllers. On the other hand, robots are, in general, non-linear multi input, multi output (MIMO) systems, for which the use of independent PID controllers for each degree of freedom (DoF) is not adequate due to coupling.

W.F. Lages (✉)
Federal University of Rio Grande Do Sul, Av. Osvaldo Aranha, 103,
Porto Alegre, RS 90035-190, Brazil
e-mail: fetter@ece.ufrgs.br
URL: http://www.ece.ufrgs.br/~fetter

© Springer International Publishing Switzerland 2016
A. Koubaa (ed.), *Robot Operating System (ROS)*, Studies in Computational
Intelligence 625, DOI 10.1007/978-3-319-26054-9_26

671

Although the use of MIMO and non-linear joint controllers is not new in robotics, their use under ROS is not common and it is very hard to find an example of such an implementation. Probably the reason for that is that researchers with background on control systems and used to code their controllers in Matlab are not comfortable with the software skills required to implement controllers in ROS. On the other hand, the concept of a controller in ROS is somewhat different from its concept in control theory, which confuse things further. The purpose of this tutorial chapter is to fill this gap, by presenting the details of implementation of a general controller in ROS. Hence, the main contribution of this chapter is the description of how to use the ROS infrastructure to implement a generic (nonlinear, MIMO) control laws in a ROS controller and how to make it run in real-time.

The computed torque control law is used. Although it is a classical controller, it is non-linear and MIMO. Hence, its implementation in ROS is generic and representative of almost any other controller and is a good example to serve as reference for further implementation of any other control law (linear, non-linear, SISO, MIMO, whatever), while the low-level controllers available from standard ROS packages, such as `ros_controllers`, and used as examples on available tutorials are only representatives of SISO controllers. To explore the generality of the implementation of the proposed controller, the inner working of the real-time loop of ROS is explained in detail, motivating the further implementation of advanced controllers in ROS such as in [17], where a computed torque controller is implemented to control a biped robot.

More specifically, the remainder of this chapter will cover the following topics:

- a background on control systems
- ROS packages for implementation of controllers
- configuring the system for real-time
- testing the installed packages
- implementing controllers in ROS

2 Background on Control Systems

Control systems can operate either in open-loop or in closed-loop. A control system operates in open-loop when the control actions do not depend on the plant (the part of the system being controlled) output or state. A typical example of an open-loop control system is the position control by using a stepper motor: The position of the motor axis is commanded by the controller which activates the motor windings in an appropriate sequence to drive the motor the number of required steps. Note that this process does not require that the controller receives any information about the effectiveness of its control actions. It is just assumed that the control actions produce the desired effect.

Figure 1 shows a block diagram of a typical open-loop control system. Besides the controller and the plant, there are three signals: $y(t)$ which is the plant output,

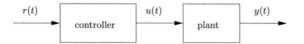

Fig. 1 Open-loop control system

$u(t)$ which is the plant input, also called the control action and $r(t)$ which is the reference. The purpose of any control system is to force $y(t)$ to follow $r(t)$.

Open-loop control systems, as shown in Fig. 1, are nice if a very good mathematical model of the plant is available and it is not subject to disturbances, parameter variations due to temperature, aging and any other imperfection of the real world. Just imagine what would happen in the above example of the stepper motor if the axis is stuck for some moment and the motor can not advance the commanded step. The controller would not be aware that the motor axis is not at the desired position.

In order to improve system capabilities to reject disturbances and increase its stability and robustness to parameter variations, a closed-loop control system is used. In a closed-loop control system, the control action depends on the system output (which is called output feedback, Fig. 2) or state (which is called state feedback, Fig. 3). Usually, an error signal ($e(t)$) is computed from the reference and the plant output and then is used by the compensator to compute the control action $u(t)$. Note that the controller is composed by the adder (actually a subtractor) and the compensator. However, it is common to use the term controller to refer to the function implemented by the compensator alone, excluding the adder.

An example of a closed-loop output feedback control system is the classical Proportional+Integral+Derivative (PID) controller, which computes the control action as:

$$u(t) = K_p e(t) + K_i \int_0^t e(t)dt + K_d \frac{de(t)}{dt} \tag{1}$$

where K_p, K_i and K_d are the gains associated to the proportional, integral and derivative terms, respectively.

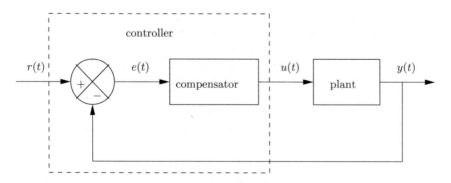

Fig. 2 Closed-loop control system with output feedback

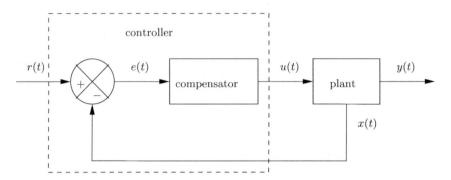

Fig. 3 Closed-loop control system with state feedback

An expression like (1), which prescribes how to compute the control action, is called a control law.

The plant may be a single motor, an entire robot, or even a team of robots. However, it is important if it is a single input, single output (SISO) system, like a motor, where the input is the voltage applied to it and the output is its axis position or a multi input, multi output (MIMO) system, like a robot, where the inputs are the torques for each joint and the outputs are the positions of each joint. In a SISO system $u(t)$ and $y(t)$ are scalar signals, while in a MIMO system $u(t)$ and $y(t)$ are vectors (not necessarily with the same dimensions). Note that $r(t)$ has the same dimension as $y(t)$.

In a SISO system, it is trivial that the single $u(t)$ should be manipulated by the controller to force the single $y(t)$ to the desired $r(t)$. However, in a MIMO system, it is not obvious which element of $u(t)$ drives what element of $y(t)$. Possibly each element of $y(t)$ depends on all elements of $u(t)$. Furthermore, the dimensions of $y(t)$ and $u(t)$ may not be the same. In a manipulator robot with n joints, $y(t)$ (the position of each joint) has n elements and $u(t)$ (the torque applied to the joints) has n elements as well. Nonetheless, for a differential-drive mobile robot, the dimension of $y(t)$ (the robot pose) is 3 while the dimension of $u(t)$ is 2 (the torque or velocities in each wheel).

It is important to note that, in general, a MIMO system can not be regarded as a set of SISO systems. That is only possible when the system is decoupled which means that each $y_i(t)$ depends on one and just one $u_i(t)$. Usually, that is not the case for robotic systems, as effects such as inertia, centrifugal and Coriolis forces almost always create couplings among all joints.

Some closed-loop control systems do not compute their control action from the plant output, but from a set of system internal variables, called state variables ($x(t)$), as shown in Fig. 3. Those systems are called a state feedback systems.

Another aspect of the plant is its linearity. A linear plant is one in which the relation between $y(t)$ and $u(t)$, which is called its model, is a linear function,[1] implying that its model can be written in the form:

---

[1] A linear function is one in which $f(\alpha_1 x_1 + \alpha_2 x_2) = \alpha_1 f(x_1) + \alpha_2 f(x_2)$ for any constant α_1 and α_2 and any x_1 and x_2.

$$\begin{cases} \dot{x}(t) = Ax(t) + Bu(t) \\ y(t) = Cx(t) + Du(t) \end{cases} \tag{2}$$

where $x(t)$ is the state of the system and A, B, C and D are parameters of the system.

A non-linear plant is one for which the linearity does not hold. In this case, its model can not be written as (2) but in a more general form given by:

$$\begin{cases} \dot{x}(t) = f(x(t), u(t)) \\ y(t) = h(x(t), u(t)) \end{cases} \tag{3}$$

There is a huge amount of knowledge and methods on how to design controllers for linear systems [1, 5, 7, 20]. However, robots are almost always non-linear plants. Although not as massive as for linear systems, there are also some methods for control of non-linear systems [11, 12, 19, 25]. In many cases, robot models fall in a special class of non-linear systems, whose models can be written as:

$$\begin{cases} \dot{x}(t) = f(x(t)) + g(x(t)) u(t) \\ y(t) = h(x(t)) \end{cases} \tag{4}$$

which are called affine systems and have some properties making them more easier to control than general non-linear systems.

It is not in the scope of this chapter to discuss how to design control laws for robots. The purpose of this chapter is to explain how to implement those control laws in a ROS controller. Hence, the remain of this chapter will be based on a classical, but yet general enough, control law for robots known as the Computed Torque Control Law [8]. That control law is representative of a general non-linear, MIMO, totally coupled controller and its implementation is a good example for implementing any control law in ROS, while the usual examples of implementation of controllers in ROS are based on linear, SISO, totally decoupled PID controllers.

2.1 Computed Torque Controller

The dynamic model of a manipulator robot (and actually of almost any robot) can be obtained by the Lagrange-Euler formulation [8] and is given by:

$$\tau = M(q)\ddot{q} + V(q, \dot{q}) + G(q) \tag{5}$$

where q is the vector of joint positions, $M(q)$ is the inertia matrix, $V(q, \dot{q})$ is the vector of centrifugal and Coriolis forces, $G(q)$ is the Vector of gravitational forces and τ is the vector of torques applied to the joints.

By defining $x = \begin{bmatrix} q & \dot{q} \end{bmatrix}^T$ and $u = \tau$ it is possible to write (5) as:

$$\begin{cases} \dot{x} = \begin{bmatrix} \dot{q} \\ -M(q)^{-1}\left(V(q,\dot{q})+G(q)\right) \end{bmatrix} + \begin{bmatrix} 0 \\ M(q)^{-1} \end{bmatrix} u \\ y = \begin{bmatrix} I & 0 \end{bmatrix} x \end{cases} \tag{6}$$

which has the form of (4), with $f(x) = \begin{bmatrix} \dot{q} \\ -M(q)^{-1}\left(V(q,\dot{q})+G(q)\right) \end{bmatrix}$, $g(x) = \begin{bmatrix} 0 \\ M(q)^{-1} \end{bmatrix}$ and $h(t) = \begin{bmatrix} I & 0 \end{bmatrix} x$.

The computed torque control law uses the dynamic model of the robot to implement a feedback linearization [25]. Then, a PD controller is used to control the resulting linear system. The resulting control law is then given by:

$$\tau = M_n(q)\left[\ddot{q}_r + K_d(\dot{q}_r - \dot{q}) + K_p(q_r - q)\right] + V_n(q,\dot{q}) + G_n(q) \tag{7}$$

where q_r is the position reference, K_p is the proportional gain matrix, K_d is the differential gain matrix, $M_n(q)$ is the nominal inertia matrix, $V_n(q,\dot{q})$ is the nominal vector of centrifugal and Coriolis forces, $G_n(q)$ is the nominal vector of gravitational forces.

The structure of the computed torque control system is shown in Fig. 4. Note that there is an abuse with respect to the usual semantics of a block diagram. The variables are represented in time domain and while the continuous lines denote inputs which are multiplied by the block "gain", the dashed lines denote inputs which are just necessary for the block computation.

By supposing that there is no model mismatch, then $M(q) = M_n(q)$, $V(q,\dot{q}) = V_n(q,\dot{q})$, $G(q) = G_n(q)$, and by applying (7) to (5), it is possible to obtain:

$$\ddot{e} + K_d\dot{e} + K_p e = 0 \tag{8}$$

where $e = q_r - q$.

Expression (8) shows that by choosing the matrices K_p and K_d in a diagonal form, it is possible to obtain a decoupled closed-loop system, where the behavior of each joint error is given by a second order differential equation. The natural frequency ω_n and the damping coefficient ξ of each equation are determined by choosing the gain matrices: $K_p = \text{diag}\left(\omega_n^2\right)$ and $K_d = \text{diag}(2\xi\omega_n)$. See [20] for a discussion of how to choose ω_n and ξ for a desired performance of the control system. In robotics, it is usual to set $\xi = 1$ for a step response without overshoot and then compute the ω_n for a desired settling time (T_s) as $\omega_n \approx \frac{4}{\xi T_s}$ for a settling to within a margin of error of 2% around the reference.

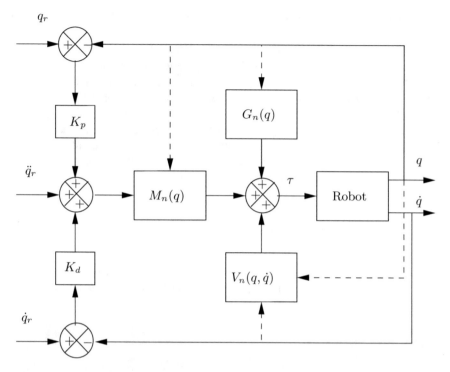

Fig. 4 Block diagram of the computed torque control

Note that to compute the computed torque control law (7) it is possible to use the classical form of the Newton-Euler formulation[2] by letting:

$$v = \ddot{q}_r + K_d(\dot{q}_r - \dot{q}) + K_p(q_r - q) \tag{9}$$

and computing $\tau = f(q, \dot{q}, v)$.

The computations required by the kinematic and dynamic models of a robot can be done by using the OROCOS Kinematics and Dynamics Library (KDL) [3]. In particular, the `ChainIdSolver_RNE` class, implements the well-known Newton-Euler algorithm [6].

3 ROS Packages for Implementation of Controllers

This section describes the installation of some packages useful to implement controllers in ROS. Some of them are not present in a standard Desktop installation of ROS and should be installed. Besides that, a few custom packages, containing

---

[2]The Newton-Euler formulation is a well-known recursive procedure to compute the torque in (5) as $\tau = f(q, \dot{q}, \ddot{q})$.

our example implementation of the computed torque controller, should be installed as well.

Of course, it is possible to implement controllers without using those packages, as long as a ROS node is a process in the host operating system and therefore can do anything that a process can do. Also, many commercial robots implement the joint controllers in their firmware or use servo-motors with built-in controllers, and hence they can be used in ROS without joint controllers implemented in ROS. However, the assumption here is that the goal is to implement a controller compliant with the ROS definition of a controller. Furthermore, note that the architecture of ROS nodes is not suitable for real-time processing, while the architecture of ROS controllers was created with hard real-time processing in mind.

3.1   Setting up a Catkin Workspace

The packages to be installed for implementing ROS controllers assume an existing catking workspace. If it does not exist, it can be created with the followind commands (assuming a ROS Indigo version):

```
source /opt/ros/indigo/setup.bash
mkdir -p ~/catkin_ws/src
cd ~/catkin_ws/src
catkin_init_workspace
cd ~/catkin_ws
catkin_make
source ~/catkin_ws/devel/setup.bash
```

3.2   `ros_control`

The `ros_control` meta-package includes a set of packages to implement generic controllers. It is a rewrite of the `pr2_mechanism` packages to be used with all robots and no just with PR2. This packages implements the base architecture of ROS controllers and hence is required for creating controllers compliant with the ROS definition of a controller.

It includes the following packages:

`control_toolbox`: contains modules that are useful for all controllers.

`controller_interface`: implements a base class for interfacing with controllers, the `Controller` class.

`controller_manager`: implements the controller manager, the `Controller Manager` class.

`hardware_interface`: base class for implementing hardware interface, the `RobotHW` and `JointHandle` classes.

`joint_limits_interface`: base class for implementing the joint limits.

transmission_interface: base class for implementing the transmission interface.

realtime_tools: contains a set of tool that can be used from a hard real-time thread.

The ros_control meta-package is not included in the standard ROS desktop installation, hence it should be installed. On Ubuntu, it can be installed from Debian packages with the command:

```
sudo apt-get install ros-indigo-ros-control
```

3.3 *ros_controllers*

This meta-package is not strictly necessary for implementing controllers in ROS, but it is useful for testing the installation as it contains simple controllers based on the PID control law. Those are the controllers to use if PID controllers are enough. In this chapter, those controllers are not used. However, it is convenient to install this package to use the joint_state_controller controller, which by its name seems an state-space controller in the joint space, but actually is just a publisher for the values of the position and velocities of the joints. Then, that topic can be inspected to check for the response of the implemented controllers.

More specifically, ros_controllers include the following:

forward_command_controller: just a bypass from the reference to the control action as they are the same physical variable.

effort_controllers: implements effort controllers, that is, SISO controllers in which the control action is the torque (or an equivalent physical variable) applied to the robot joint.

  joint_effort_controller: just a bypass from the reference to the control action as they are the same physical variable.

  joint_position_controller: a joint position controller in which the reference is joint position and the control action is torque. The PID control law is used.

  joint_velocity_controller: a joint velocity controller in which the reference is joint velocity and the control action is torque. The PID control law is used.

position_controllers: implements position controllers, that is, SISO controllers in which the control action is the position (or an equivalent physical variable) applied to the robot joint.

  joint_position_controller: just a bypass from the reference to the control action as they are the same physical variable.

velocity_controllers: implements position controllers, that is, SISO controllers in which the control action is the position (or an equivalent physical variable) applied to the robot joint.

joint_velocity_controller: just a bypass from the reference to the control action as they are the same physical variable.

joint_state_controller: implements a sensor controller which publishes the joint state as a sensor_msgs/JointState message, the JointState Controller class.

The ros_controllers meta-package is not included in the standard ROS desktop installation, hence it should be installed. On Ubuntu, it can be installed from Debian packages with the command:

```
sudo apt-get install ros-indigo-ros-controllers
```

3.4 robot_model

This meta-package is usually already installed in a standard ROS desktop installation. Although many other standard packages are used for implementing controllers, robot_model deserves special attention as it includes packages that are used to parse robot models described in many formats such as URDF, Collada and KDL.

This package is used here for two reasons: first it is used to parse the URDF description of the robot into a KDL tree description. This KDL tree description is used to compute the dynamic model of the robot necessary for the implementation of the computed torque controller. Second, this package includes a node called robot_state_publisher, which publishes the Cartesian transforms between the links of the robot. Although not necessary for implementation of joint controllers, that functionality is very useful for higher level layers of software.

Among others, this package includes the following:

kdl_parser: parses an URDF description into a KDL tree description of the robot.

robot_state_publisher: publishes the state of a robot to the transform library topic.

3.5 orocos_kdl

This package is usually installed in a standard ROS desktop installation which deserves special attention. It includes the KDL library which has classes for computing kinematic and dynamic models of robots. It is used here to compute the dynamic model of the robot in the computed torque controller.

3.6 *gazebo_ros_pkgs*

This is a collection of ROS packages for integrating the `ros_control` controller architecture with the Gazebo simulator [13], containing the following:

`gazebo_ros_control`: Gazebo plugin that instantiates the `RobotHW` class in a `DefaultRobotHWSim` class, which interfaces with a robot simulated in Gazebo. It also implements the `GazeboRosControlPlugin` class.

The `gazebo_ros_pkgs` meta-package is not included in the standard ROS desktop installation, hence it should be installed. On Ubuntu, it can be installed from Debian packages with the command:

```
sudo apt-get install ros-indigo-gazebo-ros-pkgs ros-indigo-gazebo-ros-control
```

3.7 *ufrgs_wam*

This is a custom meta-package with an example of how to implement a complex controller in ROS. It contains an URDF description of the Barrett WAM robot [2] and the implementation of the computed torque controller. More specifically it includes the following packages:

`wam_description`: URDF description of the Barrett WAM robot.
`wam_controllers`: implementation of a computed torque controller for the Barrett WAM robot.

There are other ROS packages available for the Barrett WAM, but this chapter is based on the UFRGS custom version because the parameters used in the URDF description are tuned to the robot existing at UFRGS laboratories and its more simple, since it contains just what is used here.

The `ufrgs_wam` meta-package can be downloaded and installed in the ROS workspace with the commands:

```
cd ~/catkin_ws/src
git clone \url{https://github.com/ufrgs-ece/ufrgs_wam}
cd ~/catkin_ws
catkin_make
source ~/catkin_ws/devel/setup.bash
```

4 Configuring the System for Real-Time

Real-time tasks should be scheduled by a preemptive strict priority based scheduler policy. However, the default Linux scheduler policy is not strict priority based and only privileged processes can chance their scheduler policy to a real-time one. To enable processes from normal users to do this, it is necessary to configure the user limits.

In Linux distributions using Pluggable Authentication Modules (PAM), such as Debian, Ubuntu and OpenSuSE, the configuration of user limits can be done by creating, as the root user, the `/etc/security/limits.d/wam_controllers.conf` file with the contents:

```
username soft cpu       unlimited
username -    rtprio 99
username -    nice      -20
username -    memlock unlimited
```

where `username` should be the login of the user executing the real-time tasks.

It is necessary to logout and login again for the configuration to take effect. Note that it is not enough to close the terminal and open it again.

For Linux distributions not using PAM, such as Slackware, the user limits are configured in the `/etc/limits` file by including a line with:

```
username T- O99 I39 M-
```

However, the configuration in the `/etc/limits` file is valid only for logins through the `login` program. That means text console logins only. Unfortunately, the KDM login manager does not understand the configuration in the `/etc/limits` file. This is a bug in KDM and can be overcome by issuing the following command:

```
ulimit -t unlimited -r 99 -e 39 -l unlimited
```

before running KDM, for example in the beginning of the `/etc/rc.d` `rc.4` script. However, this will set the limits for all users and anyone would be able to change scheduler policy to a real-time one.

The correct configuration of the user limits can be tested with the command:

```
ulimit -a
```

The result should show a real-time priority limit of 99 and an unlimited max locked memory.

Even by changing the scheduler policy to a real-time one, at most soft-real-time would be possible because the stock Linux kernel is not fully preemptible. To obtain hard-real-time it is necessary to install the PREEMPT_RT kernel patch. This patch turns the Linux kernel into a fully preemptible kernel. See [22] for an interesting table of real-time features added by the PREEMPT_RT patch.

The installation of the PREEMPT_RT patch requires the compilation of the Linux kernel. Hence, it should only be attempted by experienced users who already have successfully compiled the kernel before. Here it is assumed that the skills necessary for compiling and installing a stock kernel are already mastered. The computed torque controller presented in subsequent sections should execute without problems even without the PREEMPT_RT patch if executed in a machine with low load.

The general instructions for installing the PREEMP_RT patch are (execute everything as the root user):

1. Download the `PREEMPT_RT` from ⟨`ftp://ftp.kernel.org/pub/linux/kernel/projects/rt`⟩. There is a directory for the patch for each supported kernel version. For example, for kernel version 4.0.5:

```
cd /usr/src
wget ftp://ftp.kernel.org/pub/linux/kernel/projects/rt/4.0/patch-4.0.5-rt4.patch.xz
```

2. Download and untar the source code for the Linux kernel from ⟨`ftp://ftp.kernel.org/pub/linux/kernel`⟩:

```
wget ftp://ftp.kernel.org/pub/linux/kernel/v4.x/linux-4.0.5.tar.xz
tar -xJvf linux-4.0.5.tar.xz
```

3. Patch the kernel.

```
cd linux-4.0.5
xzcat ../patch-4.0.5-rt4.patch.xz | patch -p1
cd ..
mv linux-4.0.5 linux-4.0.5-rt4
cd linux-4.0.5-rt4
```

4. If necessary, install the ncurses library:

```
apt-get install ncurses-dev
```

5. Configure the kernel by executing the configuration script:

```
make menuconfig
```

Configuring a kernel requires some experience and it is recommended to start with a working configuration. To enable the `PREEMPT_RT` patch it is necessary to set the **Preemption Model** in the **Processor type and features** submenu to **Fully Preemptible Kernel (RT)**

6. Save the configuration and leave the configuration script.
7. Compile the kernel with the command (it takes a lot of time):

```
make -j 4
```

8. Compile the kernel modules:

```
make -j 4 modules
```

9. Install the kernel modules and the kernel itself (this should also create an `initrd` file and congigure GRB for the new kernel:

```
make modules_install
make install
```

10. Reboot and select the new kernel in the GRUB menu.
11. Check for "PREEMPT RT" in the kernel version string with the command:

```
uname -a
```

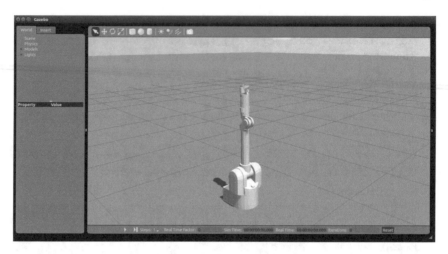

Fig. 5 Gazebo with the Barrett WAM robot

For some machines, specially laptops, it may be necessary to set the kernel parameter `processor.max_cstate=1` for the correct working of `PREEMPT_RT` patch. This parameter limits the maximum `CSTATE` entered by the processor while in idle. The `CSTATE` value is associated to power saving. For larger values, more components of the hardware are powered-off to save energy. In general for `CSTATE` above 1, the clock for the APIC chip is stopped, which can block the scheduler.

5   Testing the Installed Packages

A simple test for the installation of the packages described in Sect. 3 is performed here.

In order to test the installation of the ROS packages lets load the Barrett WAM modem in Gazebo and launch the computed torque controller. This can be done with the commands:

```
source /opt/ros/indigo/setup.bash
source ~/catkin_ws/devel/setup.bash
roslaunch wam_controllers computed_torque.launch
```

The robot should appear in Gazebo as shown in Fig. 5. Note that the robot is launched with all joints at zero degrees.

Then, start the simulation by clicking in the play button in the Gazebo panel and issue the following commands to move the robot to its home position as shown in Fig. 6:

```
source /opt/ros/indigo/setup.bash
source ~/catkin_ws/devel/setup.bash
rosrun wam_controllers move_home.sh
```

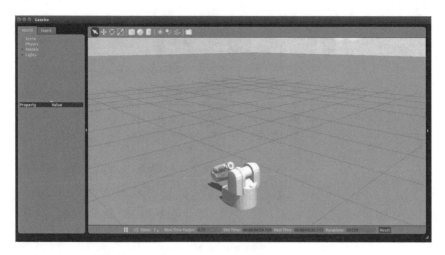

Fig. 6 Gazebo with the Barrett WAM robot on its home position

The WAM robot can be returned to the starting position, as shown in Fig. 5, with the commands:

```
rosrun wam_controllers move_zero.sh
```

The `move_home.sh`, `move_zero.sh` are scritps with examples of how to set the reference position, velocity and acceleration for the controller. Here, the scripts just publish the required values by using the `rostopic` command. In a real application, those references would be generated by a planning package, such as MoveIt! [26] or a robot navigation package.

6 Implementing Controllers in ROS

In a first look it appears that is a good idea to associate the block diagram of a control system such as the one in Fig. 2 to a ROS computation graph, with blocks implemented as ROS nodes and signals implemented as ROS topics. At least for real-time joint controllers that is not a good idea. ROS nodes do not operate in real-time and ROS topics are an asynchronous communication mechanism, for which there are no guarantees on timing of message delivery. ROS nodes and topics are adequate for higher level tasks.

For low-level joint control there are the ROS controllers and the ROS real-time loop, where computations are real-time safe and communication is based on synchronous function calls. Of course, that is not enough to ensure real-time. For real-time, controller execution should be scheduled by a preemptive strict priority based scheduler policy. Here, ROS relies on the underling operating system. The default Linux scheduler policy is not strict priority based and only privileged processes can chance their scheduler policy to a real-time one. To enable processes from normal users to do this, it is necessary to configure the user limits, as detailed in Sect. 4. Even then,

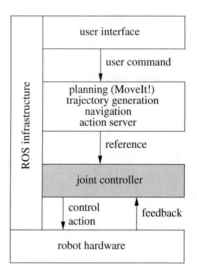

Fig. 7 Software stack of a robot

at most soft-real-time would be possible because the stock Linux kernel is not fully preemptible. To obtain hard-real-time it is necessary to install the PREEMPT_RT kernel patch as discussed in Sect. 4.

Note also, that a real-time capable system (be it soft or hard) does not ensure, by itself, that the timing requirements of the tasks are satisfied. For that, a schedulability analysis of the tasks on the system is necessary. See [4] for details on how to assure the timing requirements of the tasks.

The joint controller is one of the many software layers necessary to make a robot useful. In particular, it is responsible for ensuring that the joints of the robot are driven in a such a way that the reference commands from higher level layers are followed. Figure 7 shows the place of joint controllers in the software stack of a robot. The joint controllers are first software layer above the robot hardware and communicate directly with actuators and sensors. They receive the reference from higher level layers such as MoveIt! [26] which makes planning, trajectory generation and/or navigation. Usually those packages implement an action server for receiving commands from the user and then generates the desired reference for each sampling period of the controller.

6.1   Controllers in ROS

From the control engineering point of view, ROS presents some particularities regarding controllers. One of them is related to nomenclature, as what is called a controller in ROS is not necessarily a controller in control systems nomenclature. A controller in ROS is a plugin for the controller manager which implements the

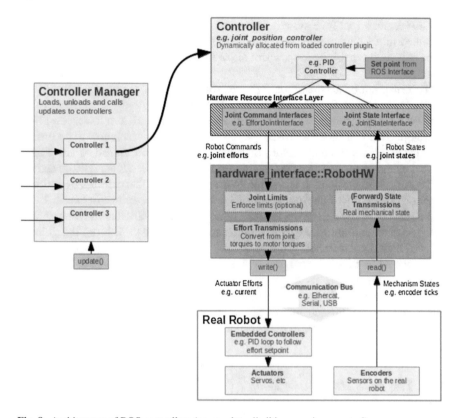

Fig. 8 Architecture of ROS controllers (source: http://wiki.ros.org/ros_control)

`Controller` interface and typically actuates the joint through some object exposing a `JointCommandInterface` and obtains data from sensors through an object exposing a `JointStateInterface`. Note that a ROS controller can perform a function which not necessarily the function of a controller in a control system. An interesting example is the `JointStateController`, which by its name seems an state-space controller in the joint space, but actually is just a publisher for the values of the position and velocities of the joints.

Another problem is that it appears that the ROS infrastructure was conceived for single-input, single-output (SISO) controllers. Advanced control laws for robots are intrinsically MIMO and can not be decomposed in a set of SISO control laws. Furthermore, there is the synchronization problem. In a MIMO controller all outputs are driven at the same time, while a set of SISO controllers would need some sort of synchronization to achieve the same effect.

This chapter explains how to implement generic control laws in a ROS controller. Note that contrariwise to most ROS controllers available in standard ROS packages [9, 18, 21], where independent PID controllers for each joint are described, here, the implementation of general, possibly MIMO, non-linear, control laws are described.

ROS is not a real-time system. However, it has some real-time capability when executed in a Linux system with the PREEMP_RT kernel patch [22]. Even then, there is a single real-time loop where all controllers are executed by the controller manager.

This may not be adequate for every real-time task in a complex system. In this cases, an alternative for executing many real-time tasks with diverse rates is to use the OROCOS (Open Robot Control Software) [3] framework as a lower-level layer for running the real-time tasks. See [10, 14, 24] for examples on how to do this.

Figure 8 shows the architecture of controllers in ROS. Controllers are plugins loaded by the controller manager. The controller manager can load, unload, activate and deactivate controllers. It is also the controller manager that calls the update() function of each controller in the real-time loop of ROS, described in Sect. 6.2.

In each control cycle, the update() function of each active controller is sequentially invoked and used by each controller to perform his task. From the point of view of digital control theory, the paradigm is that of a continuous time controller implemented in a digital computer with a sampling rate fast enough to neglect the digitization effects.

Anyway, the paradigm of a continuous time controller is appropriate for nonlinear controllers as the one proposed here. The update() function of the controller implements the control law, as described in Sect. 6.5.

6.2   The ROS Real-Time Loop

Figure 9 (adapted from [17]) shows a diagram of the real-time loop in ROS. The rectangular blocks represent the classes of objects used to implement the loop and the arrows represent function calls or access to variables through pointers. For functions, the arrow points to the class implementing the function while its base is at the class that calls the function. For pointers, the arrow points to the class that holds the variable. The ellipsis represent topics published or subscribed by the classes and are used, basically, for communication with nodes outside the real-time loop.

The RobotHW class implements the ROS interface with the robot hardware. In the case of a simulation, this class is derived to the RobotSim class, which implements the interface with the Gazebo simulator. Here, the RobotSim class is derived to the RobotSimWam class, which implements the details of the interface of ROS with the simulation of the Barrett WAM robot.

The GazeboRosControlPlugin class implements the time synchronization between the Gazebo simulator and ROS, more specifically, with the controller manager. When the robot is simulated that is very important because the time is also simulated and then the ROS should operate with basis on simulated time and not on real-time. When driving the real robot, this class should be replaced by a custom class that generates the synchronization for sampling sensors, running the controller and outputting the control action to the actuator.

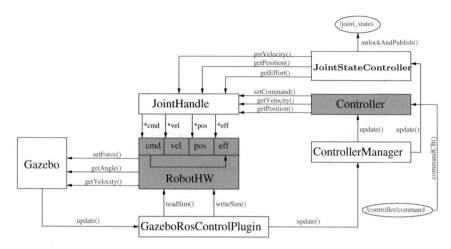

Fig. 9 Real-time loop in ROS

The ROS controller manager is implemented by the `ControllerManager` class. The function of the `ControllerManager` in the real-time loop is to execute the active controllers at each sampling time. The active controllers are executed one-by-one, in sequence. Again, it is important to note that in ROS nomenclature a controller is an object which is loaded by the controller manager as a plugin and has its `update()` function called at each sampling period. Not necessarily this object implements a control law. It can just sample a sensor and publish its data or perform any other function in the system. In the diagram shown in Fig. 9, the `JointStateController` and `Controller` classes are ROS controllers, but just the latter one implements a control law. The `JointStateController` class just publishes the robot state on the `/joint_states` topic and does not have a controller function in the control systems sense.

The reference is received through the `/controller/command` topic. Actually, the way in which the reference is received depends on the class that implements the `Controller` class. Here, the `Controller` class will be derived to the `ComputedTorqueController` class to implement the computed torque controller control law (7), as shown in Sect. 6.5.

The `JointHandle` class implements an abstraction of the robot joints, exposing functions for reading joint positions, velocities and efforts and to apply efforts to the joints. Those functions use pointers to access the respective variables in the objects of the `RobotHW` class, which implements the real access to the robot hardware or simulation and maintain copies of the values read from the hardware or simulator.

In ROS nomenclature, the structure shown in Fig. 9 is called the real-time loop and executes at a constant sampling period. It is important to note that there is only one real-time loop where all controllers are executed in sequence and therefore, at the same rate. If a lower frequency is required by some controller, it has to implement a sub-sampling by itself.

The execution of the real-time loop follows the sequence (see Fig. 9):

1. The Gazebo simulator simulates the time as well. Hence, at each sampling time it calls the `GazeboRosControlPlugin::update()` function.
2. The `GazeboROSControlPlugin::uptate()` function calls the `Robot HWSim::readsim()` function. The `RobotHWSim` class is the implementation of the `RobotHW` class for a simulated robot, thus interfacing with Gazebo.
3. The `RobotHWSim::readSim()` function reads the data from robot sensors, through the `getAngle()` and `getVelocity()` functions from the Gazebo library. The values are stored in private variables for latter use.
4. The `GazeboROSControlPlugin::update()` function calls `Cont rollerManager::update()` function.
5. The `ControllerManager::update()` function calls, in sequence the `update()` function of each active controller.
6. The `update()` function of each active controller, obtains (if necessary) the readings of the sensors, typically through `JointHandle::getPosition()` and `JointHandle::getVelocity()` functions, computes the control law and drives the actuators, typically, through the `JointHandle::setCommand()` function.
7. The `JointHandle::getPosition()` and `JointHandle::getVelocity()` functions just return the values which were stored by the `Robot HWSim::readSim()` function. The `JointHandle::setCommand()` function just stores the command value to be used by the `RobotHWSim::writeSim()`. If a real robot were used those functions would access the actual sensors and actuators.
8. The `GazeboROSControlPlugin::update()` function calls the `RobotHWSim::writeSim()` function.
9. The `RobotHWSim::writeSim()` function, obtains the effort to be applied to the joint from a private variable and calls the `setForce()` function from the Gazebo library to apply the effort to the joint, thus concluding the control cycle.

Note that this same architecture, including the very same classes would be used for driving a real robot instead of a simulator. The difference is that the Gazebo and `GazeboRosControlPlugin` blocks would not exist and the `RobotHW` class would be derived to a class with functions to directly drive the robot hardware instead of being derived to the `RobotHWSim` class, which implements the interface with the robot simulated in Gazebo.

6.3 Implementation of a Computed Torque Controller

In this section, the `ufrgs_wam` ROS meta-package is detailed. This meta-package consists of an URDF description of the Barrett WAM robot and an implementation of a computed torque controller.

6.4 The `wam_description` Package

The `wam_description` package, has the URDF description of the Barrett WAM. The files in `xacro` directory describe the geometric and inertia parameters of the many parts of the robot and the `meshes` directory holds the STL (STereoLithography) files with the meshes for those parts.

The files in the `launch` directory are used to load the robot model in the ROS parameter server. The `wam.launch` file loads the robot model in the ROS parameter server, while the `wam_sim.launch` file loads the robot model in the ROS parameter server and calls the Gazebo simulator.

It is beyond the scope of this chapter to discuss the modeling of robots in URDF. The reader is directed to the introductory ROS references for learning the details about URDF modeling in general. However, one key point for simulating ROS controllers in Gazebo is to tell it to load the plugin for connecting with `ros_control`. In the `wam_description` package this is done in the top level URDF file, within the ⟨gazebo⟩ tag, as shown in Listing 1.

Listing 1. Plugin description in `wam.urdf.xacro`

```
<gazebo>
  <plugin name="gazebo_ros_control" filename="libgazebo_ros_control.so" >
    <robotNamespace>/wam</robotNamespace>
    <robotSimType>gazebo_ros_control/DefaultRobotHWSim</robotSimType>
    <controlPeriod>0.001</controlPeriod>
  </plugin>
</gazebo>
```

The ⟨gazebo⟩ tag is used to add the `gazebo_ros_contol` plugin, which is the connection between the Gazebo simulator and the controller implemented in ROS. The parameters are the name of the plugin and the name of the library implementing it. The child elements are:

⟨robotNamespace⟩:   the namespace to be used for this instance of the plugin
⟨robotSimType⟩:   the name of the robot interface to be used
⟨controlPeriod⟩:   the period for controller update, in seconds

The default robot interface is `DefaultRobotHWSim` and is also implemented in the `gazebo_ros_control` library. It implements a simulated `ros_control` `hardware_interface::RobotHW`. For most robots, whose joints are actuated by effort (torque or force) and have position and/or velocity sensors, there is no reason to not use the default robot interface. For robots with more sophisticated types of actuation or sensors, a custom plugin may be necessary and would be specified by the ⟨robotSimType⟩ tag.

6.5 The `wam_controllers` ROS Package

The `wam_controllers` package implements controllers for the Barrett WAM robot. In particular, the computed torque controller, described in Sect. 2.1 is implemented. The files in the `config` directory specify the parameters for the controllers such as joint of the robot, gains and update rate. The `script` directory has some useful scripts for setting the reference for the controller for some interesting positions and can be used for testing the controller.

The `wam_controllers_plugins.xml` file specifies that the classes implementing the controllers are plugins for the ROS controller manager. The files in the `launch` directory are used to load the controllers with their respective configuration files.

The `package.xml` file for the `wam_controllers` package should use the ⟨export⟩ tag to export the name of the file which describes the plugin implementing the controller, as shown in Listing 2. In this case, this file is called `wam_controllers_plugins.xml` and resides on the root directory of the package.

Listing 2. Exporting the plugin configuration in `package.xml`

```
<export>
  <controller_interface plugin="${prefix}/wam_controllers_plugins.xml"/>
</export>
```

The `wam_controllers_plugins.xml` file is shown in Listing 3. The first line specifies the path to the shared library with the binary of the plugin. Then, the classes implemented in this plugin are described. In this case there is only the class named `wam_controllers/ComputedTorqueController`, which has the type `wam_controllers::ComputedTorqueController` and is derived from the `controller_interface::ControllerBase` class.

Listing 3. `wam_controllers_plugins.xml`

```
<library path="lib/libwam_controllers">

  <class name="wam_controllers/ComputedTorqueController"
    type="wam_controllers::ComputedTorqueController"
    base_class_type="controller_interface::ControllerBase">
    <description>
      The ComputedTorqueControllers linearizes the Barrett WAM dynamic
      model.  The linearized inputs are joint accelerations.
      It expects an EffortJointInterface type of hardware interface.
    </description>
  </class>

</library>
```

The `ComputedTorqueController` class is derived from the `Controller` class template (see Fig. 9) and its declaration is shown in Listing 4. The argument to the template is the type of the interface used by the controller. In this case, it is a `EffortJointInterface` because the computed torque controller is a controller whose control action is torque.

Listing 4. ComputedTorqueController class

```
class ComputedTorqueController: public controller_interface::
        Controller<hardware_interface::EffortJointInterface>
{

    ros::NodeHandle node_;

    hardware_interface::EffortJointInterface *robot_;
    std::vector<hardware_interface::JointHandle> joints_;

    ros::Subscriber sub_command_;

    KDL::ChainIdSolver_RNE *idsolver;

    KDL::JntArray q;
    KDL::JntArray dq;
    KDL::JntArray v;

    KDL::JntArray qr;
    KDL::JntArray dqr;
    KDL::JntArray ddqr;

    KDL::JntArray torque;

    KDL::Wrenches fext;

    Eigen::MatrixXd Kp;
    Eigen::MatrixXd Kd;

    void commandCB(const trajectory_msgs::JointTrajectoryPoint::ConstPtr
            &referencePoint);

    public:
    ComputedTorqueController(void);
    ~ComputedTorqueController(void);

    bool init(hardware_interface::EffortJointInterface *robot,
            ros::NodeHandle &n);
    void starting(const ros::Time& time);
    void update(const ros::Time& time,const ros::Duration& duration);
};
```

The private variable members of the ComputedTorqueController class are:

node_: the ROS node in which the controller runs, this variable is used to store the node handle received as argument to the init() function.

robot_: handle for the hardware interface received as argument to the init() function

joints_: vector with the handles for the joints

sub_command_: ROS topic subscriber to receive the controller reference

idsolver: handle for the recursive Newton-Euler solver implemented in the KDL library

q: vector of joint positions

dq: vector of joint velocities

v: vector of virtual joint accelerations

qr: vector of reference joint positions
dqr: vector of reference joint velocities
ddqr: vector of reference joint accelerations
torque: vector of torques to be applied to the joints
fext: external wrenches applied to the robot
Kp: proportional gain matrix
Kd: differential gain matrix

The commandCB() function is the callback for the ROS topic subscriber that receives the references for the controller.

A controller should define the init(), starting() and update() functions. The init() function is called by the ROS controller manager when the controller is loaded, the starting() function is called when the controller is started and the update() function is called at each execution of the real-time loop, as described in Sect. 6.2.

The implementation of the init() function shown in Listing 5. Its arguments are a pointer to a hardware interface and a reference to the ROS node where the controller is running. Those parameters are stored for latter use.

Listing 5. ComputedTorqueController::init() function

```cpp
bool ComputedTorqueController::
        init(hardware_interface::EffortJointInterface *robot,ros::NodeHandle &n)
{
        node_=n;
        robot_=robot;

        XmlRpc::XmlRpcValue joint_names;
        if(!node_.getParam("joints",joint_names))
        {
                ROS_ERROR("No 'joints' in controller. (namespace: %s)",
                        node_.getNamespace().c_str());
                return false;
        }

        if(joint_names.getType() != XmlRpc::XmlRpcValue::TypeArray)
        {
                ROS_ERROR("'joints' is not a struct. (namespace: %s)",
                        node_.getNamespace().c_str());
                return false;
        }

        for(int i=0; i < joint_names.size();i++)
        {
                XmlRpc::XmlRpcValue &name_value=joint_names[i];
                if(name_value.getType() != XmlRpc::XmlRpcValue::TypeString)
                {
                        ROS_ERROR("joints are not strings. (namespace: %s)",
                                node_.getNamespace().c_str());
                        return false;
                }

                hardware_interface::JointHandle j=robot->
                        getHandle((std::string)name_value);
                joints_.push_back(j);
        }
```

```
        sub_command_=node_.subscribe("command",1000,
                &ComputedTorqueController::commandCB, this);

        std::string robot_desc_string;
        if(!node_.getParam("/robot_description",robot_desc_string))
        {
                ROS_ERROR("Could not find '/robot_description'.");
                return false;
        }

        KDL::Tree tree;
        if (!kdl_parser::treeFromString(robot_desc_string,tree))
        {
                ROS_ERROR("Failed to construct KDL tree.");
                return false;
        }

        KDL::Chain chain;
        if (!tree.getChain("wam_origin","wam_tool_plate",chain))
        {
                ROS_ERROR("Failed to get chain from KDL tree.");
                return false;
        }

        KDL::Vector g;
        node_.param("/gazebo/gravity_x",g[0],0.0);
        node_.param("/gazebo/gravity_y",g[1],0.0);
        node_.param("/gazebo/gravity_z",g[2],-9.8);

        if((idsolver=new KDL::ChainIdSolver_RNE(chain,g)) == NULL)
        {
                ROS_ERROR("Failed to create ChainIDSolver_RNE.");
                return false;
        }

        q.resize(chain.getNrOfJoints());
        dq.resize(chain.getNrOfJoints());
        v.resize(chain.getNrOfJoints());
        qr.resize(chain.getNrOfJoints());
        dqr.resize(chain.getNrOfJoints());
        ddqr.resize(chain.getNrOfJoints());
        torque.resize(chain.getNrOfJoints());

        fext.resize(chain.getNrOfSegments());

        Kp.resize(chain.getNrOfJoints(),chain.getNrOfJoints());
        Kd.resize(chain.getNrOfJoints(),chain.getNrOfJoints());

        return true;
}
```

Joint names are recovered from the ROS parameter server and then used to obtain joint handles which are stored in the `joints_` vector for latter use.

A subscriber to the ROS topic `command` is created to receive the references for the controller through the `commandCB()` call-back function.

For the implementation of a computed torque controller it is necessary to compute the inverse model of the robot (5). This can be done by using the recursive Newton-Euler solver implemented in the KDL library. In order to create that solver, the URDF robot description is obtained from the parameter server and converted to a KDL tree description. Then, a single chain from `"wam_origin"` to `"wam_tool_plate"` is extracted from the KDL tree. KDL trees are used to represent a general mechanisms which may present bifurcations. A KDL chain represents a mechanism without bifurcations. The Newton-Euler solver implemented in KDL works only with open chains.

One important point here is that URDF and KDL use different conventions for representing the inertia of the links. KDL specifies the inertia parameters in the reference frame of the link, while URDF specifies the inertia in the center of inertia reference frame. This difference is handled automatically considered by the kdl_parser. However, the programmer should take care of it if creating a KDL tree or chain directly from scratch.

The gravity vector is recovered from Gazebo through the parameter server or, if not created by Gazebo, it is set to default values of 0 in X and Y axis and $-9.8\,\text{m/s}^2$ in the Z axis.

In the end of the init() function, the vectors created with zero length in the constructor of the ComputedTorqueController class are resized to the proper size as a function of the number of joints of the robot.

The init() function should return true to indicate success.

The starting() function, shown in Listing 6, is called when the controller is started and should set everything for controller operation. It is important to keep in mind that a controller may be stopped and started again, hence this function should reset the controller to its default initial condition. Here, the gain matrices Kp and Kd are set and the current values for joint positions and velocities are stored in q and dq vectors, respectively. The reference position and velocities are set to be equal to the current position and velocity and the reference acceleration is set to zero. That is important to avoid large values of control actions when the controller starts to operate. By setting the references to the actual position and velocity, the position and velocity errors would be zero, thus avoiding a large control signal.

The real-time properties of the controller are determined by the scheduler policy and priority configured at the end of the starting() function. Note that the update() function, which actually implements the control law, is called by the same thread as the starting() function, therefore, the settings made in here are in effect when the update() function is called. The scheduling policy is changed to SCHED_FIFO, which is a real-time policy enforcing strict priority scheduling. Task with lower priority are run only while there are no higher priority task ready run. The priorities in this policy range from 0 (the lowest priority) to 99 (the highest priority). In this example the maximum priority is obtained through a call to the sched_get_priority_max() function for compatibility. Note that the priority of a task in a real-time system has no relation with the importance of the task and that generally it is not a good idea to set the maximum priority for the task. In general, it its more appropriate to assign the task priorities by using a rate monotonic or deadline monotonic strategy [4, 15, 16].

Also note that setting a high priority by using the sched_setscheduler() function is not enough to ensure hard real-time scheduling. A real-time scheduling still depends on proper support from the underlying operating system. In Linux, a hard real-time patch, such as PREEMPT_RT, RTAI or Xenomai, is required

Listing 6. `ComputedTorqueController::starting()` function

```
void ComputedTorqueController::starting(const ros::Time& time)
{
        Kp.setZero();
        Kd.setZero();
        for(unsigned int i=0;i < joints_.size();i++)
        {
                Kp(i,i)=Wn*Wn;
                Kd(i,i)=2.0*Xi*Wn;
                q(i)=joints_[i].getPosition();
                dq(i)=joints_[i].getVelocity();
        }
        qr=q;
        dqr=dq;
        SetToZero(ddqr);

        struct sched_param param;
        param.sched_priority=sched_get_priority_max(SCHED_FIFO);
        if(sched_setscheduler(0,SCHED_FIFO,&param) == -1)
        {
                ROS_WARN("Failed to set real-time scheduler.");
                return;
        }
        if(mlockall(MCL_CURRENT|MCL_FUTURE) == -1)
                ROS_WARN("Failed to lock memory.");
}
```

to ensure hard real-time behavior. For plain Linux or other operating systems without hard real-time support, such as OS X, at most a soft real-time behavior is obtained. The `sched_setscheduler()` function is not available in OS X and some other operating systems, but most of those systems supports the `pthread_setschedparam()` function, which can be used to change the priority of a task as well.

Another important point for a setting up a real-time task is to lock its memory so that it is not paged-out to disk by the virtual memory system. This is done by the `mlockall()` function. It is obvious that if low-latencies are required, it is not possible to afford the extra time to page-in from disk the code or data of a real-time task.

The actual implementation of the control law is in the `update()` function, shown in Listing 7. It reads the robot sensors by calling the `getPosition()` and `getVelocity()` functions through the joint handles stored in the `joints_` vector. Then, it sets the external wrenches `fext` to zero, assuming that there are no external forces acting on the robot. Anyways, in our setup there are no sensors to measure those forces, therefore any external force would be considered by the controller as a perturbation. Then, the virtual acceleration (9) and the torque to be applied to the joints are computed by using the recursive Newton-Euler solver. The torque is applied to the joints by calling the `setCommand()` function.

The macro `PLUGINLIB_EXPORT_CLASS()` should be used to make the controller available as a plugin.

Listing 7. `ComputedTorqueController::update()` function

```
void ComputedTorqueController::update(const ros::Time& time,
    const ros::Duration& duration)
{
    for(unsigned int i=0;i < joints_.size();i++)
    {
        q(i)=joints_[i].getPosition();
        dq(i)=joints_[i].getVelocity();
    }
    for(unsigned int i=0;i < fext.size();i++) fext[i].Zero();

    v.data=ddqr.data+Kp*(qr.data-q.data)+Kd*(dqr.data-dq.data);
    if(idsolver->CartToJnt(q,dq,v,fext,torque) < 0)
        ROS_ERROR("KDL inverse dynamics solver failed.");

    for(unsigned int i=0;i < joints_.size();i++)
        joints_[i].setCommand(torque(i));
}
```

Configuration files for the controllers are placed in the `config` directory. Here, the `computed_torque_control.yaml` file, shown in Listing 8, is used to configure two controllers.

Listing 8. `computed_torque_control.yaml`

```
wam:
    joint_state_controller:
      type: joint_state_controller/JointStateController
      publish_rate: 100

    computed_torque_controller:
      type: wam_controllers/ComputedTorqueController
      joints:
        - wam_joint_1
        - wam_joint_2
        - wam_joint_3
        - wam_joint_4
        - wam_joint_5
        - wam_joint_6
        - wam_joint_7
```

The first one is not a controller in the control systems sense, but a sensor that publishes the position, velocity and effort at the robot joints. However, it is a controller in ROS sense because it is loaded as a plugin by the controller manager and has its `update()` function called at the ROS real-time loop. The second one is the computed torque controller explained above.

In the first line of `computed_torque_control.yaml` is the name of the robot, then there are blocks for configuring the controllers: `joint_state_controller` and `computed_torque_controller`. The configuration of each controller includes its name, its type and its parameters. For the `joint_state_controller`, there is only a single parameter, the `publish_rate`. In this case it means that the joint states will be published 100 times per second. For the `computed_torque_controller` there is also a single parameter called `joints`, but it is a vector where each element is the name of one joint of the robot.

For loading the controller, a launch file as shown in Listing 9 can be created. The ⟨arg⟩ tag is used to start the Gazebo simulator in paused mode. This way, it is possible to set the robot to a desired initial position before running the controller. This is very interesting for experiments evaluating the controller dynamic behavior. The robot_state_publisher node subscribes to the joint_states published by the joint_state_controller and publishes the homogeneous of the robot links.

Listing 9. computed_torque.launch

```
<launch>
        <arg name="paused" default="true"/>

        <include file="$(find wam_description)/launch/wam_sim.launch">
                <arg name="paused" value="$(arg paused)"/>
        </include>

        <rosparam file="$(find wam_controllers)/config/computed_torque_control.yaml"
                command="load"/>

        <node name="controller_spawner" pkg="controller_manager" type="spawner"
                respawn="false" output="screen" ns="/wam"
                args="joint_state_controller computed_torque_controller"/>

        <node name="robot_state_publisher" pkg="robot_state_publisher"
                type="state_publisher"
                ns="/wam" />
</launch>
```

Figure 10 shows the ROS computation graph for the system launched with the computed_torque.launch. Note that the controllers do not appear. That is because they are not ROS nodes, but just plugins that are loaded by the controller manager. The /wam/computed_torque_controller/command topic is where the reference (position, velocity and acceleration) for the controller should be published. Usually, that comes from a trajectory generation, navigation or planning package such as MoveIt! [26].

Some scripts useful for operating the Barrett WAM robot with the computed torque controller are in the scripts directory. The move_home.sh script is used to move the robot to the home position recommended by the manufacturer. The script is shown in Fig. 10.

Basically, the move_home.sh script calls sets the reference position for the controller to the recommended values by publishing in the /wam/computed_torquecontroller/command topic. The parameters for publishing using rostopic are the topic name, the topic type, the vectors for reference position, reference velocity and reference acceleration and the duration of the trajectory in seconds and nanoseconds.

The move_zero.sh script moves the robot to the position where all joints are at zero. Although that is the initial position used by the Gazebo simulator it is not the initial position recommended by the Barrett WAM manufacturer.

Listing 10. `move_home.sh`

```
#!/bin/bash

rostopic pub /wam/computed_torque_controller/command \
trajectory_msgs/JointTrajectoryPoint \
"[0.0,-2.0,0.0,3.1,0.0,0.0,0.0]" \
"[0.0, 0.0, 0.0, 0.0, 0.0, 0.0, 0.0]" \
"[0.0, 0.0, 0.0, 0.0, 0.0, 0.0, 0.0]" \
"[0.0, 0.0, 0.0, 0.0, 0.0, 0.0, 0.0]" \
"[0.0, 0.0]" "-1"
```

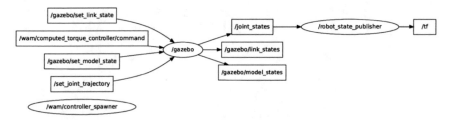

Fig. 10 ROS computation graph for the system launched with the `computed_torque.launch`

7 Conclusion

This chapter presented the implementation of an MIMO, non-linear controller in ROS, which is an important departure from the traditional low-level ROS controllers which consider a SISO system using PID controllers. The internal working of the ROS real-time loop was described in details and a computed torque controller was used as example and has its implementation discussed. The controller was implemented for running in hard-real-time under the PREEMPT_RT kernel patch.

References

1. K.J. Åström, B. Wittenmark, *Computer Controlled Systems—Theory and Design* (Prentice-Hall, Englewood Cliffs, 1984)
2. WAM User Manual (Barrett Technology Inc., Cambridge, 2011)
3. H, Bruyninckx, Open robot control software: the orocos project, in *Proceedings of the 2001 IEEE International Conference on Robotics and Automation*, (IEEE Press, Seoul, 2001), pp. 2523–2528
4. A. Burns, A. Wellings, *Real-Time Systems and Programming Languages*, 3rd edn. (Addison-Wesley, Reading, 2001)
5. C.T. Chen, *Linear System Theory and Design* (Holt, Rinehart & Winston, New York, 1984)
6. R. Featherstone, *Rigid Body Dynamics Algorithms* (Springer, New York, 2008)
7. G.F. Frankin, J.D. Powell, M.L. Workman, *Digital Control of Dynamic Systems*, 2nd edn., Addison-Wesley Series in Electrical and Computer Engineering: Control Engineering (Addison-Wesley, Boston, 1989)
8. K.S. Fu, R.C. Gonzales, C.S.G. Lee, *Robotics Control, Sensing, Vision and Intelligence*, Industrial Engineering Series (McGraw-Hill, New York, 1987)

9. P, Goebel, ROS by Example. Lulu, Raleigh, NC (Abr 2013), http://www.lulu.com/shop/r-patrick-goebel/ros-by-example-hydro-volume-1/paperback/product-21460217.html
10. Ioris, D., Lages, W.F., Santini, D.C.: Integrating the OROCOS framework and the barrett WAM robot, in *Proceedings of the 5th Workshop on Applied Robotics and Automation. Sociedade Brasileira de Automática*, (Bauru, 2012)
11. A. Isidori, *Nonlinear Control Systems*, 3rd edn. (Springer, Berlin, 1995)
12. H.K. Khalil, *Nonlinear Systems*, 2nd edn. (Prentice-Hall, Upper Saddle River, 1996)
13. N, Koenig, A, Howard, Design and use paradigms for gazebo, an open-source multi-robot simulator, in *Proceedings of the 2004 IEEE/RSJ International Conference on Intelligent Robots and Systems (IROS 2004)*, Sep 2004, vol. 3 (IEEE Press, Sendai, 2004), pp. 2149–2154
14. W.F. Lages, D. Ioris, D. Santini, An architecture for controlling the barrett wam robot using ros and orocos, in *Proceedings for the Joint Conference of 45th International Symposium on Robotics and 8th German Conference on Robotics* (VDE Verlag, Munich, Germany, 2014). ISBN: 978-3-8007-3601-0
15. J.Y.T. Leung, J. Whitehead, On the complexity of fixed-priority scheduling of periodic, real-time tasks. Perform. Eval. **2**(4), 237–250 (1982)
16. C.L. Liu, J.W. Layland, Scheduling algorithms for multiprogramming in a hard-real-time environment. J. ACM **20**(1), 46–61 (1973)
17. Maciel, E.H., Henriques, R.V.B., Lages, W.F.: Control of a biped robot using the robot operating system, in *Proceedings of the 6th Workshop on Applied Robotics and Automation* (Sociedade Brasileira de Automática, São Carlos, SP, Brazil, 2014)
18. A. Martinez, E. Fernández, *Learning ROS for Robotics Programming* (Packt Publishing, Birmingham, 2013)
19. H. Nijmeijer, A. van der Schaft, *Nonlinear Dynamical Control System* (Springer, New York, 1990)
20. K. Ogata, *Modern Control Engineering* (Prentice-Hall, Englewood Cliffs, 1970)
21. J.M. O'Kane, A Gentle Introduction to ROS. CreateSpace Independent Publishing Platform (2013), http://www.cse.sc.edu/~jokane/agitr/. Accessed Oct 2013
22. Open Software Automation Development Lab: Osadl project: Realtime linux (2012), https://www.osadl.org/Realtime-Linux.projects-realtime-linux.0.html
23. M. Quigley, B. Gerkey, K. Conley, J. Faust, T. Foote, J. Leibs, E. Berger, R. Wheeler, Ng, A.: ROS: an open-source robot operating system, in *Proceedings of the IEEE International Conference on Robotics and Automation, Workshop on Open Source Robotics*, May 2009 (IEEE Press, Kobe, Japan, 2009)
24. D.C. Santini, W.F. Lages, An architecture for robot control based on the OROCOS framework, in *Proceedings of the 4th Workshop on Applied Robotics and Automation.* (Sociedade Brasileira de Automática, Bauru, SP, Brazil, 2010)
25. J.J.E. Slotine, W. Li, *Applied Nonlinear Control* (Prentice-Hall, Englewood Cliffs, 1991)
26. I.A. Sucan, S. Chitta, MoveIt! (2015), http://moveit.ros.org

Authors' Biography

Walter Fetter Lages graduated in Electrical Engineering at Pontifícia Universidade Católica do Rio Grande do Sul (PUCRS) in 1989 and received the M.Sc. and D.Sc. degrees in Electronics and Computer Engineering from Instituto Tecnológico de Aeronáutica (ITA) in 1993 and 1998, respectively. From 1993 to 1997 he was an assistant professor at Universidade do Vale do Paraíba (UNI-VAP), from 1997 to 1999 he was an adjoint professor at Fundação Universidade Federal do Rio Grande (FURG). In 2000 he moved to the Universidade Federal do Rio Grande do Sul (UFRGS) where he is currently an associate professor. In 2012/2013 he held an PostDoc position at Universität Hamburg. Dr. Lages is a member of IEEE, ACM, the Brazilian Automation Society and the Brazilian Computer Society.

LIDA Bridge—A ROS Interface to the LIDA (Learning Intelligent Distribution Agent) Framework

Thiago Becker, André Schneider de Oliveira, João Alberto Fabro
and Rodrigo Longhi Guimarães

Abstract This chapter presents a tutorial on how to build a cognitive robotic system with the LIDA Framework. In order to ease this development, a new ROS module (the LIDA Bridge, made available at https://github.com/lidabridge/lidabridge) is presented. The LIDA Framework is a Java implementation of the LIDA conceptual model, which is a cognitive model of artificial consciousness. This work performs an in-depth discussion about LIDA conceptual model, its components and how they interact in order to manage a general-purpose cognitive system. These concepts are applied in a step-by-step tutorial to create a fully cognitive robot based on this ROS wrapper to LIDA Framework, that is able to learn with new experiences or different perceptions.

1 Introduction

In the last few years there was a big leap forward both in industrial and service robotics [1, 2]. In the service robotics field, a wide range of robots were projected to operate in human circulation environments [3]. These robots can be used in offices, hospitals, museums and many others environments where interaction with people is necessary. These robots can perform tasks like delivery, cleaning, entertainment and education. There is also the development of robotics towards accessibility like

T. Becker (✉) · A.S. de Oliveira · J.A. Fabro · R.L. Guimarães
LASER - Advanced Laboratory of Embedded Systems and Robotics,
Federal University of Technology - Parana,
Av. Sete de Setembro, Curitiba 3165, Brazil
e-mail: beckerthiago@gmail.com

A.S. de Oliveira
e-mail: andreoliveira@utfpr.edu.br

J.A. Fabro
e-mail: fabro@utfpr.edu.br

R.L. Guimarães
e-mail: rolongui@yahoo.com.br

© Springer International Publishing Switzerland 2016
A. Koubaa (ed.), *Robot Operating System (ROS)*, Studies in Computational
Intelligence 625, DOI 10.1007/978-3-319-26054-9_27

autonomous wheeled chairs [4]. Other application that has been widely developed is the passengers transport by autonomous cars. In these conditions, the robots must perform their tasks in dynamic environments, while in presence of moving objects, people and even animals. In such environments, the navigation difficulties raise substantially, when comparing with robots that act only in controlled environments (such as the industrial ones). Due to its complexities, many techniques have been developed to solve the autonomous navigation problem.

The cognitive navigation is a field of research that associate the navigation problem with artificial intelligence, combining mobile robotics and cognitive processing of high level perceptions of the environment. One widely used architecture for cognitive navigation uses a two-layer navigator, where the upper layer is responsible for identifying the high level aspects (such as semantic interpretation of the sensor readings, and planning). An updated plan of action is then passed as input to the lower control layer, which then execute the plan.

The low-level layer comprise the numeric and geometric attributes that the robot must process to execute its navigation. In order to achieve the navigation, the robot must have a precise estimation of its localization and a consistent environment map. These goals can be achieved by using the Simultaneous Localization and Mapping (SLAM) technique. This technique consists in accomplish the two tasks simultaneously with the use of probabilistic distributions of the location of the robot given a set of measurements performed by the robots sensors.

In the high level layer, the semantic interpretation of the environment is responsible for creating a high level model of the current situation of the robot inside the environment, and to realize the planning of the actions that must be executed by the robot, coordinating the low level layer in the execution of its tasks. Within the tasks performed by the high level layer are the recognition of obstacles previously seen, recognition of new obstacles, the identification of dangerous situations, and the definitions of the actions to be taken.

This chapter presents a way to implement a dual layer cognitive robot navigator for any robot using ROS: the LIDA Bridge ROS module. The implementation of this module is just a connector module, that allows the use of the features made available by the the LIDA Framework [5], which is an implementation of the Baars-Franklins cognitive architecture [6].

Artificial consciousness is a sub-field of artificial intelligence, that applies the functional features of the human consciousness to develop computer systems. The consciousness is quoted by many researches as an attribute of the human being, that enables us to get along with complex and diverse situations. By applying consciousness models to develop robotic control systems, a larger degree of flexibility and adaptability to the environment is expected.

The next section explains the Global Workspace Theory which is the cognitive theory that servers as a basis for the LIDA Framework. Section 3 discuss the LIDA cognitive model, which is the theoretical model of consciousness that makes use the Global Workspace Theory and other cognitive models in order to define a mechanism that emulates several aspects of human consciousness. Section 5 presents the LIDA Bridge, a tool developed by our research group to ease the process of connecting a

ROS Robot with the LIDA Framework in order to build a `conscious` robot. This section presents a tutorial on how to set up and define a simple LIDA Agent that acts as a navigator for a Robot.

2 Global Workspace Theory

The Global Workspace Theory, as proposed by Baars [7], presents the consciousness as a global access that assists the recruitment and integration of independent brain functions. Baars suggests that the nervous systems has many specialized processors that execute its functions unconsciously and autonomously. These processors are highly specialized and efficient in the execution of its own tasks, working in parallel and creating a system with big processing capabilities. These specialized processors can cooperate, forming coalitions in order to perform tasks cooperatively, as proposed by Crick and Koch in their model [8]. The coalitions can be composed of specialized processors, or composed of other processors coalitions. A processor can bring its contents to the consciousness when its level of activation is greater than that of all the other processors. One processor raises its activation level when it brings new, unexpected information. When a processor brings routine information, its activation level remains low. The activation level of one processor collaborates with the activation of the coalition it belongs, augmenting the chances of bringing its contents to the consciousness.

The Global Workspace is the central structure of the Baars theory. It is at this point that the conscious experience is realized. It works as a working memory that intermediates the information exchange between the processors. The Global Workspace, differently from the specialized processors, is a purely serial structure, with low processing capacity. Because of these two properties (serial and low processing), only one coalition can be in the workspace at a certain time (only one thing—though—can be considered conscious at any time). A competition between the coalitions occurs, in order to select which coalition will gain access to the global workspace. The winning coalition can then use the broadcast capabilities of the global workspace to transmit its contents (in a conscious way) to all the other processors.

To explain his theory, Baars uses a theatre metaphor, called the "Interactive Theatre Metaphor". In this theatre there is the stage, a spotlight, the actors, the public and the backstage. The spotlight leads the publics attention to the actors, and these compete for the attention of the spotlight. The public is composed of specialized processors and in the backstage are the auxiliary processors that can influence who is in the stage.

During the play, many activities occur at the stage, but only what is being illuminated by the spotlight is conscious. The events continue to happen at the stage and the assistants (backstage unconscious processors) influence the events that are going under the light of the spotlight. Only the more important actors are illuminated, so they can transmit their message do the audience and to the backstage. In the interactive theatre, when the audience identifies themselves with the message that is being transmitted, they can go up the stage and realize their performance together with the actors of the spotlight, or even in the dark part to the stage.

This metaphor represent the most import coalition as being the actors under the spotlight, transmitting its messages to the other specialized processors, expressing the limitation of the conscious contents. The coalitions that are in the global workspace (receiving the attention of the processors) are doing tasks related to new or conflicting situations. The unconscious processors execute their specific tasks continuously, whenever their find a relevant situation.

The LIDA Model uses the Global Workspace Theory as the core of its consci-ousness system, this cognitive model is the theoretical base of the LIDA Framework and will be explained in next section.

3 The LIDA Model

The LIDA (Learning Intelligent Distribution Agent) Model [7] is a cognitive model that tries to specify how to computationally implement the concepts of artificial consciousness, based on the Global Workspace Theory [7]. The LIDA model realizes a series of cognitive process, including, perception, action selection and learning capabilities. In order to achieve these goals, the architecture uses many technologies previously developed, like the Copycat architecture [9], behaviour network [10], sparse distributed memory [11] and the Pandemonium theory [12]. The LIDA Model is composed of many modules as can be seen in Fig. 1, where each module is represented

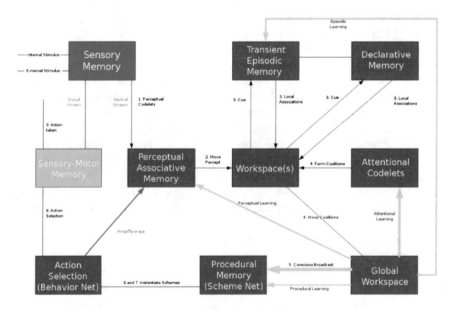

Fig. 1 The LIDA cognitive cycle (Baars and Franklin 2007: Fig. 4)

by a box. Blue and green boxes are the modules that interface with the outside world, receiving sensory data and sending `motor` actions. Arrows represent the communication between the modules. Yellow arrows are communications related to the learning process. The model uses a cognitive cycle in order to execute its processes, in a way that the system runs multiple cycles of sensing the environment, deciding and acting. Thus the system iterates continuously on this cycle, activating its modules.

The LIDA Framework is a implementation based on the conceptual model of LIDA, being the result of a series of research works developed since 1996 [13–15]. It was developed in Java, and the objective of the first applications was to autonomous elaborate the distribution of job rotations to the sailors of the US Navy. In the next sub-sections, the concepts implemented in each module are briefly described.

3.1 Codelets

The term Codelet comes from the Copycat architecture [9]. Each Codelet correspond to a processor that is specialized in one task, that is independently executed. In the Global Workspace Theory, these specialized processors form the base of the cognitive process. Each codelet is implemented as a small portion of code, being executed in a independent computer thread. Besides of being specialized in the execution of a simple task, one codelet generally works in group with other codelets, forming coalitions in order to build high level behaviours. The codelets normally act as daemons [16] waiting for the appropriate condition to act.

Each codelet has an activation level. When noticing a convenient situation, a codelet will increase its activation level according to a correspondence of the situation and its preferences. The activation level of a coalition is increased when the activation level of the codelets that belong to the coalition raises.

3.2 Perception

The perception module from the LIDA Model utilizes a Perceptual Associative Memory (PAM) as a knowledge base, combining messages received with pre-established models, through a semantic network called `slipnet`. The `slipnet` is a correlated concept network, where each concept is represented by a node and each relationship between concepts is represented by a link, which has a numeric value that represents the `conceptual distance` between the involved nodes. The lower the distance between the concepts, the higher is the activation level of the relationship. The concepts represented in the form of nodes, act like symbols representing the perceptions of the agent. A node can represent an individual, a category, or even high level ideas. The activation level of a note reflects its level of relevance to the current situation.

External and internal sensorial stimulus are received and interpreted by the perception module, in an unconscious way. In the perception phase, the sensory entries
are captured by specialized perception codelets. The codelets that find relevant features to its capacity, activate the appropriate nodes in the semantic network, then the
network is stabilized, bringing convergence to many streams of different meanings
and grouping meaning parts in bigger parts. These bigger parts in the network form
what is called a `Percept`.

3.3   Episodic and Long Term Memories

The long term associative memory and the Transient Episodic Memory are implemented in LIDA by a sparse distributed memory (SDM). The SDM was proposed
by Kanerva [11] as a model to the long term human memory. Kanerva's architecture
enables the storage of patterns and its posterior retrieval based in partial combination
of the current sensorial entries. In other words, it is a content addressable memory,
in a way that part of its content is send as entry (instead of its storage address) in the
retrieval of an item. The sparse distributed memory is capable of making associations
to the entries with the previously stored content.

Human beings have associative memories, that are transient, and content addressable. Human memory usually have a decay rate in the order of hours, which enables
us to remember with great detail the events that occurred during a day. These details
stay present in the memory only for a short period of time, for example, when we
remember what we ate for lunch or what was discussed with somebody some hours
ago. We have richness of detail in this memories, but these details are forgiven as
time passes. The Transient Episodic Memory in LIDA model represent this short term
memory. When a situation that is present in this short term memory is experienced
again, it is reinforced instead of decaying. Memories that haven't been forgiven in the
given period are stored in the long term associative memory, in the same way human
beings can keep memory of something important or that was experienced several
times, LIDA implements this different memories by having two Episodic memories,
the short term memory with a fast decay rate and the long term memory with a much
slower decay rate.

3.4   Consciousness Mechanism

The mechanism to produce the `consciousness` is described by [17], in four
components: a coalition manager, a spotlight controller, a broadcast manager and
a set of organized codelets. The components of the consciousness module can be
comprised with a implementation of some aspects of the process executed by the
theatre metaphor.

3.4.1 Attention Codelets

Attention codelets recognize problematic or new situations. The task of the attention codelets is to bring information to the consciousness. The attention codelets constantly observes a particular situation that may occur. When it finds such situation, the attention codelet will be associated with a small group of codelets that describe the situation. This association must bring the set of information codelets, together with the attention codelet that collected them, creating a coalition. The attention codelet raises its activation according to its fitness to the current situation.

3.4.2 Coalition Manager

The coalition manager creates and monitors coalitions of codelets. A codelet is initialized without associations and in its own coalition. Active codelets enter the `stage` and build associations with other codelets. Associations grows in co-activity and result in coalitions of one or more codelets [18].

3.4.3 Spotlight Controller

The relevance of a coalition is based on the mean of the activation level of the codelets that compose it. The spotlight controller computes the mean of activation of the coalitions and chooses the coalitions with the greater mean to go to the consciousness and be illuminated by the spotlight, according to the theatre metaphor.

3.4.4 Broadcast Manager

The broadcast manager realizes the broadcast of the consciousness contents to all system codelets, gathering the content of all the codelets in the coalition chosen by the Spotlight controller. This content is then transmitted to all the codelets, and each codelet then decides if this information is relevant for itself or not.

3.5 Action Selection

The Action Selection mechanism, uses a Behaviour Network [10], with improvements developed by [19]. This mechanism and the consciousness mechanism are the core elements of the LIDA model. LIDA selects and execute the actions to attend its internal purposes, but there may be several purposes running in parallel, which can vary in urgency along the time and environment changes. An active Behaviour Network is composed of structures called Behaviour Chains. These Behaviour Chains have goal nodes and behaviour nodes. Behaviours and goals in a behaviour network are interconnected through a series of links. Behaviours are typically mid level actions, and they can have many behaviour codelets for its execution. Goals are similar to behaviours, with the exception that its actions consists in finish the goals under specific conditions. The Behaviour Network acts together with the consciousness mechanism to select the actions. If a behaviour codelet finds a relevant information in a `conscious broadcast` a behaviour chain is instantiated, then it becomes part of the active Behaviour Network. Behaviours compete and also cooperate with each other, and the dynamics of the network eventually selects a relevant behaviour to act.

The previous sections presented the conceptual models used in LIDA, the next section will present in detail the computational framework based on these concepts that we will use in Sect. 5 to build a cognitive ROS Robot with all of these concepts inside.

4 LIDA Framework

The conceptual model of LIDA, as presented in the previous section, shows the proposal of a cognitive model that tries to cover a great portion of the human cognitive system, combining sub-systems of action selection, attention, episodic memory, sensory memory, perceptual memory and sensorimotor memory. This section presents the LIDA Framework, that is the computational implementation of the conceptual LIDA model, using the JAVA language.

4.1 Structure

The LIDA framework defines several data structures and algorithms, and it is composed of a variety of components. The main components are the modules, that consist of interconnected elements which represent the modules of the conceptual model. Each of the conceptual model modules has a correspondent on the framework.

Most of the frameworks modules are independent of the problem domain. For each of these modules, the framework has a standard implementation. However, some modules are dependent of the problem domain and have only a basic implementation, to be extended with specific features for the domain.

4.1.1 Communication

For the communication to be possible between modules, the LIDA framework utilizes the Observer design pattern. A listener module receives information from a producer module. The listener can subscribe to the producer to listen what the producer has to send. Each time the producer has something to send, it transmits the information to all its listeners. There are several listener instances on LIDA Framework, and a module can be registered as a listener to several other modules. It is also possible for a module to be a producer and a listener of other modules at the same time.

4.1.2 Tasks

Since all the modules and codelets run in parallel on the conceptual model of the Baars-Franklin architecture, the LIDA Framework implements a system of task management. The tasks on LIDA wrap small processes. A module can create multiple tasks to perform its functions. A task can execute only once or it can execute its functions repeatedly. The task manager controls the execution of all the tasks and performs their execution on separated threads, in order to achieve parallel execution in a way that is transparent to the user.

4.1.3 Data Structures

Nodes and *Links* are the main data structures of the LIDA Framework. Both of them have an activation, which represents the measure of their importance on the current context. This activation can be excited in several situation and it decays over time. A node can represent features, objects, concepts, events, actions, etc. Each node is based on a *PerceptualAssociativeMemory* node (PamNode). The nodes are instances of PamNode, and each node has a reference to the PamNode that originated it, as well as a unique *id* that identifies it.

4.1.4 Activation

Nodes, links and other elements like coalitions, codelets, schemes and behaviours, all have an activation level. The activation is represented by a real number between 0.0 and 1.0. In general, the activation represent the relevancy of an element. Other elements have and additional activation called *base-level activation*, which is used for the learning implementation. The framework uses two interfaces to the implementation of these functionalities: *Activatible* and *Learnable*.

One element that implements the *Activatible* interface has excitation and decay methods to regulate the activation. An *Activatible* interface should specify two strategies to determine how its activation level changes when the decay and excite methods are called.

One element that implements the *Activatible* interface has, in addition, the *base-level activation* for the implementation of the learning. The current version of the LIDA Framework does not have any learning algorithm implemented. To implement learning, currently, it is possible to develop custom modules that implement the learning method. These methods shall modify the *base-level activation* of the *Learnable* elements.

4.1.5 Graphical User Interface

The LIDA Framework has a customizable user interface that allows the online exhibition of the module contents, parameter values, tasks being executed and other variables of interest during the LIDA execution. A properties file allows the addition of standard or custom panels to configure the appearance of the graphical user interface.

5 Using the LIDA Framework on ROS Through LIDA Bridge

The LIDA Bridge is a Java library that provides the tools to build a navigation strategy to a ROS robot using the LIDA Framework. It consists in a set of classes that provides abstraction to the communication between LIDA and the ROS system. This section will show a step by step tutorial on how to set up a LIDA agent for navigating a ROS managed robot. This section will present a example of a simple mobile robot navigator for the following situation: there is a goal position the robot must must reach; in the environment, there is a moving object (representing a person or other mobile robot) which will serves as an dynamic obstacle for the robot; the navigator must decide when to follow its path (pre-defined by the navigation stack) or stop and wait for the person to pass (if the person is too close to the robot). This is obvious a unrealistic scenario, but it can demonstrate how to implement a cognitive agent with LIDA.

5.1 Prerequisites

Before beginning with this tutorial, you must have a ROS environment already configured for your robot. The example used in this tutorial will be of a mobile robot, so the Navigation Stack must be also working in your environment. LIDA Framework and the LIDA Bridge library are written in Java, so is necessary to have a Java Development Kit configured in your environment. The example was tested in Oracle JDK, but it's likely to work in OpenJDK either. The LIDA Framework can be downloaded in the Cognitive Computing Research Group from the University of Memphis (http://ccrg.cs.memphis.edu/framework.html). The user can download the framework after completing a registration form. The LIDA Bridge library can be downloaded in https://github.com/lidabridge/lidabridge. The LIDA Bridge library communicates with ROS through websockets by the Rosbridge package, which can be downloaded in http://wiki.ros.org/rosbridge_suite.

Once you have a ROS system with the Navigation Stack and Rosbridge running, we are able to proceed to the next step.

5.2 Setting up a New Project

To develop a LIDA agent you need to create a Java Project with the IDE of your choice and add the jar files for the LIDA Framework and LIDA Bridge in the Java Classpath, in the LIDA Bridge page you can get a list of all the dependencies needed to start up the system.

The LIDA Framework uses a set of configuration files in order to set up the system, it is a very flexible and customizable system, due to the fact that it was developed for building any kind of intelligent system. We will begin by creating a folder in our project to store all the configuration files. The file structure for the new project will be somewhat like this:

- myRobotAgent

  - src—Folder to store all the source code of the agent.
  - lib—Folder to store the library dependencies.
  - configs—Folder to store the configuration files.

There are two main types of configuration files used by the system, the first of them is the properties file (.properties). In these files we will set up the basic configuration for the system like loggings features and graphical interface. The other type of file is XML. The core LIDA configurations are defined in XML files that define almost every property of the agent.

Lets begin by creating a property file called `lidaConfig.properties` in our `configs` folder. This file will be the base for defining our system. The file must have the following content:

Listing 1.1. lidaConfig.properties

```
1  #Agent properties
2  lida.agentdata=configs/basicAgent.xml
3  lida.elementfactory.data=configs/factoryData.xml
4
5  #Gui properties
6  lida.gui.panels=configs/guiPanels.properties
7  lida.gui.commands=configs/guiCommands.properties
8  lida.gui.enable=true
9
10 #Logging properties
11 lida.logging.configuration=configs/logging.properties
```

The first line is a commented line (all the lines that begin with a # in the properties files are ignored by the LIDA Framework). Line 2 defines a `basicAgent.xml` file in configs folder associated to the `lida.agentdata` property. This file is defines the agent specific configurations. Line 3 defines a `factoryData.xml` associated to the *lida.elementfactory.data* property. This file specifies which classes will be instantiated by the LIDA Framework for the many components of the system. The LIDA Framework uses the `Element Factory` design pattern and this file contains the factories definitions. Lines 6 and 7 define user interface property files and line 8 tells if we will be using the LIDA user interface or not. Finally, line 11 defines a logging properties file, for setting up the system logs. With all these files created we have the following project structure:

- myagent
    - src—Folder to store all the source code of the agent.
    - lib—Folder to store the library dependencies.
    - configs—Folder to store the configuration files.
        basicAgent.xml
        factoryData.xml
        guiPanels.properties
        guiCommands.properties
        logging.properties

We will be editing these files through this tutorial, but you will get a example version of each file together with the LIDA Framework, now lets go to the Java source code.

In order to get our agent running we need to define a static *main* method to serve as the entry point of our system. We will create a file called Run.java with a proper class, to handle our emphmain method.

Listing 1.2. Run.java

```
1  package myagent;
2
3  import edu.memphis.ccrg.lida.framework.initialization.
       AgentStarter;
4  public class Run {
5
6      public static void main(String[] args ){
7          AgentStarter.main(args);
8      }
9  }
```

In line 7, the method AgentStarted.main is invoked in order to start the LIDA Framework agent, this method will read the configuration files and set the agent running. Our agent won't run right now, because there are some code and configurations missing, lets create them.

Earlier in this chapter, it was said that the LIDA Framework's Environment module is problem dependent and we need to implement this module now. LIDA Bridge comes with an implemented environment module that will provide an environment almost ready to use in our ROS/LIDA navigator. Although the LIDA Bridge environment module provides us a ROS Environment, we still need to specify some specific settings for our robot. We will create a class called MyRobot that inherits from lidabridge.ROSEnvironment. Inside this class, we will set up the ROS Topics we will be subscribing and publishing within our system, as well as the actions that, when decided to be taken by LIDA, will affect the ROS system. We will do all this customization in the constructor method of our main class, because all of this configurations must be ready before our agent starts.

The code bellow shows the example of the Environment module, inherited from ROSEnvironment with all the customizations needed to our example. The first thing that appears in our constructor (line 21) is the rosSetAddr method from the ROSEnvironment class. It sets the URI (Uniform Resource Indicator) which points to the ROS Bridge node, so it can communicate with it, through web sockets.

Lines 13–18 define several objects, each one representing a ROS Topic that is advertised or subscribed by the agent. The PoseStampedTopic class that can manage a ROS topic with the PoseStamped message type. The StringTopic class that can manage a ROS topic with the String message type.

Line 13 defines a PoseStampedTopic named robotPosition, that will be used to handle the position of our robot. Line 23 instantiates the object robot-Position, of the PoseStampedTopic class, that has the following arguments: the ROS topic, an alias to be used by LIDA Bridge, and a TopicAccessType that tells if the topic will be published or subscribed. Besides the robotPosition we create for our example, we have other three topics for managing positions, mainGoal that handles the final goal that the robot must pursuit, the goal topic, that handles temporary positions that LIDA uses to send the goal for the ROS Navigation Stack, and finally the person topic representing the position of the person (or any other mobile obstacle). Lines 17 and 18 create two objects of the StringTopic class, the status topic will be used to publish the last action taken by the agent, and the message topic will publish a simple message depending on the action taken. The StringTopic constructor has the same parameters of the PoseStampedTopic. Lines 26–40 instantiate all the other topics, just like the robotPosition.

In lines 46–65 two Actions are created and added to the Actions list of the ROSEnvironment. An Action is a LIDA Bridge class used to define what the agent must do when a decision is taken by LIDA. The actions are created in this example with two anonymous classes that inherits from the Action class. The constructor takes two parameters, the first is the ROSEnvironment object related to this action, and the second is the LIDA command that is result of LIDA's action selection mechanism (commands will be discussed later in this chapter). Every Action must override the doWork method. In this example the two actions correspond to the algorithm.goal and algorithm.stop commands. The first is selected when LIDA decides the robot must continue going to the goal. Line 49 sets a simple message, and line 51 sets the goal topic with the mainGoal value, as the goal topic is sent every cycle to the Navigation Stack, the robot will continue going to the goal. Line 52 sets the status of the robot as goal, which tells LIDA that the last action taken by the robot was to move towards the goal. The second Action sets the robot position to the current robot position, so the robot will stop.

Listing 1.3. MyRobot.java

```java
 1  package myagent;
 2
 3  import java.util.Random;
 4
 5  import lidabridge.Action;
 6  import lidabridge.ROSEnvironment;
 7  import lidabridge.topic.PoseStampedTopic;
 8  import lidabridge.topic.StringTopic;
 9  import lidabridge.topic.TopicAccessType;
10
11  public class MyRobot extends ROSEnvironment {
12
13      private PoseStampedTopic robotPosition;
14      private PoseStampedTopic mainGoal;
15      private PoseStampedTopic goal;
16      private PoseStampedTopic person;
17      private StringTopic status;
18      private StringTopic message;
19
20      public MyRobot() {
21          setRosAddr("ws://localhost:9090");
22
23          robotPosition = new PoseStampedTopic("odom",
24              "agentpose", TopicAccessType.SUBSCRIBED);
25          this.getTopics().add(robotPosition);
26
27          mainGoal = new PoseStampedTopic("/goal",
28              "maingoal", TopicAccessType.SUBSCRIBED);
29          this.getTopics().add(mainGoal);
30
31          goal = new PoseStampedTopic("/move_base_simple/
32              goal", "goal", TopicAccessType.ADVERTISED);
32          this.getTopics().add(goal);
33
34          person = new PoseStampedTopic("/person_pose",
35              "person", TopicAccessType.SUBSCRIBED);
36          this.getTopics().add(person);
37
38          message = new StringTopic("/message", "message",
39              TopicAccessType.ADVERTISED);
39          this.getTopics().add(message);
40
```

```
41      status = new StringTopic("/status", "status",
            TopicAccessType.ADVERTISED);
42      status.setValue("stop");
43      this.getTopics().add(status);
44
45
46      // Actions
47
48      // Action for the robot follow the main goal
49      this.getActions().add(new Action(this,
50          "algorithm.goal") {
51          @Override
52          public void doWork() {
53              message.setValue("Following torwards the
                    goal");
54
55              goal = mainGoal;
56              status.setValue("goal");
57          }
58      });
59
60      // Action for the robot stop
61      this.getActions().add(new Action(this,
62          "algorithm.stop") {
63          @Override
64          public void doWork() {
65              message.setValue("Stop...");
66              goal.setX(robotPosition.getX());
67              goal.setY(robotPosition.getY());
68              goal.setOrientation(robotPosition.
                    getOrientation());
69              status.setValue("stop");
70          }
71      });
72  }
73 }
```

Now thanks to the LIDA Bridge library our environment communication is ready, and we just need to set up the standard LIDA settings.

The next thing we need to do to accomplish this task is to implement our perception codelets. These codelets will monitor the environment and if it recognizes a particular situation it will activate the corresponding nodes in the Perceptual Associative Memory. The following code shows a codelet to activate the node ofar (obstacle is far) or onear (obstacle is near) in PAM, depending on the distance of the robot from the person in our environment. In the init method (lines 21–27) the nodes are taken from the PAM using the `getNode` method by their names. The `detectLinkables` method is responsible for executing the detection and the activation of the nodes. Lines 32 and 35 read the values of the position of the person and the agent respectively, then a simple euclidian distance is computed and if the distance is less than 1.2 meters, the `onear` node is excited (line 50) otherwise the `ofar` node is excited (line 53).

Listing 1.4. ObstacleDistanceDetector.java

```
1  package myagent.featuredetectors;
2
3  import java.util.HashMap;
4  import java.util.Map;
5
6  import lidabridge.topic.PoseStampedTopic;
7  import edu.memphis.ccrg.lida.pam.PamLinkable;
8  import edu.memphis.ccrg.lida.pam.tasks.
       DetectionAlgorithm;
9  import edu.memphis.ccrg.lida.pam.tasks.
       MultipleDetectionAlgorithm;
10
11 public class ObstacleDistanceDetector extends
       MultipleDetectionAlgorithm implements
       DetectionAlgorithm {
12
13     private PamLinkable near;
14     private PamLinkable far;
15     private PamLinkable obstaclenode;
16
17     private final String modality = "";
18     private Map<String, Object> detectorParams = new
           HashMap<String, Object>();
19
20     @Override
21     public void init() {
22         super.init();
23
24         near = (PamLinkable) pam.getNode("onear");
25         far = (PamLinkable) pam.getNode("ofar");
26         obstaclenode = (PamLinkable) pam.getNode
27             ("obstacle");
28     }
29
30     @Override
31     public void detectLinkables() {
32         detectorParams.put("mode", "person");
```

```
33          PoseStampedTopic obstacle = (PoseStampedTopic)
                sensoryMemory.getSensoryContent(modality,
                detectorParams);
34
35          detectorParams.put("mode", "agentpose");
36          PoseStampedTopic agent = (PoseStampedTopic)
                sensoryMemory.getSensoryContent(modality,
                detectorParams);
37
38          if ((obstacle == null) || (agent == null))
39              return;
40
41          // Compute the euclidian distance between the
                agent and the obstacle
42
43          double distance;
44          Double dx = Math.pow(obstacle.getX() - agent.
                getX(), 2);
45          Double dy = Math.pow(obstacle.getY() - agent.
                getY(), 2);
46          distance = Math.sqrt(dx + dy);
47
48          pam.receiveExcitation(obstaclenode, 1);
49
50          if (distance < 1.2 )
51              pam.receiveExcitation(near, 1);
52
53          else
54              pam.receiveExcitation(far, 1);
55
56      }
57
58  }
```

One more perceptual codelet is needed to our example to detect the the status of the robot, and excite the nodes corresponding to the two possible actions of the robot, stop and goal. You can do it yourself or download the full code of this example in the LIDA Bridge page.

The last piece of code needed is a main class to start our agent, this is accomplished by invoking the main method of the AgentStarter class, as shown in Listing 1.5. This class reads all the configuration from the XML files and runs the LIDA Framework with the proposed setup.

Listing 1.5. Run.java

```
1  package myagent;
2
3  import edu.memphis.ccrg.lida.framework.initialization.
     AgentStarter;
4  public class Run {
5
6      public static void main(String[] args ){
7          AgentStarter.main(args);
8      }
9  }
```

Finally, now we need to set up things in LIDA's XML files. In `lidaConfig.properties` file we defined a `basicAgent.xml`. This file is the core of LIDA's agent declaration, where we can tell LIDA how every module will work. The full documentation of the file can be found on LIDA project documentation and a example file in also be found at LIDA Bridge's page. Here we will focus only on what is specifically associated to the problem proposed to our example.

Listing 1.6. basicAgent.xml (Environment)

```
1      <module name="Environment">
2          <class>myagent.MyRobot</class>
3          <param name="environment.ticksPerRun" type="
             int">1</param>
4          <taskspawner>defaultTS</taskspawner>
5      </module>
6      <module name="SensoryMemory">
7          <class>lidabridge.modules.ROSSensoryMemory
8             </class>
9          <associatedmodule>Environment
10            </associatedmodule>
11         <taskspawner>defaultTS</taskspawner>
12         <initialTasks>
13             <task name="backgroundTask">
14                 <tasktype>SensoryMemoryBackgroundTask
                     </tasktype>
15
16                 <ticksperrun>5</ticksperrun>
17             </task>
18         </initialTasks>
19     </module>
```

In the modules section we need to specify the environment module we developed, to do so in the Environment module configuration we set the class property to our `MyRobot` class, then when the agent start it will use our class as the environment module. Another custom module that must be configured is the `SensoryMemory` module, because we will be using the LIDA Bridge's sensory memory. To do this you must set the class as `lidabridge.modules.ROSSensoryMemory` as shown in Listing 1.6.

In the Perceptual Associative Memory module configuration, we define which will be the nodes that will represent the perceptual model of the environment;. We do so with the `nodes` section in the `PerceptualAssociativeMemory` module configuration. Listing 1.7 shows the section `nodes` filled with 6 nodes: `obstacle, onear, ofar, robot, rstop, rgoal`. The node obstacle represent the person moving in the environment, the nodes `onear` and `ofar` represent the distance of the agent from the obstacle (as we defined before in the `Obstacle DistanceDetector`), the node robot as the name says represent the robot itself, the nodes `rstop` and `rgoal` represent the two possible states, the robot has stopped or is moving towards the goal. Still in the PAM configuration, we have a section named `links`, this section defines which are the links between the nodes in PAM, and they are represented using the following notation: `<node>:<node>`. In Listing 1.7 we defined four links: `obstacle:near`, `obstacle:far, robot:rstop, robot:rgoal`. These nodes and links defined here will be enough for our simple example.

In order to monitor the sensory data and excite the PAM nodes, we developed two codelets or feature detectors. The PAM configuration also as to include those feature detectors information. The perception codelets are defined by adding a TASK attribute for each codelet.

Listing 1.7. basicAgent.xml (PerceptualAssociativeMemory)

```
1  <module name="PerceptualAssociativeMemory">
2      <class>edu.memphis.ccrg.lida.pam.
           PerceptualAssociativeMemoryImpl</class>
3      <associatedmodule>TransientEpisodicMemory
4          </associatedmodule>
5    <param name="pam.upscale" type="double">
6      0.5 </param>
7      <param name="pam.downscale" type="double">0.6  </
         param>
8    <param name="pam.perceptThreshold" type="double">
9      0.8 </param>
10     <param name="pam.excitationTicksPerRun" type="int
         ">1</param>
11   <param name="pam.propagationTicksPerRun" type="int">
12   1</param>
13     <param name="pam.propagateActivationThreshold" type
         ="double">0.5</param>
14
15     <param name="nodes">
16         obstacle, onear, ofar,
17         robot, rstop, rgoal
```

```
18        </param>
19        <param name="links">
20            obstacle:onear, obstacle:ofar,
21            robot:rstop, robot:rgoal
22        </param>
23
24        <taskspawner>defaultTS</taskspawner>
25        <initialTasks>
26            <task name="ObstacleDistanceDetector">
27              <tasktype>ObstacleDistanceDetector</tasktype>
28                <ticksperrun>3</ticksperrun>
29                <param name="nodes" type="string">onear,
                      ofar</param>
30            </task>
31            <task name="RobotStatusDetector">
32                <tasktype>RobotStatusDetector</tasktype>
33                <ticksperrun>3</ticksperrun>
34                <param name="nodes" type="string">rgoal,
                      rstop, rexcuse</param>
35            </task>
36        </initialTasks>
37        <initializerclass>edu.memphis.ccrg.lida.pam.
              BasicPamInitializer</initializerclass>
38    </module>
```

The attention module is where the Attention Codelets are defined, just like the perceptual codelets in the Perceptual Associative Memory, the Attention Codelets in the Attention Module are defined by TASKS. In this example we set up three attention codelets:

- AttentionCodelet-01—observes the nodes robot, ofar and rstop;
- AttentionCodelet-02—observes the nodes robot, rgoal and onear;
- AttentionCodelet-03—observes the nodes robot, rstop and obear;

In Listing 1.8 is shown the example of the Attention Module configuration.

Listing 1.8. basicAgent.xml (AttentionModule)

```
1  <module name="AttentionModule">
2              <class>edu.memphis.ccrg.lida.
                  attentioncodelets.AttentionCodeletModule
                  </class>
3              <associatedmodule>Workspace
                  </associatedmodule>
4              <associatedmodule>GlobalWorkspace
                  </associatedmodule>
```

```
5              <param name="attentionModule.
                  defaultCodeletType" type="string">
                  NeighborhoodAttentionCodelet</param>
6              <param name="attentionModule.
                  codeletActivation" type="double">1.0
                  </param>
7            <param name="attentionModule.
                  codeletRemovalThreshold" type="double">
8              -1.0</param>
9            <param name="attentionModule.
                  codeletReinforcement" type="double">0.5</
                  param>
10
11           <taskspawner>defaultTS</taskspawner>
12
13           <initialTasks>
14               <task name="AttentionCodelet-01">
15                   <tasktype>
                         NeighborhoodAttentionCodelet</
                         tasktype>
16                   <ticksperrun>5</ticksperrun>
17                   <param name="nodes" type="string">
                         robot, ofar, rstop</param>
18                   <param name="refractoryPeriod" type="
                         int">5</param>
19                   <param name="initialActivation" type
                         ="double">1.0</param>
20                   <param name="learnable.
                         baseLevelActivation" type="double
                         ">1.0</param>
21               </task>
22               <task name="AttentionCodelet-02">
23                   <tasktype>
                         NeighborhoodAttentionCodelet
                         </tasktype>
24                   <ticksperrun>5</ticksperrun>
25                   <param name="nodes" type="string">
                         robot, rgoal, onear</param>
26                   <param name="refractoryPeriod" type="
                         int">5</param>
27                   <param name="initialActivation" type="
28                   double">1.0</param>
29                   <param name="learnable.
30                     baseLevelActivation" type="double">
                         1.0</param>
```

```
31                      </task>
32                      <task name="AttentionCodelet-03">
33                          <tasktype>
                                NeighborhoodAttentionCodelet
                                </tasktype>
34                          <ticksperrun>5</ticksperrun>
35                      <param name="nodes" type="string">robot,
                            rstop, onear</param>
36                      <param name="refractoryPeriod" type="int
                            ">5</param>
37                      <param name="initialActivation"type="
                            double">1.0</param>
38                      <param name="learnable.
                            baseLevelActivation" type="double">
                            1.0</param>
39                      </task>
40
41                  </initialTasks>
42              </module>
```

In the Procedural Memory configuration we define the action schemes used by the action selection mechanism (Listing 1.9).

Listing 1.9. Action schemes syntax

```
1   <scheme_name>|(<context>)()|<action>|(<result>)()
        <base_activation>
```

Listing 1.10 shows the configuration of the module, where the parameters named scheme.1a and scheme.2a define the action schemes of our procedural memory.

Listing 1.10. basicAgent.xml (ProceduralMemory)

```
1           <module name="ProceduralMemory">
2               <class>edu.memphis.ccrg.lida.
                    proceduralmemory.ProceduralMemoryImpl
                    </class>
3               <param name="proceduralMemory.ticksPerStep"
                    type="int"> 14 </param>
4               <param name="proceduralMemory.
                    conditionDecayStrategy">conditionDecay
                    </param>
5               <param name="proceduralMemory.
                    schemeSelectionThreshold" type="double
                    ">0.1</param>
6               <param name="proceduralMemory.contextWeight"
                    type="double">1.0</param>
```

```
7            <param name="proceduralMemory.
                addingListWeight" type="double">1.0
                </param>
8            <param name="proceduralMemory.schemeClass">
                edu.memphis.ccrg.lida.proceduralmemory.
                SchemeImpl</param>
9            <param name="scheme.1a">Move|(robot, ofar,
                rstop)()|action.seguir|(rgoal)()|0.2
                </param>
10           <param name="scheme.2a">Stop|(robot, rgoal,
                onear)()|action.parar|(rstop, ofar,
                omoving)()|0.2</param>
11
12           <taskspawner>defaultTS</taskspawner>
13           <initializerclass>edu.memphis.ccrg.lida.
                proceduralmemory.
                BasicProceduralMemoryInitializer
                </initializerclass>
14           <!--<initializerclass>myagent.initializers.
                ProceduralMemoryInitializer
                </initializerclass>-->
15        </module>
```

Finally we will set up the Sensory Motor Memory, in which we will map an action (given by the action schemes) to a command (the ones we used to define out actions in the environment class). Listing 1.11 shows the configuration of the SensoryMotorMemory.

Listing 1.11. basicAgent.xml (SensoryMotorMemory)

```
1    <module name="SensoryMotorMemory">
2            <class>edu.memphis.ccrg.lida.
                sensorymotormemory.
                BasicSensoryMotorMemory</class>
3            <associatedmodule>Environment
                </associatedmodule>
4            <param name="smm.processActionTaskTicks"
                type="int">1</param>
5            <param name="smm.mapping.1">action.stop,
                algorithm.stop</param>
6            <param name="smm.mapping.4">action.move,
                algorithm.move</param>
7            <!-- <param name="smm.3">action.releasePress
                ,algorithm.releasePress</param> -->
8            <taskspawner>defaultTS</taskspawner>
9            <initializerclass>edu.memphis.ccrg.lida.
                sensorymotormemory.
```

```
                    BasicSensoryMotorMemoryInitializer
                    </initializerclass>
10             </module>
```

These are the main configurations that must be set up in order to run LIDA. A complete set of default configuration files can be found inside the LIDA Framework. Now with our robot configured to read and publish the proper topics we can run our agent and monitor all the LIDA modules with the LIDA GUI.

The following topics are published by the robot and subscribed by the agent, they will provide the data that LIDA will use to create the perception model of the environment:

- /pose
- /person
- /goal

While the agent executes, the following topics are published by the agent:

- /status
- /move_base_simple/goal
- /message

The `status` and `message` topics are used to show actions are being selected by the agent, the `move_base_simple/goal` is the topic used by the Navigation Stack and it points to the position the robot must go.

We can see in Fig. 2 the rqt interface, showing all the topics used by the agent in the ROS system. Figure 3 presents the LIDA Framework Interface showing the graph build by LIDA's current situational model, which represents the actual state of the environment in the form of a graph.

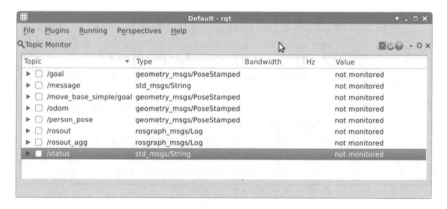

Fig. 2 RQT Interface showing the topics managed by the agent

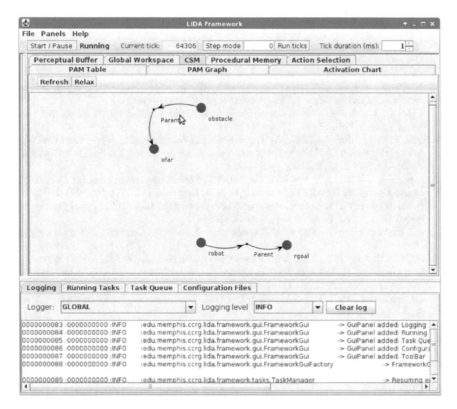

Fig. 3 LIDA GUI showing the current situational model graph

6 Conclusions

The main points in developing a LIDA powered robot are due to the its refined attentional system, its learning capabilities and association potential. The core feature of the `conscious` system is the attentional system which is the implementation of the Global Workspace Theory. This feature makes our system more similar to a biological brain in some ways, and it can rapidly change its tasks in the conscious and unconscious contexts just like a human or animal does when a novel or unexpected situation is experienced. The LIDA's graph representation of the situations is a powerful tool, in our example (due to the space limitations for this chapter) the graph which represent the perceptions is very simple and may somehow not enough to show LIDA's association capabilities. The various memory systems working together gives LIDA a high association power, which can infer new information as long as it experiences diverse environment situations. LIDA Bridge is a very helpful tool because the LIDA Framework is a general purpose cognitive system, it does not have any ROS integration nor any robot specific settings. This simple but useful tool can ease the process of building LIDA controlled robots, and allow for an easier starting point to an exciting new area of study: conscious robotics.

References

1. G. Landis, Teleoperation from Mars orbit: a proposal for human exploration. Acta Astronautica **61**(1), 59–65 (2008)
2. E. Guizzo, How Googles self-driving car works (2011), http://spectrum.ieee.org/automaton/robotics/artificial-intelligence/how-google-self-driving-car-works
3. L. Iocchi, J. Ruiz-del-Solar, T. van der Zant, Advances in domestic service robots in the real world. J. Intell. Robot. Syst. **76**(1), 3–4 (2014)
4. B.M. Faria, L.P. Reis, N. Lau, A Survey on Intelligent Wheelchair Prototypes and Simulators, in *New Perspectives in Information Systems and Technologies* vol. 1 (Springer International Publishing, Switzerland, 2014), pp. 545–557
5. J. Snaider, R. McCall, S. Franklin, The LIDA framework as a general tool for AGI, in *The Proceedings of the Fourth Conference on Artificial General Intelligence (AGI-11)* (2011)
6. B.J. Baars, S. Franklin, Consciousness is computational: the LIDA model of global workspace theory. Int. J. Mach. Conscious. **1**(1), 23–32 (2003)
7. B.J. Baars, *A Cognitive Theory of Consciousness* (Cambridge University Press, Cambridge, 1988)
8. F. Crick, C. Koch, A framework for consciousness. Nat. Neurosci. **6**(2):119–126 (2003)
9. D.R. Hofstader, M. Mitchell, The Copycat Project: A model of mental fluidity and analogy-making, in *Advances in Connectionist and Neural Computation Theory Volume 2: Analogical Connections*, ed. by K. Holyoak, J. Barnden (Ablex Publishing Corporation, Norwood, 1994), pp. 31–112
10. P. Maes, How to do the right thing. Connect. Sci. J. **1**, 291–323 (1989)
11. P. Kanerva, *Sparse Distributed Memory* (MIT Press, Cambridge, 1988)
12. J.V. Jackson, Idea for a mind. ACM SIGART Bull. **xx**(101):23–26 (1987)
13. S. Franklin, A. Graesser, O. Brent, H. Song, N. Aregahegn, Virtual mattie—an intelligent clerical agent
14. J. Newman, B.J. Baars, S.-B. Cho, A neural global workspace model for conscious attention. Neural Netw. **10**(7), 1195–1206 (1997)
15. S. Franklin, A. Kelemen, L. Mccauley, IDA: a cognitive agent architecture, in *IEEE Conference on Systems, Man and Cybernetics*, pp. 2646–2651 (1998)
16. O.G. Selfridge, Pandemonium: a paradigm for learning, ed. by D.V. Blake, A.M. Uttley, *Proceedings of the Symposium on Mechanisation of Thought Processes*, London, pp. 511–529 (1959)
17. A.S. Negatu, Cognitively inspired decision making for software agents: integrated mechanisms for action selection, expectation, automatization and non- routine problem solving. Ph.D. thesis, The University of Memphis (2006)
18. R. Capitanio, R.R. Gudwin, A conscious-based mind for an artificial creature, in *Artificial Life XII: Proceedings of the Twelfth International Conference on the Synthesis and Simulation of Living Systems*, Odense, Denmark, pp. 616–623 (2010)
19. A.S. Negatu, S. Franklin, An action selection mechanism for "conscious" software agents. Cogn. Sci. Q. **2**, 363–386 (2002)

Printed in the United States
By Bookmasters